Particles in the Dark Universe

Yann Mambrini

Particles in the Dark Universe

A Student's Guide to Particle Physics and Cosmology

Springer

Yann Mambrini
Laboratory of the Physics of the Two
Infinities Irène Joliot-Curie (IJCLab)
CNRS/University Paris-Saclay
Orsay, France

ISBN 978-3-030-78138-5 ISBN 978-3-030-78139-2 (eBook)
https://doi.org/10.1007/978-3-030-78139-2

This Springer imprint is published by the registered company Springer Nature Switzerland AG
The registered company address is: Springer Nature Switzerland AG

'If you are doing everything well, you are not doing enough'
Howard Georgi.

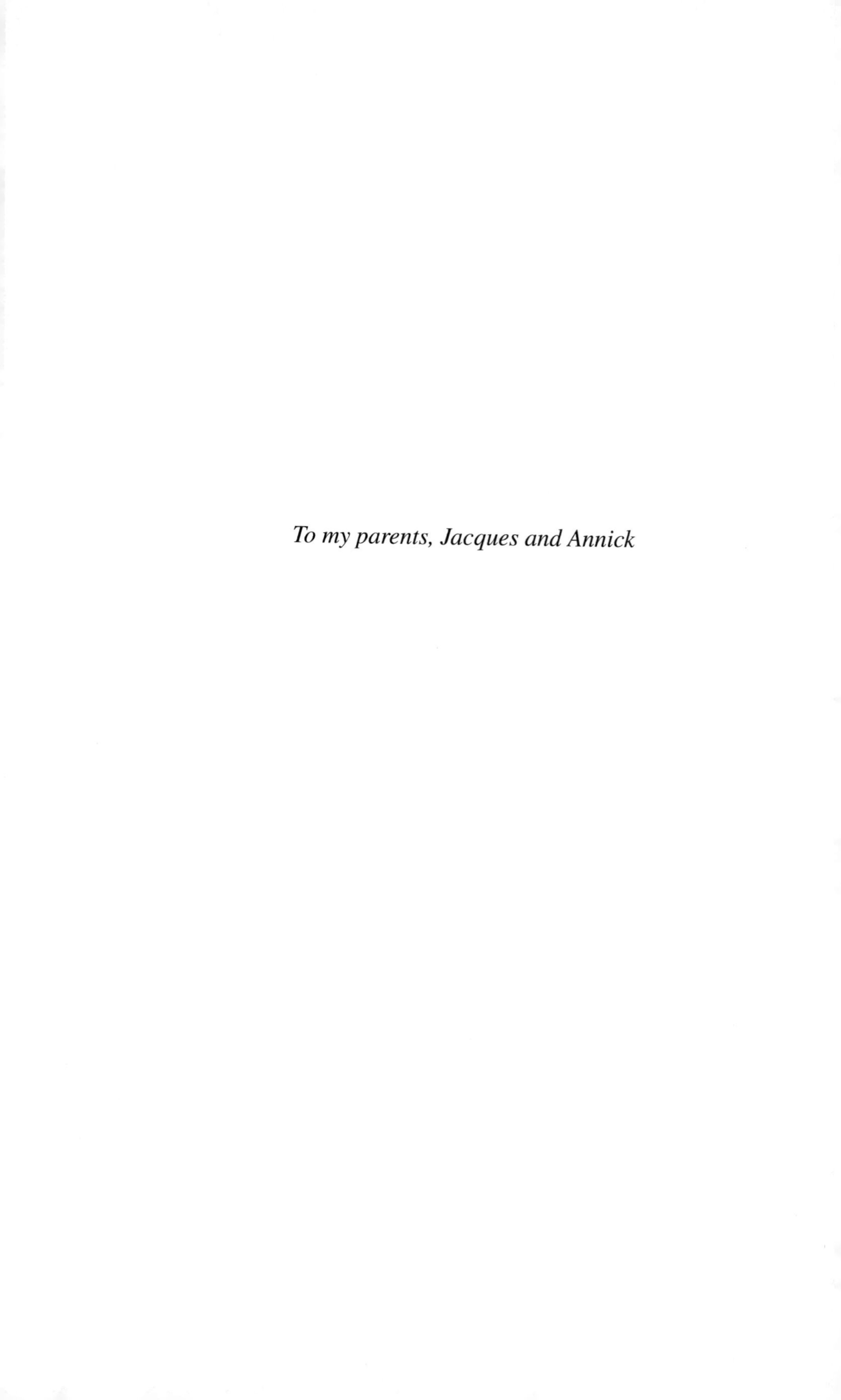

To my parents, Jacques and Annick

Foreword

Like all successful areas in physics, cosmology is based on an interplay between theory and experiment. In cosmology, experiment typically means observation, and what we observe is light. We assume that the spectral properties of light are literally universal, and as a result, the observed shifted spectra, allow us to determine the dynamics of the galaxies and clusters of galaxies hosting the light-producing gas and stars. These dynamics indicated the presence of significantly more matter than could be accounted for by the observed luminosity. What began as the problem of missing mass (or more aptly phrased as the problem of missing light—as the presence of mass without accompanying light) became to be known as the problem of dark matter.

Modern cosmology or big bang cosmology originated in the late 1940s from work principally by Alpher, Gamov, and Herman in their attempts to understand the pattern of element abundances observed in the Universe. Recognizing that nuclear reactions needed sufficient temperatures and densities to be effective, they envisioned the currently expanding Universe at early times as a primeval nuclear reactor. While this theory of big bang nucleosynthesis was only successful in accounting for the abundance of elements through Li, as a by-product it led to the prediction by Alpher and Herman of the cosmic microwave background with a temperature of 1–5 K. This was of course famously discovered in 1964 by Penzias and Wilson, with the interpretation by Dicke, Peebles, Roll, and Wilkinson in 1965 as a direct remnant of the big bang. The work on element abundances however shifted from cosmological to stellar, as most elements are produced in stars or as a result of stellar explosions. Nuclear astrophysics began its boom in the 1950s, highlighted by the work of Burbidge, Burbidge, Fowler, and Hoyle.

The 1950s also saw a boom in particle physics, with the discovery of new particles. By the late 1960s, there emerged a new Standard Model for particle physics, one that has held up remarkably well culminating in the discovery of the Higgs boson at the LHC at CERN in 2012. The role of particle physics in cosmology also began in the 1960s, although the role of neutrinos in cosmology was known in the early work on nucleosynthesis. At the time, it was not known if neutrinos possessed mass or if they were truly massless particles. Indeed, the Standard Model was first described in terms of massless neutrino states, and the determination of neutrino mass, in some sense, became the first evidence of physics beyond the Standard Model.

Even a tiny neutrino mass, of order a few eV, would have an enormous impact on cosmology. Cosmological limits on neutrino masses began in the 1960s and excelled in the 1970s with limits on particle masses and lifetimes. Thus was born the field of particle astrophysics or as it is more commonly called: astro-particle physics.

As it was becoming clear that some form of dark matter was a necessity to explain several very distinct observations, its nature was and remains unknown. Could it be normal matter in a non-luminous state (very difficult as normal matter couples to light and easily shines in one wavelength or another), or could it represent some change in Einstein gravity? Another possibility is that dark matter resides in particle form. As just noted, neutrinos with eV masses could in fact supply all of the dark matter in the Universe. There was one catch though, it is the wrong kind of dark matter. In the early 1980s, work on the formation of structure in the Universe led to the realization that all forms of dark matter were not equal. Structure in the Universe begins to form when the Universe becomes dominated by matter, that is, dark matter, which can be classified as either hot or cold depending on whether or not it is relativistic at the time structure formation begins. Neutrinos, as very light dark matter candidates, would be hot, and as a result, they would erase most all small-scale structures, leaving the Universe very different from the one we observe. In contrast, cold dark matter (CDM) preserves structures on small scales and is in effect in excellent agreement with observations. Hence, a Standard Model of cosmology, known as ΛCDM, where Λ, refers to a dark energy component which may simply be Einstein's cosmological constant.

Is dark matter then a new particle? Particle physics has a long history of solving problems by introducing a new particle, necessitated by either theory or experiment. Examples are plentiful. The positron was needed to complete Dirac's theory of relativistic quantum mechanics. The neutrino was needed to explain the missing energy in Tritium decay. In later examples , the charm quark is needed to explain the suppression of flavor-changing neutral currents. Indeed, one can argue that the entire third generation of quarks and leptons was needed to account for CP violation in weak interactions. The Higgs boson was proposed to explain the breaking of the electroweak gauge symmetry. Thus, to a particle physicist, there is nothing unusual about proposing the existence of a new particle which solves a problem.

Two examples of proposed dark matter candidates are worth mentioning. Both were proposed to solve other problems in particle physics, but yet could play an important role as dark matter. The smallness of the neutron electric dipole moment is an indication that CP violation in strong interactions is very small. A priori, there is no reason for it to be small, and a symmetry was proposed as an explanation. With this symmetry, there is a new very light scalar particle called the axion which could account for the dark matter. Though the axion is light, unlike the neutrino, it is in fact cold due to the mechanism leading to presence in the Universe. A second example comes from an extension of the Poincare algebra known as supersymmetry. Supersymmetric transformations lead to (roughly) a doubling of the number of particle types, and the lightest of these is expected to be stable. Often assumed to be the partner of either the photon, Z, or Higgs bosons, the lightest supersymmetric particle is another well-studied dark matter candidate.

This book is about particle dark matter. It is also not for the faint hearted. It is a serious exposé of particle dark matter, and in particular the mechanisms which lead to its production in the Universe. After a brief introduction to the history and motivation for dark matter, Yann Mambrini jumps straight to a description of the Friedmann-Robertson-Walker Universe and Inflation. Apart from the resolution of classic cosmological problems, inflation is key to producing density fluctuations, which lead to structure formation, and reheating, producing a thermal bath from which the radiation-dominated era in standard cosmology is born. Yann Mambrini spends considerable effort on the latter and begins with a somewhat non-traditional approach by examining dark matter production during the period of reheating after inflation.

Whether produced immediately after inflation or through equilibrium production and subsequent thermal freeze-out, the thermodynamics of the early Universe is a necessary component. The book covers the radiation-dominated era in great detail, including a detailed discussion of thermalization and decoupling. When applied to dark matter, Yann Mambrini provides a complete derivation for the computation of the relic density for cold dark matter. All this is done generically with a few simple examples.

The second part of the book examines the state of dark matter today. Split into two chapters, the reader will find all that's necessary for computations of direct detection rates as well as the complications involved in indirect detection.

The book could be titled "Everything you need to know about particle dark matter." However, Yann Mambrini shies away from the particular dark matter candidate. There is no lengthy discussion of supersymmetry or axion dark matter, the discussion throughout is very generic and can be applied to nearly any dark matter model, though many examples are supplied.

As an added bonus, the Appendices in the book are mini-text books in themselves. All of the basic relativity, particle physics, neutrino physics, and statistics are placed in separate appendices with a final one on useful values in cosmology and particle physics.

Minneapolis, MN, USA
New York, NY, USA
Paris, France
January 2021

Keith Olive
P. J. E. Peebles
Joseph Silk

Preface

I wrote this book out of frustration. Several excellent works and reviews are on the market. I cite some of them at the beginning of the appendix with their particularity. However, a vast majority of them has been written by astrophysicists, and as a student, I always needed the equivalent of the Particle Data Booklet to compute cross sections in specific processes, especially fighting with the "2π-like" factors. On the other hand, particle physics textbooks were very light in the treatment of thermodynamics of the Universe or radiative effects like the loss of energy of a charged particle in astrophysical framework. In other words, I was always juggling between the *Kolb and Turner* book [1] to solve subtleties in Boltzmann equation, the *Jackson* [2] textbook to clarify the radiative effects of a model at the Galactic scale, the *Jungman, Kaminowski and Griest* [3] review for some amplitudes computations and *Shifmann* et al. papers for details in the nucleus composition. This work is a humble attempt of a particle physicist to propose a textbook which can be useful for any phenomenologist who is interested in astroparticle physics, especially in its dark matter aspects.

I also tried to unify units. In such a vast field as dark matter research, we have to deal with cosmological scales (when solving Boltzmann equation or computing reheating temperatures), astrophysical scales (when looking at propagation of cosmic rays or indirect detection of dark matter) and of course microscopic scale when computing interaction cross section. It is always possible to convert Jansky to Joule, $\mathrm{cm}^2\,\mathrm{s}^{-1}$ to GeV^{-2}... However, as a particle physicist by formation, I decided to work in a unified GeV-framework. From this point of view, comparison between energy loss, expansion rates or annihilation cross section is straightforward. In any case, all conversion factors are given in the Appendix and I also give the results in the more "natural" units for people who are used to their own scale-related units.

I want to prevent reading of the manuscript from A to Z, but want the reader to look at the table of contents, or even better, the index, picking a word they would like to understand better and read the corresponding section. Indeed, I wanted a chronological structure, starting from the Big Bang, inflation, reheating, thermal Universe, CMB, and Current Universe, presenting at each step the possible mechanisms of dark matter production. This is also the structure I use for the courses I give at the University. At first sight, this structure seems elegant and logical, but it may be that some elements necessary to a step are at the next step (need to know a thermal distribution even if the Universe is not yet thermalised. for example). The reader should therefore not hesitate to skip from chapter to chapter. The advantage of this presentation is that one can begin reading the book at any stage. No need to read the first chapters to understand the mechanisms involved in the direct or indirect detection of dark matter, for example. I thought this book as an efficient tool above all.

This book should stir up curiosity in reader more than teaching them a complete history of the Universe. I also advice the reader to keep this book in a place easily accessible: anytime they read an article related to dark matter, or physics beyond the standard model, they should find the answer or at least a hint of their answer in this book. If not, they should not hesitate to send me an email concerning a subject not treated or too lightly treated here. This textbook is far from complete and will evolve thanks to the comments of all the readers. To appreciate at its best the reading of the manuscript, some symbol * and ** are added at the beginning of each section of chapter depending on its technical difficulty. So we advice the readers to skip them at a first lecture.

Orsay, France
February 2021

Yann Mambrini

Acknowledgements

I want to especially thank all the people who contributed directly or indirectly to the production of this book. Especially Genevieve Belanger, Soo-Min Choi, Emilian Dudas, Marcos Garcia, Andreas Goudelis, Lucien Heurtier, Kunio Kaneta, Keith Olive and Mathias Pierre.

Above all, I must thank those who, through my discussions, my meetings and by reading their writings, have always been able to fuel my flame. First, Keith Olive who guided me throughout my career, for whom physics is an art of living. Keith taught me to feel the phenomena behind the equations and managed to transmit to me a part of his intuition. His background in particle physics, astrophysics or cosmology is just incredible. Thank you, Keith.

I would also like to thank Jim Peebles. He doesn't know it, but for years I closed my lectures or seminars insisting that, from my point of view, I couldn't understand how the Nobel Committee hadn't yet awarded him the ultimate prize in our field. This was done in 2019, and I was lucky enough to be in Stockholm at the time of the nomination. I read and reread his seminal articles, and their clarity of presentation and calculation make them works of art in scientific literature.

Another physicist that was often present, indirectly, during my career was Joe Silk. Joe was always accessible "next door" at the Institute of Astrophysics of Paris every time I had a question, a doubt, or needed an explanation about an observation or theoretical subtleties. His kindness is matched only by his incredible scientific talent.

I would also like to warmly thank Lisa Scalone, from Springer Publishing, for her incredible patience, her boundless kindness and her constant support in this adventure that I did not imagine to be so trying at the beginning of the project. This book would definitely not be in your hands without her.

Finally, I would like to dedicate this book to my mentor, Pierre Binetruy, without whom none of this would have been possible. He introduced me to the fabulous world of theoretical physics, from my university years to the heart of my scientific career. He is missed by all of us, and I hope that wherever you are, you enjoy the journey as you read this book.

Contents

About the Author

Yann Mambrini is research director at the French CNRS (Centre National de la Recherche Scientifique) in the Irene Joliot Curie Laboratory, University Paris-Saclay. His research interests concern the field of particle physics and fundamental interactions. He works actively on extensions of the standard model, from supergravity to grand unified theories, and more specifically on their cosmological consequences, especially involving dark matter aspects. He was awarded with the Prix d'Excellence Scientifique of the National Research Council in 2010, 2014 and 2018. Currently, he is lecturing at Ecole Normal Supérieure. One of his passions is the concept of time, from its measurement to its nature, and the history of ideas in physics. He wrote several popular books on the subject: Histoires de Temps and Le Siècle des révolutions Scientifiques, Ed. Ellipses and Newton à la Plage, Ed. Dunod. He shares this passion with magic, with which he presents shows for the general public mixing illusions, time and mysteries of the Universe.

Introduction

1

Abstract

The twentieth century studies evidenced the existence of a new form of matter which have inspired interest in modern physics scenario. It has been named "Dark Matter" (DM), exotic name but with a clear meaning: a component of matter that does not emit luminous radiation. Beginning from a study presented by Zwicky in 1933 [4] who analyzed the motion of individual galaxies in the Coma cluster, subsequently other observations have indicated the presence of dark matter from the kinematics of gravitationally bound systems and rotating spiral galaxies, the effects of gravitational lensing of background objects, various evidences among which the observation of the Bullet Cluster, until recent results from the PLANCK satellite. Furthermore, the dark matter appears to have an important role in the formation of the structures, in the evolution of galaxies and also has effects on non-uniformity observed cosmological microwave of background radiation. Before going into detailed analyses of each step structuring the dark matter presence and interaction in our Universe, we will first introduce in this chapter the most important evidences, explaining where the dark matter may intervene to resolve the oddities observed before listing the general features of dark matter particles.

1.1 The First Dark Matter Paper

There are many books that retrace in a more or less faithful and more or less exhaustive way the history of dark matter research and its interpretations. I advise the reader to refer to three very accessible references. First of all, the summary by G. Bertone and D. Hooper [5] is an excellent introduction to the field. They retrace the steps that led to the hypothesis of a dark matter in the form of a particle. R.H. Sanders in [6] summarizes very well, on the astrophysical side, the disappointments and difficulties that had to be overcome before admitting the presence of missing

Y. Mambrini, *Particles in the Dark Universe*,
https://doi.org/10.1007/978-3-030-78139-2_1

mass in our Universe. In [7, 8] you will find a complete summary of the dark matter candidates, their properties, and the status of the detection prospect. Finally, the excellent review by J. Peebles [9] reminds us of the steps that had to be taken in order to have an image of an expanding cosmos dominated by predominantly black components. Concerning inflationary models, there is no more complete review than the one proposed by Keith Olive [10] and the seminal articles by A.H. Guth [11] and A. Linde [12].

Each of my colleagues working in the so-called dark matter field will have their own interpretation of facts and history. I remember a question I was asked at a seminar in Moscow: "What is, *for you*, the first paper dealing with dark matter?" This question, which seemed trivial, turned out to be much more difficult to answer. And I turned it into a seminar that I still manage to give for Master's students. First of all, we have to agree on the term "Dark Matter." What we consider in this textbook is a dark matter in the form of a particle or a field. For a very long time the presence of missing mass has been debated. Then, once its presence was irrefutable, the problem was not to explain its presence, but to understand why it was not seen (light path deflected, obscured by interstellar gas? Stars too old to be visible?). It took several years, and the work of Peebles and Ostriker on the stability of galaxies, to realize that 'something more' was needed. The neutrino was obviously the first candidate, quickly discarded because of its too light mass and relativistic character. It was not until the 1970s that the first paper mentioning the need for a new particle appeared, and its authors calculated the expected relic density, based on data from the cosmological diffuse background and astronomical observations of the time.

Prehistory

The first mention of the word "Dark Matter" appeared in a work by Poincaré in 1908. In "*The Milky Way and the Theory of Gases*," Poincaré follows Lord Kelvin's 1904 hypothesis and proposes to treat stars in galaxies as atoms in a gas. He then deduced, by comparing the movement of the Earth around the Sun with that of the Sun around the nearest star Proxima, an approximation of the mass of the Milky Way. Assuming an approximately constant star density, he deduced that our galaxy should be made up of a billion stars. This value was quite close to the number of stars then visible in our galaxy (we know today that it contains 200 to 400 times more stars). So he concluded

> *There is not dark matter, or at least not so much as there are of shining matter.*

Of course, his constant density hypothesis finally made his calculation wrong, most of the dark matter being present around the galactic center which is a strong gravitational well. But here, we must appreciate the revolutionary idea of treating a galaxy as a gas, and applying a kinetic theory to it, to calculate its mass. The second appearance of the word «dark matter» in the literature is in a paper of the physicist

Jan Oort from Netherland in 1932. While he was analyzing the radial velocities, he noticed a discrepancy with Newton's law. He computed that only one-third of the dynamically inferred mass was present in bright visible stars. It is clear from the context that, as characterizing the remainder as «dark» («Dunkle Materie»), Oort was describing all matter which is not in the form of visible stars with luminosity comparable or larger than that of the Sun. Gas and dusts between the stars were the main constituents of his «invisible mass» that should be found (for him) soon. The main reason evoked at this time was the presence of low luminosity objects (dead stars) or large absorbing gas. Imagining a new dark component took a very long time to physicists, who even preferred to modify the law of gravity at large scale before invoking a new particle. In this sense, the first real work underlining that the missing mass could be problematic is Fritz Zwicky in 1933.

It is by observing the Coma cluster that Zwicky realizes a problem in the movement of the galaxies that make it up. The Coma cluster is a highly regular gravitationally bound system of thousands of galaxies at a distance of about 100 Mpc. The astrophysicist noticed that each of its galaxies was following a movement that seemed fast compared to the movements of closer galaxies such as Andromeda. He then applied the virial theorem to this cluster of 800 galaxies of $\sim 10^9$ solar masses and deduced that the average speed of each of them must be of the order of 80 km/s. What was his surprise when he measured the individual velocities of the galaxies in this cluster, and that they were close to 1000 km/s. The only possible explanation was the presence of invisible matter which increased the gravity potential. Hence his conclusion at the end of his article was

> *In order to obtain the observed value of an average Doppler effect of 1000 km/s or more, the average density in the Coma system would have to be at least 400 times larger than that derived on the grounds of observations of luminous matter. If this would be confirmed we would get the surprising result that dark matter is present in much greater amount than luminous matter.*

This result was then completely forgotten and no one took Zwicky's comment seriously. Indeed, large scale astrophysics had only just emerged following the discovery of Hubble, and many physicists believed that the problem of *missing mass* will be solved when they better understand the mechanisms of light absorption in the interstellar/internebulae medium. In fact, the *missing mass* problem has long been viewed as a *missing luminosity* problem: why do not we see the astrophysical bodies that should be responsible for Newtonian dynamics. On the other hand, several scientists tried rather to modify (already in the 30s) the law of attraction in $\frac{1}{r^2}$. It is not surprising. The theory of general relativity was then only in its observational beginnings and was much less untouchable than today. Introducing a new component of matter into the Universe was much more esoteric. It was then that the analysis of galaxies began.

The Galactic Scale

The history of measurements of the rotation curves of galaxies dates back to 1914 when Vesto Slipher at the Lowell Laboratory observed the Andromeda galaxy. It is the closest galaxy to us, at a distance of ~800 kpc, but estimated at 210 kpc at this time due to the approximate determination of the Hubble constant. Slipher noticed that the speeds of the stars measured to the left of the galaxy bulge were approaching us at higher speeds (~320 km/s) than those on the right side of the central bulge (~280 km/s). This observation corresponds to that of a disc spinning in front of us. In 1918, Pease at the Mount Wilson Observatory measured the speed of rotation over a radius of 600 pc (the central part of Andromeda). Its result could be expressed by the formula

$$V_c = -0.48r - 316, \tag{1.1}$$

where V_c is the measured circular speed (in km/s) at a distance r from the central Andromeda bulge. This central part seems to rotate at a constant angular speed. It is Horace Babcock who, in his PhD thesis in 1939, extended the study to larger scale, up to 24 kpc from the center. Measuring still a constant angular velocity, he concluded

> *..constant angular velocity discovered for the outer spiral arms is hardly to be anticipated from current theories of galactic rotations.*

From the computation of the density, he deduced a total mass for Andromeda of 10^{11} solar mass, equivalent to a mass-to-luminosity ratio $M/L = 50$. He then concluded

> *This last coefficient is much greater than that for the same relation in the vicinity of the Sun.*

As we can see, for Babcock at the scale of galaxies, just like for Zwicky at the scale of galaxy clusters, a problem remains. But for the moment, no explanation is really satisfactory. When Karl Jansky observes radio waves of synchrotron radiation from the galactic center in 1932, and Oort and Van de Hulst study the sky by observing the spectral line 21cm from the hydrogen atom in 1957, astrophysicists realize that the presence of an invisible matter, responsible for the rotation of the closest galaxies, extends well beyond the visible domain. Van de Hulst does not insist too much in his article on the flatness of the rotation curve. But, by calculating the mass of M31 (Andromeda), he deduces that it is much heavier than the Milky Way. The "dark matter" hypothesis does not (yet) hit the galactic scale.

Stabilization of the Structures

In the 70s, Moore's law of exponential development describing the evolution of computing power over time affected astrophysical studies: the computing power doubling every two years, it was possible at the end of the 60s to develop electronic machines for the digital resolution of complex problems (technically, it was the replacement of vacuum tubes by transistors that gave a big leap in the field). Franck Hohl carried out in 1971 one of the first "N-body" simulations (100,000 stars!!) to test the stability of galactic structures, with a disk of particles supported in equilibrium almost entirely by rotation. He noticed that an elongated spiral shape forms after 2 revolutions, but quickly the kinetic energy diffuses the particles and the final state is a pressure dominated gas with large elongated axisymmetric ellipses. To keep the observed spiral shape, Miller, Pendergast, and Quirk tried to stabilize the model by artificially adding energy loss. Nevertheless, the heating of the gas destroyed the spiral structure after a few revolutions. It was at this point that the proposition of the presence of a dark halo came to the rescue and is mentioned for the first time in an article.

In 1973, P.J.E. Peebles and J.P. Ostriker [13] noticed that the individual random velocities of stars in our galaxies (around 30–40 km/s) are much smaller than the average circular motion (around 200 km/s). So not only is the system unstable as noted by Hohl et al., but it also shows that the galaxies seem to be dominated by a cold gravitational system and not by kinetic pressure. Peebles and Ostriker noticed that if at least 28% of the kinetic energy is stored in the rotation, the system is unstable and destabilized very quickly. However, in our Milky Way, the rotational speed is around 200 km/s, while the individual velocities approach 40 km/s, which gives a ratio of 50% if one takes into account only the visible stars, well over the limit of stability. The ingenious idea of Peebles and Ostriker is then to add an additional component to the galaxy, a dark halo U_{dark}, which contributes at least 50% of the mass around the position of the Sun:

$$U \rightarrow U + U_{dark}, \tag{1.2}$$

where U represents the total gravitational potential of the galaxy. This spheroidal system increases the gravitational potential energy but adds nothing to the rotational energy. The ratio (kinetic energy)/(gravitational energy) would then be reduced and perhaps stability restored. In their 1973 seminal paper, they conclude

Adding an extended component corresponding to the "halo" [. . .] apparently will stabilize the system if the halo mass is equal to or somewhat greater than the disk mass.

Vera C. Rubin and W. Kent Ford [14] also analyzed the rotation curve of the Andromeda galaxy, up to 120 minutes of arc, and proposed also the presence of a halo surrounding all the nebulae, even if their conclusion states

.. extrapolation beyond that distance is a matter of taste.

A study by D.H. Rogstad and G.S. Shostak in 1972 showed that the presence of a dark halo is also necessary in several galaxies surrounding the Milky Way. In 1985, Dekel and Silk went further, showing in a work ahead of its time [16] how dark matter could explain the formation of dwarf galaxies, finding that the relation between their mass and radius can fit in a cosmological model which includes a right component of cold dark matter.

After several independent works by Dolgov, Zeldovich, Cowsik, Mc Clelland, Hut, Lee, and Weinberg, which give limits on the dark matter mass from observations, one has to wait for the paper by Steigman et al. in 1978 [17] to see a first particle candidate, beyond the Standard Model, including the computation of its relic abundance and a complete analysis of the equilibrium issue. Preceding important works proposed neutrinos as dark matter, showing they cannot fill the entire missing mass, or even massive neutrino–like particles, but [17] is the first to generalize to a yet unknown electroweak particle. I am sure that a lot of colleagues will dispute my choice. However, I consider it as the first dark matter paper in the sense that they study, for the first time, the cosmological consequences of the existence of *any* stable, massive, neutral lepton, particle that is now more known under the acronym "WIMP" for Weakly Interacting Massive Particle. Since then, several models have been proposed to fit a candidate in a more general framework. The more solid proposition (until now) stays the supersymmetric version of the Standard Model, which includes two serious candidates: the neutralino (the lightest partner of the weak gauge bosons) and the gravitino (supersymmetric partner of the graviton). The first complete study on the subject was made by Ellis, Hagelin, Olive, Nanopoulos, and Srednicki in their seminal work [18], where they proposed scenarios with Higgsino (partner of the Higgs), photino (partner of the photon), and gravitino dark matter, computing their relics from the Big Bang. After this brief historical introduction, one needs to see in more detail how the local abundance has finally been measured, and how we can extrapolate its distribution all over the large scale structures of the Universe.

1.2 Local Dark Matter

The dynamical density of matter in the Solar vicinity can be estimated using vertical oscillations of stars around the galactic plane. The orbital motions of stars around the galactic center play a much smaller role in determining the local density. Oort in 1932 [15] indicated, in his analysis, as members of a "star atmosphere," a statistical ensemble in which the density of stars and their velocity dispersion define a "temperature" from which one obtains the gravitational potential. The result contradicted grossly the expectations: the potential provided by the known stars was not sufficient to keep the stars bound to the galactic disk because the density of visible stellar populations by a factor of up to 2, and so the galaxy should rapidly be losing stars [19]. Since the galaxy appeared to be stable, there had to be some missing matter near the galactic plane, Oort thought, exerting gravitational attraction. This limit is often called the Oort limit. This used to be counted as the first indication for the possible presence of dark matter in our galaxy: the amount of invisible matter in the Solar vicinity should be approximately equal to the amount of visible matter. A modern calculation of the local relic abundance using several observables can be found in the work of R. Catena and P. Ullio in [20]. Their result gives a local amount of dark matter of $\rho_0 \simeq 0.39\ \mathrm{GeV/cm^3}$, which corresponds mainly to 100,000 particles passing through us per cm^2 per second.

It is important to notice that such a value for a local amount of dark matter ($0.39\,\mathrm{GeV/cm^3}$) is very small. To collect one gram of dark matter, we would have to put together all the particles contained in a volume comparable to that occupied by the entire Earth. Indeed, the dark matter is much more present near the galactic center. That also explains why we can safely use the Newton laws in the solar system. Of course, when one looks at larger scale, the situation is completely different: between the Sun and the next star, for instance, the void is huge. A quick calculation shows that the density of the planets in the solar system in a sphere that extends halfway to the next star, Proxima Centauri, is about the same as the density of dark matter.

1.3 Anomalies in Rotation Curves of Galaxies

Radiowave radiation from interstellar gas, in particular that of neutral hydrogen, is not strongly absorbed or scattered by interstellar dust [21]. It can therefore be used to map and to study the motion of neutral hydrogen clouds concentrated in spiral arms. We can therefore determine the angular velocity of the gas at different distances from the galactic center r, and plot the corresponding rotation curve $v = v(r)$. The most convincing and direct evidence for dark matter on galactic scales comes mainly from the observations of the rotation curves of galaxies. Rotation curves are usually obtained by combining observations of the 21 cm line with optical surface photometry: if θ is the angle between the velocity of the star and the line of sight, the velocity components can be written as $v_r = v \cos\theta$ and $v_t = v \sin\theta$. The tangential velocity v_t results in the proper motion, which can be measured by taking photos at intervals of several years or decades. The radial velocity v_r can be measured from Doppler shift of the stellar spectrum, in which the spectral lines are often displaced toward the blue or red. The blueshift means that the star is approaching, while the redshift indicates that it is receding. From 1940 (Oort's) numerous observations in spiral galaxies showed, in outer regions of galaxies, an anomaly in rotation velocity that can be translated in a high M/L, mass-to-luminosity ratio. Observed rotation curves usually exhibit that the central part of the galaxy rotates like a rigid body, i.e. $v \propto r$, and then the velocity reaches a maximum value, or plateau. At this point we would expect a decreasing velocity outward, as the third Kepler law suggests, instead there is a characteristic flat behavior until edges of galaxies where few light is emitted. Let us consider for simplicity a spherical distribution of matter in the galaxy. In Newtonian dynamics, the virial theorem determines the circular velocity, which is expected to be

$$\frac{v^2}{r} = \frac{GM(r)}{r^2} \Rightarrow v = \sqrt{\frac{GM(r)}{r}}, \tag{1.3}$$

where $M(r) = 4\pi \int_0^r \rho(a)a^2 da$ is the mass of the matter of density $\rho(a)$ contained in a sphere of radius r. From Eq. (1.3), $v(r)$ should fall following $1/\sqrt{r}$ outside the optical disk where M should be constant without the presence of dark matter. The fact that the observation gave $v(r) \simeq$ constant implies the existence of a halo with $M(r) \propto r$ and then $\rho(r) \propto 1/r^2$ confirming that a large part of the mass is present in the outer part of the galaxy and not in the visible disk. This is well illustrated in Fig. 1.1. The mass distribution is obtained from rotational curves, determined from light distribution of luminous components in the galaxy. By photometry the estimated mass in our galaxy between the galactic center (GC, Sagitarius A*) and the Sun (at a distance $r_\odot = 8$ kiloparsec (kpc) from the GC) is $M = 9 \times 10^{10} M_\odot$, while for the outer edge of galaxy, where the luminosity decreases exponentially, the component of luminous is negligible. At this distance from the galactic center, the rotational velocities verify approximately the Keplerian law as the visible matter dominates the matter distribution at such distances from the GC.

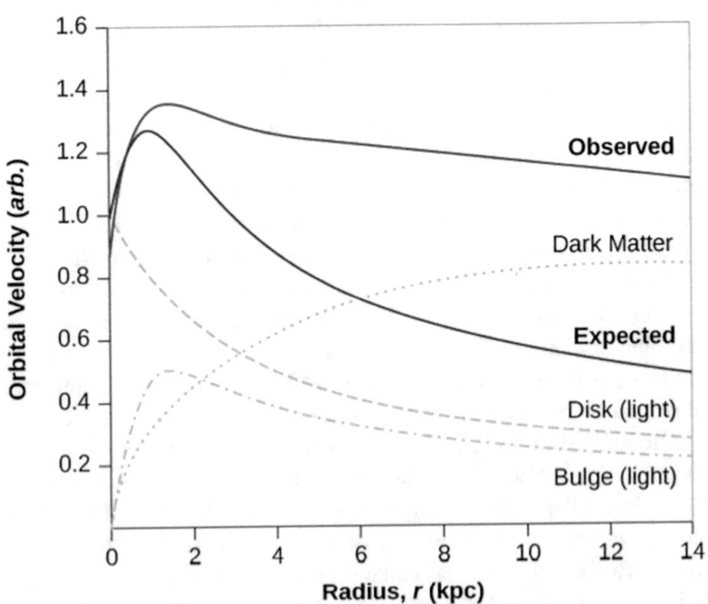

Fig. 1.1 Rotation curve of a galaxy with the different components extracted from observations: bulge, disk, and dark halo. OpenStax University Physics, extracted from the book *University Physics Volume 1* under *Creative Commons Attribution 4.0 International* license

Below $r_\odot$ some discrepancies in the rotational velocity can be explained by the presence of an invisible mass halo, called dark matter halo or dark halo, around our galaxy. This dark component of matter would be spherically distributed in a halo extended until 230 kpc from galactic center and having a density profile following

$$\rho(r) = \frac{\rho_0}{\left(\frac{r}{a}\right)\left(1+\frac{r}{a}\right)^2}, \tag{1.4}$$

a being a typical scale of the halo depending on the galaxy. With this profile the galaxy behaves like $1/r$ at the center ($r \ll a$) and $1/r^3$ in the edges $r \gg a$. With this calculation the mass of halo of dark matter must be $5.4 \times 10^{11} M_\odot$ within 50 kpc and $2.5 \times 10^{12} M_\odot$ at 230 kpc [22]. The profile proposed in Eq. (1.4) is called NFW profile and was proposed by Navarro Frenk and White in [23], whereas the first measurement of the rotation curve was made by Ford and Rubin in [14].

1.4 Cluster Dark Matter

Another mass discrepancy was found by Zwicky in 1933 [24]. He measured redshifts of galaxies in the Coma cluster and found that the velocities of individual galaxies with respect to the cluster mean velocity are much larger than those expected from the estimated total mass of the cluster, calculated from masses of individual galaxies. His article was in fact more a review on the status of observational cosmology in 1933. It is just at the end of its article that he concludes on the remark concerning the probable presence of dark matter in the Coma cluster. Stars move in galaxies and galaxies in clusters along their orbits, and those are virialy bound systems: the orbital velocities are balanced by the total gravity of the system, similar to the orbital velocities of planets moving around the Sun in its gravitational field. In the simplest dynamical framework, one treats clusters of galaxies as statistically steady, spherical, self-gravitating systems of N objects of average mass m, and average orbital velocity v. The total kinetic energy E of such a system is then

$$E = \frac{1}{2} N m v^2.$$

If the average separation is r, the potential energy of $N(N-1)/2$ pairings is

$$U = -\frac{1}{2} N(N-1) \frac{Gm^2}{r}.$$

The virial theorem states that for such a system,

$$E = -\frac{U}{2}.$$

The total dynamic mass M can then be estimated from v and r from the cluster volume

$$M = Nm = \frac{2rv^2}{G}.$$

Zwicky was the first to use the virial theorem to infer the existence of unseen matter. He found that the orbital velocities are almost a factor of ten larger than expected from the summed mass of all galaxies belonging to the clusters, and this implies that the average mass of galaxies within the cluster has a value about 400 times greater than expected from their luminosity. The gravity of the visible galaxies in the cluster would be far too small for such fast orbits, so something extra was required. This is known as the "missing mass problem," and he proposed that most part of the missing matter was a dark, non-visible form of matter, which would provide enough of the mass and gravity to hold the cluster together.

There exists another method to determine the mass of cluster: the temperature of the hot intracluster gas, like the galaxy motion, traces the cluster mass. Indeed, hot gas inside the clusters emits X-ray radiation through bremsstrahlung process (see Sect. 5.6.4 for more details). Observations show that the gas is in hydrodynamic equilibrium ($dF_{grav} = dF_{press} = \frac{dP}{dr} = -\frac{GM_r\rho}{r^2}$ with M_r the inner total mass inside the radius r), and it moves in the gravitational field of cluster in orbits with velocities depending on the mass of the cluster. Through spectroscopic analysis of hot gas, we can obtain density and temperature of gas as a function of galactic distance r. With these parameters, we can get mass distribution of cluster. For example, gas mass of the Coma cluster is $M_{gas} = 1.05 \times 10^{14} M_{\odot}$, which is larger than the visible mass $M_{vis} = 1.5 \times 10^{13} M_{\odot}$, but not sufficiently to explain the value extracted from the virial theorem, that is, $M_{vir} = 3.3 \times 10^{15} M_{\odot}$. We illustrate in Fig. 1.1 the different components of the galactic structure and their relative abundance as a function of their distance from galactic center.

N-body simulations also give information on the possible distribution of dark matter in our galaxy. This is a daunting task especially when one considers that even the three-body problem—the problem of describing the orbits of three celestial bodies under their reciprocal gravitational attraction—is extraordinarily difficult and can only be solved in some simplified cases. How can we therefore hope to solve the problem of computing the reciprocal interactions of *all particles in the Universe?* The problem is that, in principle, one has to calculate for each particle the attraction of every other particle in the universe. Eric Holmberg, an ingenious Swedish scientist, found an original solution to the problem in 1941. He decided to simulate the intersection of two galaxies using 74 light bulbs, together with photocells and galvanometers, using the fact that light follows the same inverse square law as the gravitational force. He then calculated the amount of light received by each cell and manually moved the light bulbs in the direction that received the most intense light.

The first application of computer calculations to gravitational systems was probably by John Pasta and Stanislaw Ulam in 1953. Their numerical experiments were performed on the Los Alamos computer, which by then had already been applied to a variety of other problems, including early attempts to decode DNA sequences and the first chess-playing program. Two young astrophysicists, the Toomre brothers, had access in the early 1970s to one of the NASA's two IBM 360-95 computers, completed with high-resolution graphics workstations and auxiliary graphics–rendering machines (computing facilities far in advances of any other astrophysics laboratory). They set up a series of simulations of galaxy grazings and collisions using a simple code that described the galaxies as two massive points surrounded by a disk of test particles. The outcome of the analysis was a very influential paper, published in 1972, that contained a detailed discussion of the role of collisions in the formation of galaxies. Together with their paper, the Toomre brothers also created a beautiful 16 mm microfilm movie "Galactic Bridges and Tails" that you can find at http://kinotonik.net/mindcine/toomre. A photo of the simulation is represented in Fig. 1.2.

Numerical simulations have improved immensely since these pioneering attempts, thanks to a dramatic increase in computing power. Modern supercomputers allow us to simulate entire universes by approximating their constituents with up to ten billion particles, as in the case of the Millennium simulation shown in Fig. 1.2, which was run at the Max Planck Society's Supercomputing Center in Garching, Germany. This has a computing power ten million times larger than the old IBM 360-95 used by the Toomre brothers. We usually parameterize our "ignorance" of the exact distribution obtained by the N-body simulation systems with an empirical formula for the dark matter halo in the cluster or in our Milky Way:

$$\rho_{DM}(r) = \frac{\rho_0}{\left(\frac{r}{r_s}\right)^{\alpha}\left(1+\frac{r}{r_s}\right)^{3-\alpha}},$$

where r_s is a typical scale (usually the distance from the Sun to the galactic center), and ρ_0 is the normalization constant, determined from the observation of the local density of dark matter in the vicinity of the Sun. The values of α are determined by observations or after running simulations and are approximately between $0 \leq \alpha \leq 3/2$. For a more detailed analysis on the profile structure and how one obtains them, have a look at Sect. 5.8.2. It is however important to notice that N-body simulations deal with "particle" mass of the order of several solar masses and that one should extrapolate their results down to electroweak masses, which can infer some errors. Moreover, the smaller dimension scale that these simulations can reach is of the order of the kiloparsec, so all spatial extrapolations around the parsec galactic center can be dubious, especially, when we do not know exactly which kind of ingredient and initial conditions are taken by the simulations. That explains the

Fig. 1.2 N-body simulation in 1972 with 120 particles (top) and in 2014 with ten billions of particles (bottom), from the Illustris project, https://www.illustris-project.org/, licensed under the Creative Commons Attribution-Share Alike 4.0 International license

orders of magnitude existing between prediction in indirect detection of dark matter when most of the annihilation processes occur around the galactic center, where the precisions of the simulations are the worst, without real visible observables to test them.

1.5 Gravitational Lensing

The theory of general relativity teaches us that gravitational field curves the space-time metric, whereas the particles or photons travel in geodetic trajectory. An observable consequence of this effect is the gravitational lensing: a photon in a gravitational field moves as if it possessed mass, and light rays therefore bend around gravitating masses. Thus celestial bodies can serve as gravitational lenses probing the gravitational field, whether baryonic or dark without distinction, and thus can probe the dark massive component of any celestial object. If we consider that a trajectory of light ray in a gravitational filed with a spherical symmetry (r, θ, ϕ) is represented as

$$\frac{d^2}{d\phi^2}\left(\frac{1}{r}\right) + \frac{1}{r} = 3\frac{GM}{r^2}.$$

The solution of this equation can be thought as a perturbation of special relativity (without gravitational field), and the deflection angle is derived in the Appendix, Eq. (A.106):

$$\delta = 4\frac{GM}{r_0 c^2},$$

r_0 being the closest distance from the light ray to the massive body causing the deflection. The deflexion provided for a light ray that enters in gravitational field of the Sun is $\delta \simeq 1.75''$ and has been measured by Eddington in 1919. Since photons are neither emitted nor absorbed in the process of gravitational light deflection, the surface brightness of lensed sources remains unchanged. Changing the size of the cross section of a light bundle only changes the flux observed from a source and magnifies it at a fixed surface brightness level. We can categorize three classes of gravitational lensing as follows:

- Strong lensing, the photons move along geodesics in a strong gravitational potential, which distorts space as well as time, causing larger deflection angles and requiring the full theory of general relativity. The images in the observer plane can then become quite complicated because there may be more than one null geodesic connecting source and observer. Strong lensing is a tool for testing the distribution of mass in the lens rather than purely a tool for testing general relativity. The masses of clusters of galaxies determined using this method confirm the results obtained by the virial theorem and the X-ray data.
- Weak lensing, which refers to deflection through a small angle when the light ray can be treated as a straight line, and the deflection as if it occurred discontinuously at the point of closest approach (the thin-lens approximation in optics). One then only invokes special theory of relativity, which accounts for the distortion of clock rates. This kind of lensing allows to determine the distribution of dark matter in clusters as well as in superclusters: the lensing mass estimate is almost twice as high as that determined from X-ray data.

- Microlensing, if the mass of the lensing object is very small, one merely observes a magnification of the brightness of the lensed object. Microlensing of distant quasars by compact lensing objects (stars, planets) has also been observed and used for estimating the mass distribution of the lens-quasar systems. A fraction of the invisible baryonic matter can lie in small compact objects. To find the fraction of these objects in the cosmic balance of matter, special studies have been initiated, based on the microlensing effect. This process is used to find Massive Compact Halo Objects (MACHOs), small baryonic objects as planets, dead stars, or brown dwarfs, which emit so little radiation that they are invisible most of the time. A MACHO may be detected when it passes in front of a star and the MACHO's gravity bends the light, causing the star to appear brighter. Some authors claimed that up to 20% of the dark matter in our galaxy can be in low-mass stars (Fig. 1.3).

1.6 Bullet Cluster

The Bullet Cluster (1E 0657-558) consists of two colliding clusters of galaxies. It is at a co-moving radial distance of 1.141 Gpc (3.721 Gly). Gravitational lensing studies of the Bullet Cluster are claimed to provide the best evidence to date for the existence of dark matter [25]. At a statistical significance of 8σ, it was found that the spatial offset of the center of the total mass from the center of the baryonic mass peaks cannot be explained with an alteration of the gravitational force law. In Fig. 1.4, we show the Chandra X-ray Observatory image of this cluster taken in 2004 [26]. This cluster was formed after the collision of two large clusters of galaxies. Hot gas detected by Chandra in X-rays is seen as two pink clumps in the image and contains most of the "normal" or baryonic matter in the two clusters. The bullet-shaped clump on the right is the hot gas from one cluster, which passed through the hot gas from the other larger cluster during the collision. An optical image from Magellan and the Hubble Space Telescope shows the galaxies in orange and white. The blue areas in this image show where astronomers find most of the mass in the clusters. The concentration of mass is determined using the effect of gravitational lensing, where light from the distant objects is distorted by intervening matter. Most of the ordinary visible matter in the clusters is clearly separate from the matter responsible of the gravitational lensing, giving direct evidence that nearly all of the matter in the clusters is dark: the hot gas in each cluster was slowed down by a force like air resistance, whereas the dark matter was not slowed by the impact because it does not directly interact with the gas itself or if not through gravity. Therefore, during the collision, the lumps of dark matter from the two clusters moved ahead of the hot gas, producing the separation of dark matter and baryonic matter. If hot gas was the most massive component in the clusters, as proposed by alternative theories of gravity, this effect would not be seen: this result shows that dark matter is required at least in the Bullet Cluster.

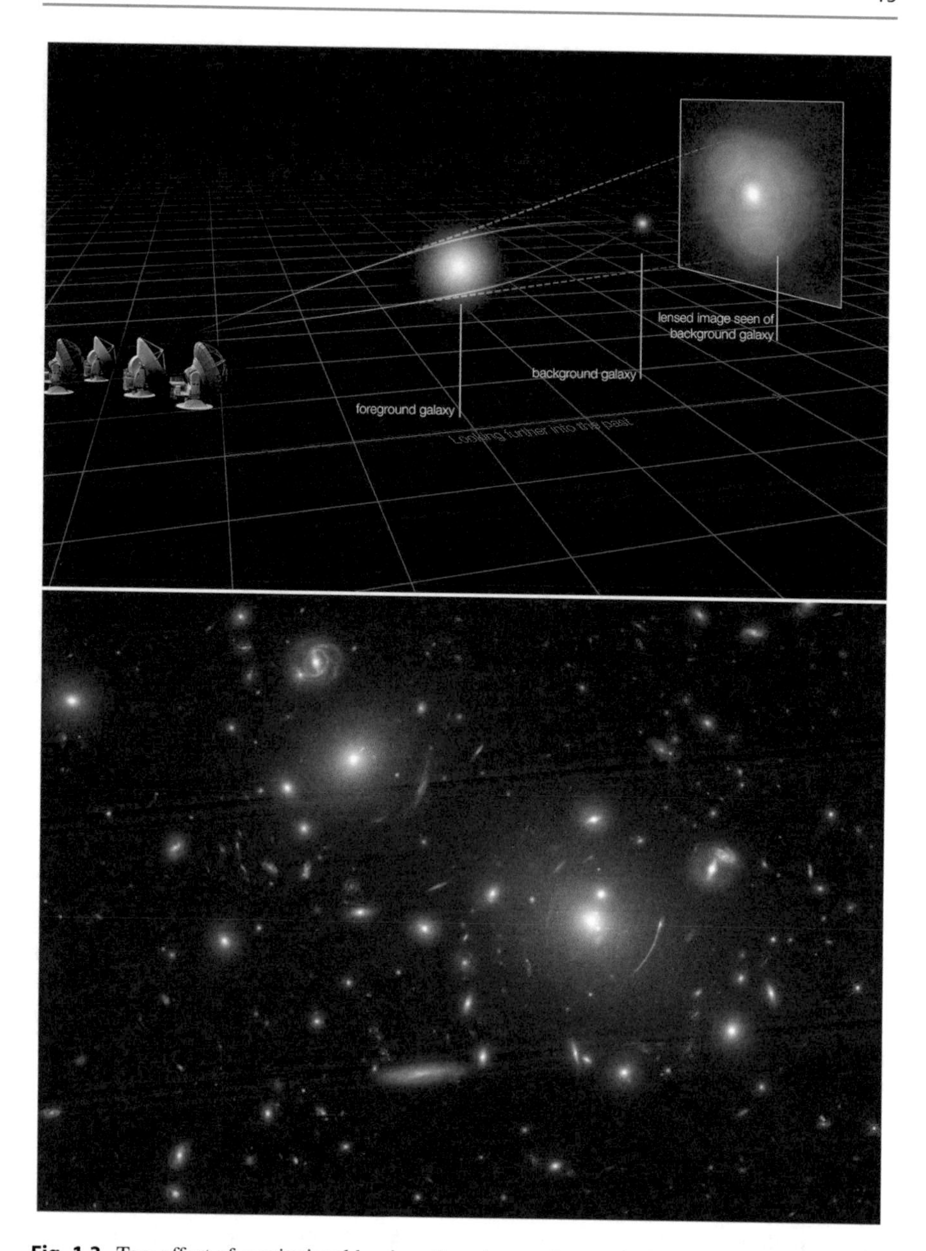

Fig. 1.3 Top: effect of gravitational lensing of a galaxy acting on the light emitted from a galaxy on its way to the Earth, ALMA (ESO/NRAO/NAOJ), L. Calçada (ESO), Y. Hezaveh et al. Bottom: distorted images due to the gravitational lensing system called SDSS J0928+2031 observed by the Hubble telescope. ESA/Hubble image released by the ESA under the Creative Commons Attribution 4.0 Unported license

Fig. 1.4 Bullet Cluster photo in X-ray (red) superimposed with the gravitational lensing (blue), exposition time of about 140 h and megaparsec scale [26] (NASA/CXC/M. Weiss)

1.7 Comparison of Three Matter Abundance

From all the measurements described above, one can estimate the relative mass contribution of the galaxies $\Omega_g = \frac{\rho_g}{\rho_{crit}}$, ρ_g being the density of mass in the galaxies and ρ_{crit} the critical density of the Universe (see Sect. 2.1.6 for more details) to be $\Omega_g \simeq 0.03 - 0.07$. This estimation comes from combining the mean luminosity per unit volume produced by galaxies $\mathcal{L}_g$ and the mean mass to light ratio $\langle M/L \rangle$

$$\rho_g = \mathcal{L}_g \times \langle \frac{M}{L} \rangle \simeq 6 \times 10^{-31} h^2 \text{ g cm}^3,$$

which gives[1] $\Omega_g \simeq 0.03 - 0.07$. At the level of the cluster of galaxies, estimations of the gravitational lens produced by Abell 2218 or Abell 1689 give a similar result, with a mean $\langle M/L \rangle$ around 10 times the one in galaxies (almost constant for scales above the Mpc [27]), and coherent with other observations, which gives $\Omega_{cluster} \simeq 0.2-0.4$. The discrepancy between the three values of Ω_i can then also be attributed to the presence of non-luminous dark matter, which may play an important role in

[1] $\rho_c = 1.78 \times 10^{-29} h^2$, see Sect. 2.1.6.

structure formation. For a mass scale $R > 1.5\,h^{-1}$ Mpc (typical galaxy radius), the mass-to-light ratio of superclusters of galaxies confirms that there does not exist an additional quantity of dark matter at higher scale, $R = 6\,h^{-1}$ Mpc.

We can also estimate the contribution from baryonic material by comparing the observed abundances of light elements (deuterium, ^{3}He, ^{4}He, and ^{7}Li) with the predictions of primordial nucleosynthesis computations, which gives us $\Omega_b \simeq 0.04$. This value is obtained from the standard Big Bang nucleosynthesis and (except from the lithium "problem") corresponds to recent observations. For the curious people, they can go to have a look at Sect. 3.4 for more details. In few words, according to the Big Bang model, the Universe began in an extremely hot and dense state. For the first few seconds, it was so hot that atomic nuclei could not form, and space was filled with a hot soup of protons, neutrons, electrons, photons, and other short-lived particles. Occasionally a proton and a neutron collided and sticked together to form a nucleus of deuterium (a heavy isotope of hydrogen), but at such high temperatures they were broken immediately by high-energy photons. When the Universe cooled down, these high-energy photons became rare enough that it became possible for deuterium to survive, during a short period before the expansion freezes out the production process. This narrow time window is called the deuterium bottleneck. These deuterium nuclei could keep sticking to more protons and neutrons, forming nuclei of ^{3}He, ^{4}He, lithium, and beryllium. This process of element formation is called nucleosynthesis. The denser proton and neutron gas is at this time, the more of these light elements will be formed. As the Universe expands, however, the density of protons and neutrons decreases and the process slows down. Neutrons are unstable (with a lifetime of about 15 min) unless they are bound up inside a nucleus. After a few minutes, therefore, the free neutrons are gone and nucleosynthesis stops. There is only a small window of time in which nucleosynthesis can take place, and the relation between the expansion rate of the Universe (related to the total matter density) and the density of protons and neutrons (the baryonic matter density) determines how much of each of these light elements are formed in the early Universe. Figure 1.5 shows the computed abundance of deuterium D (^{2}H), ^{3}He, ^{4}He, and ^{7}Li (compared with (H) hydrogen). The abundances are all shown as a function of η_b, the baryon-to-photon ratio, which is related to Ω_b by $\Omega_b = 0.004\,h^{-2}\frac{\eta_b}{10^{-10}}$. The estimates of the primordial values of the relative abundances of these elements obtained by WMAP [28] and PLANCK [29] appear[2] to be in agreement with nucleosynthesis predictions, but only if the density parameter in baryonic material is $\Omega_b h^2 = 0.02$.

[2] $\eta_b = 6.19 \times 10^{-10}$ as measured by WMAP.

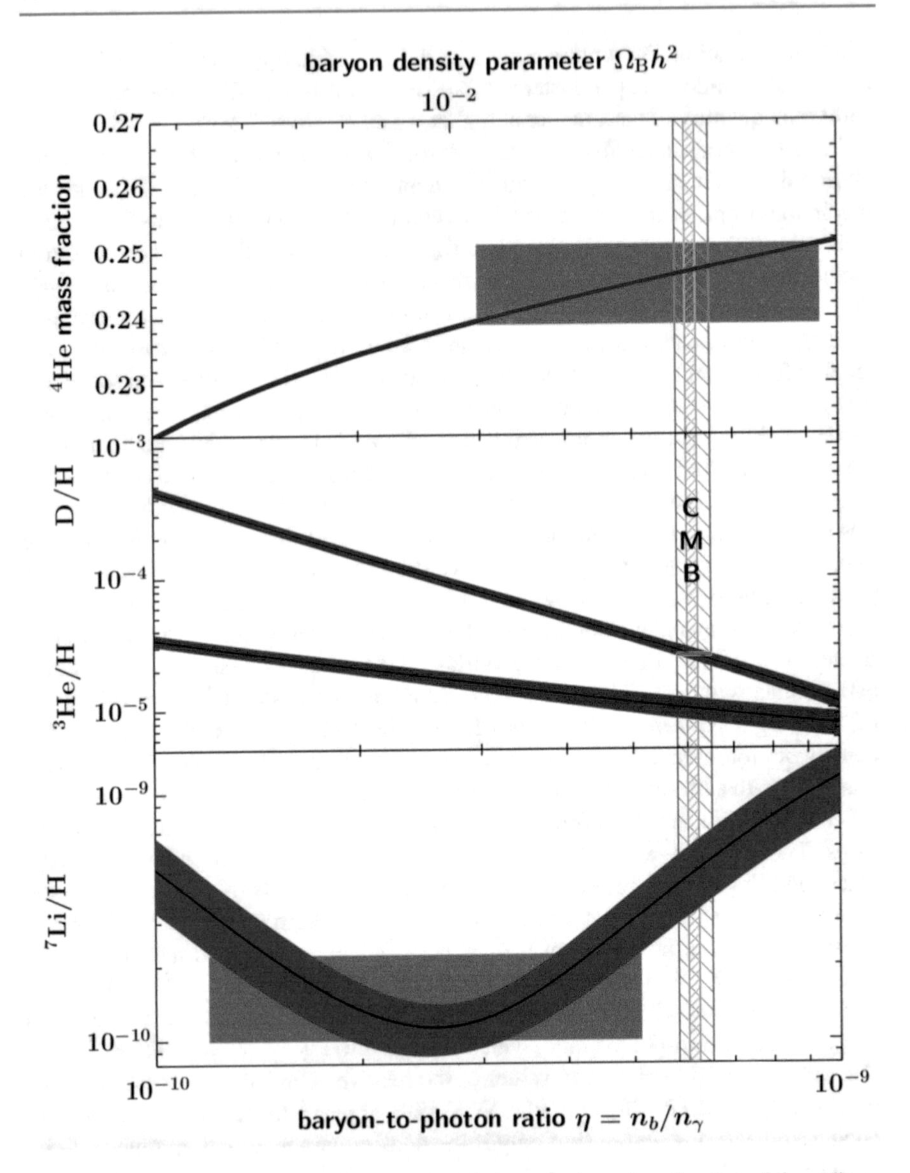

Fig. 1.5 Relative abundance of light elements (relative to hydrogen) as function of the ordinary matter relative to photon (η_b). Licensed under the Creative Commons Attribution-Share Alike 4.0 International license

1.8 Cosmic Microwave Background (CMB)

The relic abundance of dark matter can also be extracted from the analysis of the Cosmic Microwave Background (CMB), radiation originating from the propagation of photons in the early Universe once they decoupled from matter. This is the recombination era (see Sect. 3.3.4 for more details). In 1965, this radiation was detected by Penzias and Wilson, and this discovery was a powerful confirmation of the Big Bang theory. WMAP [28] and more recently the European PLANCK satellite [29] gave us the more precise photo of the Universe and its ingredients. After many decades of experimental efforts, the CMB is known to follow with extraordinary precision a black-body spectrum corresponding to a temperature $T = 2.726$ K. It is also quasi-isotropic withbtemperature fluctuations (called anisotropy) of the order $\frac{\delta T}{T} \simeq 10^{-3} - 10^{-5}$. The variations of temperature in the CMB can be expressed as a sum of spherical harmonics Y_{lm}

$$\frac{\delta T}{T}(\theta, \phi) = \sum_{l=2}^{\infty} \sum_{m=-l}^{l} a_{lm} Y_{lm}(\theta, \phi), \tag{1.5}$$

where a_{lm} gives us the variance $C_l = \langle |a_{lm}|^2 \rangle = \frac{1}{2l+1} \sum_{m=-l}^{l} |a_{lm}|^2$. If the temperature fluctuations are assumed to be Gaussian, as it appears to be the case, all of the information contained in CMB maps can be encoded into the power spectrum, essentially giving the behavior of $l(l+1)C_l/2\pi$ as a function of l. WMAP and PLANCK data could map universal fluctuations after removing the dipole anisotropy ($l = 1$) and galactic and extragalactic contaminations. To extract information from CMB, we must consider a cosmological model with fixed number of parameters. Comparing each position of the peak and its height, one can deduce the deep composition of the Universe. We represent the result of PLANCK in Fig. 1.6. This spectrum was fitted by a model that considers a Universe with a cosmological constant Λ and a cold dark component of matter (Λ_{CDM}). The position of the first peak determines $\Omega_m h^2$. Combining the measurements of the temperature power spectrum with a determination of the Hubble constant h, the PLANCK team found the total mass density parameter $\Omega_m = 0.3175$. The ratio of amplitudes of the second-to-first Doppler peaks determines the baryonic density parameter $\Omega_b = 0.048$; the dark matter component is then $\Omega_{DM} = 0.2695$.

1.9 Alternatives

To be complete, we should say that there exist some alternatives to the dark matter scenario. Indeed, all the evidences of dark matter that we have discussed above relied on the strong assumption that we know the law of gravity at all scales. Is it possible that by changing the laws of gravity we can avoid this mysterious component of the universe? In 1983, Mordehai Milgrom proposed to get rid of dark matter altogether and to replace the known laws of gravitation with the so-

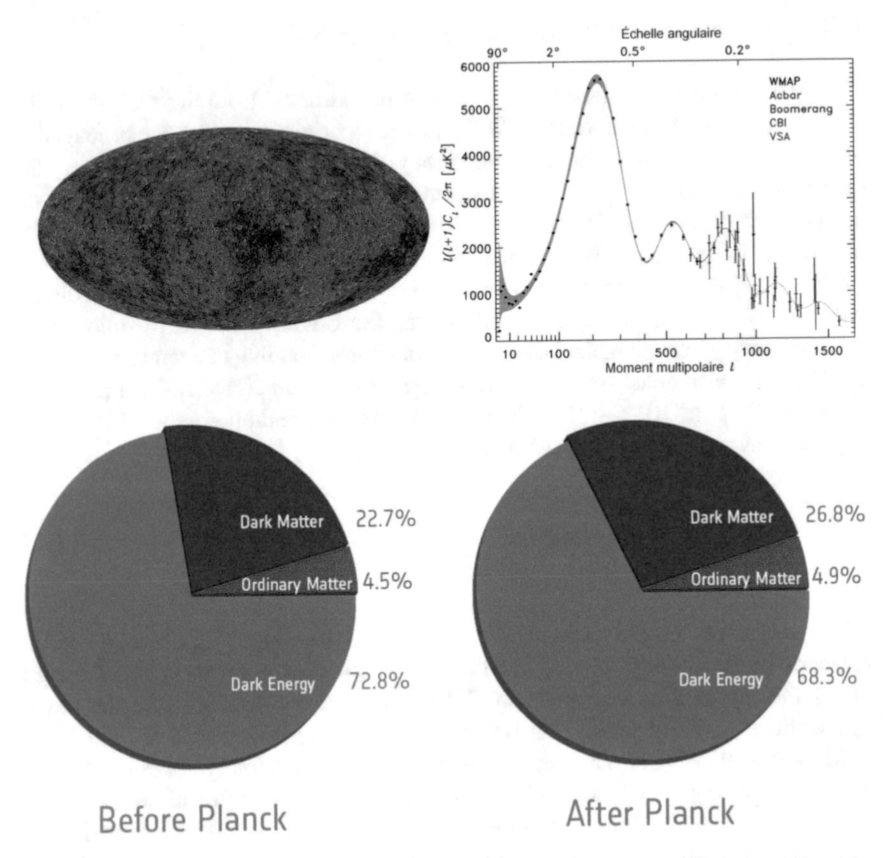

Fig. 1.6 The sky observed by WMAP satellite (top) and composition of the Universe deduces by PLANCK (bottom). NASA/WMAP Science Team, ESA

called MoND paradigm, short for "modified Newtonian dynamics." The price to be paid was to abandon general relativity, a theory that is particularly appealing to many physicists because of its elegance and formal beauty. But there is no dogma in physics, and history has taught us that our theories can always be refined and improved. Milgrom proposed that the law of gravity is modified below a certain *acceleration*, that is, when the gravitational force becomes very weak.

This proposal is very clever, because it bypasses one of the main difficulties of theories of modified gravity: the easiest way to construct them is to introduce a new distance scale, above which the gravitational force is modified from its characteristic inverse square law. But observations tell us that modifications of gravity (or, alternatively, the presence of dark matter) are observed on different scales in different systems. MoND is surprisingly accurate on the scale of galaxies, and it even addresses some mysterious correlations between properties of galaxies that find no explanation in the standard dark matter paradigm. A few years later, Jacob Bekenstein even embedded Milgrom's proposal into a more relativistic theory called

TeVeS, for "tensor-vector-scalar" theory, promoting it from a phenomenological model to a more fundamental theory. It is without doubt an interesting proposal which attracted and is still attracting substantial interest.

There is a problem with these theories, however. Indeed, as soon as we move from the scale of galaxies to the scale of galaxy clusters, they fail to reproduce the observational data. Perhaps the biggest challenge to MoND-like theories, and one of the most direct proofs of the existence of dark matter, is provided by systems like the so-called Bullet Cluster discussed previously in Fig. 1.4, the system of two clusters of galaxies that have recently collided, with one of the two passing through the bigger one like a bullet (hence the name). To explain this observation with MoND-like theories, one has to postulate the existence of additional matter, in the form of massive neutrinos, for instance. But there is an ensign between the properties required of the neutrinos and current data, and, in general, it is not very appealing to require at the same time a modification of gravity *and* the presence of some form of dark matter.

It is however still perfectly possible that it is through a modification of the laws of gravity that we will be able to explain the motion and the shape of cosmic structures, and it is important that part of the research effort of the scientific community should focus on this possibility. Fortunately, as Bekenstein says,

> *The increasing sophistication of the measurements in [gravitational lensing and cosmology] should eventually clearly distinguish between the various modified gravity theories, and between each other and General Relativity.*

The dark matter paradigm will remain a conjecture until we finally put our hands on the particles by measuring their properties in our laboratories. Before describing the techniques that have been devised so far to detect dark matter however, we need to understand a little bit our thermal history and the behavior of the dark matter in the primordial Universe.

References

1. E.W. Kolb, M.S. Turner, Early Universe. Front. Phys. **69**, 1 (1990)
2. J.D. Jackson, *Classical Electrodynamics*, 3rd edn. (Wiley, New York, 1999)
3. G. Jungman, M. Kamionkowski, K. Griest, Phys. Rept. **267**, 195 (1996)
4. F. Zwicky, Helv. Phys. Acta **6**, 10 (1933) [Gen. Rel. Grav. **41**, 207 (2009)]
5. G. Bertone, D. Hooper, History of dark matter. Rev. Mod. Phys. **90**(4), 045002 (2018). https://doi.org/10.1103/RevModPhys.90.045002. [arXiv:1605.04909 [astro-ph.CO]]
6. "The Dark Matter Problem: A Historical Perspective"; The Dark Matter Problem: A Historical Perspective
7. G. Bertone, D. Hooper, J. Silk, Phys. Rept. **405**, 279–390 (2005). https://doi.org/10.1016/j.physrep.2004.08.031. [arXiv:hep-ph/0404175 [hep-ph]]
8. G. Arcadi, M. Dutra, P. Ghosh, M. Lindner, Y. Mambrini, M. Pierre, S. Profumo, F.S. Queiroz, Eur. Phys. J. C **78**(3), 203 (2018). https://doi.org/10.1140/epjc/s10052-018-5662-y. [arXiv:1703.07364 [hep-ph]]
9. P.J.E. Peebles, *Cosmology's Century: An Inside History of Our Modern Understanding of the Universe* (Princeton University Press, Princeton, 2020)

10. K.A. Olive, Inflation. Phys. Rept. **190**, 307–403 (1990). https://doi.org/10.1016/0370-1573(90)90144-Q
11. A.H. Guth, Adv. Ser. Astrophys. Cosmol. **3**, 139–148 (1987). https://doi.org/10.1103/PhysRevD.23.347
12. A.D. Linde, Adv. Ser. Astrophys. Cosmol. **3**, 149–153 (1987). https://doi.org/10.1016/0370-2693(82)91219-9
13. J.P. Ostriker, P.J.E. Peebles, Astrophys. J. **186**, 467–480 (1973). https://doi.org/10.1086/152513
14. V.C. Rubin, W.K. Ford, Jr., Astrophys. J. **159**, 379–403 (1970). https://doi.org/10.1086/150317
15. J. Oort, Bull. Astro. Inst. Neth. **6**, 289–294 (1932)
16. A. Dekel, J. Silk, Astrophys. J. **303**, 39–55 (1986). https://doi.org/10.1086/164050
17. J.E. Gunn, B.W. Lee, I. Lerche, D.N. Schramm, G. Steigman, Astrophys. J. **223**, 1015–1031 (1978). https://doi.org/10.1086/156335
18. J.R. Ellis, J.S. Hagelin, D.V. Nanopoulos, K.A. Olive, M. Srednicki, Nucl. Phys. B **238**, 453–476 (1984). https://doi.org/10.1016/0550-3213(84)90461-9
19. M. Roos, arXiv:1001.0316 [astro-ph.CO]
20. R. Catena, P. Ullio, JCAP **1008**, 004 (2010) [arXiv:0907.0018 [astro-ph.CO]]
21. H. Karttunen et al. *Fundamental Astronomy* (Springer, Berlin, 2007)
22. B. Carrol, D. Ostlie, *An Introduction of Modern Astrophysics* (Pearson, London, 2007)
23. J.F. Navarro, C.S. Frenk, S.D.M. White, Astrophys. J. **462**, 563 (1996) [astro-ph/9508025]
24. S. Smith, Astrophys. J. **83**, 23 (1936)
25. D. Clowe, A. Gonzalez, M. Markevitch, Astrophys. J. **604**, 596 (2004) [astro-ph/0312273]
26. http://chandra.harvard.edu/photo/2006/1e0657/, http://chandra.harvard.edu/press/06releases/press082106.html
27. N.A. Bahcall, Phys. Scripta T **85**, 32 (2000) [astro-ph/9901076]
28. E. Komatsu et al. [WMAP Collaboration], Astrophys. J. Suppl. **192**, 18 (2011) [arXiv:1001.4538 [astro-ph.CO]]
29. P.A.R. Ade et al. [Planck Collaboration], arXiv:1303.5076 [astro-ph.CO]

Part I
The Primordial Universe

Et fiat lux. . . Not exactly. The Universe was not first dominated by radiation but by matter in the form of an inflation, 13.8 billion years ago. This field is at the same time responsible of the very fast inflationary phase (in the first 10^{-37} s), followed by a reheating phase (until $\sim 10^{-20}$ s) and then a thermal phase the following 380,000 years, before matter dominates during 6 billion years. Then the expansion of the universe is globally (but not locally) driven by the cosmological constant. In the first chapter, we propose to review in detail the three phases preceding a thermal world (inflation, thermalization, and reheating) and look at how dark matter can be produced in each of these epochs. Then we will study the evolution of the primordial plasma and how particles can be decoupled from it, neutrino as well as dark matter.

Inflation and Reheating $[M_P \to T_{RH}]$

2

Abstract

The inflation and reheating phases of the Universe concern a period where the Universe changes very quickly from a vacuum/constant density domination to an oscillation/matter domination and then a radiation domination. These transitions are not only fast but also violent. We will analyze in detail each of these phases, insisting on the possibility of producing dark matter before reaching the thermal equilibrium. But before, one needs to understand the equation that will lead our expanding Universe: the Hubble law.

2.1 The Context

When dealing with the physics of the primordial Universe, one needs to study high energy processes in a specific space-time metric and evolution, which is determined by its content (matter or radiation) through the Hubble scaling. The interplay between radiation and matter also plays a role in the inflaton decay and the reheating. These entanglements can seem at first extremely complex on the formal side (Special and General Relativity, Quantum Field Theories, Unification, etc.) as well as on the physics side (inflation, thermodynamics in an expanding Universe, finite temperature effects, etc.). Nevertheless, paradoxically, the whole framework can be summarize by a set of "Mann's" equations: Boltzmann's and Friedmann's equations are summarized in Eqs. (2.206–2.208) and can be directly used by the researcher who already knows the context. They both can be derived from General Relativity principles, as we show in Appendix A.2. However, a pure Newtonian approach can also lead to the same system of equations (forgetting the cosmological constant, which is a consequence of a non-flat space-time). This can be seen as an intermediate step. The reader who is not directly interested in the General Relativity approach can directly jump to Sects. 2.1.1 and 2.1.2 to have a Newtonian

Y. Mambrini, *Particles in the Dark Universe*,
https://doi.org/10.1007/978-3-030-78139-2_2

approach of the Hubble law and Friedmann equations, respectively, whereas a classical description of the Boltzmann equation can be found in Sect. 2.3.1.

2.1.1 The Hubble Law

According to cosmological principle, the Universe is homogeneous and isotropic over distances of the order Gpc but becomes highly inhomogeneous for scale below 100 Mpc. It is also an observed fact that the Universe expands (and cools down) following the Hubble law since around 13.8 billions of years. The Hubble law that governs the expansion of the Universe is fundamental in many aspects. Indeed, it governs the physics in the primordial Universe and the decoupling of particles from the thermal bath, acts on the structure formation of galaxies and clusters of galaxies, and determines the mean free path of relativistic particles in our present Universe. Understanding this law is then a critical step to understand all the processes involved in the evolution of the Universe. In this section, we will show how the basic Newtonian concepts give a relatively clear picture of the phenomena.

In an expanding, homogeneous, and isotropic Universe, the relative velocities of observers obey the *Hubble law*: the recession velocity of observer **B** with respect to **A**, $\mathbf{v}_{AB}$ is

$$\mathbf{v}_{AB} = H(t)\mathbf{r}_{AB}, \tag{2.1}$$

where $\mathbf{r}_{AB}$ is the relative position of **B** with respect to **A**.

Homogeneous and isotropic

Homogeneity and isotropy are fundamental because it means that our physics and the results we obtain from our computation are valid in the whole Universe. In other words, cosmology as a science can exist.[1] We can then distinguish a *Homogeneous Universe*, which is the same wherever you are, and an *Isotropic Universe*, which looks the same on every angle you look at (see Fig. 2.1). Notice that there exists a kind of *anthropic* link between the two notions. Indeed, if a Universe is isotropic but *not* homogeneous, the only possibility is that we are at the center of the Universe (Fig. 2.1, right). This is an Aristotelian conception of the world. In other words, combining the *homogeneous* and *isotropic* constraints tells us that we are not in a unique position in the Universe. A space that is isotropic about every point is necessary homogeneous.

(continued)

[1] Hubble never had the Nobel prize because cosmology was not recognized as a scientific discipline until his death in 1953.

It is possible to show that *the unique law that respects a Homogeneous and Isotropic Universe is in fact the Hubble law*. Indeed suppose three observers **A**, **B**, and **C** (Fig. 2.2) define the velocity function of an observer i, $v^i = f^i(r, t)$. The observer **A** sees the point **C** moving away following $v_C^A = f^A(r_{AC}, t)$, whereas the observer **B** sees **C** receding at a velocity $v_C^B = f^B(r_{BC}, t)$. The observer **A** then *believes* that **B** sees **C** moving following its own law. Then for **A**, the velocity of **C** observed by **B** is the velocity of **C** observed by himself (**A**) minus the velocity of **B** relative to himself: $v_C^B = f^A(r_{AC}, t) - f^A(r_{AB}, t)$. If the Universe is homogeneous and isotropic, **A** and **B** should observe the same law in any direction. One can then write $f^B = f^A$, and so, equalizing the velocity for both observers, $v_C^B = f(r_{BC}, t) = f(r_{AC}, t) - f(r_{AB}, t)$. For **C**= **A**, on obtains $f(r_{BA}, t) = -f(r_{AB}, t)$. Noting $r_{BC} = r_{BA} - r_{CA}$, one deduces $f(r_{BA} - r_{CA}, t) = f(r_{BA}, t) - f(r_{CA}, t)$. Developing $f(x) = \sum_i H_i(t) x^i$, it is straightforward to see that only the linear term survives this condition. We have just proven that the homogeneity and isotropy of the Universe impose f to be a linear function of r, $f(r, t) = v(r, t) = H(t) r$. The Hubble law is indeed the unique law which respects a homogeneous and isotropic Universe.

This law means that for an object **B** situated at a distance twice as large as the object **C** from the observer **A**, the velocity of **B** is twice larger than the one of **C**. If we go back in time and consider constant velocity, it will take the same time for **C**

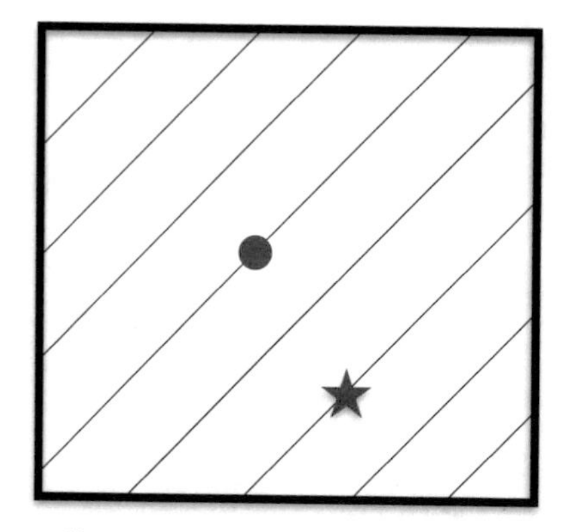

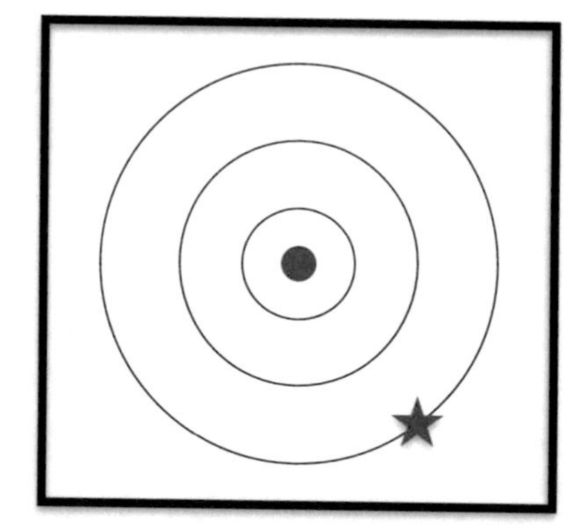

Fig. 2.1 *Left*: a homogeneous but not isotropic Universe: the dot and the star see the same Universe, but not isotropic ones because one direction is preferred. *Right*: An isotropic but non-homogeneous Universe: The dot sees the same Universe whatever direction he looks at, but the star does not see the same Universe as the dot

to join **A** as **B** because even if being further away, its velocity compensates its larger distance. Measuring the Hubble parameter nowadays ($H_0 \simeq 67$ km/s/Mpc), we can compute the time t_0 needed for **B** and **C** to join **A**, $t_0 \simeq 1/H_0 \simeq 14.2$ billions years, which as a good approximation corresponds to the age of the Universe (13.8 billion years). One can also understand why the Hubble law is valid only at scales above 100 Mpc. Indeed, at such distances, the velocity due to the expansion is of the order of 6700 km/s, which is larger than the peculiar velocities of galaxies (of the order of hundreds of km/s). For instance, the Andromeda galaxy, which is at a distance of 0.7 Mpc from the Milky Way (see Sect. 5.1), possesses a negative recession velocity: Andromeda falls *toward us* at a velocity of around 300 km/s: at such low distance, the Hubble expansion has little influence on the Newton law of attraction.

The computation is in fact a little bit more subtile as the Hubble parameter is not constant and depends on the Universe content (matter, radiation, or cosmological constant). This law is easy to interpret if one considers points on a radius of an expanding sphere of radius $a(t)$ as illustrated in Fig. 2.2. One can define $r_{AB}(t) = \chi_{AB} a(t)$, and then $v_{AB} = \chi_{AB} \dot{a}(t)$. χ_{AB} is called the *comoving coordinate*. This corresponds to a system of coordinates in which matter is at rest all the time, which

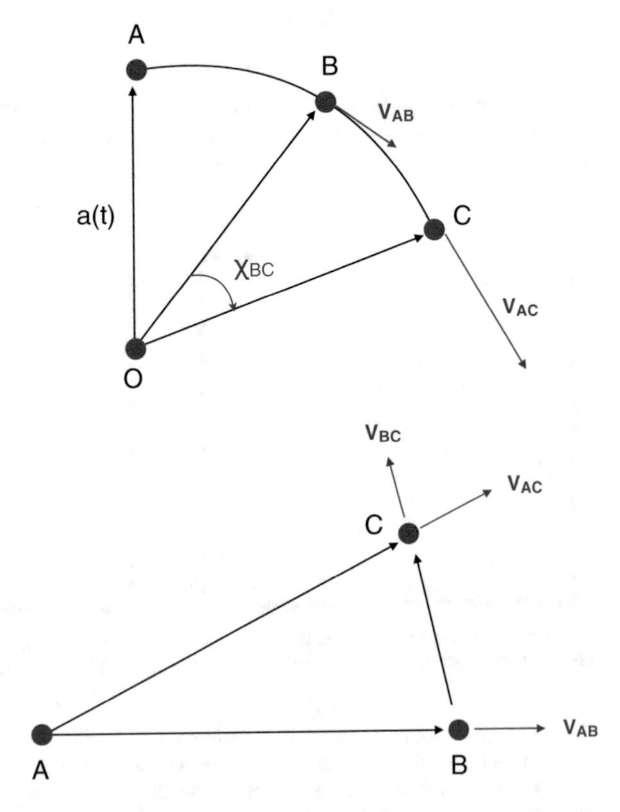

Fig. 2.2 Illustration of the Hubble law (see text for details). *Up*: Universe with expansion parameter a(t) and comoving coordinate χ. *Bottom*: velocity of a point **C** viewed by observers **A** and **B**

does not evolve with time either and can be considered as an initial condition. Then

$$v_{AB} = H(t) r_{AB} \;\Rightarrow H(t) = \frac{\dot{a}(t)}{a(t)}. \tag{2.2}$$

2.1.2 The Friedmann Equations in a Dust Universe

2.1.2.1 The Hubble Parameter

It becomes interesting to solve the equation of motion of an observer **B** at a distant r_{AB} of **A**, in a homogeneous and isotropic space in expansion dominated by matter (dust dominated[2]), i.e. respecting the Hubble law. If one defines ρ the mass density that we assume homogeneous between **A** and **B**, one can write

$$M = \frac{4\pi}{3} r^3 \rho(t), \quad \Rightarrow \rho(t) = \rho_0 \left(\frac{a_0}{a(t)}\right)^3, \tag{2.3}$$

where $r = r_{AB} = \chi a(t)$, and M is the mass inside the radius r. On the other hand, the Newtonian equation of motion for the observer **B** of mass m_B is

$$m_B \ddot{r} = m_B \chi \ddot{a}(t) = -\frac{G m_B M}{\chi^2 a^2(t)}, \quad \Rightarrow \ddot{a}(t) = -\frac{4\pi G}{3}\rho(t) a(t) = -\frac{4\pi G}{3}\rho_0 \frac{a_0^3}{a^2(t)}. \tag{2.4}$$

Multiplying by $\dot{a}$ on both sides of Eq. (2.4) and integrating with constant of integration k, one obtains

$$\frac{1}{2}\dot{a}^2 = \frac{4\pi G}{3}\rho_0 \frac{a_0^3}{a(t)} + k, \tag{2.5}$$

$$\Rightarrow \left(\frac{\dot{a}(t)}{a(t)}\right)^2 = H^2(t) = \frac{8\pi G}{3}\rho(t) + \frac{k}{a^2(t)} = \frac{8\pi}{3M_{Pl}^2}\rho(t) + \frac{k}{a^2(t)},$$

or

$$H^2(t) = \frac{\rho(t)}{3M_P^2} + \frac{k}{a^2(t)}, \tag{2.6}$$

[2] A matter dominated Universe is called a "dust" Universe. It is a Universe composed of matter whose pressure p is negligible compared to its energy density ρ.

where we used $H(t) = \dot{a}(t)/a(t)$ and $G = \frac{1}{M_{Pl}^2} = \frac{1}{8\pi M_P^2}$, $M_P = \frac{M_{Pl}}{\sqrt{8\pi}} \simeq 2.4 \times 10^{18}$ GeV being the reduced Planck mass. This is the Friedmann equation that we obtained in a pure Newtonian formalism and which determines entirely the physics of an expanding mass dominated ("dust dominated") Universe (up to the cosmological constant). We will see that the equation is still valid in the General Relativity framework, ρ being the density of energy.

In fact, Eq. (2.6) can also be seen as the conservation of energy equation, k being the initial energy. Indeed, if we rewrite Eq. (2.6)

$$\frac{1}{2}\dot{a}^2 - \frac{4\pi G\,\rho_0\,a_0^3}{3a} = \frac{1}{2}\dot{a}^2 + V(a) = k, \tag{2.7}$$

we recognize a familiar expression we used to tool with in high school when dealing with the classical rocket problem and escape velocity: for a given gravitational potential (V), what is the escape velocity ($\dot{a}$) needed to escape the attraction ($k \geq 0$). This velocity is also called the *critical* velocity. In a sense, a positive k means that at the beginning of the expansion, the kinetic energy was larger than the potential attraction: the Universe evolves in a continuous expansion with time. On the opposite side, a negative k drives the Universe, dominated by the gravitational attraction, to a future collapse on itself. In a relativistic approach, we will see that k corresponds to the curvature of the Universe in a Friedmann–Robertson–Walker metric. We can go further in the "rocket" analogy. Indeed, historically speaking, astronomers gathered much more data concerning the expansion of the Universe (especially from Hubble's observations) before knowing about its mass composition. Even in the 1960s, the value of the mass density in the Universe was mainly given by approximations of the star component in the galaxies, and the common accepted value was $\rho_{now} \simeq 10^{-31}$g cm^{-3}. The measurements of star velocities were best known before having access to the Cosmological Microwave Background. The problem was then the opposite: knowing the escape velocity $\dot{a}$ from the Hubble constant, what is the *critical* density ρ_c that allows such an expansion. From Eq. (2.7), we obtain

$$k \geq 0 \;\Rightarrow\; \rho \leq \rho_c \equiv \frac{3H^2}{8\pi G}, \tag{2.8}$$

where we used $\rho = \rho_0 \left(\frac{a_0}{a}\right)^3$. We can then rewrite Eq. (2.6)

$$k = \frac{4\pi G\,\rho_c\,a^2}{3}\left[1 - \frac{\rho}{\rho_c}\right] = \frac{4\pi G\,\rho_c\,a^2}{3}\left[1 - \Omega(t)\right], \tag{2.9}$$

where $\Omega(t) = \frac{\rho}{\rho_c}$.

It is interesting to compute another useful cosmological parameter called the *deceleration* parameter $q(t)$

$$q(t) = -\frac{1}{H^2(t)}\frac{\ddot{a}(t)}{a(t)}. \tag{2.10}$$

Historically speaking, this was the main source of information on the matter content of the Universe before the precise measurement of the CMB and its anisotropies. From Eq. (2.4), one obtains

$$q(t) = \frac{4\pi G}{3H^2(t)}\rho(t) = \frac{1}{2}\frac{\rho(t)}{\rho_c(t)} = \frac{1}{2}\Omega(t). \tag{2.11}$$

In the early 60s until the 80s, the matter content of the present Universe was limited by the measurement of $q_0 = q(t_0) \lesssim 2.5$ from the observation of the deceleration of nearby clusters of galaxies. This limit gives $\Omega(t_0) \lesssim 5$, far above the present limit given by the measurement of the anisotropies of the CMB ($\Omega^{CMB}(t_0) \lesssim 0.3$).

2.1.2.2 The Continuity Equation

To compute any physical processes in the early Universe, we need to have the expression of the radius, or scale factor a as a function of time, which means solving Eq. (2.6). For that, we first need an expression of ρ as a function of a and t. This is obtained through the continuity equation, tightly related to the conservation of energy. In the case of a Universe dominated by dust, which means a non-relativistic matter of density ρ_m, the density is given by

$$\rho_m \propto \frac{M}{a^3}, \quad \text{or} \quad \rho_m = \rho_m^0 \left(\frac{a_0}{a}\right)^3, \tag{2.12}$$

which gives

$$\dot{\rho}_m = -3\rho_m^0 \left(\frac{a_0}{a}\right)^3 \frac{\dot{a}}{a} = -3H\rho_m, \tag{2.13}$$

implying

$$\boxed{\dot{\rho}_m + 3H\rho_m = 0.} \tag{2.14}$$

Notice that the right-hand side of the previous equation is null because we considered an isolated system. If energy is injected or even lost (by the decay of a particle, for instance), "0" should be replaced by the source term.

2.1.3 The Friedmann Equations in a Radiative Universe

The treatment described in the previous section is elegant but, being Newtonian, cannot deal with a Universe dominated by relativistic species (as it is the case during the reheating phase, dominated by a gas with pressure $= 1/3$ of the energy density, see Eq. (2.204) for the precise calculation) or by a cosmological constant (as it is the case nowadays, corresponding to an effective negative pressure). It also does not take into account the curvature of space due to the gravitational deformation of its metric. Moreover, the notion of absolute time is absent, whereas lengths depend on the referential. We can, however, also describe a Universe dominated by radiative (relativistic) species from a classical perspective.

A system of relativistic particles is fundamentally different from a system of static (massive) particles due to its pressure. Indeed, if we consider a system of dust made of particles "i" and compress it, its total energy is not modified, as we do not create or extract particles from the system. Their individual energies (only given by their mass because they do not move) are constant. Thus, the total energy, E_m, is given by $E_m = \rho_m \times V = \sum_i m_i$, in a volume V. This is not the case anymore for relativistic particles because they exert a pressure p in the system. In the same way that a bicycle pump heats when we compress its piston because of the increasing pressure of the air inside the pump, we understand quite well that the same phenomenon appears in an expanding Universe: the pressure does work, modifying the internal energy of the system:

$$dE = -pdV. \tag{2.15}$$

The negative sign reflects the fact that the internal energy decreases for an increasing volume and vice versa (the same way than a bicycle pump.[3]) To imagine physically the origin of such energy, we need to understand that in a closed system, which is the case of the whole Universe as no particle can escape from it, diluting the space will decrease the energy of each relativistic particle and thus of the entire system (by redshift,[4] for instance). The pressure is just defined as the proportionality coefficient between the energy lost and the volume gained. In other words, "how much energy I lose per unit of volume I gain." It is then common and natural to express it as a function of the total energy density

$$p = w\rho, \tag{2.16}$$

w being a constant depending on the nature of the content of Universe (0 for dust, 1/3 for a relativistic gas, -1 for the cosmological constant, etc., see Sect. 2.3.1.2).

[3] A more complete statement can be found in Sect. 3.1.2.

[4] The redshift is the propensity for a relativistic particle to have its wavelength changed in an expanding Universe or if it is emitted by a moving source. See Sect. 2.1.4.3 for more details.

V being proportional to a^3, we can rewrite Eq. (2.15) for a gas with density of energy ρ_R

$$d\rho_R = -3(\rho_R + p)d\ln a \tag{2.17}$$

or

$$\dot{\rho}_R + 3H(\rho_R + p) = 0. \tag{2.18}$$

Notice that this equation is not only valid for a relativistic gas of particles but also for any kind of energy density content of the Universe, as long as one knows the relation between the pressure p and the density ρ. We also considered a completely isolated system. If, by any means, the Universe can be populated by another source of energy, the "0" on the right side should be replaced by its density rate of injection. Equation (2.18) should then be replaced by

$$\sum_i \dot{\rho}_i + 3H(\rho_i + p_i) = 0. \tag{2.19}$$

At the end, it seems that we have a perfect set of equations, and people can ask "why the need to go further toward a complex General Relativity approach?" First of all, it is true that the Newton theory being a limit of Einstein theory in a flat space-time, away from extreme deformations (like a black hole or the very early Universe) the classical approach is valid. However, for a curved space-time, the integration constant k has a deep meaning and can be measured. There were two periods in our history where the curvature was dominating the Universe: nowadays and during the violent expansion phase of inflation. In both cases, a General Relativity treatment is not at all a refinement but an obligation, as we will see when dealing with the inflationary sector. Indeed, for matter or radiation, the energy–momentum is well defined and we can treat them classically. For a less conventional potential, in a curved space-time, the metric and energy content should be analyzed with care as we detailed in Appendix A.2. We will present in the following section the basics one needs to understand to deal with the Friedmann equations in a General Relativity framework. Let us begin first with the difference between the metric in a flat space and in a curved space.

2.1.4 The Friedmann–Lemaitre–Robertson–Walker (FLRW) Metric

2.1.4.1 Generalities

The construction of a general space-time metric of the Universe is based on the hypothesis that the Universe is homogenous and isotropic. This is called the *cosmological principle* and is empirically justified on scales larger than 100 Mpc. So one should first build a 3D spatial metric plunged in a 4D space. Using Cartesian coordinates (x, y, z, w) but replacing (x, y, z) by spherical coordinates (r, θ, ϕ), we have for the infinitesimal space interval dl

$$dl^2 = d\rho^2 + \rho^2 d\Omega^2 + dw^2,$$

where $d\Omega^2 = d\theta^2 + \sin^2\theta d\phi^2$ is shorthand for the solid angle. For a positive curvature, one can also write

$$x^2 + y^2 + z^2 + w^2 = \rho^2 + w^2 = R^2,$$

where R is the curvature radius, independent of the position (x, y, z) by the hypothesis of homogeneity. We then have

$$\rho d\rho + w dw = 0.$$

Therefore,

$$dw^2 = \frac{\rho^2}{w^2} d\rho^2 = \frac{\rho^2}{R^2 - \rho^2} d\rho^2$$

and so

$$dl^2 = d\rho^2 + \frac{\rho^2 d\rho^2}{R^2 - \rho^2} + \rho^2 d\Omega^2$$

giving

$$dl^2 = \frac{d\rho^2}{1 - (\rho/R)^2} + \rho^2 d\Omega^2.$$

This is a homogeneous, isotropic 3D space of (positive) curvature $1/R^2$. Notice that if $R \to \infty$, we recover the Euclidean 3D metric $dl^2 = d\rho^2 + \rho^2 d\Omega^2$. Setting $r = \frac{\rho}{R}$, we can express

$$dl^2 = R^2 \left(\frac{dr^2}{1 - r^2} + r^2 d\Omega^2 \right)$$

with $0 \leq r \leq 1$. This is the expression of the metric in a 3D sphere, corresponding to a positive curvature. The coordinates (r, θ, ϕ) will play an important role as "comoving" coordinate we discussed in the previous section. This corresponds to the metric in a unit 3D-sphere or, in other words, in sphere of radius $R = 1$ constant with time. Negative and zero curvatures are also possible. For the negative case, the radius condition should be written as

$$dl^2 = d\rho^2 + \rho^2 d\Omega^2 - dw^2 \text{ and } \rho^2 - w^2 = -R^2.$$

This is the equivalent of the hyperboloid surfaces in 2D. We then obtain

$$dl^2 = R^2 \left(\frac{dr^2}{1+r^2} + r^2 d\Omega^2 \right).$$

We can then combine all the possible curvatures with the generic expression

$$dl^2 = R^2 \left(\frac{dr^2}{1-kr^2} + r^2 d\Omega^2 \right)$$

with $k = +1$ for a positive curvature, $k = 0$ for a flat one, and $k = -1$ for a negative curvature. In general we must allow for R to be an arbitrary function of time $R(t)$ (not position since that would destroy homogeneity). The coordinates (r, θ, ϕ) are called comoving coordinates, in the sense that, in this system or coordinates, even if the Universe expands, the distance between two fixed observers does not change. In this referential, only proper movements have dynamic. If we apply the definition of the invariant metric of special relativity defined in Eq. (A.11)

$$ds^2 = c^2 dt^2 - dl^2,$$

we obtain, in a curved space,

$$ds^2 = c^2 dt^2 - R^2(t) \left(\frac{dr^2}{1-kr^2} + r^2 d\theta^2 + r^2 \sin^2\theta d\phi^2 \right). \tag{2.20}$$

This is the Friedmann–Robertson–Walker metric. It was first derived by Friedmann in 1922 and then more generally by Robertson and Walker in 1935. It applies to *any* metric theory of gravity, not just General Relativity.

2.1.4.2 Geometry of the Universe

We can analyze in more detail the properties of this metric. We have just seen that we have three cases:

- $k = 1$: positive curvature, closed Universe,
- $k = 0$: zero curvature, flat Universe (flat space, not flat space-time), and
- $k = -1$: negative curvature, open Universe.

An alternative form of the metric is often useful. For $k = 1$, setting $r = \sin\chi$, the interval becomes

$$ds^2 = c^2dt^2 - R^2(t)(d\chi^2 + \sin^2\chi d\Omega^2),$$

which can be written by generalization

$$ds^2 = c^2dt^2 - R^2(t)(d\chi^2 + S_k^2(\chi)d\Omega^2),$$

where

$$S_1(\chi) = \sin\chi, \quad S_0(\chi) = \chi, \quad S_{-1}(\chi) = \sinh\chi. \tag{2.21}$$

This convention has the great advantage of describing the metric in a "flat" way for $d\Omega = 0$ (in a straight geodesic line, as the line of sight of a photon) as we will see when computing the redshift or expressing the Hubble law.

2.1.4.3 Redshift

This convention of variables (t, χ, Ω) is less obvious to understand the physics lying beyond the equations, but much more practical to deal with formal calculations. This becomes clear when one needs to compute processes like the redshift, for example. Indeed, the wavelength of light from astronomical sources is a crucial, easily measured observable. Consider two pulses of light emitted at times $t = t_e$ and $t = t_e + \delta t_e$ by an object at χ toward an observer at the origin who picks them up at $t = t_0$ and $t = t_0 + \delta t_0$. The light is emitted in a solid angle dΩ=0 and follows a geodesic ($ds = 0$).

For photons traveling toward the origin, since $ds = 0$

$$cdt = R(t)d\chi.$$

Therefore, because χ represents a comoving coordinate, which is fixed and independent of time,

$$\chi = \int_{t_e}^{t_0} \frac{cdt}{R(t)} = \int_{t_e+\delta t_e}^{t_0+\delta t_0} \frac{cdt}{R(t)}.$$

Subtracting the first integral from the second,

$$\int_{t_0}^{t_0+\delta t_0} \frac{cdt}{R(t)} - \int_{t_e}^{t_e+\delta t_e} \frac{cdt}{R(t)} = 0.$$

For small intervals, $R(t)$ is almost constant,[5] so

$$\frac{\delta t_0}{R(t_0)} = \frac{\delta t_e}{R(t_e)}.$$

Therefore the redshift z, which is defined as the relative difference between the observed wavelength and the emitted one ($z = \frac{\lambda_0 - \lambda_e}{\lambda_e}$), is given by

$$1 + z = \frac{\lambda_0}{\lambda_e} = \frac{\nu_e}{\nu_0} = \frac{\delta t_0}{\delta t_e} = \frac{R(t_0)}{R(t_e)}.$$

2.1.4.4 The Hubble Law

We can now try to find the Hubble law expression in a general space-time metric. The universal "fluid" (=galaxies) is at rest in *comoving* coordinates χ, θ, and ϕ. Expansion of the Universe is encoded in the scale factor $R(t)$. Consider the instantaneous physical distance (or proper distance) to a galaxy at radius χ

$$d_P = \int_0^{\chi} R(t) d\chi = R(t)\chi.$$

Since χ is fixed, the rate of recession of the galaxy is

$$v = \frac{d}{dt}(d_P) = \dot{R}\chi = \frac{\dot{R}}{R} d_P.$$

Identifying

$$H(t) = \frac{\dot{R}}{R} = \frac{\dot{a}}{a},$$

where we have defined a dimensionless scale factor $R(t) = R_0 a(t)$, R_0 being the present Universe radius, we then have

$$v = H(t) d_P,$$

which is the Hubble law, while $H(t)$ is the Hubble "constant" $= H(t_0) = H_0$ today. Hubble's law is thus a direct outcome of homogeneity and isotropy and has the same expression in a curved space than in a flat space.

[5] Whereas t_e and t_0 are usually very spaced: $R(t_e) \neq R(t_0)$.

2.1.4.5 Measuring the Size of the Universe

This is a little exercise I use to give to master students for oral exam:

Knowing that the Universe began to be dominated by a cosmological constant at $t \simeq 10$ *Gyrs, after being dominated by matter, what is the size of the Universe?*

Naively speaking, students want to answer "13.8 billion light-years," forgetting that, since the emission of the CMB radiation, which indeed occurred 13.8 billion years ago[6] at a time $t = t_{CMB}$, the Universe has expanded. The point source of the radiowave we receive now (at time $t = t_0$) is presently at a distance further than ct_{CMB}. To compute the distance of this point source, one needs to compute the distance traveled by the light, while the Universe is expanding continuously. The exercise is quite similar to the one with an ant on an inflating balloon, who tried to reach the north pole while following a meridian, like the light follows a geodesic on its way from the CMB to us. But the curvature is not the point here. The important fact is that the structure of the space dilates with time.

Exercise Considering an ant walking on the meridian of an inflating balloon, walking at v=1 cm/s toward the north pole of the balloon which is 2 cm away. The balloon inflate at a rate of H = 0.4 cm/s per centimeters. Show that the time to reach the north pole is given by

$$t = \frac{\ln(5)}{0.4} \simeq 4 \text{ s}. \tag{2.22}$$

Hint: during a time dt, the distance that the ant still needs to travel is

$$d\Delta = \Delta \times H \times dt - vdt. \tag{2.23}$$

Solving the equation, $\Delta = 0$ gives you the time of the travel. Compute also the distance of the point of origin when the ant arrived to the north pole.

We can then do the same exercise for our Universe. Depending on its composition, the scaling factor a evolves differently with time. For instance, in the case of a matter or radiation dominated Universe,

$$a(t) \propto t^\alpha = a_0 \left(\frac{t}{t_0}\right)^\alpha \tag{2.24}$$

[6]Or 13.8 Gyrs −380,000 years to be more precise because the CMB took place 380,000 years after the Big Bang as we will discuss in Sect. 3.3.4.

with $\alpha = \frac{2}{3}$ for a matter domination and $\alpha = \frac{1}{2}$ for a radiation domination. We can then compute the distance λ_i^0 traveled by the light between t_i and t_0:

$$R\, d\chi = R_0\, a(t)\, d\chi = cdt \;\Rightarrow\; R_0 \chi_i^0 = R_0 \int_i^0 d\chi = \frac{ct_0}{a_0(1-\alpha)} \tag{2.25}$$

$$\Rightarrow\; \lambda_i^0 = R_0\, a_0\, \chi_i^0 \simeq \frac{ct_0}{1-\alpha}, \tag{2.26}$$

where we have supposed, in the last equation, $t_0 \gg t_i$. We can then write the distance λ traveled by the light from $t_i \ll t_0$ to t_0

$$\lambda = 3ct_0 \text{ [matter domination] ;} \quad \lambda = 2ct_0 \text{ [radiation domination].} \tag{2.27}$$

Exercise Show that in the case of a cosmological constant Λ dominated Universe (in other words, with a constant Hubble parameter H_0), the distance traveled by the light, from t_i to t_0 is

$$\lambda = c(t_0 - t_i)\left[\frac{r-1}{\ln(r)}\right], \quad \text{with} \quad r = \frac{a_0}{a_i} \quad [\Lambda \text{ domination}]. \tag{2.28}$$

As a first approximation, if we consider that the Universe was dominated by the radiation during almost 380,000 years, then 13.8 billion years of matter domination, we can write

$$\lambda_{tot} = c \times \left(2 \times 380{,}000 + 3 \times 13.8 \times 10^9\right) = 41.4 \times 10^9 \text{ light} - \text{years} \tag{2.29}$$

for the radius of the observable Universe or a diameter of ~83 billion light-years. Fortunately, this naive approach is quite ok and gives a result not so far from the reality, which is 92 billion light-years. For the exact calculation, one needs to take into account 2 important points: the Universe has been dominated by a cosmological constant at a redshift $z = 0.326$ *and* the evolution between the two dominations is smooth. As it is almost always the case, it is easiest to integrate the free path with respect to the scale factor a and not the time t. We compute the exact solution in Sect. 2.1.7.2.

2.1.5 Friedmann's Equation in General Relativity

After having studied in detail the subtleties of a curved space-time metric, we can generalize it to a metric with local deformations. This is the aim of General Relativity, where the gravity is the source of local deformations of space-time.

The reader will easily understand that it is impossible to give a complete and fair treatment of such a complex subject in a book devoted on the dark Universe. A lot of fantastic textbooks exist on the subject. I personally love the D'Inverno one [1] for its way to avoid complex machinery, keeping concentrated on the essential, and the Hartle book [2] for some refinements. The subject is developed in greater detail in the book of Weinberg [3]. In this section, we will give the necessary tools needed to understand where and how do the fundamental Friedmann equations are affected by a locally curved metric. To find the Friedmann solution to Einstein equations, we will just need (obviously) the Einstein equations (A.88) and the Robertson–Walker metric (2.20):

$$G^{\mu\nu} = R^{\mu\nu} - \frac{1}{2}\mathcal{R}g^{\mu\nu} = 8\pi G\, T^{\mu\nu} + \Lambda g^{\mu\nu} \tag{2.30}$$

$$ds^2 = c^2dt^2 - R^2(t)\left(\frac{dr^2}{1-kr^2} + r^2d\theta^2 + r^2\sin^2\theta d\phi^2\right), \tag{2.31}$$

where $G = 1/M_{Pl}^2$ is the gravitational coupling (M_{Pl} is the Planck mass = 1.22×10^{19} GeV) and Λ the cosmological constant. I can understand the need to clarify the origin of Eq. (2.30), that is, the reason why we devoted a complete appendix on the subject for any student who feels the need to recover the Einstein equations of General Relativity. Even if most readers are surely familiar with these equations, Appendix A.2 is available for the untrained reader. However, for the study of a primordial Universe with an (almost) flat geometry, I think the reader can accept, at a first step, this set of equations that will lead to the fundamental Friedmann relations.

2.1.5.1 The Friedmann Equations

Concretely speaking, the idea is to write the set of equations (2.30) for all possible values of (μ, ν) parameters, given a definite metric like the one given by (2.31). Naively speaking one could think that 16 equations should be taken into account, but in fact, because of the homogeneous and isotropic conditions, only two relations are of interest for us: the G_{00} and G_{ii} ones.[7] Using the Robertson–Walker metric normalized to $c = 1$,

$$g_{00} = g_{tt} = 1\,; \quad g_{rr} = -\frac{R^2(t)}{1-kr^2}\,; \quad g_{\theta\theta} = -R^2(t)r^2\,; \quad g_{\phi\phi} = -R^2(t)r^2\sin^2\theta, \tag{2.32}$$

[7] In fact, all the G_{ii} conditions are the same due to the isotropic principle: none of the 3 directions can be distinguished from the others.

we obtain the following Christoffel symbols:[8] $\Gamma^{\mu}_{\nu\rho} = \frac{1}{2} g^{\mu\alpha}\left[\partial_\nu g_{\alpha\rho} + \partial_\rho g_{\alpha\nu} - \partial_\alpha g_{\nu\rho}\right]$

$$\Gamma^t_{rr} = \frac{R\dot{R}}{1-kr^2};\quad \Gamma^t_{\theta\theta} = r^2 R\dot{R};\quad \Gamma^t_{\phi\phi} = r^2\sin^2\theta R\dot{R};$$
$$\Gamma^r_{rr} = \frac{kr}{1-kr^2};\quad \Gamma^r_{\theta\theta} = -r(1-kr^2)\Gamma^r_{\phi\phi} = -r(1-kr^2)\sin^2\theta;\quad \Gamma^r_{tr} = \frac{\dot{R}}{R};$$
$$\Gamma^\theta_{r\theta} = \frac{1}{r};\quad \Gamma^\theta_{\phi\phi} = -\cos\theta\sin\theta;\quad \Gamma^\theta_{t\theta} = \frac{\dot{R}}{R}$$
$$\Gamma^\phi_{r\phi} = \frac{1}{r};\quad \Gamma^\phi_{\theta\phi} = \cot\theta;\quad \Gamma^\phi_{t\phi} = \frac{\dot{R}}{R}, \tag{2.33}$$

all the other components vanishing.

Exercise Derive the above expression, and compare the Christoffel symbols for a Cartesian metric.

We can then deduce the Riemann and the Ricci tensors $R^\alpha_{\mu\beta\nu}$ and $R_{\mu\nu} = R^\alpha_{\mu\alpha\nu}$:

$$R^\alpha_{\mu\beta\nu} = \Gamma^\alpha_{\sigma\beta}\Gamma^\sigma_{\mu\nu} - \Gamma^\alpha_{\sigma\nu}\Gamma^\sigma_{\mu\beta} + \partial_\beta\Gamma^\alpha_{\mu\nu} - \partial_\nu\Gamma^\alpha_{\mu\beta} \tag{2.34}$$
$$R_{\mu\nu} = R^\alpha_{\mu\alpha\nu} \Rightarrow R_{00} = R_{tt} = -3\frac{\ddot{R}}{R}\ ;\ \mathcal{R} = R^\mu_\mu = -6\frac{\ddot{R}}{R} - 6\frac{(\dot{R})^2}{R^2} - 6\frac{k}{R^2}.$$

Exercise With the help of the results (2.33), prove that $R^t_{ttt} = 0$, $R^r_{trt} = -\frac{\ddot{R}}{R}$, $R^\theta_{t\theta t} = -\frac{\ddot{R}}{R}$, $R^\phi_{t\phi t} = -\frac{\ddot{R}}{R}$, and then deduce Eq. (2.34). Do the same with the other components of the Ricci tensor to show that $R_{rr} = \frac{R\ddot{R}+2(\dot{R})^2+2k}{1-kr^2}$, $R_{\theta\theta} = r^2\left[R\ddot{R} + 2(\dot{R})^2 + 2k\right]$ and $R_{\phi\phi} = r^2\sin^2\theta\left[R\ddot{R} + 2(\dot{R})^2 + 2k\right]$, giving the Ricci scalar (2.34).

Combining Eqs. (2.30), (2.32), and (2.34), we obtain

$$G_{00} = G_{tt} = R_{tt} - \frac{1}{2}\mathcal{R}g_{tt} = 3\left(\frac{\dot{R}}{R}\right)^2 + 3\frac{k}{R^2} = 8\pi G T_{tt} + \Lambda g_{tt} = 8\pi G\rho + \Lambda \tag{2.35}$$

[8]See Appendix A.2.4 for details.

or

$$\left(\frac{\dot{R}}{R}\right)^2 = H^2 = \frac{8\pi G}{3}\rho - \frac{k}{R^2} + \frac{\Lambda}{3}, \tag{2.36}$$

where we have considered the stress-energy tensor of a perfect fluid of density of energy ρ (see Eq. A.96), as it should be the case in the early Universe.[9] This first time-like component of the Einstein equations is in fact the general expression of the Hubble law we obtained by a Newtonian approach in Eq. (2.6), where the "k" term is directly linked to the metric (curvature) of the space-time, and a new term, the cosmological constant, is present. Even if not necessary, this term was at first, not included by Einstein himself in his solutions. But observations of certain types of supernovae in 1995 confirmed the presence of the cosmological constant Λ, which in fact dominates the expansion of the Universe since almost 4 billion years.

2.1.5.2 The Deceleration Equation

To be complete, we have now to compute the G_{ii} component of the Einstein equations.

If one considers the case of a perfect fluid of energy density ρ and pressure P, we can use the expression (A.97) for $T^{\mu\nu}$

$$T^{\mu\nu} = (\rho + P)\frac{dx^\mu}{d\tau}\frac{dx^\nu}{d\tau} - g^{\mu\nu}P, \tag{2.37}$$

and noticing that (we let the reader prove it)

$$G_{ii} = \left[-2\frac{\ddot{R}}{R} - \left(\frac{\dot{R}}{R}\right)^2 - \frac{k}{R^2}\right]g_{ii},$$

we obtain

$$\left[-2\frac{\ddot{R}}{R} - \left(\frac{\dot{R}}{R}\right)^2 - \frac{k}{R^2}\right]g_{ii} = 8\pi G\left[(\rho + P)\frac{dx_i}{d\tau}\frac{dx_i}{d\tau} - g_{ii}P\right] + \Lambda g_{ii}.$$

[9] During the phase of inflation, the energy–momentum tensor should depend on the dynamics of the scalar inflaton as we will see in Sect. 2.2.

If we place ourselves in the rest frame (comoving frame) of the perfect fluid $\frac{dx_r}{d\tau} = \frac{dx_\theta}{d\tau} = \frac{dx_\phi}{d\tau} = 0$, we then can write[10]

$$2\frac{\ddot{R}}{R} + \left(\frac{\dot{R}}{R}\right)^2 + \frac{k}{R^2} = -8\pi G P + \Lambda. \tag{2.38}$$

It is not easy to understand the meaning of Eq. (2.38) by itself. However, eliminating the Hubble rate $\frac{\dot{R}}{R}$ given by Eq. (2.36), we obtain an equation for the acceleration.

$$\frac{\ddot{R}}{R} = -\frac{4\pi G}{3}(\rho + 3P) + \frac{\Lambda}{3}, \tag{2.39}$$

also called the the Raychaudhuri equation.

This equation is in fact a deceleration equation, as $\ddot{R}$ is always negative if one neglects the cosmological constant. It is interesting to go back a century ago, when Einstein tries to find a static solution of his General Relativity equations (2.36) and (2.39). Asking for $\ddot{R} = 0$, we obtain $\Lambda = 4\pi G(\rho + 3P) = 4\pi G\rho$ for a dust matter, and $\dot{R} = 0$ implies then $4\pi G\rho = \frac{k}{R^2}$. In other words, the conditions

$$\Lambda = 4\pi G\rho = \frac{k}{R^2} \tag{2.40}$$

are the conditions for a static Universe, which should *not* be flat. Eddington already noticed that this equality represents a curious Universe where a dynamical variable (ρ) should be *exactly* equal to a constant of Nature (Λ) to ensure a static Universe. Moreover, one should also remark that modifying slightly locally the density of matter will render all the system unstable. On the other hand, supposing a homogeneous density of matter (as proposed by Friedmann and Lemaitre) to solve the cosmological principle of Einstein was not a priori so obvious, especially when we observe such an inhomogeneous sky every nights. The main result of the Hubble discovery was that the Universe was in fact homogeneous and isotropic at largest scale, and this was far to be obvious in the 20s. The deceleration was the parameter measured by astrophysicists, giving constraints to the density of the Universe nowadays, before the discovery of the Cosmological Microwave Background (CMB). Indeed, the beauty of this equation is the absence of the curvature k, rendering it easily testable. To be more precise, physicists were using

[10]There are subtleties in this argument. One way to see it is to think that in the rest frame of a perfect fluid, a particle with a velocity $\frac{dx_\theta}{d\tau}$, for instance, will have a counterpart of another particle of velocity $-\frac{dx_\theta}{d\tau}$, canceling the velocity part of the stress–energy tensor T_{ii} (2.37).

the measurement of the deceleration parameter

$$q = -\frac{\ddot{R}R}{(\dot{R})^2} = -\left(\frac{\ddot{R}}{R}\right)\frac{1}{H^2}, \tag{2.41}$$

the minus sign having been added in the definition to render q positive because historically physicists believed the Universe should contract under the gravitational forces. This was of course much before the discovery of the cosmological constant.

Another combination of (2.38) and (2.36) by elimination of the term $\frac{\ddot{R}}{R}$ gives

$$\dot{\rho} = -3H(\rho + P), \tag{2.42}$$

which is a generalization of the energy conservation equations (2.14) and (2.18)

Exercise Recover (2.42) from the energy conservation equation $D_\mu T^{\mu\nu} = 0$.

In summary, the set of Friedmann equations that we will need to study the evolution of the early Universe can be expressed as

$$\begin{cases} H^2 = \frac{8\pi G}{3}\rho - \frac{k}{R^2} + \frac{\Lambda}{3} \\ \dot{\rho} = -3H(\rho + P) = -3H(1+w)\rho \text{ with } P = w\rho. \end{cases} \tag{2.43}$$

Remark To obtain this set of equations, we have supposed a "stable" source of energy density ρ. If there exist decay processes with a width Γ (which is the case for the inflaton), one needs to add a $-\Gamma \times \rho$ term on the right-hand side of Eq. (2.42). In the same manner, if a source of energy is injected in the volume under consideration, a term of the form $+\Gamma \times \rho_{source}$ should be included. We also want to point out that the relation between pressure and energy density, $P = w\rho$, also called equation of state, has a classical physical interpretation called the Laplace law as we show in Appendix A.5 and more specifically in Eq. (A.115).

Teaching Friedmann equations

When I give lectures at a bachelor level, or when I know that students do not have sufficient training in General Relativity, I present things differently, the goal always being to arrive at Friedmann–Lemaitre's equations. First of all, the virial theorem gives, by equalizing, the mean kinetic energy and the mean potential energy in a (non-relativistic) gas system of density ρ, made of

(continued)

identical particles of masses m:

$$\frac{1}{2}m\langle v^2\rangle = \frac{Gm\rho \times \frac{4\pi}{3}\langle R^3\rangle}{\langle R\rangle}, \tag{2.44}$$

or, getting rid of the mean,

$$H^2 = \left(\frac{\dot{R}}{R}\right)^2 = G\frac{8\pi\rho}{3} + \frac{\Lambda}{3}, \tag{2.45}$$

where we added a cosmological term Λ, as Einstein did to counterbalance the gravitational attraction and keep a static Universe. The internal energy $U = \rho V$ of the gas at a pressure P respects

$$dU = d\rho V + 3\rho\frac{dR}{R}V = -PdV \;\Rightarrow\; \dot{\rho} + 3H(\rho + P) = 0. \tag{2.46}$$

Deriving the expression for H (2.45), using (2.46), one obtains

$$\frac{\ddot{R}}{R} = -G\frac{4\pi(\rho + 3P)}{3} + \frac{\Lambda}{3}. \tag{2.47}$$

In the classical convention we use in this textbook, writing $R = a \times R_0$ (R_0 being the present radius of the Universe) and $G = \frac{1}{8\pi M_P^2}$, $M_P = 2.4 \times 10^{18}$ GeV, we finally have

$$\left(\frac{\dot{a}}{a}\right)^2 = H^2 = \frac{\rho}{3M_P^2} + \frac{\Lambda}{3};$$
$$\dot{\rho} + 3H(\rho + P) = 0;$$
$$\frac{\ddot{a}}{a} = -\frac{\rho + 3P}{6M_P^2} + \frac{\Lambda}{3}. \tag{2.48}$$

An interesting point, noted of course by Einstein, is that, if you want to force a static Universe, the condition $\ddot{a} = 0$ requires

$$\Lambda = \frac{\rho}{2M_P^2}. \tag{2.49}$$

(continued)

Notice also that we recover in Eq. (2.48) the fact that the attractive force is proportional to $1/a^2$ and the repulsive one, driven by Λ, to a: an object twice as far as an observer will be 4 times less attracted by the matter, but twice more repulsed by the cosmological constant term Λ. This can give a hint to the student why at small distances, the attractive gravity dominates and we can apply Newton laws, whereas at large scales, one needs to look at the forces induced by the cosmological term.

Exercise Supposing that in the Universe today, matter dominates over radiation, compute the value of Λ needed to have a static Universe ($\ddot{a} = 0$). Compute then the Hubble constant and compare with its initial value after the inflation. What can you conclude?

2.1.5.3 The Cosmological Constant Case

The other possibility to consider is a Universe dominated by the cosmological constant Λ. Far from being artificial or an exercise, this is in fact the present situation we observe since almost 4 billion years. If we look at Eq. (2.30), we see that it corresponds to an energy–momentum tensor $T^{\Lambda}_{\mu\nu} = g_{\mu\nu}\Lambda$, or in other words, in a flat metric ($k = 0$) from (A.96):

$$8\pi G\, T^{\Lambda}_{\mu\nu} = g_{\mu\nu}\Lambda = \begin{pmatrix} \Lambda & 0 & 0 & 0 \\ 0 & -\Lambda & 0 & 0 \\ 0 & 0 & -\Lambda & 0 \\ 0 & 0 & 0 & -\Lambda \end{pmatrix} = 8\pi G \begin{pmatrix} \rho_\Lambda & 0 & 0 & 0 \\ 0 & P_\Lambda & 0 & 0 \\ 0 & 0 & P_\Lambda & 0 \\ 0 & 0 & 0 & P_\Lambda \end{pmatrix}, \tag{2.50}$$

which corresponds to an equation of state $P_\Lambda = w\rho_\Lambda$ with $w = -1$, where ρ_Λ is constant and can be identified to $\frac{\Lambda}{8\pi G}$, and $\ddot{a} > 0$: we are indeed in the presence of a Universe in an accelerating expansion. Notice that Lemaitre in 1934 remarked that this vacuum energy density is not changed by a velocity transformation. Indeed, $T^{\Lambda}_{\mu\nu}$ is proportional to the Minkowski metric tensor $\eta_{\mu\nu}$ and therefore is unchanged by a Lorentz transformation. In other words, ρ_Λ *does not* define a preferred frame of motion, and it is *not* from the same nature that the ether in special relativity.

2.1.6 Another Look on the Hubble Expansion

We can take a more detailed look to the Hubble expansion rate predicted by the Friedmann equations

$$H^2 = \left(\frac{\dot{a}}{a}\right)^2 = \frac{8\pi G}{3}\rho - \frac{k}{a^2 R_0^2} + \frac{\Lambda}{3} \tag{2.51}$$

with k the curvature factor $(0, \pm 1)$. Compared to the formulae of the preceding section, a is a dimensionless parameter defined as $R(t) = R_0 \times a(t)$, R_0 being the present time radius of the Universe (and then $a \leq 1$), and $\rho = \rho_r + \rho_m$ is the energy density of radiation and matter in the Universe. The presence of the cosmological constant Λ, which appeared in the 90s through observational measurements, changed completely the fate of the Universe. Indeed, without it, the Universe had 3 options:

- If $k = -1$, the expansion slows down all the time, but without stopping.
- If $k = 0$, the expansion slows down and stops at $t = \infty$.
- If $k = +1$, the expansion slows down, stops, and then turns over to contraction.

All these different destinies are related by a slowing down Universe. Indeed, in all the cases, the ρ ($\propto 1/a^3$ if matter dominated and $1/a^4$ if radiation dominated, see Sect. 3.1.6) term or curvature terms were *decreasing* as a is increasing. The presence of a positive Λ term inverts all the process, giving nowadays an accelerating expanding Universe.

If the Universe is matter dominated, $\rho = \rho_0/a^3$, ρ_0 being the density of matter today. We can also define

$$\rho_c = \frac{3H^2}{8\pi G}, \qquad \Omega_M = \frac{\rho_M}{\rho_c}, \qquad \Omega_R = \frac{\rho_R}{\rho_c},$$
$$\Omega_k = -\frac{k}{R^2 H_0^2} \quad \text{and} \quad \Omega_\Lambda = \frac{\Lambda}{3H_0^2}, \tag{2.52}$$

where ρ_c is the critical density. It corresponds to the density we would expect if the Universe is flat ($k = 0$). Any deviation on ρ_c can be interpreted as a measurement of the space curvature. Nowadays, for a value of $H_0 = 100h$ km s^{-1} Mpc$^{-1} = 2.13 \times 10^{-42}\, h$ GeV (corresponding to the speed of a galaxy 1 Mpc away, h being measured to be in 2019, $h \simeq 0.74$, whereas 2018 Planck measurement gave $h \simeq 0.68$), one obtains, reminding that $G = 1/M_{Pl}^2$,

$$\rho_c^0 = \frac{3H_0^2}{8\pi G} = 1.05 \times 10^{-5} h^2 \text{ GeV cm}^{-3} = 1.88 h^2 \times 10^{-29} \text{ g cm}^{-3}. \tag{2.53}$$

The CMB measurements as well as the type Ia supernovae observations seem to favor a flat Universe with a matter component composing a fraction $\Omega_M = \rho_M/\rho_c^0 = 0.3$ of the critical density. Notice that we can define densities ρ_k and ρ_Λ to have a uniform definition of H

$$H^2 = \frac{8\pi G}{3}(\rho_R + \rho_M + \rho_k + \rho_\Lambda)$$
$$\Rightarrow \rho_k = -\frac{3k}{8\pi G\, R^2}, \quad \rho_\Lambda = \frac{\Lambda}{8\pi G}. \tag{2.54}$$

One can then compute which fraction of this matter is composed of baryonic ($\simeq$1 GeV as it is mainly proton/hydrogenic clouds) matter from the data given by WMAP or PLANCK of the ratio of baryon to photon number density [4]

$$\eta = \frac{n_B}{n_\gamma} = 6.12 \times 10^{-10}. \tag{2.55}$$

Knowing the temperature of photons nowadays (2.725 K), one can deduce the number density of photon Eq. (3.42): 411 cm^{-3}, and so the number density of baryon is $n_b = \eta \times n_\gamma = 2.4 \times 10^{-7}$cm^{-3}. Considering 1 GeV baryon, and $h = 0.71$, from the value of ρ_c^0 computed in (2.53), we deduce

$$\Omega_b = \frac{1\ \text{GeV} \times n_b}{\rho_c^0} \simeq 0.044. \tag{2.56}$$

As we notice, this value is far to be sufficient to account for the matter component of the Universe ($\Omega_M = 0.3$), which implies the need for a dark component. Substituting this into the Friedmann equation, Eq. (2.51), and replacing a with $a = 1/(1+z)$, where z is the redshift defined in Sect. 2.1.4.3, gives

$$H^2(z) = H_0^2 \left[\Omega_R^0 (1+z)^4 + \Omega_M^0 (1+z)^3 + \Omega_k^0 (1+z)^2 + \Omega_\Lambda^0 \right], \tag{2.57}$$

or as function of a,

$$H(a) = H_0 \sqrt{\Omega_R^0 a^{-4} + \Omega_M^0 a^{-3} + \Omega_k^0 a^{-2} + \Omega_\Lambda^0} \tag{2.58}$$

Exercise Considering a Universe dominated by matter and a cosmological constant, show that the age of the Universe can be written as

$$t_U = \int_0^1 \frac{\sqrt{a}}{H_0 \sqrt{\Omega_M^0 + \Omega_\Lambda a^3}} da. \tag{2.59}$$

Deduce the age of the Universe for the Einstein–de Sitter model ($\Omega_M^0 = 1, \Omega_\Lambda^0 = 0$) and the Universe corresponding to the observed content ($\Omega_M^0 = 0.311, \Omega_\Lambda^0 = 0.689$).

As we already discussed in Sect. 2.1.5.2, before having access to the CMB measurement, the deceleration parameter was the best mean to compute the relic abundance of matter. Indeed, from (2.41), we can deduce in a flat space,

$$q_0 = \frac{1}{2}\Omega_M^0 - \Omega_\Lambda^0. \tag{2.60}$$

It was then common to extract $\Omega_M^0 = 2q_0$, q_0 being measured by astrophysicists (in the hypothesis of null cosmological constant).

In 1932, Einstein and de Sitter proposed a model with $\Omega_M = 1$ and $\Omega_R = \Omega_k = \Omega_\Lambda = 0$. The simple measurement of the *sign* of q_0 could distinguish their proposition from models dominated by Ω_Λ. This became known as the Einstein–de Sitter cosmological model. It is interesting to see how Einstein, following the experimental discovery of a non-static Universe, totally rejected the idea of the cosmological constant that he himself had introduced. One can feel his bitterness in a letter written in 1947 to Georges Lemaitre:

" *Since I have introduced this term, I had always a bad conscience... I cannot help to feel it strongly and I am unable to believe that such an ugly thing should be realized in nature.*"

It is also interesting to see that in 1931, Georges Lemaitre proposed a model where a part of the energy density is taken by the cosmological constant. The main advantage is that, at this time, the age of the Earth was measured to be $t_{\text{Earth}} \simeq 1.6 \times 10^9$ years (from radioactive decay sources), whereas the expansion timescale was believed to be $H_0^{-1} = 1.8 \times 10^9$ years. Lemaitre concluded that the Einstein–de Sitter model predicting an age of a matter dominated Universe $t = \frac{2}{3}H_0^{-1}$ would conflict with the radioactive decay age. This is the reason why he proposed to introduce the cosmological constant. In a letter to Einstein dated July 30, 1947, he wrote

"*that the cosmological constant is necessary to get a time-scale of evolution which would definitely clear out from the dangerous limit imposed by the known duration of geodesic ages,*"

letter to which Einstein replied that "*it offers a possibility, it may even be the right one.*" Einstein was finally even thinking to reintroduce the cosmological constant to address the issue of the age of the Earth. The proposition of Lemaitre is easy to understand by a quick look to Eq. (2.39). We clearly see that the role of the matter is to decelerate the Universe. The higher is the recession velocity, the further we are from the asymptotic limit, the younger is the Universe, whereas for a cosmological constant Universe, the higher is the recession velocity, the older the Universe is. Both phenomena counterbalance, which explains why adding a little bit of cosmological constant in Ω ages the Universe. Paradoxically, more precise measurements of H_0 in the 1950s will once again bring the age of the Earth within the limits of the age of the Universe, once again rendering the cosmological constant obsolete, until the revolution of 1998, which will bring Λ back into the limelight.

Exercise From Eq. (2.57), and taking $\Omega_M = 0.3$, $\Omega_\Lambda = 0.7$, $\Omega_R = \Omega_k = 0$, show that z_Λ, the redshift when the cosmological constant density began to dominate the evolution of the Universe is $z_\Lambda = 0.32$ (∼9 Gyrs). Show that the condition $\ddot{R} = 0$ from Eq. (2.39) gives $z_\Lambda = 0.67$, corresponding age of the Universe is 6.4 Gyrs. This phase of the Universe was called by Lemaitre, the "hoovering Universe", when, he believed, the slowdown of the expansion was favorable to the formation of galaxies. Comment. Now, suppose that the Universe is composed of $\Omega_M = 0.1$ and $\Omega_\Lambda = 0.9$, whereas $\Omega_R = 10^{-4}$, show that at the redshift $z = 10^{10}$, where the first light elements are produced, the ratio $\frac{\Omega_\Lambda(z=10^{10})}{\Omega_R(z=10^{10})} = 10^{-36}$. Same question if the Universe is composed today of 10% matter and 90% curvature. Show in this case that $\frac{\Omega_k(z=10^{10})}{\Omega_R(z=10^{10})} = 10^{-16}$. What you can conclude from these numbers? Do you understand the Weinberg proposition of multiverse?

2.1.7 The Comoving Distance or Codistance

2.1.7.1 Generalities

Sometimes, when computing a physical process, one needs to know the phenomena characteristic (spectra, lifetime, temperature, etc.) at a given distance and a given time t (or equivalently, redshift z). At that redshift z, the Universe was quite different than today, and its metric was also reduced by a factor $1/(1+z)$. One should then be careful when computing the signal observed now, from a source producing it at a high redshift. For instance, let us think about the spectrum of a source, which emits a monochromatic signal continuously in the Universe. One has to take into account that the energy emitted at the distance $d(z)$ has been redshifted along its way from the source to the Earth. As a consequence, the observed signal on the Earth is not monochromatic anymore, but a sum of redshifted signal, on a distance which evolves also with time (as z evolves with time). Let us put it in numbers. The first step is to compute the distance of the source from the Earth at a redshift z, which we call the comoving distance χ. In other words, the distance x that a photon has crossed if it has been produced at a redshift z. For a massless particle like the photon, we can write

$$dt^2 - a^2(t)dx^2 = 0 \;\Rightarrow\; dx = \frac{dt}{a(t)}. \tag{2.61}$$

In the meantime, we know that by definition, $H(t) = \frac{da(t)/dt}{a(t)}$, which implies

$$dx = \frac{da(t)}{a^2(t)H(t)}, \quad a = \frac{a_0}{(1+z)} \quad \Rightarrow dx = -\frac{dz}{a_0 H(t)},$$

where we kept $a_0(=1)$ for a better understanding.[11] The codistance $d\chi$ *from* the Earth can then be written $d\chi = -dx$ or

$$\frac{d\chi}{dz} = \frac{1}{a_0 H(z)}.$$

We now need to find the expression of $H(z)$, which obliviously depends on the redshift as, if the cosmological constant stays the same, the *density* of matter evolves with time due to the expansion of the metric. Using Eq. (2.57), one obtains

$$H(z) = H_0\sqrt{(1+z)^4\Omega_R^0 + (1+z)^3\Omega_M^0 + (1+z)^2\Omega_k^0 + \Omega_\Lambda^0} \;\Rightarrow \tag{2.62}$$
$$\frac{d\chi}{dz} = \frac{1}{a_0 H_0 (1+z)^{\frac{3}{2}}\sqrt{(1+z)\Omega_R^0 + \Omega_M^0 + \Omega_k^0/(1+z) + \Omega_\Lambda/(1+z)^3}}.$$

Depending on the problem, one then can integrate from $z_0 = 0$ to any redshift z to compute the consequences of an event on the Earth, which is occurring regularly since the redshift z, like a decaying dark matter, for instance.

2.1.7.2 The Size of the Universe (bis)

Using the comoving distance is also the easiest way to compute the exact size (and age) of the Universe. Indeed, in Sect. (2.1.4.5), we gave a solution, under the form of an exercise for students. The real computation of the size of the Universe is not so much complicated but should be done with the scale factor (or redshift) as the variable to integrate on. It is indeed the easiest dynamical variable, as the time is more difficult to define because it depends itself already on the composition of the Universe. As in (2.25), we need to compute the distance λ traveled by the light within an expanding Universe:

$$d\lambda = R_0\, a(t)\, d\chi = c\, dt$$
$$\Rightarrow R_0\chi = c\int \frac{dt}{a(t)} = c\int_0^1 \frac{da}{a^2 H(a)}, \tag{2.63}$$

where we used $dt = \frac{da}{aH}$. Rewriting Eq. (2.62) as function of a,

$$H(a) = H_0\sqrt{\Omega_R^0 a^{-4} + \Omega_M^0 a^{-3} + \Omega_k^0 a^{-2} + \Omega_\Lambda^0} \tag{2.64}$$

[11] We remind the reader that by convention, the radius of the Universe at a time t is written $R(t) = a(t)R_0$, where R_0 is the present radius of the Universe, of the order of 46 Gpc.

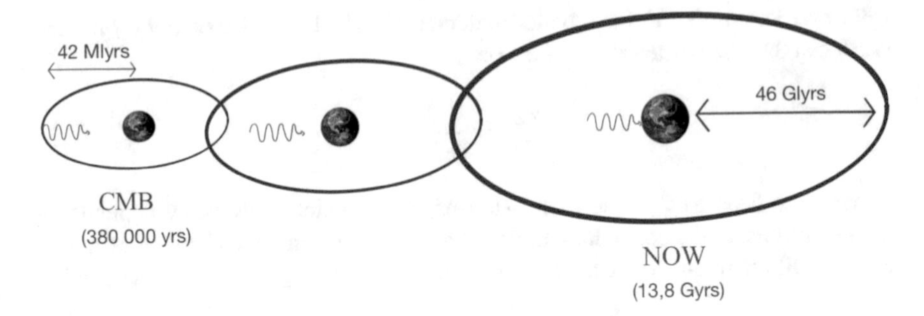

Fig. 2.3 Evolution of the Universe from the CMB emission till now

giving

$$\lambda = R_0\chi = \frac{c}{H_0}\int_0^1 \frac{da}{a^2\sqrt{\Omega_R^0 a^{-4} + \Omega_M^0 a^{-3} + \Omega_k^0 a^{-2} + \Omega_\Lambda^0}} \simeq 44.7\times 10^9 \text{ lyrs.} \tag{2.65}$$

Using the values $\Omega_\Lambda = 0.69$, $\Omega_M^0 = 0.31$, $\Omega_R = 8 \times 10^{-5}$, $H_0^{-1} = 13.8$ Gyrs, and a flat Universe[12] ($k = 0$). This result of 89.4×10^9 light-years is very near from the one computed with our naive assumptions (2.29). This is coming from the fact that light has (almost) not traveled before the CMB, and the Λ-domination time happened in a very late time ($z \simeq 0.66$), so the Universe can be considered as a mater dominated Universe in its whole history from the CMB. To compute the age of the Universe T, one needs to integrate

$$T = \int dt = \int_0^1 \frac{da}{aH(a)} \simeq 13.2 \text{ Gyrs,} \tag{2.66}$$

not so far from the measured value $T \simeq 13.8$ Gyrs. The difference comes from the naive integration we do. We illustrate in Fig. 2.3 the evolution of the Universe from the time the light was emitted from the CMB till now, with the respective size of the Universe in both cases (46×10^9 years now and $\frac{46}{1100} = 42 \times 10^6$ light-years at the CMB epoch.

[12] A slightly more precise computation commonly used in the literature gives $\lambda = 46.3 \times 10^9$ lyrs.

2.2 Inflation [10^{-43} – 10^{-37} s]

The cosmological standard model, from the reheating process to the galaxies formation is consistent and can explain the presence of dark matter, nucleosynthesis, the relative abundance of hydrogen, lithium and helium, structures formations, and even the cosmological constant and the expansion rate of the Universe can be included with minimum changes. We have to admit that few models have so many predictions with so few hypothesis and parameters: almost all the physics is included in Eq. (2.36). However, when putting some numbers, it seems that the vanilla scenario exhibits some issues in the very early stage of the Universe, around the initial singularity of the theory.

2.2.1 The Horizon Problem

The first problem that was noticed is usually labeled "horizon" problem, but also "homogeneity" or "isotropy" issue. The idea is very simple. Indeed, the observable Universe today ($t_0 = 13.8$ Gyr) has a horizon, defined by[13]

$$d_H^0 = d_H(t_0) = c \times t_0 \simeq 1.3 \times 10^{26} \text{ m},$$

whereas this horizon was, at the very initial time, the Planck time $t_i = t_{Planck} = 10^{-43}$ s

$$d_H^i = d_H(t_i) = d_H(t_0) \times \frac{a_i}{a_0}, \tag{2.67}$$

where $a_i = a(t_i)$ and $a_0 = a(t_0)$ are the scale factors of the Universe at the time t_i and t_0, respectively. At t_i, with a good approximation, the light has covered a distance $ct_i = 3 \times 10^{-35}$ m, or in other words,

$$\frac{d_H^i}{c \times t_i} \simeq 5 \times 10^{28}. \tag{2.68}$$

That means that it would exist almost $(5 \times 10^{28})^3 \simeq 10^{85}$ causally disconnected "bubbles" in the very first instant of the Universe,[14] rendering impossible to observe such a homogeneity $\frac{\delta E}{E} \simeq \frac{\delta T}{T} \lesssim 10^{-4}$ in the CMB spectrum. Another way to see it is to say that our present horizon would be composed of 10^{85} initially disconnected patches. If we apply the same reasoning to compute the number of

[13]To be more precise, one should take into account the *physical* horizon distance, i.e. the actual distance traveled by the light, which is given by Eq. (2.65), but this approximation is quite valid for the argument.

[14]The same result can be obtained by computing the entropy of the Universe in the very early time.

causally disconnected bubbles at the CMB time, we obtain (we let the reader to prove it)

$$\frac{d_H^{CMB}}{c \times t_{CMB}} \simeq 30 \tag{2.69}$$

corresponding to more than 1000 regions that could not be connected.

To understand better the phenomena, we can rewrite Eq. (2.67) as

$$\frac{d_H^i}{c \times t_i} = \frac{t_0}{t_i}\frac{a_i}{a_0} \simeq \frac{\dot{a}_i}{\dot{a}_0} \gg 1, \tag{2.70}$$

where we have supposed in the last equation that $a(t) \propto t^\alpha$, α being a constant, which is effectively the case in every type of density domination we will encounter (except during the inflation phase of course as we will demonstrate). But, as $\dot{a}$ decreases with time, the horizon distance will always be larger than the causally connected region "$c \times t$" when going back to time. That is, completely unavoidable as we show in Fig. 2.4. However, if we suppose an expending phase, where $\dot{a}$ increases in a very short period of time before the radiative Universe, we can make the horizon "re-entering" in the causally connected bubbles. How is it possible?

If we do not consider the presence of a cosmological constant, the evolution of $\dot{a}$ is given by the Hubble parameter $H \propto \sqrt{\rho}$, Eq. (2.43), where we supposed a flat Universe (we will discuss the flatness problem in the following section). That means that neither a radiative domination ($\rho \propto a^{-4} \Rightarrow \dot{a} \propto \frac{1}{a}$) nor a matter dominated Universe ($\rho \propto a^{-3} \Rightarrow \dot{a} \propto \frac{1}{\sqrt{a}}$) can induce a phase of increasing $\dot{a}$. However, if we suppose the presence of a field whose density ρ_Λ is constant just after the Planck time t_i, we have

$$H(t_i) = \frac{\sqrt{\rho_\Lambda}}{\sqrt{3}M_P} \Rightarrow \dot{a} = a\frac{\sqrt{\rho_\Lambda}}{\sqrt{3}M_P} > 0. \tag{2.71}$$

To be more precise, if the Universe is dominated by ρ_Λ between the Planck time t_i and a time t_f, and we suppose that at t_i all the horizon is causally connected ($d_H^i = ct_i$), we can compute how stretched is the horizon d_H^f at the time t_f supposing the Universe dominated by ρ_Λ. Making use of (2.63), we can write

$$R_0 a(t) d\chi = cdt \Rightarrow R_0 \chi_f = \int_i^f \frac{cda}{H_\Lambda a^2} \Rightarrow d_H^f = R_0 a_f \chi_f = \frac{c}{H_\Lambda}\frac{a_f}{a_i},$$

where we supposed $a_i \ll a_f$. Asking for d_H^f to fit the observable contracted horizon from now to t_f:

$$d_H^f = c \times t_0 \frac{a_f}{a_0}, \tag{2.72}$$

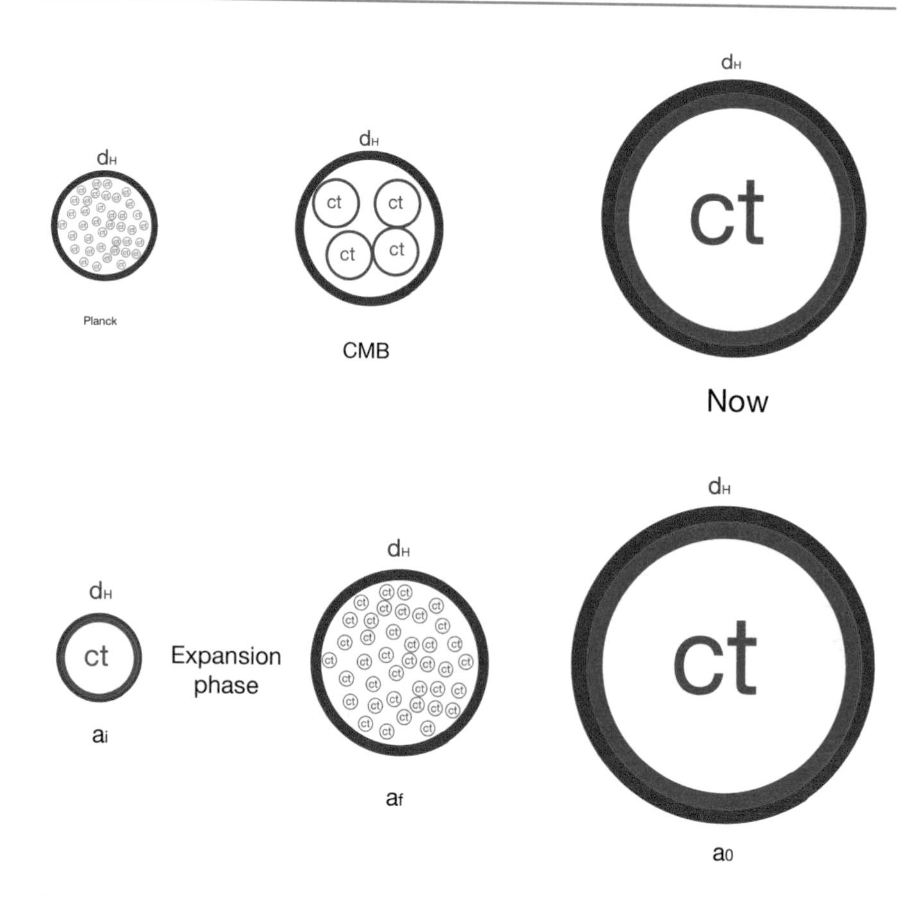

Fig. 2.4 Top: an illustration of the evolution of the causally connected regions (ct in blue) with the horizon (d_H in red) in the case of a pure radiative Universe (without a phase of inflation). Bottom: same but with the hypothesis of an inflationary phase prior to the radiative Universe

we obtain

$$\frac{c}{H_\Lambda}\frac{a_f}{a_i} = c \times t_0\,\frac{a_f}{a_0}, \tag{2.73}$$

which gives

$$\frac{a_f}{a_i} \gtrsim \frac{H_\Lambda}{H_0}\frac{a_f}{a_0} = \frac{a_0}{a_f}, \tag{2.74}$$

where we used in the last relation, $H \propto \frac{1}{a^2}$ in a radiation dominated Universe,[15] $H_0 \simeq \frac{1}{t_0}$, and $H_\Lambda = H(a_f)$ because H is constant during the inflation phase. Notice that we transformed an equality into a $\gtrsim$ relation because the relation holds as long as the horizon distance *after* the inflation phase is greater than the one computed from now to t_f. We summarize all the process in Fig. 2.4.

Exercise Integrating the expression (2.61), recover the preceding result by computing the distance traveled by the light from $t_i \sim 0$ to $t = t_{CMB}$ and then the same horizon distance from $t = 0$ to t_0. Show that the distance of the CMB horizon *today* d_H^{CMB} can be expressed as a function of the present horizon d_H^0 by $d_H^0 = \sqrt{1+z} \times d_H^{CMB}$, which gives (2.69) for $z = z_{CMB} = 1000$.

2.2.2 The Flatness Problem

Even if we do not consider the horizon problem, another issue arises in the vanilla thermal Universe model, and it concerns its curvature. From the CMB measurement, we know that Ω_k (2.52) is $\lesssim 0.02$ (and is probably much less) at present time t_0, while the radiation density is $\Omega_R = \frac{\rho_R}{\rho_c} \simeq 10^{-4}$. However, as we can see from Eq. (2.51), the curvature density scales as a^{-2} whereas the radiation density scales as a^{-4}. That means that the hierarchy between the radiation and curvature densities increases drastically as we go back in time. If we assume by simplicity that both densities are at most equal today ($\rho_k \leq \rho_R$), we obtain at the Planck time t_i

$$\frac{\rho_k(t_i)}{\rho_R(t_i)} = \frac{\rho_k^0 \left(\frac{a_0}{a_i}\right)^2}{\rho_R^0 \left(\frac{a_0}{a_i}\right)^4} \leq \left(\frac{a_i}{a_0}\right)^2 = \sqrt{\frac{\rho_R^0}{\rho_R^i}} \simeq \left(\frac{10^{-13}}{10^{19}}\right)^2 \simeq 10^{-64}, \tag{2.75}$$

where we considered $\rho_R^i = \rho_R(t_i) \simeq M_{Pl}^4$ GeV^4 and took the measured value $\rho_R^0 = \rho_R(t_0) \simeq (10^{-4}\ \text{eV})^4$. That means that the curvature should be very near zero at the very early stage of the Universe, whereas it would be natural to consider a homogeneous Universe with an equal amount of density of each kind in the very beginning (that is, $\rho_k(t_i) = \rho_R(t_i)$), especially if one considers a Universe emerging from a quantum gravity phase. It corresponds to justify a fine tuning of more than 64 orders of magnitude.

The main problem comes from the fact that the ratio $\frac{\rho_k}{\rho_R}$ increases with time, proportionally to $\left(\frac{a}{a_i}\right)^2$, which very quickly reaches huge values. We can then play the same game we did when we tried to find a solution to the horizon problem. Indeed, if one supposes at the beginning of time, a reasonable density of curvature, a

[15]That is the phase during which the Universe evolves the more, compared to the matter or dark energy period.

phase of rapid expansion will dilute it sufficiently such that nowadays, the curvature takes reasonable values again. Concretely speaking, if between the Planck time t_i and the end of this inflationary phase t_f the Universe is dominated by a constant density ρ_Λ, without yet any density of radiation, supposing the natural initial condition $\rho_k^i = \rho_\Lambda$, we obtain

$$\rho_k^f = \rho_k^i \left(\frac{a_i}{a_f}\right)^2 = \rho_\Lambda \left(\frac{a_i}{a_f}\right)^2 \Rightarrow \rho_k^0 = \rho_\Lambda \left(\frac{a_i}{a_0}\right)^2. \tag{2.76}$$

Then, if all the energy contained in ρ_Λ is converted into radiation at t_f, one has

$$\rho_R^f = \rho_\Lambda \Rightarrow \rho_R^0 = \rho_\Lambda \left(\frac{a_f}{a_0}\right)^4. \tag{2.77}$$

Combining the two previous equations, we obtain the condition to respect the observation $\rho_k^0 \leq \rho_R^0$:

$$\frac{a_i}{a_f} \leq \frac{a_f}{a_0}, \tag{2.78}$$

which is surprisingly the same relation we obtained to solve the horizon problem, Eq. (2.74). What tells Eqs. (2.74) or (2.78) is that the volume of the Universe should have the same expansion rate between t_i and t_f that between t_f and t_0. A convenient measure of expansion is the so-called *e-fold number* defined as

$$N \equiv \ln a. \tag{2.79}$$

If we suppose (as we will see later on) that the energy scale of the radiation at t_f is $\rho_R^f \simeq (10^{16}\ \text{GeV})^4$, and knowing the present value $\rho_R^0 = (10^{-4}\ \text{eV})^4$, we can deduce

$$N_{f0} = N_0 - N_f = \ln \frac{a_0}{a_f} = \frac{1}{4} \ln \frac{\rho_R^f}{\rho_R^0} \simeq 67. \tag{2.80}$$

We can then conclude that if inflation takes place at a scale of $\sim 10^{16}$ GeV, it should last for a minimum of 67 e-folds. We then have a one-to-one correspondence between the energy scale of the inflation and the number of e-folds necessary to achieve it. The larger is the scale, the more e-folds are needed to dilute sufficiently and counterbalance the evolution of the curvature density between t_f and t_0.

Exercise Compute the number of e-folds necessary if the inflation scale is of the order of TeV (solution: 37).

Computing the necessary number of e-folds

For the students, there is a shorter formulation that gives, at first order, a good approximation of the number of e-folds necessary to solve the horizon problem. The first step is to compute the size of the horizon, from the initial scale factor a_i (which we will take at $t_i = t_{planck} \simeq 10^{-43}$ s) to the final stage of the inflation a_f (at $t = t_f$). With the same method we used to compute the size of the Universe, the distance traveled by the light from t_i to t_f can be written as

$$d_f^H = a_f R_0 \chi_f = a_f \int_{t_i}^{t_f} \frac{cdt}{a} = a_f \int_{a_i}^{a_f} \frac{cda}{H_\Lambda a^2} \simeq \frac{a_f}{a_i} \frac{c}{H_\Lambda}, \tag{2.81}$$

where we applied Eq. (2.63) and considered a constant Hubble rate $H_\Lambda = \frac{\sqrt{\rho_\Lambda}}{\sqrt{3} M_P}$. We also supposed $a_f \gg a_i$. To solve the horizon problem, we want that, at t_f, the size of the horizon corresponds to the size of the Universe. In other words, we want that at t_f, all the points inside the volume $(R_0 a_f)^3$ are causally connected:

$$d_f^H = R_0 a_f \chi_f \;\; \Rightarrow \;\; \ln\left(\frac{a_f}{a_i}\right) \simeq \ln\left(\frac{H_\Lambda}{H_0}\right) + \ln a_f, \tag{2.82}$$

where we used $R_0 \simeq ct_0 \simeq \frac{c}{H_0}$. Taking a radiation dominated Universe (which is the period where the Universe evolved the more), we can write

$$a_f = \frac{a_f}{a_0} = \sqrt{\frac{t_f}{t_0}} \simeq \sqrt{\frac{10^{-43}}{10^{17}}} = 10^{-30}, \tag{2.83}$$

which gives at the end, if one considers a potential $\rho_\Lambda = V(\phi) \simeq (10^{16})^4$ GeV^4 at unification scale,

$$N_e = \ln\left(\frac{a_f}{a_i}\right) \simeq \ln 10^{26} \simeq 60, \tag{2.84}$$

where we took $H_0 \simeq 10^{-42}$ GeV.

2.2.3 The Inflaton

The issues we discussed above can be solved by the introduction of a scalar field, called inflaton.[16] It will be responsible of the rapid expansion phase between t_i and t_f but will also reheat the thermal bath through its decay into Standard Model particles from t_f during all its lifetime. We talk about *reheating* and not *heating* because all the initial radiation that could have been produced before the inflationary phase has been largely diluted. However, before going in detail into the thermalization process, one needs to understand in greater detail the evolution of the inflaton field ϕ with time. To find its equation of motion, we should minimize its action. As we will see, we can minimize it with respect to the metric or with respect to the field (and its derivative) itself. If we consider the inflaton as a classical field with a Lagrangian density

$$\mathcal{L}_\phi = \frac{1}{2} g^{\mu\nu} \partial_\mu \phi \partial_\nu \phi - V(\phi),$$

its action can be written as

$$S_\phi = \int d^4x \sqrt{-g} \mathcal{L}_\phi \tag{2.85}$$

asking the invariance of S_ϕ under a metric transformation gives

$$\delta S_\phi = \int d^4x \left[\delta(\sqrt{-g}) \mathcal{L}_\phi + \sqrt{-g} \delta(\mathcal{L}_\phi) \right]. \tag{2.86}$$

Noticing that $\delta g = g g^{\mu\nu} \delta g_{\mu\nu}$ (A.77) and $g^{\mu\nu} \delta g_{\mu\nu} = -g_{\mu\nu} \delta g^{\mu\nu}$ (A.83) implying

$$\delta(\sqrt{-g}) = -\frac{\sqrt{-g}}{2} g_{\mu\nu} \delta g^{\mu\nu}, \tag{2.87}$$

we can rewrite Eq. (2.86) as

$$\delta S_\phi = \frac{1}{2} \int d^4x \sqrt{-g} \left(-g_{\mu\nu} \mathcal{L}_\phi + \partial_\mu \phi \partial_\nu \phi \right) \delta g^{\mu\nu} = \frac{1}{2} \int d^4x \sqrt{-g}\, T^\phi_{\mu\nu} \delta g^{\mu\nu},$$

which gives

$$T^\phi_{\mu\nu} = \partial_\mu \phi \partial_\nu \phi - g_{\mu\nu} \mathcal{L}_\phi. \tag{2.88}$$

[16] Even if the original model of Starobinsky had no scalar, it was shown to be equivalent to a scalar theory.

Exercise From

$$\mathcal{L}_S = \frac{1}{2} g^{\mu\nu} \left[\partial_\mu S^* \partial_\nu S + \partial_\nu S^* \partial_\mu S \right] - V(S)$$

$$\mathcal{L}_\psi = \frac{i}{4} g^{\mu\nu} \left[\bar{\psi} \gamma_\mu \partial_\nu \psi + \bar{\psi} \gamma_\nu \partial_\mu \psi - \partial_\mu \bar{\psi} \gamma_\nu \psi - \partial_\nu \bar{\psi} \gamma_\mu \psi \right] - m \bar{\psi} \psi$$

$$\mathcal{L}_V = -\frac{1}{4} g^{\mu\alpha} g^{\nu\beta} F_{\alpha\beta} F_{\mu\nu},$$

where S is a complex scalar, ψ a Dirac fermion, and V_μ a vector with $F_{\mu\nu} = \partial_\mu V_\nu - \partial_\nu V_\mu$, show that in the approximation of flat metric,

$$T^S_{\mu\nu} = (\partial_\mu S^* \partial_\nu S + \partial_\nu S^* \partial_\mu S) - g_{\mu\nu} \mathcal{L}_S \tag{2.89}$$

$$T^\psi_{\mu\nu} = \frac{i}{4} \left[\bar{\psi} \gamma_\mu \partial_\nu \psi + \bar{\psi} \gamma_\nu \partial_\mu \psi - \partial_\mu \bar{\psi} \gamma_\nu \psi - \partial_\nu \bar{\psi} \gamma_\mu \psi \right] - g_{\mu\nu} \mathcal{L}_\psi \tag{2.90}$$

$$T^V_{\mu\nu} = -g^{\alpha\beta} F_{\alpha\mu} F_{\beta\nu} + \frac{1}{4} g_{\mu\nu} F_{\alpha\beta} F^{\alpha\beta}. \tag{2.91}$$

Developing $g_{\mu\nu} = \eta_{\mu\nu} + \frac{1}{M_P} h_{\mu\nu}$, where $h_{\mu\nu}$ is the graviton field, show that the coupling of the matter to the graviton can be written as

$$\mathcal{L}_{hi} = \frac{1}{2M_P} h^{\mu\nu} T^i_{\mu\nu}. \tag{2.92}$$

Notice that in General Relativity, the derivatives ∂_μ should be thought as covariant derivatives D_μ defined in Eq. (A.51). However, for a scalar, $D_\mu \phi = \partial_\mu \phi$, and for a vector, $D_\mu V_\nu - D_\nu V_\mu = \partial_\mu V_\nu - \partial_\nu V_\mu$.

The fermionic case is much more tricky due to the fact that one should use the geometric γ^μ matrices (that you can think as being also bent by the metric $g^{\mu\nu}$) and defined them as $\gamma^\mu = e^\mu_a \gamma^a$, γ^a being the classical dirac matrices in the Minkowski flat space and e^μ_a the *vierbein*, see Eq. (A.43) and Eq. (A.59). A detailed and pedagogical computation of the stress–energy tensor of a free dirac field can be found in [5]. The connection term present in Eq. (A.59) disappears in the flat space approximation. To compute $\delta \mathcal{L}_\psi$, do not forget to compute $\delta \gamma_\mu$:

$$\delta \gamma_\mu = \delta(e^a_\mu \gamma_a) = \delta(e^a_\mu) \gamma_a = -\frac{1}{2} g_{\mu\alpha} \delta g^{\alpha\nu} (e^a_\nu \gamma_a), \tag{2.93}$$

where we used from $g_{\mu\nu} = e^a_\mu e^b_\nu \eta_{ab}$: (prove it)

$$\delta e^a_\mu = \frac{1}{2} \delta g_{\mu\nu}\, e^{\nu a}, \quad \delta e^{\mu a} = -\frac{1}{2} \delta g^{\mu\nu} e^a_\nu. \tag{2.94}$$

The density of energy of the stress–energy tensor is given by the {00} component of $T^{\phi}_{\mu\nu}$,

$$\rho_\phi = T^{\phi}_{00} = (\partial_t\phi)^2 - g_{00}\mathcal{L}_\phi = \frac{1}{2}\dot{\phi}^2 + V(\phi), \tag{2.95}$$

where we have supposed a flat and homogeneous metric for the space-time, whereas the pressure is the spatial $\{ii\}$ component of the tensor[17]

$$P_\phi = T^{\phi}_{ii} = (\partial_i\phi)^2 - g_{ii}\mathcal{L}_\phi = \frac{1}{2}\dot{\phi}^2 - V(\phi), \tag{2.96}$$

our matrix $T^{\phi}_{\mu\nu}$ can then be written (A.96)

$$T^{\phi}_{\mu\nu} = \begin{pmatrix} \rho_\phi & 0 & 0 & 0 \\ 0 & P_\phi & 0 & 0 \\ 0 & 0 & P_\phi & 0 \\ 0 & 0 & 0 & P_\phi \end{pmatrix} = \begin{pmatrix} \frac{1}{2}\dot{\phi}^2 + V(\phi) & 0 & 0 & 0 \\ 0 & \frac{1}{2}\dot{\phi}^2 - V(\phi) & 0 & 0 \\ 0 & 0 & \frac{1}{2}\dot{\phi}^2 - V(\phi) & 0 \\ 0 & 0 & 0 & \frac{1}{2}\dot{\phi}^2 - V(\phi) \end{pmatrix}.$$

Notice that for a non-homogenous field ϕ, the energy density ρ_ϕ is

$$\rho_\phi = \frac{1}{2}\left[(\partial_t\phi)^2 + (\partial_i\phi)^2\right] + V(\phi). \tag{2.97}$$

This situation appears when one needs to deal with constraints from inhomogeneities present in the CMB spectrum. This remark is also valid at $t_i = t_{Planck}$, where $\partial_i\phi$ need not be zero but just "sufficiently" small to get washed out by expansion.

2.2.4 The Equation of Motion

Implementing ρ_ϕ and P_ϕ in Eq. (2.42), we obtain the equation of motion for ϕ:

$$\begin{aligned} &\dot{\rho}_\phi + 3H(\rho_\phi + P_\phi) = 0 \\ &\Rightarrow \ddot{\phi} + 3H\dot{\phi} + V'(\phi) = 0, \end{aligned} \tag{2.98}$$

where $V'(\phi)$ stands for $\partial_\phi V(\phi)$.

[17] We remind the reader that in our convention of the flat metric, $g_{00} = g^{00} = +1$ and $g_{ii} = g^{ii} = -1$.

Exercise Noticing that $g_{\mu\nu}$ is independent of ϕ, recover Eq. (2.98) applying the Euler Lagrange equation on $\delta S = \sqrt{-g}\mathcal{L}_\phi$ with respect to $(\phi, \partial_\mu\phi)$. Hints: you will need the relation (prove it) $\partial_t\sqrt{-g} = 3H\sqrt{-g}$, and Eq. (2.166) can help you.

A quick look at Eq. (2.98) shows that ϕ behaves like a rolling-down ball, slowed down by a friction term represented by H, the expansion of the Universe. That is indeed the classical exercise given at high school to compute the limit velocity reached by a falling body submitted by a friction force $F_f = -k\mathbf{v} = -k\dot{\mathbf{x}}$, H playing the friction role of $\frac{k}{3}$. The equation of motion in a potential $V(x)$ is then $m\mathbf{a} = m\ddot{\mathbf{x}} = -k\dot{\mathbf{x}} - \frac{d}{d\mathbf{x}}V(\mathbf{x})$, which is exactly the expression we obtained for ϕ (2.98).

Notice also that this equation assumes a stable field ϕ. If, as we will see when discussing the reheating phase, the ϕ field decays into lighter particles, to give rise to the Standard Model bath, one needs to add to Eq. (2.98) a term of the form $\tilde{\Gamma}_\phi\dot{\phi}$, $\tilde{\Gamma}_\phi$ being the width of the inflaton. The evolution of ϕ then becomes

$$\ddot{\phi} + 3H\dot{\phi} + V'(\phi) = -\tilde{\Gamma}_\phi\dot{\phi}. \tag{2.99}$$

Each term of the equation above has a clear physical meaning. The evolution of the density of the kinetic energy, frictional energy, and potential energy is converted into the lost of energy by decay. It is always possible to solve numerically Eq. (2.99), but we can find quite accurate analytical solution in two regimes of interests: the slow roll regime and the coherent oscillating regime.

2.2.5 The Equation of Motion (Generalization)

We can generalize Eq. (2.98) to fields that interact with the inflaton. This is useful when one needs to compute production of dark matter in the preheating phase. Let us consider first the case of a scalar field S of mass m_S interacting with the inflaton ϕ through a dimensionless coupling proportional to λ:

$$\mathcal{S}_S = \int d^4x\sqrt{-g}\left[\frac{1}{2}g^{\mu\nu}\partial_\mu S\partial_\nu S - \frac{1}{2}m_s^2S^2 - \frac{1}{2}\lambda\phi^2S^2\right] = \int d^4s\sqrt{-g}\mathcal{L}_{\phi S}. \tag{2.100}$$

The Euler–Lagrange equation can be written as

$$\partial_\mu\sqrt{-g}\frac{\partial\mathcal{L}_{\phi S}}{\partial_\mu S} - \sqrt{-g}\frac{\partial\mathcal{L}_{\phi S}}{\partial S} = 0, \tag{2.101}$$

which gives, for a flat metric,

$$ds^2 = c^2dt^2 - dx^2 - dy^2 - dz^2 = c^2dt^2 - a(t)^2d\chi^2, \tag{2.102}$$

and noticing that $\sqrt{-g} \propto a(t)^3$,

$$\ddot{S} + 3H\dot{S} + \frac{\partial V(S)}{\partial S} + \frac{|\mathbf{k}|^2}{a^2}S = 0, \tag{2.103}$$

where we used $\partial_t\sqrt{-g} = 3\frac{\dot{a}}{a} = 3H$ and $V(S) = \frac{1}{2}m_S^2S^2 + \frac{1}{2}\lambda\phi^2S^2$ and implicitly worked in Fourier space. The $\frac{|\mathbf{k}|^2}{a^2}$ term can be understood as a redshifted kinetic energy and comes from

$$\partial^i\partial_iS = -\frac{\partial}{a^2\partial\chi^2}S = -\frac{|\mathbf{k}|^2}{a^2}S. \tag{2.104}$$

2.2.6 The Slow-Roll Regime

2.2.6.1 The Context

To understand the evolution of the inflaton field, one needs to solve Eq. (2.98) or (2.99) if one takes into account the possibility of decay. Of course, it is always possible to do it numerically, but it is good to find analytical solutions to feel the behavior of ϕ. The first stage is called the slow-roll regime, which name is quite explicit. At the very beginning, we can first neglect the acceleration part $\ddot{\phi}$. Physically we are in the presence of a field "falling" in a potential $V(\phi)$ with a friction term $3H$. The only point, which differs from the classical falling body Newtonian analogy we took, is that the friction parameter is not constant but depends strongly on "the height" (ϕ).

Remember that, to realize inflation, we need a phase of constant density $\rho = \rho_\phi$ during a very short period of time, between the Planck times t_i and t_f. A look at Eq. (2.43) or (2.50) shows that we need the equation of state $P_\phi = w\rho_\phi$ with $w = -1$. From Eqs. (2.95) and (2.96),

$$w = \frac{P_\phi}{\rho_\phi} = \frac{\frac{1}{2}\dot{\phi}^2 - V(\phi)}{\frac{1}{2}\dot{\phi}^2 + V(\phi)}, \tag{2.105}$$

we deduce that the scalar field ϕ has the desired equation of state $w = -1$ only if $\dot{\phi}^2 \ll V(\phi)$. In other words, during a short period of time, the kinetic energy of the inflaton should be subdominant to realize the inflation phase. Writing $P_\phi = -\rho_\phi + \dot{\phi}^2$ helps us to understand that the inflation phase should last as long as the kinetic energy $\dot{\phi}^2$ is kept sufficiently small compared to the potential energy $V(\phi)$, where "*as long as*" should be understood as, as long as the 67ish e-folds of Eq. (2.80) have not yet been reached. This condition is possible if $V(\phi)$ evolves very slowly with ϕ between t_i and t_f, this is called the *slow-roll regime* and t_f can be considered as the time when inflation ends. From now on, we will then call it t_{end}.

Exercise Combining Eqs. (2.39), (2.95) and (2.96), and neglecting Λ show that the inflation condition $\ddot{R} \gg 0$ corresponds to $V(\phi) \gg \dot{\phi}^2$.

2.2.6.2 The $V = \frac{1}{2}m^2\phi^2$ Case

We will first concentrate in the simplest quadratic example for $V(\phi)$. Indeed, even if this model is in tension with data, this will help us to understand the dynamics of the inflation. Concretely speaking, the idea is quite straightforward. We need to solve the set of equations Eq. (2.43) for $\phi(t)$ and then extract $H(t)$, which in turn will give us the solution $a(t)$. To begin with, one should first rewrite the set of equations (2.43) as a function[18] of ϕ:

$$\ddot{\phi} + 3H\dot{\phi} + V'(\phi) = 0 \tag{2.106}$$

$$H^2 = \frac{\frac{1}{2}\dot{\phi}^2 + V(\phi)}{3\ M_P^2} \tag{2.107}$$

$$V(\phi) = \frac{1}{2}m^2\phi^2. \tag{2.108}$$

As we discuss in the previous section, at the beginning of the "falling" of the inflaton, its acceleration $\ddot{\phi}$ and its velocity $\dot{\phi}$ can be neglected. Equation (2.107) then gives

$$H \simeq \frac{m\phi}{\sqrt{6}M_P}, \tag{2.109}$$

which we can replace in (2.106) to write

$$\dot{\phi} = -\sqrt{\frac{2}{3}}m\ M_P \quad \Rightarrow \quad \phi(t) \simeq \phi_i - \sqrt{\frac{2}{3}}m\ M_P(t - t_i), \tag{2.110}$$

and as a consequence

$$H = H(t) = \frac{m\phi}{\sqrt{6}M_P} \simeq \frac{m\phi_i}{\sqrt{6}M_P} - \frac{m^2}{3}(t - t_i), \tag{2.111}$$

[18]We will not consider a decaying ϕ at this stage. This is justified because if $\tilde{\Gamma}_\phi$ is comparable to the Hubble rate, then the slow-roll regime is not valid anymore: the inflaton decays too fast to obtain the needed 67 e-folds.

where t_i is the Planck time. Defining the end of inflation t_{end} by $m\ \phi(t_{end}) < |\dot{\phi}|(t_{end})$, or in other words, by the time when the kinetic energy begins to reach values similar to the potential energy, Eq. (2.110) gives

$$t_{end} - t_i \simeq \sqrt{\frac{3}{2}}\frac{\phi_i}{m\ M_P}, \quad |\phi_{end}| = \sqrt{\frac{2}{3}}M_P. \tag{2.112}$$

Another way to understand the condition for the inflation to end is to see that if $\phi(t_{end}) = 0$, the inflation effectively ends because $V(\phi)$ vanished for $\phi = 0$, and the kinetic part $\dot{\phi}$ dominates on $V(\phi)$, giving an end to the slow-rolling inflation regime. Indeed, the negative solution for $\dot{\phi}$ shows that during inflation, ϕ will decrease, inducing automatically a decrease in $V(\phi)$, whereas $\frac{1}{2}\dot{\phi}^2 \simeq m^2 M_P^2$ stays almost constant. It is then unavoidable that there will exist a moment when $\dot{\phi}^2 < V(\phi)$

We have now all the tools in hand to compute $a(t)$ and apply the condition (2.80) to obtain a condition on ϕ_i in order to obtain sufficient e-folds for the inflation to occur. Integrating $H(t)$ in Eq. (2.111) from t_i to t_{end}, and supposing $\phi_i \gg M_P$, we have

$$a(t_{end}) \simeq a_i e^{\frac{\phi_i^2}{4M_P^2}}, \tag{2.113}$$

which means we need

$$\phi_i = \sqrt{4 \times 67} M_P \simeq 16\ M_P \tag{2.114}$$

to obtain the necessary 67 e-folds. It is worth noticing that having a field *above*, the Planck mass is not problematic (in the sense, one does not need to look for a quantum theory of gravity) as long as the energy density stays below the Planck scale. For instance, for an inflaton mass of $m = 10^{13}$ GeV and $\phi_i = 10^6 M_P$, the potential $V(\phi) = \frac{1}{2}m^2\phi_i^2 = \frac{1}{2}(10^{19})^4$ GeV^4 is still sub-Planckian at the Planck time t_i. We can also see that

$$P_\phi = -\rho_\phi + \frac{2}{3}m^2 M_P^2, \tag{2.115}$$

which means that when ϕ has decreased to a value $\sim M_P$, the kinetic part begins to be important in the equation of state: the inflaton will enter in an oscillatory regime and will not behave anymore like a vacuum energy, but like a dust.[19]

Exercise Noticing that the total number of e-folds N_e can be generalized by

$$a(t_e) = a(t_{Pl}) \times e^{\int_{t_e}^{t_{Pl}} H(t')dt'} \Rightarrow N_e = \ln \frac{a(t_e)}{a(t_{Pl})} = \int_{t_e}^{t_{Pl}} H(t')dt', \quad (2.116)$$

where t_{Pl} and t_e represent, respectively, the Planck time and the time when inflation ends and compute the number of e-folds for a potential $V(\phi) = \lambda\phi^4$ and $V(\phi)$ generic.

Conditions for the slow-roll regime

It is interesting to write the general condition for the slow-roll regime, which corresponds to neglecting the term $\ddot{\phi}$ in Eq. (2.106). In other words,

$$3H\dot{\phi} \simeq -V'(\phi) \quad (2.117)$$

$$\ddot{\phi} \ll -3H\dot{\phi} \simeq V'(\phi). \quad (2.118)$$

Remarking that, during the inflation era, ϕ is almost constant ($\dot{\phi}^2 \ll V(\phi)$),

$$H^2 = \frac{\rho}{3M_P^2} = \frac{\frac{1}{2}\dot{\phi}^2 + V(\phi)}{3M_P^2} \simeq \frac{V(\phi)}{3M_P^2}, \quad (2.119)$$

Equation (2.117) can be written as

$$\dot{\phi}^2 \simeq \frac{V'(\phi)^2 M_P^2}{3V(\phi)} \ll V(\phi) \Rightarrow \epsilon \equiv \frac{M_P^2}{2}\left(\frac{V'(\phi)}{V(\phi)}\right)^2 \ll 1. \quad (2.120)$$

In the same way, writing $\frac{d}{dt} = \frac{d\phi}{dt}\frac{d}{d\phi}$, we can express condition (2.118)

$$\ddot{\phi} = \frac{d}{dt}\dot{\phi} = -\dot{\phi}\frac{d}{d\phi}\frac{V'(\phi)}{3H} = \frac{V''(\phi)V'(\phi)}{9H^2} \ll V'(\phi) \Rightarrow \eta \equiv \frac{M_P^2 V''(\phi)}{V(\phi)} \ll 1, \quad (2.121)$$

where we supposed $H \simeq$ cst.

(continued)

[19] To be more precise, as we will see in the next section, the inflaton will behave like a dust for a quadratic potential $V(\phi)$, whereas it will behave like a radiation for a quartic potential.

Exercise Show that Eq. (2.121) can be recovered as a consequence of (2.120) and (2.117).

2.2.7 The Coherent Oscillation Regime

In this regime, the field ϕ has "rolled down" the potential and oscillates. Another way is to notice that $V(\phi)$ has decreased down to the value $\sim \dot{\phi}_{sr}^2$ and we cannot use the approximation $\dot{\phi} \ll m\phi$ anymore. Indeed, while $\dot{\phi}$ was negative, it will become positive (and ϕ_{sr} negative, see Eq. (2.110)), pushing $V(\phi)$ toward larger values, then back until the moment where $V(\phi)$ will reach again $\dot{\phi}^2$... We clearly recognize an oscillating body, transferring its energy between kinetic ($\dot{\phi}^2$) and potential ($V(\phi)$). Rewriting the equation for the Hubble rate

$$H^2 = \frac{\dot{\phi}^2 + m^2\phi^2}{6M_P^2}, \tag{2.122}$$

we can define a new variable θ by

$$\dot{\phi} = \sqrt{6}M_P\ H \sin\theta, \quad m\phi = \sqrt{6}M_P\ H \cos\theta \tag{2.123}$$

and implementing it in Eq. (2.106) to obtain

$$\begin{aligned} \dot{H} &= -3H^2 \sin^2\theta \\ \dot{\theta} &= -m - \frac{3}{2}H \sin 2\theta. \end{aligned} \tag{2.124}$$

Exercise Recover the previous set of equations.

Noticing that for $\phi \ll M_P$ (after the end of inflation t_{end}), $H \simeq m\frac{\phi}{M_P} \ll m$, we can solve the set of equations (2.124):

$$H \simeq \frac{2}{3t}\left(1 - \frac{\sin 2mt}{2mt}\right)^{-1} \tag{2.125}$$

$$\theta \simeq -mt \tag{2.126}$$

and

$$\phi = \frac{2\sqrt{2}}{\sqrt{3}} \frac{M_P}{\left(1 - \frac{\sin 2mt}{2mt}\right)} \frac{\cos mt}{mt}. \tag{2.127}$$

We show in Fig. 2.5 the evolution of ϕ as a function of the normalized time mt, where Eq. (2.106) is solved numerically. We see clearly the two different regimes, where ϕ evolves first linearly with t, following a slope $-\sqrt{\frac{2}{3}}M_P$ as predicted in (2.110), in the slow-roll regime (for low values of mt). Then, entering in the coherent oscillation regime, ϕ oscillates with a frequency $\simeq m$, in accordance with (2.124) while still losing amplitude at a rate $\simeq 3H$. The inflation ends when $t \simeq \Gamma_\phi^{-1}$, or $mt \simeq \frac{m}{\Gamma_\phi} = \frac{8\pi}{y^2}$ if one considers an effective coupling to the standard model fermions of the form $\mathcal{L} = y\phi \bar{f} f$.

Exercise Show that the envelope of the oscillation follows a law in $a^{-3/2}$, or equivalently, $\frac{1}{t}$. Conclude that ρ_ϕ follows a dust equation of state.

2.2.8 The General Case, $V(\phi)$

We can apply the same scheme in the case of a generic potential $V(\phi)$. In this case, we need to solve the following set of equations:

$$\ddot{\phi} + 3H\dot{\phi} + V'(\phi) = 0 \tag{2.128}$$

$$H^2 = \frac{\frac{1}{2}\dot{\phi}^2 + V(\phi)}{3M_P^2}. \tag{2.129}$$

If, as before, we suppose that at the beginning the inflaton is "falling" without large acceleration $\ddot{\phi}$ nor large velocity $\dot{\phi}$, we can write

$$\dot{\phi}^2 \ll V(\phi) \quad \Rightarrow \quad H \simeq \frac{\sqrt{V(\phi)}}{\sqrt{3}M_P} \tag{2.130}$$

$$\ddot{\phi} \ll V'(\phi) \quad \Rightarrow \quad \dot{\phi} \simeq -\frac{M_P}{\sqrt{3}} \frac{V'(\phi)}{\sqrt{V(\phi)}} \tag{2.131}$$

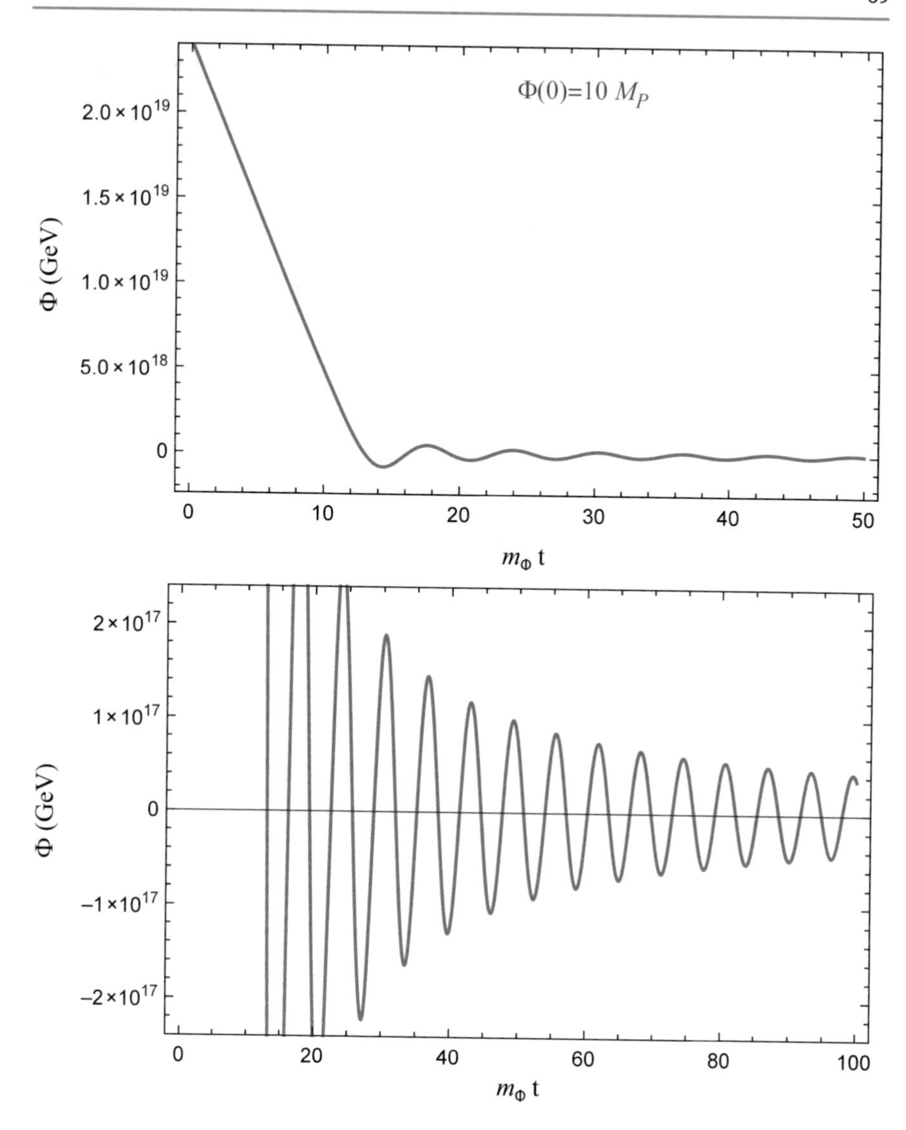

Fig. 2.5 Evolution of the inflaton field ϕ as a function of the time (normalized to m_ϕ) in the case of a quadratic potential $V(\phi) = \frac{1}{2}m_\phi^2\phi^2$, with $\phi(t_i) = 10\ M_P$

and find the solution for a:

$$H = \frac{\sqrt{V(\phi)}}{\sqrt{3}M_P} = \frac{d\ln a}{dt} = \frac{d\ln a}{d\phi}\frac{d\phi}{dt} = -\frac{M_P}{\sqrt{3}}\frac{V'(\phi)}{\sqrt{V(\phi)}}\frac{d\ln a}{d\phi} \tag{2.132}$$

$$\Rightarrow a = a_i e^{-\int \frac{V(\phi)}{M_P^2 V'(\phi)}d\phi} . \tag{2.133}$$

We then see that if one considers a potential of the form $V(\phi) \propto \phi^k$, we obtain

$$a = a_i e^{\frac{\phi_i^2 - \phi^2}{2M_P^2 k}}. \tag{2.134}$$

Asking for $N_e = 67$ e-folds, we obtain

$$\phi_i \simeq \sqrt{134\, k} M_P = \sqrt{2N_e k} M_P, \tag{2.135}$$

which is of the same order as the value of ϕ_i we found in Eq. (2.114), and we recover it for $k = 2$.

After the inflationary phase, we enter in the coherent oscillation phase. In this case, one cannot neglect anymore $\ddot{\phi}$ in Eq. (2.128). Multiplying the equation by ϕ and taking the mean (noticing that on a period, $\langle \phi\dot{\phi} \rangle = 0$), we obtain

$$-\langle \dot{\phi}^2 \rangle + 3\langle H\dot{\phi}\phi \rangle + \langle \phi V'(\phi) \rangle = 0. \tag{2.136}$$

Remarking that at the end of the inflation,

$$H \simeq \frac{\langle \dot{\phi} \rangle}{\sqrt{3} M_P} \quad \Rightarrow \quad H\langle \dot{\phi}\phi \rangle = \frac{\langle \phi \rangle}{\sqrt{3} M_P} \langle \dot{\phi}^2 \rangle \ll \langle \dot{\phi}^2 \rangle, \tag{2.137}$$

we can neglect the friction term H in Eq. (2.136), which gives

$$\langle \dot{\phi}^2 \rangle \simeq \langle \phi V'(\phi) \rangle. \tag{2.138}$$

We can then express the equation of state by the relation

$$w = \frac{P_\phi}{\rho_\phi} = \frac{\frac{1}{2}\langle \dot{\phi}^2 \rangle - \langle V(\phi) \rangle}{\frac{1}{2}\langle \dot{\phi}^2 \rangle + \langle V(\phi) \rangle} = \frac{\langle \phi V'(\phi) \rangle - 2\langle V(\phi) \rangle}{\langle \phi V'(\phi) \rangle + 2\langle V(\phi) \rangle} = \frac{k-2}{k+2}, \tag{2.139}$$

where we supposed $V(\phi) \propto \phi^k$ in the last equation. Implementing the previous relation between P_ϕ and ρ_ϕ in Eq. (2.98), we obtain the generic equation for ρ_ϕ

$$\dot{\rho_\phi} + \frac{6k}{k+2} H \rho_\phi = 0. \tag{2.140}$$

In a Universe, dominated by ϕ, Eq. (2.140) can be solved analytically and we have

$$\rho_\phi(a) = \rho_\phi(a_i) \left(\frac{a}{a_i} \right)^{-\frac{6k}{k+2}}. \tag{2.141}$$

This last equation is very interesting because we recover that for $k = 2$, $\rho_\phi \propto a^{-3}$, which means that ρ_ϕ behaves like a matter field, whereas for $k = 4$, $\rho_\phi \propto a^{-4}$ and behaves like a radiation field. Naively, it is surprising that we recover the behavior of a massive field for an oscillating field. In fact, it is easy to understand if one interprets Eq. (2.124) as the equation of a *homogeneous* field oscillating at a frequency m. ϕ can then be considered as a coherent wave constituted of "ϕ-particles," or oscillators, with zero momentum ($\mathbf{k} = 0$) and a frequency m. The density of energy for ϕ can then be written as a sum of density of energies of oscillators of number density n_ϕ, in other words,

$$\rho_\phi = n_\phi\, m, \tag{2.142}$$

which is effectively the behavior of a massive field of density n_ϕ. If one considers a density of energy $V(\phi) = \frac{1}{2}m^2\phi^2$, one obtains

$$n_\phi = \frac{1}{2}m\phi^2 \simeq 10^{92}\ \text{cm}^{-3} \tag{2.143}$$

for $m = 10^{13}$ GeV and $\phi = M_P$, which can be considered as a rough value of the density of entropy in the Universe after the inflation phase.

2.2.9 Constraint from Perturbations*

2.2.9.1 Generalities

The inflationary models we discussed were homogeneous, *i.e.* we never took into consideration any spatial distribution of the inflaton field. However, there is no doubt that the CMB sky, measured by PLANCK satellite, shows density perturbations at the level of

$$\frac{\delta\rho}{\rho} \simeq 5 \times 10^{-5}. \tag{2.144}$$

The subject of primordial perturbations is extremely complex and far beyond the scope of this book.[20] However, we would like to give some hints, and order of approximations for the reader, to help him/her understanding the philosophy and mechanics of perturbation theory.

Let us first have a rough estimate of the density perturbation we expect from a non-homogeneous field $\delta\phi$. In the absence of expansion, at the classical level, if one considers a massless scalar field ϕ, the density of energy can be written using (2.88) as

$$\rho = T^{00}_\phi = \frac{1}{2}\left[(\partial_t\phi)^2 + (\partial_i\phi)^2.\right] \tag{2.145}$$

[20]For a detailed study on the subject, see [6].

At a given time,

$$\delta\rho \sim \frac{1}{2}\delta(\partial_i \phi)^2 = \frac{1}{2}\delta(k_1^2 + k_2^2 + k_3^2)\phi^2 = \frac{1}{2}k^2\delta\phi^2. \tag{2.146}$$

Remarking that in the volume $\delta V = \lambda^3$, with $\lambda \sim \frac{1}{q}$ the wavelength of the k mode of $\delta\phi$, we should have one quanta of energy k, one deduces

$$\delta\rho \times \lambda^3 \sim \frac{\delta\rho}{k^3} = \frac{1}{2}\frac{k^2\delta\phi^2}{k^3} = k \;\Rightarrow\; \delta\phi \simeq k. \tag{2.147}$$

This means that a perturbation $\delta\phi \simeq k$ generates an energy density perturbation $\delta\rho \sim k$. In other words, if one asks for ρ to vary of a quantity of order k, one needs also ϕ to vary by the same quantity, $\delta\phi \sim k$.

Exercise For a quantum treatment of the preceding analysis, consider a field (B.77)

$$\phi(\mathbf{x}, t) = \int \frac{d^3k}{(2\pi)^{3/2}\sqrt{2E_k}}(e^{iE_k t - i\mathbf{kx}}a_k^\dagger + e^{-iE_k t + i\mathbf{kx}}a_k), \tag{2.148}$$

where the creation and annihilation operators obey

$$[a_k^\dagger, a_{k'}] = \delta(\mathbf{k} - \mathbf{k'}). \tag{2.149}$$

Show that

$$\langle\phi^2(x)\rangle = \langle 0|\phi(\mathbf{x}, t)\phi(\mathbf{x}, t)|0\rangle = \int \mathcal{P}_\phi(k)\frac{dk}{k} \tag{2.150}$$

with

$$\mathcal{P}_\phi(k) = \frac{k^2}{(2\pi)^2}. \tag{2.151}$$

From Eq. (2.150), we can understand the power spectrum $\mathcal{P}_\phi$ as the *increase* of the perturbation $\delta\phi^2 = \langle\phi^2(x)\rangle$ per decimal interval of momenta k. If we *define* the amplitude of the quantum oscillation $\delta\phi_k$ as the *variance* of the field with momentum k,

$$\delta\phi_k \equiv \Delta_\phi = \sqrt{\mathcal{P}_\phi(k)}, \tag{2.152}$$

we obtain

$$\delta\phi_k = \frac{k}{2\pi}, \qquad (2.153)$$

which is similar to the classical solution (2.147).

2.2.9.2 In an Expanding Universe

The main difference, in an expanding Universe, is that one needs to take into account the redshift of the momenta k with the scale factor $a(t)$:

$$q = \frac{k}{a(t)}. \qquad (2.154)$$

If we write the inflaton field

$$\Phi(\mathbf{x}, t) = \Phi_c(t) + \phi(\mathbf{x}, t), \qquad (2.155)$$

$\Phi_c(t)$ being the (homogeneous) classical inflaton field, the equation of motion for the perturbation ϕ is

$$\ddot{\phi} + 3H\dot{\phi} + \frac{k^2}{a^2}\phi + V''(\Phi_c)\phi = 0. \qquad (2.156)$$

Exercise Using

$$\begin{aligned} S_\phi &= \frac{1}{2}\int d^4x\sqrt{-g}[g^{\mu\nu}\partial_\mu\phi\partial_\nu\phi - V''(\Phi_c)\phi^2] \\ &= \frac{1}{2}\int dtd^3xa^3[\dot{\phi}^2 - a^{-2}(\partial_i\phi)^2 - V''(\Phi_c)\phi^2], \end{aligned} \qquad (2.157)$$

recover the expression (2.156) from the Euler–Lagrange equation (2.101).

Neglecting $V''(\phi)$ (slow-roll condition 2.117), we can briefly describe the different regimes for Eq. (2.156). At first, when a is extremely small, the first and third terms dominate. The equation is then a classical harmonic oscillator equation, with a redshifting frequency ($\sim\frac{k}{a}$) decreasing with time. When the wavelength ($\sim\frac{a}{k}$) reaches the horizon size[21] H^{-1}, then the field ϕ freezes in, and the second term of (2.117) dominates; $\dot{\phi} = 0 \Rightarrow \phi$ is constant. In the first stage, ϕ is in a *subhorizon mode*, transforming into a *superhorizon mode* in the second stage. Once

[21] That is almost constant during the whole process.

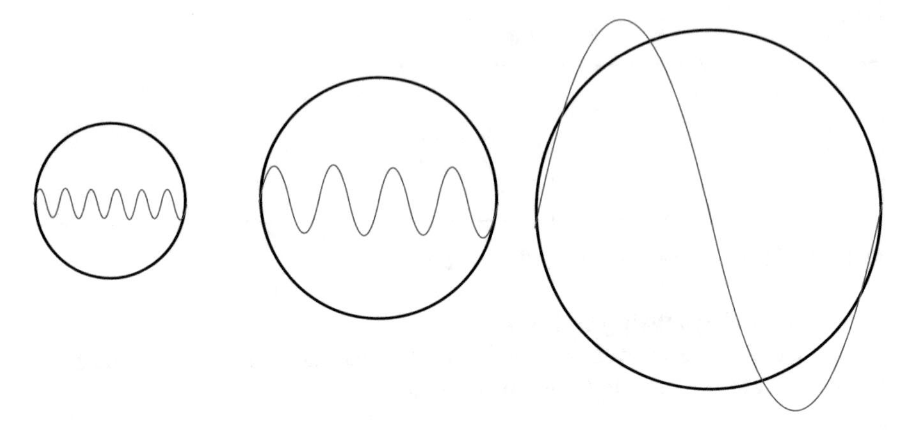

Fig. 2.6 Evolution of a fundamental mode, from the subhorizon to the superhorizon phase, when it froze in

in the superhorizon phase, the vacuum fluctuations ϕ are frozen in, even through the wavelength grows. At this frozen stage, one deduces from (2.153) (Fig. 2.6).

$$\delta\phi_k \simeq \frac{H}{2\pi}. \tag{2.158}$$

To have a feeling about the mechanics of the formation of overdense regions, we notice that some regions will possess a density $\Phi_c - |\delta\phi|$, in a slow-roll regime, where $V(\Phi_c - \delta\phi) < V(\Phi_c) = V(\langle\Phi\rangle)$. We illustrate it in Fig. 2.7. These regions, with negative fluctuations, will exit the inflation phase *before* the regions with positive fluctuations. These regions will lose energy density, whereas, during a small period δt, the region with positive $\delta\phi$ will still have the constant inflationary density. The time it will take for Φ_c to *reach* $\Phi_c - |\delta\phi|$ is given by

$$\delta\phi = \dot{\Phi}_c \delta t. \tag{2.159}$$

From the evolution of the density given in Eq. (2.43),

$$\dot{\rho} + 3H(1+w)\rho = 0 \;\Rightarrow\; \frac{\delta\rho}{\rho} \sim -H\delta t = -\frac{H}{\dot{\Phi}_c}\delta\phi = -\frac{H^2}{2\pi\dot{\Phi}_c}, \tag{2.160}$$

where we used (2.158). This condition should be evaluated at the time the k-mode exits the horizon. From the slow-roll condition (2.117), we deduce

$$\dot{\Phi}_c = -\frac{V'(\phi)}{3H} \;\Rightarrow\; \frac{\delta\rho}{\rho} \sim \frac{V^{3/2}(\phi)}{2\pi\sqrt{3}M_P^3 V'(\phi)}. \tag{2.161}$$

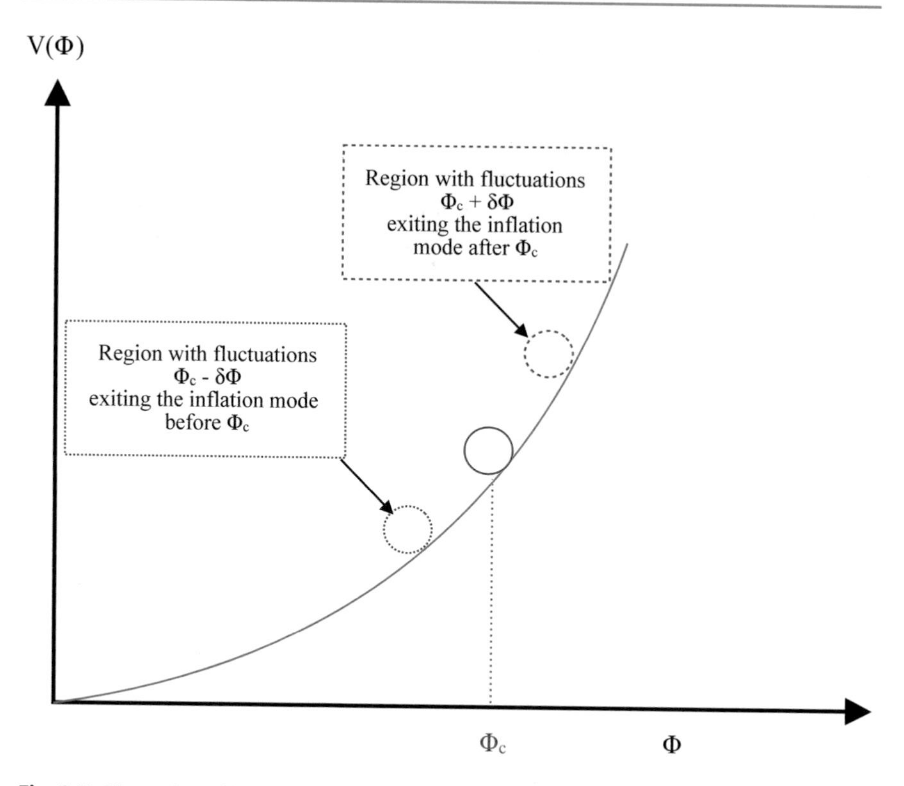

Fig. 2.7 Illustration of inflation exit for regions with different fluctuations

If one considers the chaotic inflation type of potential $V(\Phi) = \lambda\Phi^n$, one obtains

$$\frac{\delta\rho}{\rho} \sim \frac{\sqrt{\lambda}}{n} \frac{\Phi^{\frac{n}{2}+1}}{2\pi\sqrt{3}M_P^3} \simeq 5 \times 10^{-5}, \tag{2.162}$$

where the last equality holds from PLANCK measurement of the CMB. It is then easy to extract limit on $\sqrt{\lambda}$ as a function of the number of e-folds N_e using Eq. (2.135):

$$m_\phi = \sqrt{2\lambda} \simeq \frac{4}{N_e} \times 10^{-4} M_P \lesssim 1.8 \times 10^{13}\text{ GeV} \tag{2.163}$$

in the case of a quadratic potential for $N_e \gtrsim 50$, and

$$\lambda \simeq \frac{93}{N_e^3} \times 10^{-10} \lesssim 7.4 \times 10^{-14} \tag{2.164}$$

for a quartic potential.

2.2.10 Preheating and Dark Matter*

2.2.10.1 Parametric Resonance

Before treating the perturbative reheating process in the next section, we want to say few words about a preheating phenomenon called the *narrow resonance*, which is in fact a classical parametric resonance in the context of the preheating. The idea is that, if one couples the inflaton field Φ to a scalar S, there is the possibility of producing it with a large rate, increasing exponentially its occupation number, through a mechanism of resonance. For illustration, let us consider the action

$$\mathcal{S} = \int d^4x\sqrt{-g}\left(\frac{1}{2}g^{\mu\nu}\partial_\mu S\partial_\nu S - V(S,\Phi)\right) = \int d^4x\sqrt{-g}\mathcal{L}. \tag{2.165}$$

The Euler–Lagrange equation is

$$\begin{aligned}
&\frac{\partial}{\partial S}\sqrt{-g}\mathcal{L} = \partial_\mu\frac{\partial}{\partial(\partial_\mu S)}\sqrt{-g}\mathcal{L} \\
\Rightarrow\quad & -\sqrt{-g}V'(S) = \partial_t[\sqrt{-g}g^{tt}\dot{S}] + \partial_i[\sqrt{-g}g^{ii}\partial_i S] \\
\Rightarrow\quad & -\sqrt{-g}V'(S) = -\frac{\dot{g}}{2\sqrt{-g}}g^{tt}\dot{S} + \sqrt{-g}g^{tt}\ddot{S} + \partial_i[\sqrt{-g}g^{ii}\partial_i S].
\end{aligned} \tag{2.166}$$

Using

$$\sqrt{-g} \propto a^3 \;\Rightarrow\; \dot{g} \propto 6a^5\dot{a} \;\Rightarrow\; -\frac{1}{2}\frac{\dot{g}}{-g} = +3\frac{\dot{a}}{a} = 3H, \tag{2.167}$$

we obtain

$$\ddot{S} + 3H\dot{S} + \Box S + \frac{\partial V}{\partial S} = 0, \tag{2.168}$$

or, for a plane wave[22] $S \propto e^{i\mathbf{k}.\chi}$, implying $\partial_i S = \frac{1}{a}\frac{\partial}{\partial \chi_i}S = \frac{k_i}{a}S$,

$$\ddot{S} + 3H\dot{S} + \frac{k^2}{a^2}S + \frac{\partial V}{\partial S} = 0 \tag{2.169}$$

If one takes the potential $V(S,\Phi) = \mu\Phi S^2$, and suppose, at a first approximation that $a \sim$ constant during this short preheating phase, for an inflaton field of the from

[22] Keeping in mind that we can always decompose S in its fundamental modes, $S = \sum_k s_k e^{i\mathbf{k}.\mathbf{x}}$, all contributions add up to the number density n or energy density ρ.

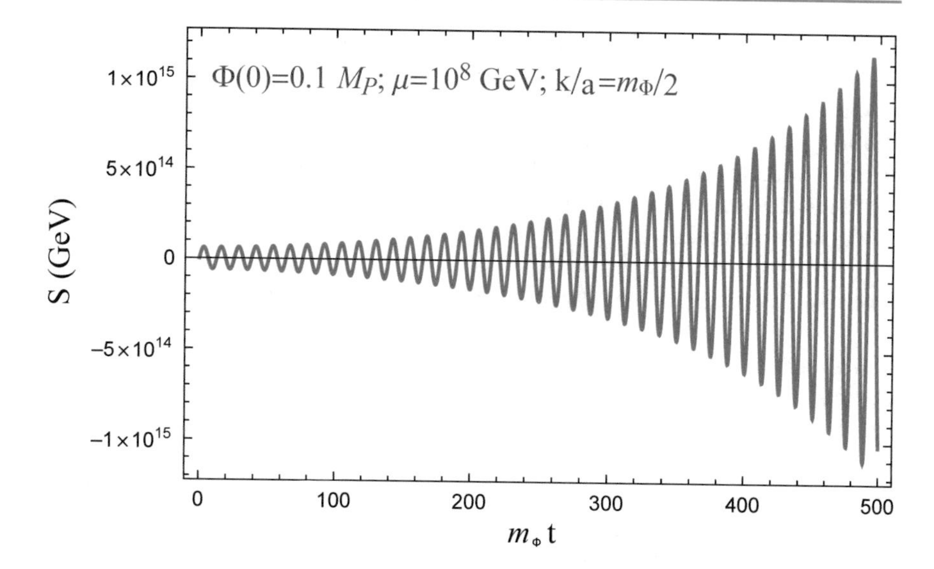

Fig. 2.8 Illustration of the parametric (also called *narrow*) resonance in the context of preheating. We can see clearly the exponential envelop of the periodic solution

$\Phi = \Phi_0 \cos(m_\Phi t)$, the equation one needs to solve is

$$\ddot{S} + 3H\dot{S} + \left[\frac{k^2}{a^2} + \mu\Phi_0 \cos(m_\Phi t)\right] S = 0. \tag{2.170}$$

Supposing $a \simeq$ constant, we can neglect H. This equation is one form of the Mathieu equation, which is the equation for an oscillator with a time dependant frequency $\omega^2(t) = \frac{k^2}{a^2} + \mu\Phi_0 \cos(m_\Phi t)$, and is present in a lot of classic phenomena involving periodical force. It can be shown that for

$$\frac{m_\Phi}{2} - \frac{\mu\Phi_0}{2m_\Phi} < \frac{k}{a} < \frac{m_\Phi}{2} + \frac{\mu\Phi_0}{2m_\Phi}, \tag{2.171}$$

we enter in a regime where the solution grows exponentially with time.[23] We can understand it easily, from the shape of the Mathieu equation, where, periodically, the coefficient $\cos(m_\Phi t)$ becomes negative and drives the evolution of S toward an exponential solution, periodically. The evolution of S is shown in Fig. 2.8. A more

[23]This situation corresponds to a narrow resonance if one considers $\mu \ll \Phi_0$. The regime $\mu \gtrsim \Phi_0$ is a broad resonance regime but exhibits similar features [7].

refined treatment necessitate to compute the Bogoliubov coefficient to extract the occupation number [7], but we give in the following section a more intuitive view of the phenomena, solving the equation for the density of the ϕ decay products. For the analytical solution of the Mathieu equation (2.170), the reader is directed to [6], which is without doubt the best textbook treating it, and [7], which is (paradoxically) the seminal *research* paper on the subject and one of the clearer and more detailed works in the literature.

Exercise Redo the previous analysis for a potential $V(S, \Phi) = \frac{1}{2}\Phi^2 S^2$.

The reason why we will not observe this highly non-perturbative effect in a classical (perturbative) vision, which we will use later, is quite delicate. It comes mainly from the fact that in perturbation theory, we consider a mother particle at rest, decaying into two daughter particles satisfying the mass shell relation $k^2 = m_S^2$. This is not exactly the case when we look at the inflaton as a background field. Indeed, even if we show in Eq. (2.141) that the inflaton field *behaves* like a set of matter fields at rest with respect to the evolution of the scale factor a, we should rather treat these fields as a set of coherently oscillating ϕ-fields and not as a physical set of independent ϕ-fields. One should in fact take into account the effects of the decay products on the width of the ϕ-fields themselves and the back reaction effects. Another way to look at it is to look at this physics from the statistical point of view of the Bose enhancement.

2.2.10.2 Narrow Resonance Interpreted as Bose Condensates

The result discussed above can be obtained numerically (as we have done) or analytically, via the Floquet method. However, there is a more intuitive interpretation in terms of Bose enhancement. Indeed, the narrow resonance effect can be calculated by considering the decay of the inflaton field into two bosons and by calculating the number of particles occupancy in the final state. In the rest frame of the ϕ-particle, the two particles produced have the same momentum, but with opposite directions. Being bosonic by nature, if the final state is already occupied, there will be an enhancement effect by a Bose factor, which we must therefore calculate.

Let us suppose an inflaton coupling to a real scalar S:

$$\mathcal{L} = \frac{1}{2} g^{\mu\nu}\partial_\mu\phi\partial_\nu\phi - V(\phi) + \frac{1}{2} g^{\mu\nu}\partial_\mu S\partial_\nu S - \frac{m_S^2}{2} S^2 - \mu\phi S^2.$$

The production rate of S should be computed combining the decay process $\phi \rightarrow S\,S$ and the inverse decay $S\,S \rightarrow \phi$. The former is proportional to

$$\Gamma_{\phi\rightarrow SS} \propto |\langle n_\phi - 1, n_{\mathbf{k}} + 1, n_{-\mathbf{k}} + 1| a_\phi a_{\mathbf{k}}^\dagger a_{-\mathbf{k}}^\dagger | n_\phi, n_{\mathbf{k}}, n_{-\mathbf{k}}\rangle|^2$$
$$= n_\phi(n_{\mathbf{k}} + 1)(n_{-\mathbf{k}} + 1),$$

where we classically defined the creation/annihilation operators $a_i^\dagger$ and a_i (see Sect. 4.9.2.4 and Eq. (4.72) in particular). The rate of the inverse decay is

$$\Gamma_{SS\to\phi} \propto |\langle n_\phi+1, n_{\mathbf{k}}-1, n_{-\mathbf{k}}-1|a_\phi^\dagger a_{\mathbf{k}} a_{-\mathbf{k}}|n_\phi, n_{\mathbf{k}}, n_{-\mathbf{k}}\rangle|^2 = (n_\phi+1)n_{\mathbf{k}}n_{-\mathbf{k}}.$$

In the following, considering a homogeneous background, the density occupation number depends only on the amplitude of the momentum and not its direction. We can then fairly define $n_k = n_{\mathbf{k}} = n_{-\mathbf{k}}$. Adding the two processes and noting that $n_\phi \equiv \frac{\rho_\phi}{m_\phi} \gg 1$, we obtain

$$\Gamma_{eff} \simeq \Gamma_\phi(1+2n_k), \quad \text{with } \Gamma_\phi = \frac{\mu^2}{32\pi m_\Phi} \tag{2.172}$$

where we used Eqs. (B.165) and (B.179). To compute the occupation number n_k, we use the definition[24] (B.85)

$$n_s = \frac{d^3k}{(2\pi)^3}n_k, \tag{2.173}$$

the volume d^3k being limited by the relation

$$E_k = \sqrt{k^2 + m_s^2 + 2\mu\phi(t)} = \frac{1}{2}m_\phi, \tag{2.174}$$

or

$$|dk| = \frac{2\mu|d\phi|}{k}. \tag{2.175}$$

For a quadratic potential, $V(\phi) = \frac{1}{2}m_\phi$, one can use the solution we obtained (2.127)

$$\phi(t) = \sqrt{\frac{8}{3}}\frac{M_P}{m_\phi t}\cos m_\phi t \equiv \Phi\cos m_\phi t, \tag{2.176}$$

approximation already valid after only one oscillation. Noticing then that $d\phi \sim 2\Phi$ and $k \sim \frac{m_\phi}{2}$, we obtain

$$|dk| \simeq \frac{8\mu\Phi}{m_\phi} \quad \Rightarrow \quad n_k = \frac{\pi^2}{\mu\Phi m_\phi}n_s, \tag{2.177}$$

[24] Be careful in the following notations; n_s represents the density number of particle S (per unit of space volume), whereas n_k represents the occupation number (no units).

and

$$\Gamma_{eff} \simeq \Gamma_\phi \left(1 + \frac{2\pi^2}{\mu \Phi m_\phi} n_s\right). \tag{2.178}$$

2.2.10.3 Production of Dark Matter in the Preheating Era

It is then straightforward to compute the production of a dark matter species (or any type of particles) by solving the Boltzmann equation

$$\frac{dn_s}{dt} = 2\Gamma_{eff} \frac{\rho_\phi}{m_\phi} = 2\Gamma_\phi \left(1 + \frac{2\pi^2}{\mu \Phi m_\phi} n_s\right) \frac{\rho_\phi}{m_\phi}, \tag{2.179}$$

the factor "2" taking into account the fact that 2 particles are produced by decay. A quick look at the above equation shows where the explosive, exponential behavior of particle production comes from. This is an effect directly related to the fact that bosons will tend to accumulate in the phase space that already contains the largest number of bosons, this term being proportional to n_s. Solving Eq. (2.179), we obtain for n_s

$$n_s(t) = \frac{\mu \Phi m_\phi}{2\pi^2} \left(e^{2\pi^2 \frac{\Phi}{\mu} \Gamma_\phi t} - 1\right) = \frac{\mu \Phi m_\phi}{2\pi^2} \left(e^{\frac{\pi}{16} \frac{\mu \Phi}{m_\phi} t} - 1\right). \tag{2.180}$$

Being at a very early time ($\Gamma_S t \ll 1$), we recover the narrow width condition ($\mu \ll \Phi$) for the explosive production to be efficient.

The natural question we are then entitled to ask is when does this exponential production end? There are several cases, depending on the value of the width of ϕ. When the perturbative decay dominates, we are in the perturbative regime, the oscillations ends, and we can jump to the next chapter. But how long should we wait to see the perturbative decay dominates, and dominates over what?

- The first (and most naive) condition is that the non-perturbative decay rate must be less than the perturbative part, i.e. from (2.180),

$$\frac{\pi}{16} \frac{\mu \Phi}{m_\phi} \lesssim \Gamma_\phi = \frac{\mu^2}{32\pi m_\phi} \quad \Rightarrow \quad \frac{\Phi}{\mu} \lesssim \frac{2}{\pi^2}, \tag{2.181}$$

or, in other words, when the narrow width condition is not satisfied anymore. However, we are in a period of time where $\Gamma_\phi \ll H$, which means, another condition should be stronger.

- Indeed, the friction term appearing in the equation of motion is of the form

$$3H + \Gamma_\phi \tag{2.182}$$

(see Eq. (2.206)), for instance). A stronger condition to stop the preheating production is then

$$\frac{\pi}{16}\frac{\mu\Phi}{m_\phi} \lesssim 3H + \Gamma_\phi \sim 3H = 3\frac{m\Phi}{\sqrt{6}M_P} \quad \Rightarrow \quad \frac{\mu M_P}{m_\phi^2} \lesssim \frac{16}{\pi}\sqrt{\frac{3}{2}}, \tag{2.183}$$

corresponding to the fact that the expansion rate dominates the production rate, and the process is frozen.
- Last but not the least, an even stronger constraint is directly related to the narrowness of the resonance. Indeed when a particle is produced with momentum k, if by redshift this momentum exits from the phase space volume defined by $|dk|$ of Eq. (2.175) at a rate larger than the produced ones, the phase space is not populated anymore. k being redshifted from time t_1 to t_{PH} (time of the end of preheating phase), we can write

$$|dk| = k\frac{a(t_1)}{a(t_{PH})}\frac{\dot{a}(t_{PH})}{a(t_{PH})}dt \simeq k\ H(t_{PH})dt \ \Rightarrow \ dt = \frac{|dk|}{k}H^{-1} = \frac{8\sqrt{6}\mu M_P}{m_\phi^3},$$

where we used $a(t_1) \sim a(t_{PH})$, the evolution being slow during the whole process, $k = \frac{m_\phi}{2}$, and $|dk| = 4\frac{\mu\Phi}{m_\phi}$ (2.175). The inflaton will then not populate efficiently the region as soon as

$$\frac{\pi}{16}\frac{\mu\Phi}{m_\phi}dt = \sqrt{\frac{3}{2}}\frac{\pi\mu^2\Phi M_P}{m_\phi^4} \lesssim 1 \ \Rightarrow \ \frac{\mu^2\Phi M_P}{m_\phi^4} \lesssim \sqrt{\frac{2}{3}\frac{1}{\pi}}. \tag{2.184}$$

This last constraint, which is the strongest of all, gives us the value of Φ at which explosive production ceases to be effective. Remembering that (2.176)

$$\Phi(t) = \sqrt{\frac{8}{3}\frac{M_P}{m_\phi t}}, \tag{2.185}$$

and implementing it in Eq. (2.184), we can then calculate the time t_{PH} from which the reheating phase starts and the preheating phase ends:

$$t_{PH} = 2\pi \frac{\mu^2 M_P^2}{m_\phi^5}. \tag{2.186}$$

To give an idea, for a typical value of $\mu = 10^5$ GeV, one obtains $t_{PH} \simeq 10^{-19}$ GeV$^{-1} \simeq 7 \times 10^{-44}$ s, where we took $m_\phi = 2 \times 10^{13}$ GeV from Eq. (2.163). This time is very close to Planck's time (5.4×10^{44} s). We can now compute the relic abundance of a bosonic particle produced from t_{Pl} to t_{RH}, the reheating temperature, and compare it to the perturbative production.

Combining Eqs. (2.180), (2.184), and (2.186), we deduce the density of S at t_{PH}, n_s^{PH}

$$n_s^{PH} = \frac{\mu \Phi m_\phi}{2\pi^2} e^{\frac{1}{16}\frac{m_\phi^2}{\mu M_P} m t_{PH}} = \frac{\mu \Phi m_\phi}{2\pi^2} e^{\frac{\pi}{4\sqrt{6}}\frac{\mu M_P}{m_\phi^2}} \simeq \frac{\mu M_P m_\phi}{2\pi^2} e^{\frac{\pi}{4\sqrt{6}}\frac{\mu M_P}{m_\phi^2}}, \tag{2.187}$$

where we have set $\Phi = M_P$ in the last approximation.[25] This density will be diluted by a factor

$$\left(\frac{t_{RH}}{t_{PH}}\right)^2 = \left(\frac{\Gamma_\phi^{-1}}{t_{PH}}\right)^2 = \frac{64 m_\phi^{12}}{\mu^8 M_P^4} \tag{2.188}$$

between its production and the end of reheating, which gives us

$$n_s^{PH}(t_{RH}) = \frac{\mu^9 M_P^5}{128\pi^2 m_\phi^{11}} e^{\frac{\pi}{4\sqrt{6}}\frac{\mu M_P}{m_\phi^2}}. \tag{2.189}$$

We can then compare this production to the one obtained by the direct perturbative decay of the inflaton at time $t_{RH} = \Gamma_\phi^{-1}$, solving Eq. (2.179) without taking into account the enhancement effect but taking into account the expansion rate:

$$n_s^{RH}(t_{RH}) = \frac{\mu^4 M_P^4}{96\pi^2 m_\phi^3}, \tag{2.190}$$

[25] To obtain the exact result, one should in fact solve numerically the combined set of Eqs. (2.276) and (2.179). Taking $\Phi \simeq M_P$ in the overall factor of n_s is a valid approximation.

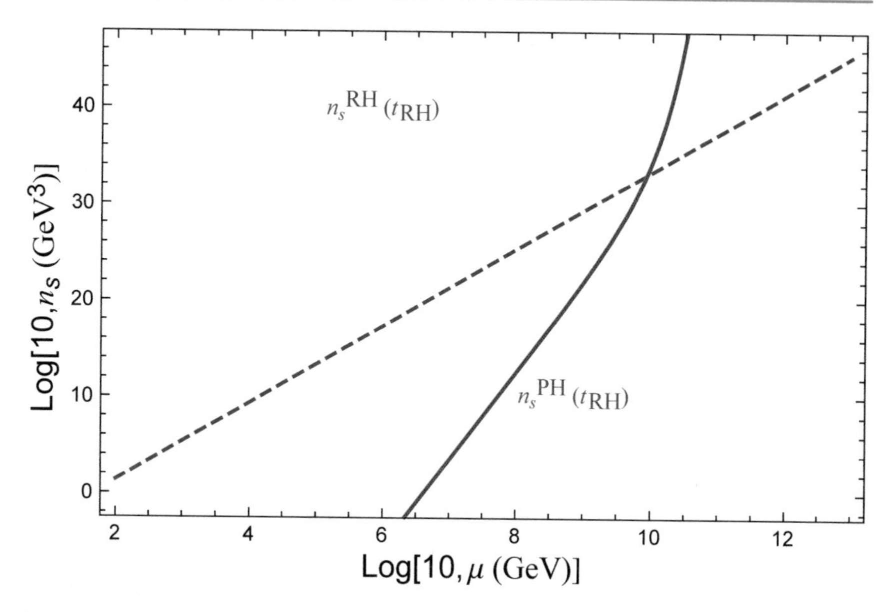

Fig. 2.9 Number density of a scalar as a function of μ produced in the preheating phase (n_s^{PH}) compared to the one produced perturbatively at reheating time (n_s^{RH})

where we used the result obtained in the reheating phase, see Eq. (2.248), with $t = \Gamma_\phi^{-1}$. This is justified because it is the dominant phase in this case. We show in Fig. 2.9 the two densities as a function of μ. For $\mu \simeq 10^{10}$, the exponential production dominates the perturbative one. The reader can find a nice detailed treatment of the bosonic condensation effect we just discussed in [8].

Exercise Redo the previous analysis for a potential $V(S, \phi) = \lambda\phi^4$.

Now that we understood how evolves the inflaton field and density from t_i to t_{end}, it is time to couple it to matter, and allow it to decay in order to reheat the Universe by the production of a thermal bath of Standard Model particles. This is the aim of the next chapter.

2.3 Reheating: Non-thermal Phase [10^{-37} – 10^{-30} s]

Thermalization of the hot and dense primordial plasma produced by the decay of the inflaton ϕ is far from being a simple subject. And, to be honest, it is far to be an understood subject. The reasons of such difficulties are multiple. We can say that computing processes in a matter dominated Universe, driven by Friedmann expansion mixed with a non-thermal statistics, renders the subject sensitive to a lot of theoretical uncertainties and/or questionable hypothesis. Paradoxically, once the Universe has thermalized, the story is much easier to describe. A thermal bath of relativistic particles is, in some sense, quite Universal. Things then become more complicated every time particles decouple, or during the Big Bang Nucleosynthesis, in other words, every time the Universe becomes colder and goes through a phase of mixed matter–radiation domination. In this chapter, we will review in detail how a set of particles, generated by the decay of the inflaton, begins to interact sufficiently strongly to overcome the expansion rate.

2.3.1 The Context

Before digging into the thermalization process, let us emphasize that before the Universe reaches its thermal bath phase, temperature has, by definition, no meaning. One should then solve any set of equations as a function of the time t, which is the dynamical parameter we have then in hands. The problem is then to compute the evolution of the content of the Universe, which is composed of a mix of non-relativistic species (behaving as a dust, like the inflaton ϕ) and relativistic species (radiation like the Standard Model particles, or any kind of relativistic matter or gauge fields that is produced by the decay of the inflaton). As we will see, dark matter can be one of them.[26] Of course, these laws can be justified by the Einstein equations we studied in the preceding chapter, however, if we neglect the space-time curvature (the parameter k), we can recover the Boltzmann–Friedmann set of equations just asking for the conservation of energy.

2.3.1.1 The Boltzmann Equation for the Dust (Inflaton or Non-relativistic Fields)

We call dust or matter field, generically, a field that obeys the relation[27] $\rho_M \propto a^{-3}$. To follow its evolution with time in an expanding Universe, let us first write the law of conservation of energy for a decaying matter field ϕ_M of mass m_M, width Γ_M and density n_M:

$$\frac{d(n_M a^3)}{dt} = -\Gamma_M (n_M a^3), \qquad (2.191)$$

and multiplying by m_M on both sides, we obtain

$$\frac{d(\rho_M a^3)}{dt} = -\Gamma_M (\rho_M a^3), \qquad (2.192)$$

[26] Due to the relative large mass of the inflaton ($m_\phi \simeq 10^{13}$ GeV), Standard Model particles produced by its decay are obviously ultra-relativistic and can be considered as a form of radiation.
[27] $\rho_M \propto a^{-3(1+w)}$ with $w = 0$ and $p = w\rho$.

where $\rho_M = n_M m_M$ is the energy density of the matter (dust) field ϕ_M. Developing the derivative, the preceding equation gives

$$\dot{\rho}_M + 3H\rho_M = -\Gamma_M \rho_M. \tag{2.193}$$

Notice that we recover exactly Eq. (2.140), which describes the evolution of the inflaton field ϕ for the case $k = 2$, but where we have added in the equation a term $\Gamma_M \rho_M$ as we allowed the possibility for the matter field ϕ_M to decay. As we have seen in the previous section, this is not surprising, as the inflaton, in a quadratic potential, behaves, during its coherent oscillation phase, as a pressureless field ($k = 2 \Rightarrow w = 0$). That will be our first ingredient to understand the thermalization:

$$\dot{\rho}_\phi + 3H\rho_\phi = -\Gamma_\phi \rho_\phi. \tag{2.194}$$

Notice that this equation is valid for *any* type of matter field which can, for instance, dominate the Universe even after the reheating has occurred.

2.3.1.2 The Boltzmann Equation for the Radiation (Relativistic Fields)

To find the evolution of the density of radiation, one needs to use the first law of thermodynamics in a system of internal energy $U = \rho_R a^3$ in contact with another system transmitting heat $dQ = -d(\rho_\phi a^3)$ (represented by the decaying inflaton ϕ) in a volume $V = a^3$. The amount of internal energy $dU = dQ - p_R dV$ (p_R being the pressure of the relativistic gas) received by the radiation sector is[28]

$$d(\rho_R a^3) = -p_R da^3 - d(\rho_\phi a^3) = -\frac{\rho_R}{3} da^3 - d(\rho_\phi a^3) \tag{2.195}$$

where we used the usual

$$p_R = \frac{\rho_R}{3} \tag{2.196}$$

relation of a relativistic gas ($w = \frac{1}{3}$), as explained in the frame below. We then obtain the equation

$$\dot{\rho}_R + 4H\rho_R = \Gamma_\phi \rho_\phi, \tag{2.197}$$

where we used Eq. (2.192):

$$\frac{d(\rho_\phi a^3)}{dt} = -\Gamma_\phi(\rho_\phi a^3). \tag{2.198}$$

---

[28] We make an important hypothesis of a constant number of degrees of freedom in the relativistic bath during all the process. That will not be the case once the Universe cools down as some particles may decouple as we will discuss in the next chapter.

Pressure of a gas: $p_R = \frac{\rho_R}{3}$

Let us recall where the relation $p_R = \frac{\rho_R}{3}$ in (2.196) comes from. For that, one needs to compute the pressure π of a homogeneous and isotropic gas of relativistic particles. Consider a surface S hit by particles with a distribution of momentum $f(\mathbf{p})$ during a time Δt. The pressure π is defined as the force applied by the collisions of the particles per unit of surface, which can be written

$$\pi = \frac{|\mathbf{F}|}{S}, \text{ with } \mathbf{F} = \frac{\Delta \mathbf{p}}{\Delta t}, \tag{2.199}$$

$\mathbf{p}$ being the transferred momentum.

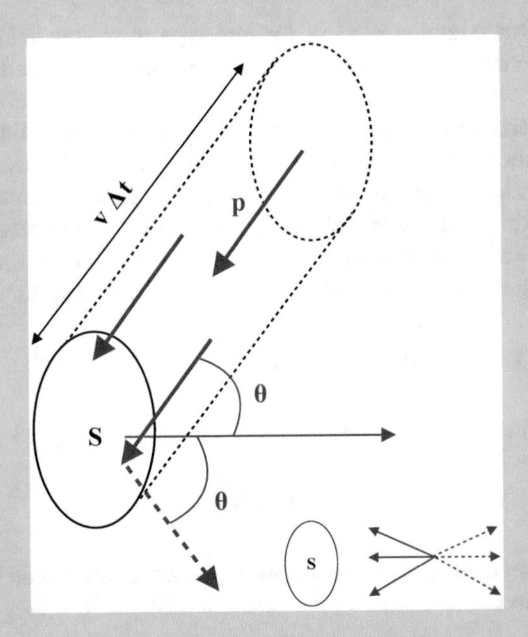

One particle arriving on the surface with an angle θ, will bounce with an angle $-\theta$ and will transfer to the surface a momentum $|\Delta \mathbf{p}| = 2p\cos\theta$. During the time Δt, this corresponds to dn particles of velocity v with distribution $f(\mathbf{p})$ in a volume $(S \times \cos\theta)v\Delta t$:

$$|\Delta \mathbf{p}|_{tot} = 2p\cos\theta \; S\cos\theta \; v\Delta t dn, \tag{2.200}$$

(continued)

which gives (considering a relativistic gas, $v = c$):

$$|\Delta \mathbf{p}|_{tot} = 2pc\ S\Delta t\ \cos\theta^2 f(\mathbf{p})\frac{d^3p}{(2\pi)^3}. \tag{2.201}$$

The pressure, given by the momentum transfer per unit of surface and unit of time, is

$$d\pi = \frac{1}{2}\frac{|\Delta \mathbf{p}|_{tot}}{S\Delta t} = \epsilon \cos^2\theta f(\mathbf{p})\frac{p^2}{(2\pi)^3}dpd\phi d\cos\theta, \tag{2.202}$$

where $\epsilon = pc$ is the energy of the relativistic particle of momentum p, and we have defined the solid angle by $d^3p = p^2dpd\Omega = p^2dpd\phi d\cos\theta$. The $\frac{1}{2}$ factor in the definition of $d\pi$ comes from the fact that only half of the distribution should be considered, the one embedding the particles with momentum directing toward the surface (see figure the above). After integrating on $\cos\theta$ between -1 and $+1$, we obtain

$$\begin{aligned}\langle\pi\rangle &= \int d\pi = \int \cos^2\theta d\cos\theta \int \epsilon f(p)\frac{p^2}{(2\pi)^3}dpd\phi \\ &= \frac{2}{3}\int \epsilon f(p)\ \frac{p^2}{(2\pi)^3}dpd\phi \\ &= \frac{1}{3}\int_{-1}^{+1} d\cos\theta \int \epsilon f(p)\ \frac{p^2}{(2\pi)^3}dpd\phi = \frac{\rho}{3},\end{aligned} \tag{2.203}$$

$\rho = \int \epsilon f(p)\ \frac{d^3p}{(2\pi)^3}$ being the energy density, which can be written for a radiation $\pi = p_R$, $\rho = \rho_R$:

$$p_R = \frac{\rho_R}{3}, \tag{2.204}$$

which is precisely Eq. (2.196).

The factor "4" in front of H in Eq. (2.197) compared to the factor "3" in Eq. (2.194) can also be understood as a redshift effect. Indeed, it is natural to write an energy density for a radiation $\rho_R \sim n_R E_R \propto a^{-3} \times a^{-1}$, the first term coming from the dilution by expansion, and the second term (a^{-1}) being the redshift of the energy E_R due to the expansion. In this case, $\frac{d\rho_R}{dt} \propto -4H\rho_R$. Quantitatively speaking, it is easy to understand that the pressure in Eq. (2.195) acts negatively, "adding" in a sense some dispersion and thus rendering then dilution faster than in the matter case.

A last way to recover Eq. (2.197) is to apply directly the General Relativity equation (2.42) adding the decay rate in the right-hand side, and replacing $P = \frac{\rho_R}{3}$ or, in other words, to use the stress–energy tensor (A.96)

$$T^R_{\mu\nu} = \begin{pmatrix} \rho & 0 & 0 & 0 \\ 0 & \frac{\rho}{3} & 0 & 0 \\ 0 & 0 & \frac{\rho}{3} & 0 \\ 0 & 0 & 0 & \frac{\rho}{3} \end{pmatrix}. \quad (2.205)$$

Combining results above, we obtain the set of equations to solve

$$\frac{d\rho_\phi}{dt} + 3H\rho_\phi = -\Gamma_\phi \rho_\phi \quad (2.206)$$

$$\frac{d\rho_R}{dt} + 4H\rho_R = \Gamma_\phi \rho_\phi \quad (2.207)$$

$$H^2 = \frac{\rho_\phi}{3M_P^2} + \frac{\rho_R}{3M_P^2} \quad (2.208)$$

with $M_P = \frac{M_{Pl}}{\sqrt{8\pi}} \approx 2.4 \times 10^{18}$ GeV (as usual in all the book) the reduced Planck mass[29] and $H = H(t) = \frac{2}{3t}$. The last expression assumes a Universe purely matter dominated (see the box below for details). A radiation dominated Universe follows the law $H = \frac{1}{2t}$. This approximation can be justified by the fact that we are dealing, for time $t \ll \Gamma_\phi^{-1}$, with a Universe where the inflaton dominates still largely the dynamic.

2.3.1.3 The Influence of the Nature of the Inflaton

It can also be interesting to see how the set of Eqs. (2.206) and (2.207) changes for the decay of an inflaton with a generic equation of state $P_\phi = w\rho_\phi$. Indeed, even if intuitive, the previous equations describe the evolution of the *density of energy* ρ_ϕ while the width appearing on the right-hand side in the width of the *field* ϕ, Γ_ϕ, and not "Γ_{ρ_ϕ}." To find the conservation of energy equation, one needs to rederive it from the original equation for ϕ, (2.99):

$$\ddot{\phi} + 3H\dot{\phi} + V'(\phi) = -\Gamma_\phi \dot{\phi}. \quad (2.209)$$

[29] Notice that for $\Gamma_\phi = 0$, we recover the classical evolutions ($\rho_\phi \propto a^{-3}$, $\rho_R \propto a^{-4}$) for independent systems with entropy conservation, the a^{-4} corresponding to the redshifted energy $E_R(a) = \frac{E_R(a_f)}{a}$, and a_f being the radius of the Universe just after the inflation.

Multiplying both sides by $\dot{\phi}$ and taking the mean, we obtain

$$\langle\dot{\phi}\ddot{\phi}\rangle + 3H\langle\dot{\phi}^2\rangle + \langle\dot{\phi}V'(\phi)\rangle = -\Gamma_\phi\langle\dot{\phi}^2\rangle. \tag{2.210}$$

Noticing that (2.95)

$$\langle\dot{\phi}^2\rangle = \rho_\phi + P_\phi, \tag{2.211}$$

and

$$\dot{\rho}_\phi = \langle\dot{\phi}\ddot{\phi}\rangle + \langle\dot{\phi}V'(\phi)\rangle, \tag{2.212}$$

we can write

$$\dot{\rho}_\phi + 3H(\rho_\phi + P_\phi) = -\Gamma_\phi(\rho_\phi + P_\phi), \tag{2.213}$$

or, if we include the contribution to the radiation,

$$\frac{d\rho_\phi}{dt} + 3H\rho_\phi = -\Gamma_\phi(1+w)\rho_\phi. \tag{2.214}$$

$$\frac{d\rho_R}{dt} + 4H\rho_R = \Gamma_\phi(1+w)\rho_\phi. \tag{2.215}$$

The Hubble parameter and time

We can have a simple expression of the Hubble parameter $H(t)$ for the two cases of Universe matter dominated and radiation dominated. Indeed, neglecting the curvature and cosmological constant (approximation completely valid in the early Universe), the Friedmann equation (2.6) can be written as

$$H^2 = \left(\frac{\dot{a}}{a}\right)^2 = \frac{8\pi G}{3}\rho = \frac{\rho}{3M_P^2}, \tag{2.216}$$

with $\rho = \rho_M = n \times M \propto a^{-3}M$ for a Universe dominated by a massive field of mass M and $\rho = \rho_R = n \times E_R \propto a^{-3}a^{-1}$ for a radiation dominated Universe (where the energy is redshifted by a factor a^{-1} due to the expansion). In this context, solving the Friedmann equation (2.216) for $a(t)$ gives $H(t) = \frac{2}{3t}$ for a matter dominated Universe and $H(t) = \frac{1}{2t}$ for a radiation dominated Universe, justifying the common approximation $H \simeq \frac{1}{t}$

2.3.2 The (Non-thermal) Distribution Function

2.3.2.1 Time Evolution of the Densities

When the reheating phase of the Universe begins, the inflaton staying in a coherent oscillation regime, we cannot yet evoke a "temperature." Whereas distribution functions will be the classical thermal Boltzmann or Bose–Einstein distributions when the Universe reaches thermal equilibrium, the situation is more complex in the early phase of reheating. The time t is then the more justified dynamical parameter to describe the evolution of the Universe.[30] The first set of equations describes the evolution of the matter field (the inflaton). During all this phase, the inflaton will dominate the Universe. It is then justified to use the relation $H(t) = \frac{2}{3t}$ that we extracted from Eq. (2.216). We are in a period much before the lifetime of ϕ, which means $t \ll \Gamma_\phi^{-1}$, or $H \gg \Gamma_\phi$. The problem is then clearly set: one has a system of 3 equations governing 3 parameters (ρ_ϕ, ρ_R, a) as a function of a dynamical parameter, the time t. We can then first find an analytical solution of Eq. (2.206).

2.3.2.2 The Matter

Let us begin by the evolution of the inflaton density in time. Defining $X_\phi = t^2 \rho_\phi$ and using $H = \frac{2}{3t}$, one can integrate Eq. (2.206) between[31] t_{end} and t

$$\frac{dX_\phi}{dt} = -t^2 \Gamma_\phi \frac{X_\phi}{t^2} \Rightarrow X_\phi(t) = X_\phi(t_{end}) e^{(t-t_{end})\Gamma_\phi}$$

$$\Rightarrow \rho_\phi = \rho_\phi(t_{end}) \left(\frac{t_{end}}{t}\right)^2 e^{-(t-t_{end})\Gamma_\phi}. \tag{2.217}$$

Notice that, at the end of the inflation, $H = \frac{2}{3t_{end}} = \frac{\sqrt{\rho_\phi(t_{end})}}{\sqrt{3}M_P}$. We can then write

$$\rho_\phi(t_{end}) t_{end}^2 = \frac{4}{3} M_P^2, \tag{2.218}$$

which implies

$$\rho_\phi(t) \simeq \frac{4}{3} \frac{M_P^2}{t^2} e^{-(t-t_{end})\Gamma_\phi}. \tag{2.219}$$

[30] It is possible to find in the scientific literature the parameter $v = t \times \Gamma_\phi$ to keep the dynamical parameter dimensionless.

[31] We will consider in all this section the time t_{end} as being the time just at the end of the inflation, at the beginning of the coherent oscillation.

2.3.2.3 The Radiation

We can make the same exercise for the radiation density of energy ρ_R. Replacing H by $\frac{2}{3t}$ in Eq. (2.207) and defining $X_R = t^{8/3}\rho_R$, one needs to solve

$$\frac{dX_R}{dt} = t^{8/3}\Gamma_\phi\rho_\phi \;\Rightarrow\; X_R = \Gamma_\phi\rho_\phi(t_{end})t_{end}^2 e^{\Gamma_\phi t_{end}} \int_{t_{end}}^{t} z^{2/3}e^{-z\Gamma_\phi}dz, \tag{2.220}$$

or, using Eqs. (A.123) and (A.124),

$$X_R = \Gamma_\phi^{-2/3}\rho_\phi(t_{end})t_{end}^2 e^{\Gamma_\phi t_{end}}\left[\gamma(\frac{5}{3},\Gamma_\phi t) - \gamma(\frac{5}{3},\Gamma_\phi t_{end})\right], \tag{2.221}$$

where $\gamma(\alpha, x)$ is the incomplete gamma function. Then,

$$\rho_R = \Gamma_\phi\rho_\phi(t_{end})t_{end}^2 \frac{e^{\Gamma_\phi t_{end}}}{t^{8/3}}\left[\gamma(\frac{5}{3},\Gamma_\phi t) - \gamma(\frac{5}{3},\Gamma_\phi t_{end})\right]. \tag{2.222}$$

In the regime $\Gamma_\phi t \ll 1$, we can use the approximation Eq. (A.124), to write

$$\rho_R \simeq \frac{3}{5}\Gamma_\phi\rho_\phi(t_{end})t_{end}^2\frac{e^{\Gamma_\phi t_{end}}}{t}\left[1-\left(\frac{t_{end}}{t}\right)^{5/3}\right] \simeq \frac{4}{5}\frac{\Gamma_\phi M_P^2}{t}\left[1-\left(\frac{t_{end}}{t}\right)^{5/3}\right], \tag{2.223}$$

where we used $\rho_\phi(t_{end})t_{end}^2 = \frac{4}{3}M_P^2$, Eq. (2.218).

2.3.2.4 The Scale Factor

To compute the evolution of the scale factor as a function of time, one needs to recall the definition of the Hubble parameter, Eq. (2.216),

$$H(t) = \frac{2}{3t} = \frac{\dot{a}}{a} \;\Rightarrow\; a(t) = a(t_{end})\left(\frac{t}{t_{end}}\right)^{2/3}. \tag{2.224}$$

2.3.2.5 Summary

Finally, the evolution of the energy densities and the scale factor in the reheating phase of Universe can be written as follows:

$$\rho_\phi(t) \simeq \frac{4}{3}\frac{M_P^2}{t^2}e^{-(t-t_{end})\Gamma_\phi} \simeq \frac{4}{3}\frac{M_P^2}{t^2} \tag{2.225}$$

$$\rho_R(t) \simeq \frac{4}{5}\frac{\Gamma_\phi M_P^2}{t}\left[1-\left(\frac{t_{end}}{t}\right)^{5/3}\right] \tag{2.226}$$

$$a(t) = a(t_{end}) \left(\frac{t}{t_{end}} \right)^{2/3} , \tag{2.227}$$

where we have considered the regime $t_{end} \ll t \ll \Gamma_\phi^{-1}$, in other words, well before the end of reheating, when the Universe is still largely dominated by the energy of the inflaton coherent oscillations.

Exercise Find the solutions for ρ_ϕ and ρ_R considering a generic equation of state for ρ_ϕ: $p_\phi = w\rho_\phi$.

We plot the evolution of $\rho_\phi(t)$, $\rho_R(t)$, and $a(t)$ in Fig. 2.10, where we fixed $t_{end} = 10^{-37}$ s (see below). The width of the inflaton is fixed by its effective Yukawa coupling y_ϕ. Indeed, it is common in the literature to write the inflaton coupling to the bath under the form (see Eq. B.182)

$$\mathcal{L}_{\phi,bath} = y_\phi \phi \bar{f} f \quad \Rightarrow \quad \Gamma_\phi = \frac{y_\phi^2}{8\pi} m_\phi, \tag{2.228}$$

where f represents Standard Model fermions.[32] The coupling in Eq. (2.228) is obviously not gauge invariant and should "include" in some way all the degrees of freedom of the Standard Model (and beyond the Standard Model) particles joining the thermal bath. This type of term can be written in a gauge invariant way through higher dimensional operators like $\phi \frac{(\bar{f}H)(H^* f)}{\Lambda^2}$, for instance.[33] In any case, naively, an invariant term $\phi|H|^2$ should effectively dominate the production rate. The expression (2.228) is then convenient because the width can be expressed as a function of a dimensionless parameter y_ϕ but should only be considered as a practical approach without real meaning, physics-wise.

We can make some comments on the Fig. 2.10. First of all, we notice that ρ_R reaches a maximum at around $t = t_{max} \simeq 5 \times 10^{-36}$ s. The value of t_{max} is in fact highly dependent on the value of t_{end}, which is itself very dependent on the inflationary model being considered. Given that $H_{end} = \frac{\sqrt{\rho_{end}}}{\sqrt{3} M_P}$ with $\rho_{end} = m_\phi^2 \langle\phi\rangle^2 \simeq m_\phi^2 M_P^2$, $t_{end} \simeq \frac{1}{m_\phi} \simeq 10^{-37}$ s for $m_\phi = 3 \times 10^{13}$ GeV. The behavior of the radiation energy density is quite easy to understand. Indeed, at the beginning, the density of radiation is extremely small due to inflation. Then it increases due to the inflaton decay. However, as time passes, the dilution and redshift effect tend to diminish the radiation density. It is then a fight between the injection of

[32] Suppose a coupling to bosons is also possible.

[33] Alternatively, one can also draw a loop diagram with two Higgs and a fermion which will give the effective decay $\phi \to \bar{f} f$.

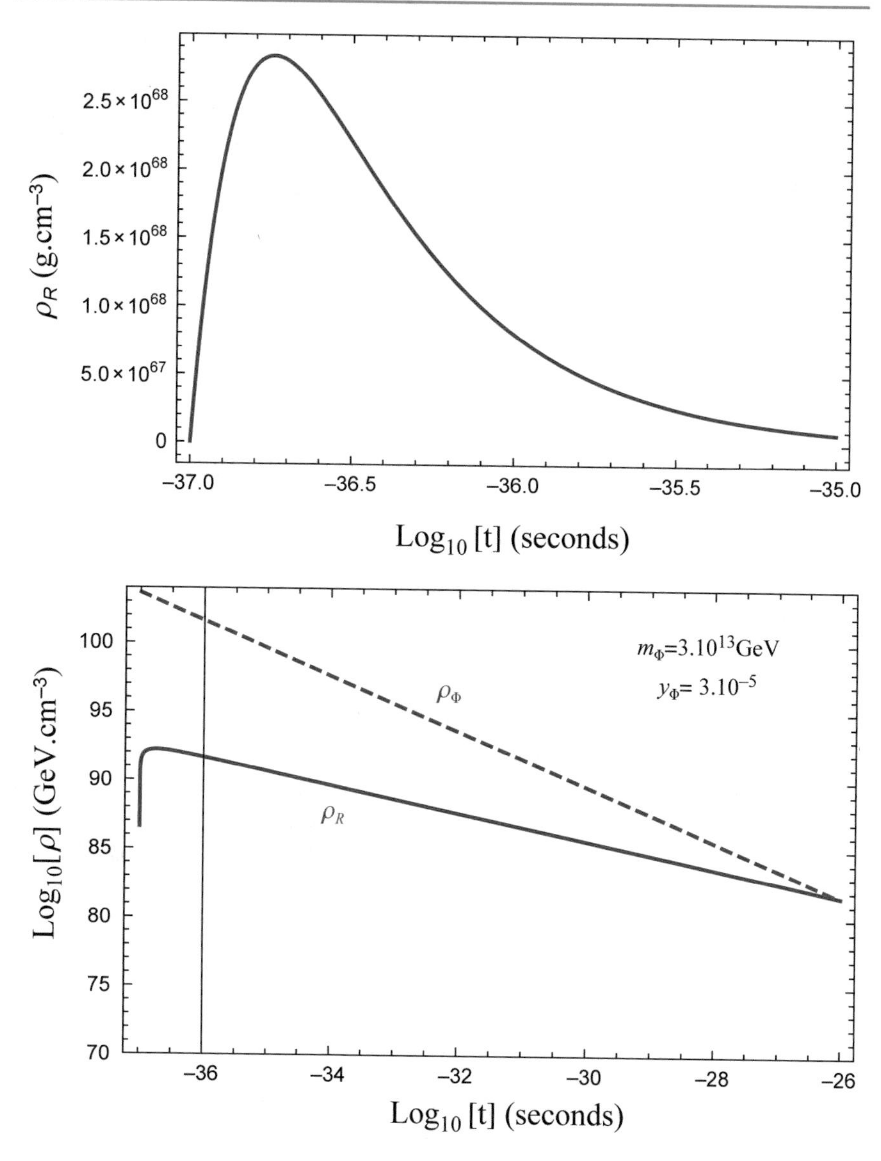

Fig. 2.10 Evolution of the density of radiation energy (in g.cm^{-3}, top) and the density of inflaton (in GeV.cm^{-3}, bottom) for $m_\phi = 3 \times 10^{13}$ GeV, $y_\phi = 10^{-5}$, and $t_{end} = 10^{-37}$ s (see the text for details)

energy through the inflaton decay (proportional to t) and its dilution (proportional to $a^{-4} \propto t^{-8/3}$, see Eq. 2.227). This eventually results in an overall decrease of ρ_R with time for $t > t_{max}$, albeit slower than a^{-4} ($\rho_R \propto \frac{1}{t}$ for $t_{max} < t < t_{reh}$). Notice also the value of the density, compared to the density of water (1 g.cm^{-3}, density of the Universe at the CMB time) or the matter density of the Universe nowadays (3×10^{-30} g.cm^{-3}).

Observing the comparison between the inflaton density and the radiation density, we notice that radiation reaches the inflaton density for a time $t = t_{reh} \simeq \Gamma_\phi^{-1} \simeq 10^{-26}$ s for $y_\phi = 10^{-5}$ (corresponding to a width $\Gamma_\phi = 119$ GeV). This is the beginning of the radiation domination era.[34] This is also logical, as the inflaton dilution is governed by the expansion rate $a^{-3}(t)$, whereas the radiation still receives energy from its decay, compensating partly the dilution factor.

Exercise Find the expression for t_{max}, the time when the radiation density reaches its maximum.

2.3.2.6 The Distribution Function

In the preceding section, we computed the *total* density of radiation ρ_R. But the key ingredient when one needs to compute any process is the energy distribution. More precisely the distribution function is defined by

$$dn = dn(p, p + dp) \equiv f(p)\frac{d^3p}{(2\pi)^3} = f(p)\frac{p^2}{(2\pi)^3}dpd\Omega, \tag{2.229}$$

where dn is the density of particles having momenta between p and $p + dp$ and $d\Omega = d\cos\theta d\phi$ the solid angle (in the momentum phase space). For the reader who is not familiar with the concept of distribution function, we encourage him/her to take a look at Sect. 3.1.3. The classical distributions $f(p) = \frac{1}{e^{p/kT} \pm 1}$ one can find in a lot of situations (Fermi–Dirac, Bose-Einstein, or even Maxwell) are only valid when the particles are in a thermal bath, *i.e.* in equilibrium at a temperature T. This is clearly not the case in the beginning of the reheating phase, where the rate of interaction radiation-radiation (decay product f on decay product f) is too weak to counterbalance the expansion ($n\langle\sigma v\rangle_{f\leftrightarrow f} < H$). It is only when the number of relativistic particles reaches a certain threshold that thermalization is achieved, at a time t_{th}. We will discuss in great detail the thermal case in the following section.

To illustrate the process, let us imagine the inflaton decaying into two relativistic particles with a rate Γ_ϕ per second. That means that every second, 2 Γ_ϕ particles are created with an energy $\frac{m_\phi}{2}$. After some time t, this energy will be redshifted by the scale factor $a(t)$, while the number density of particles produced at the time t will also be diluted by $a^3(t)$. We illustrate the phenomena in the drawing below (Fig. 2.11).

[34] It is important to point out that the radiation has thermalized well before t_{reh}.

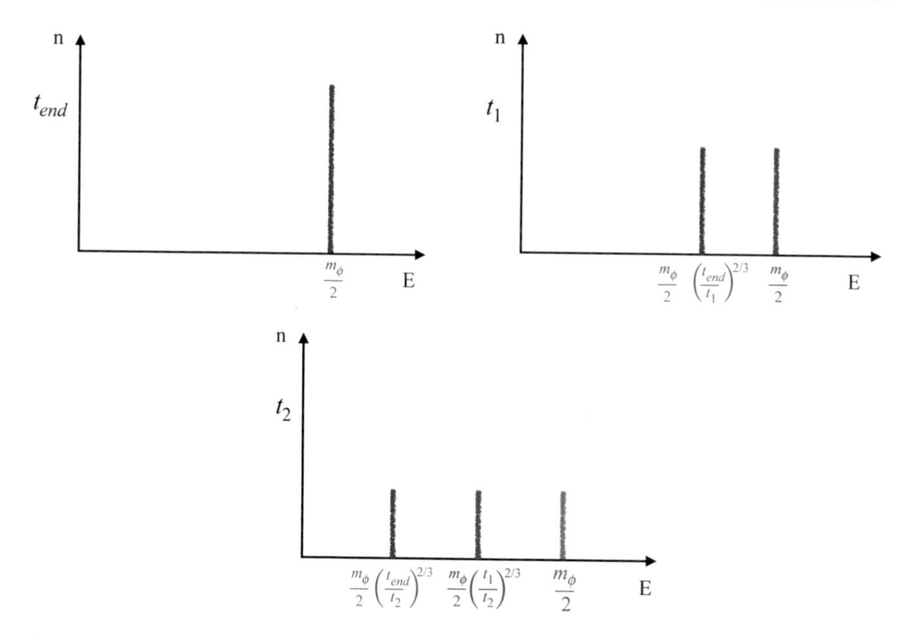

Fig. 2.11 An illustration of the evolution of a spectrum for a decaying particle of mass m_ϕ into two particles. They are produced with energy $\frac{m_\phi}{2}$ at t_{end} and then are redshifted at t_1 while being diluted by the expansion, and so on. The spectrum, which has initially a Dirac-delta form, becomes flatter and flatter with time

More concretely, to compute the spectrum, one needs to know what is the density of particles having their energies between E and $E + dE$ at a time t. These particles have been produced at a time t_i, which is defined by[35]

$$E = E(t) = E(t_i) \times \frac{a(t_i)}{a(t)} = E(t_i)\left(\frac{t_i}{t}\right)^{2/3} = \frac{m_\phi}{2}\left(\frac{t_i}{t}\right)^{2/3}, \tag{2.230}$$

which implies

$$t_i = \left(\frac{2E}{m_\phi}\right)^{3/2} t \Rightarrow dt_i = \frac{3}{m_\phi}\left(\frac{2E}{m_\phi}\right)^{1/2} dE\, t = \frac{3\sqrt{2}}{m_\phi^{3/2}} t\sqrt{E} dE. \tag{2.231}$$

In the meantime, the number density of particles produced at t_i, with the energy $E(t_i) = \frac{m_\phi}{2}$, has been diluted by a scaling factor a^{-3}:

$$dn(E, t) = dn(E, E + dE)_t = dn(\frac{m_\phi}{2}, t_i)\frac{a^3(t_i)}{a^3(t)}. \tag{2.232}$$

[35]By simplicity, we considered a 2-body decay in this section.

Then, if $\Gamma_\phi \times n_\phi(t_i) \times a^3(t_i)$ particles are produced per second at time t_i, during time dt_i given by Eq. (2.231), $dt_i \times \Gamma_\phi \times n_\phi \times a^3(t_i)$ particles are produced between t_i and $t_i + dt_i$. This means

$$dn(E,t) = dt_i \Gamma_\phi n_\phi(t_i) \frac{a^3(t_i)}{a^3(t)} = \frac{3\sqrt{2}}{m_\phi^{3/2}} \sqrt{E} dE t \Gamma_\phi n_\phi(t_i) \left(\frac{t_i}{t}\right)^2, \tag{2.233}$$

where we used $a \propto t^{2/3}$ (Eq. 2.227). Using $\rho_\phi = n_\phi \times m_\phi$, we can write

$$dn(E,t) = \frac{3\sqrt{2}\sqrt{E}\Gamma_\phi}{m_\phi^{5/2} t} \rho_\phi(t_i) t_i^2 dE. \tag{2.234}$$

With the help of Eq. (2.225), valid for $\Gamma_\phi t \ll 1$,

$$\rho(t) = 3M_P^2 H^2(t) = \frac{4}{3t^2} M_P^2,$$

one obtains

$$\frac{dn}{dE} = \frac{4\sqrt{2}\sqrt{E}\,\Gamma_\phi M_P^2}{m_\phi^{5/2} t}, \quad \text{with} \quad \left(\frac{t_{end}}{t}\right)^{2/3} \frac{m_\phi}{2} < E < \frac{m_\phi}{2}, \tag{2.235}$$

We can then extract the distribution function defined by $dn = f(p) \frac{d^3p}{(2\pi)^3}$

$$f(p) = \frac{8\sqrt{2}\pi^2 \Gamma_\phi M_P^2}{p^{3/2} m_\phi^{5/2} t}. \tag{2.236}$$

We show in Fig. 2.12 the spectrum $\frac{dn}{dE}$ as a function of the energy at different epochs, from $t = t_{end} = 2 \times 10^{-37}$ s $(\frac{1}{m_\phi})$ to $t = 10^{-36}$ s. We fixed the inflaton width through its effective Yukawa coupling y_ϕ as we discussed in Eq. (2.228).

A quick look at Fig. 2.12 teaches us interesting features of the beginning of the reheating process. First of all, if one considers the end of inflation at $t = t_{end} = 10^{-37}$ s, the spectrum becomes flat quickly. At 10^{-36} s already there is no trace of the monochromatic injection. Whereas the dependence on the width Γ_ϕ will just be an uplift of the whole spectrum, the lowest energy particles that began to populate the Universe will be the first to thermalize, much before the decay process of ϕ finishes.

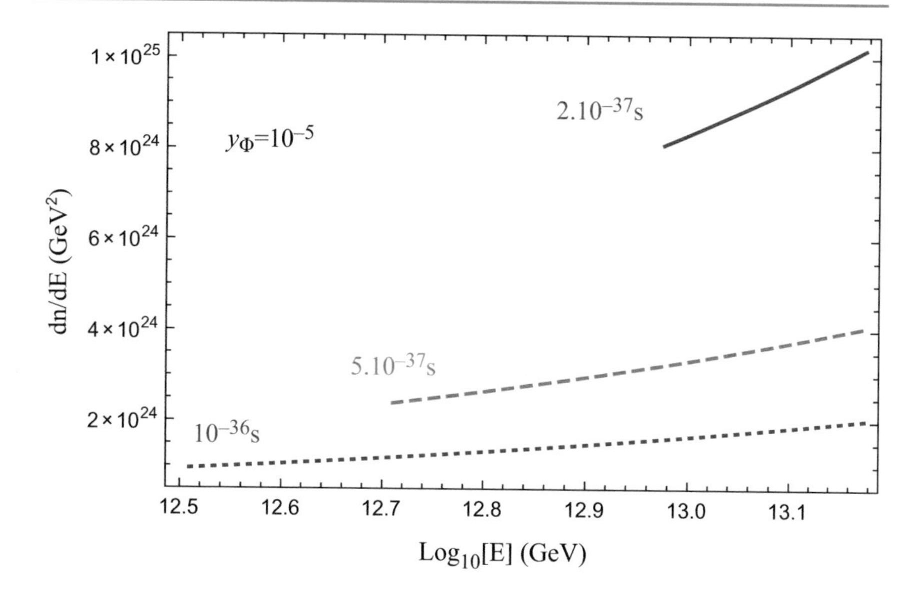

Fig. 2.12 Evolution of the spectrum of the products of inflaton decay at different times for $y_\phi = 10^{-5}$

Exercise Compute the distribution function in the case of the decay of the inflaton into n particles.

2.3.3 End of the Thermalization Process: Transition Toward a Thermal Bath

2.3.3.1 Understanding the Process

We show in Fig. 2.13 the evolution of the distribution function in the plasma as a function of time, from the end of the inflation ($t \simeq \frac{1}{m_\phi} \simeq 10^{-37}$ s) to the beginning of the thermal era ($t \simeq 10^{-30}$ s). We see clearly how the shape evolves from a delta function to a flat function (once the energies of the products begin to redshift), to the appearance of a thermal "bump" for the less energetic particles to a complete thermal distribution that will redshift also with time. We clearly see how the spectrum passes from a delta function distribution, when all the particles produced are given a momentum $p = \frac{m_\phi}{2}$ to a redshifted flat distribution as we study in Fig. 2.12 to a Boltzmann (or Fermi–Dirac/Bose–Einstein) classical distribution, as we will discuss in the following chapter devoted to the thermal bath. The complete treatment of the thermalization is very complex and deserves a book (or even several books) in itself. However, we can try to give an idea of how and why the transition physically occurs.

To understand this phenomenon, one needs to remember that the processes that are responsible for the thermalization are long range interaction, where the particles

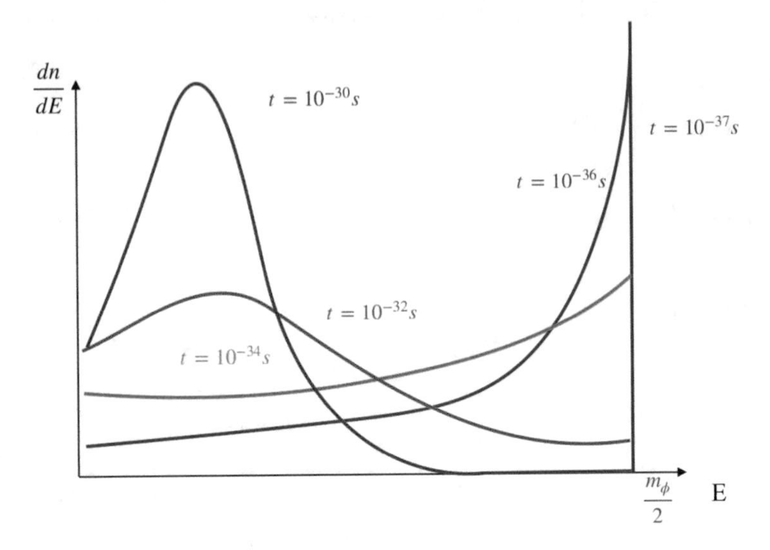

Fig. 2.13 Illustration of the evolution of the distribution function in the plasma with time, from the beginning of the oscillation ($t \simeq \frac{1}{m_\phi} \simeq 10^{-37}$ s) to the beginning of the thermalization era ($t \simeq 10^{-30}$ s)

exchange massless bosons or scalars between themselves (the electroweak phase transition has yet to occur, and all the particles in the plasma are massless at this energy scale). In this case, the interaction cross section between particles in the plasma is proportional to $\frac{1}{s}$, s being the classical center of mass energy squared. It means that the particles with the *lowest* energy will interact more efficiently between themselves to form a thermal bath, competitive with the expansion of the Universe, H, which has a tendency to make the Universe colder. In other words, at a time when the particles (all produced at an energy $E = \frac{m_\phi}{2}$) having low redshifted energies are sufficiently present, their interaction rate $n\langle\sigma v\rangle > H$, and the Universe enters in a thermal phase.

The transition phase between the thermalization and the radiation dominated Universe is then a phase where the bath *is* thermal, in the sense that any particle f produced with an energy $\frac{m_\phi}{2}$ is "caught" by particles in thermal bath because their density is sufficient to ensure $n_{bath}\langle\sigma v\rangle_{bath-f} > H$. The Universe is matter dominated because the inflaton is still the source of the energy injected in the bath, but its non-inflaton content is radiation. This is the time when the energy injected in the primordial plasma is maximal: the production rate is large enough to compensate the dilution due to the expansion rate. This feature is easy to see in Eq. (2.226), where a maximum is reached for ρ_R before a decrease proportional to $\frac{1}{t}$. The transition phase is terminated once the decay rate of the inflaton cannot compete anymore with the inflation scale ($\Gamma_\phi < H$). The Universe then enters in a radiation dominated era, until the last scattering time (the decoupling time, see Sect. 3.3.4).

2.3.3.2 Computing the End of Thermalization Process

The process of thermalization ends when any particle produced by the inflaton decay, with an energy $\frac{m_\phi}{2}$, is trapped by particles already in equilibrium with themselves, in a time shorter than the Hubble timescale H^{-1}. The physics behind the thermalization mechanism is quite complex and effects as LPM (Landau–Pomeranchuk–Migdal) effect, splitting processes, and loss of energies in inelastic scatterings have to be taken into account. The physics needed deserves a complete book by itself, and we will then just keep the result of interest in our case: the energy lost by inelastic scattering in a plasma of temperature T, for a particle of energy E, with a gauge coupling g with the particles in the plasma is given by

$$\left(\frac{dE}{dt}\right)_{\text{inelastic}} \sim g^4 T^2 \sqrt{\frac{E}{T}} = x g^4 T^2 \sqrt{\frac{E}{T}}, \tag{2.237}$$

where we have embedded the uncertainties in the complex treatment of the physics in x, which is of the order of unity. If we denote by t_{th} the time when the loss of energy ΔE is equal to $\frac{m_\phi}{2}$ (i.e. when a produced particle with energy $\frac{m_\phi}{2}$ will lose all its energy by the inelastic scattering with particles from the plasma), we obtain

$$x g^4 T^2 \sqrt{\frac{E}{T}} = \frac{\Delta E}{\Delta t} = \frac{m_\phi/2}{H^{-1}} \tag{2.238}$$

with

$$H = \frac{1}{M_P}\sqrt{\frac{\rho_\phi}{3}} \text{ and } T = \left(\frac{\rho_R}{\alpha}\right)^{1/4}, \tag{2.239}$$

where we have define the temperature by analogy with the thermal bath (see Eq. 3.35).

$$\rho_R = \frac{g_\rho \pi^2}{30} T^4 = \alpha T^4 \text{ with } \alpha = \frac{g_\rho \pi^2}{30}, \tag{2.240}$$

where g_ρ is the sum of degrees of freedom in the thermal plasma.[36] Replacing E by $\frac{m_\phi}{2}$ in Eq. (2.238), we obtain

$$x g^4 \sqrt{\frac{m_\phi}{2} \frac{\rho_R^{3/8}}{\alpha^{3/8}}} = \frac{m_\phi}{2M_P}\sqrt{\frac{\rho_\phi}{3}}. \tag{2.241}$$

[36] When calculating g_ρ, one should be careful between the fermionic and bosonic states, as one can see in Eq. (3.26): $g_\rho = 106.75$ ($\alpha \simeq 35$) for the Standard Model, and $g_\rho = 213.5$ ($\alpha \simeq 70$) in supersymmetry.

Using Eqs. (2.225) and (2.226) at time t_{th}, we obtain

$$t_{th} = \left(\frac{2}{9}\right)^{4/5} \left(\frac{g_\rho \pi^2}{24}\right)^{3/5} \left(\frac{1}{x g^4}\right)^{8/5} \frac{m_\phi^{4/5}}{M_P^{6/5} \Gamma_\phi^{3/5}}, \tag{2.242}$$

or

$$t_{th} = \frac{1.5}{x^{8/5}} \times 10^{30} \left(\frac{g_\rho}{106.75}\right)^{3/5} \left(\frac{0.1}{g}\right)^{32/5} \left(\frac{m_\phi}{3 \times 10^{13}\mathrm{GeV}}\right)^{1/5} \left(\frac{10^{-5}}{y_\phi}\right)^{5/6} \mathrm{s}, \tag{2.243}$$

where we used Eq. (2.228). We show in Fig. 2.14 the evolution of the inflaton density ρ_ϕ and radiation density ρ_R as a function of time for different values of y_ϕ. We see that the thermalization time is reached before the domination of the Universe by the radiation. This means that during a period of time, the evolution of the Universe is dominated by the inflaton energy density, while the distribution function of the bath is non-thermal (from t_{end} to t_{therm}). It then enters in a phase, still dominated by the inflaton density, but with a thermal bath (from t_{bath} to t_{rh}, the reheating time), before being ruled by the radiation density until the decoupling time (from t_{rh} to t_{cmb}). Viewing Fig. 2.14, we understand easily that increasing y_ϕ has two straight

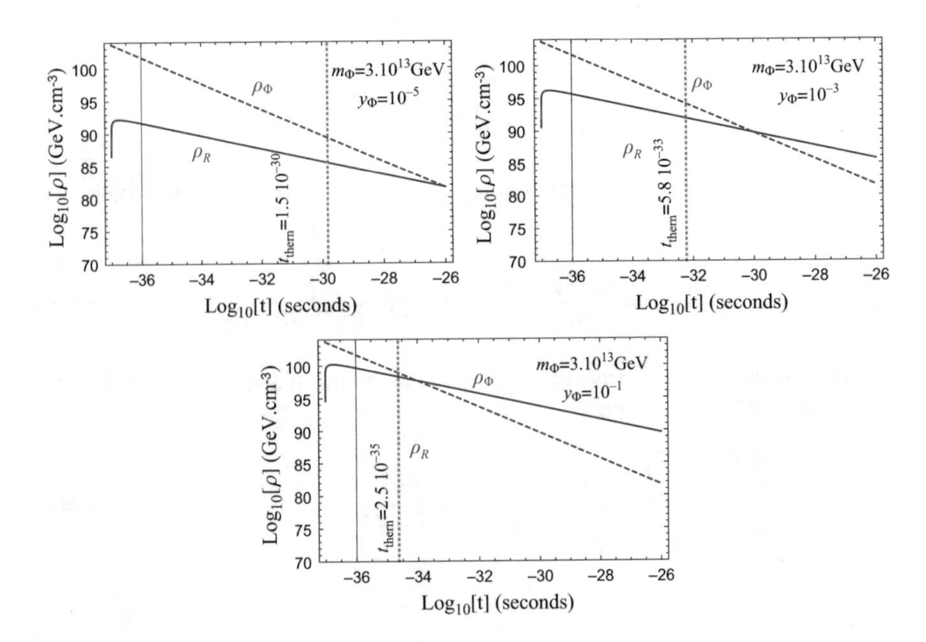

Fig. 2.14 Evolution of the inflaton density ρ_ϕ and radiation density ρ_R as a function of time for different values of the inflaton coupling to the Standard Model y_ϕ, compared to the thermalization time for the same value of y_ϕ

effects: the radiation density being larger and faster, the reheating period can be reached earlier, and the thermalization time, for the same reason, tends also to shift toward the end of inflation time and the reheating time. The interesting point is then to understand how the production of dark matter can be computed in this phase of the thermalization era.

2.3.4 Dark Matter Production During the Non-thermal Phase of the Reheating

2.3.4.1 The Context

We have now all the tools in hand to compute the amount of dark matter produced during the process of thermalization, from the end of inflation t_{end} to the thermalization time t_{th}. There are two possible sources: the inflaton decay and the scattering of Standard Model particles already produced. In any case, the exercise consists in solving the Boltzmann equation

$$\frac{dn_\chi}{dt} + 3Hn_\chi = R(t), \tag{2.244}$$

where $R(t)$ is the production rate of dark matter, from the end of inflation t_{end} to the thermalization time t_{th}. Defining the variable $X_\chi = t^2 n_\chi$, and using $H = \frac{2}{3t}$ from Eq. (2.216), the solution of Eq. (2.244) can be written as

$$n_\chi(t) = \left(\frac{t_{end}}{t}\right)^2 n_\chi(t_{end}) + \frac{1}{t^2}\int_{t_{end}}^{t} t'^2 R(t')dt'. \tag{2.245}$$

Let us apply the calculation to the production of dark matter from decaying inflaton and scattering from the (not yet thermal) particles in the plasma.

2.3.4.2 Direct production by inflaton decay

This source of production is always present, and very difficult to avoid, except by invoking specific symmetries that would seclude completely the dark sector from the inflationary sector. As we will see later, even if the dark matter is completely secluded from the inflaton sector, loop processes are able to produce dark matter in a sufficiently large amount to fill the Universe. To compute the density of a dark matter candidate χ produced by direct decay of the inflaton is straightforward. One needs to solve the Boltzmann equation for χ:

$$\frac{dn_\chi}{dt} + 3Hn_\chi = l Br_\chi \Gamma_\phi \frac{\rho_\phi}{m_\phi}, \tag{2.246}$$

where Br_χ is the branching fraction of decay of ϕ into l dark matter particles. Defining $X_\chi = t^2 n_\chi$ and using Eq. (2.225), we can write

$$\frac{dX_\chi}{dt} = \frac{4}{3} l\ Br_\chi \frac{\Gamma_\phi}{m_\phi} M_P^2 \tag{2.247}$$

from that one can extract

$$n_\chi(t) = \frac{4}{3} l\ Br_\chi \frac{\Gamma_\phi M_P^2}{m_\phi t}\left[1 - \left(\frac{t_{end}}{t}\right)\right] = l \frac{Br_\chi}{6\pi} y_\phi^2 \frac{M_P^2}{t}\left[1 - \left(\frac{t_{end}}{t}\right)\right], \tag{2.248}$$

where we used Eq. (2.228) in the last expression. With the help of Eq. (2.242), one can then compute the amount of dark matter in the Universe at $t = t_{th}$, $t_{th} \gg t_{end}$:

$$n_\chi^{dec}(t_{th}) = \frac{l\ Br_\chi}{6\pi} y_\phi^2 \frac{M_P^2}{t_{th}} = \frac{l\ Br_\chi}{(6\pi)(8\pi)^{3/5}\delta} \frac{y_\phi^{16/5}\ M_P^{16/5}}{m_\phi^{1/5}} \tag{2.249}$$

with

$$\delta = \left(\frac{1}{x g^4}\right)^{8/5} \left(\frac{2}{9}\right)^{4/5} \left(\frac{g_\rho \pi^2}{24}\right)^{3/5} \simeq \frac{7 \times 10^6}{x^{8/5}} \left(\frac{g_\rho}{106.75}\right)^{3/5} \left(\frac{0.1}{g}\right)^{32/5} \tag{2.250}$$

from Eq. (2.242), or

$$n_\chi(t_{th}) = \left(\frac{l\ Br_\chi x^{8/5}}{7 \times 10^{-32}}\right) \left(\frac{106.75}{g_\rho}\right)^{3/5} \left(\frac{g}{0.1}\right)^{32/5} \left(\frac{y_\phi}{10^{-5}}\right)^{16/5} \left(\frac{3 \times 10^{13}}{m_\phi}\right)^{1/5}. \tag{2.251}$$

We remind the reader that the coefficient x represents the uncertainties and the unknown thermalization exact process and is of the order of 1.

2.3.4.3 Production by Scattering

The second source of dark matter production is the thermal bath. The scattering of Standard Model particles can lead to a large amount of dark matter, especially if the annihilation cross section is highly energy dependent. To compute the density of particles produced this way, one needs to solve the Boltzmann equation for the dark matter field, Eq. (2.244). The rate $R(t)$ can be computed directly from the expressions (B.87) and (B.95). Indeed, the interaction rate (the number of interactions per unit of time) dw_{fi} is given by (B.86)

$$dw_{fi} = V \times (2\pi)^4 \delta^{(4)}(p_f - p_i)|\mathcal{T}_{fi}|^2 \prod_{f=1}^{n_f} \frac{V}{(2\pi)^3} d^3 p_f.$$

To understand each term of the equation above, notice that $\mathcal{T}_{fi}$ has been defined from the $\mathcal{S}$ matrix element $\mathcal{S}_{fi} = 1 + (2\pi)^4\delta^{(4)}(p_f - p_i)\mathcal{T}_{fi}$, and $w_{fi} = \frac{|\mathcal{S}_{fi}|^2}{T}$, which gives, in a phase space of volume[37] $V \times d^3p$, $dw_{fi} = w_{fi} \times \prod_{f=1}^{n_f} \frac{V\, d^3 p_f}{(2\pi)^3}$ (and using the relation $[(2\pi)^4\delta^{(4)}(p_f - p_i)]^2 = V \times T \times (2\pi)^4\delta^{(4)}(p_f - p_i)$).

However, the precedent reasoning was made in a non-covariant form (quantum mechanics). As explained in Sect. B.4.1.4, one needs to normalize the wave function by $\psi_i \to \frac{1}{\sqrt{2E_i V}}\psi_i$ to render the theory Lorentz invariant (normalization corresponding to $2E_i$ particles i in volume V). We then define the matrix element $\mathcal{M}_{fi}$ by (see Eq. B.87)

$$\mathcal{T}_{fi} = \left(\prod_{i=1}^{n_i} \frac{1}{\sqrt{2E_i V}}\right)\left(\prod_{f=1}^{n_f} \frac{1}{\sqrt{2E_f V}}\right)\mathcal{M}_{fi}.$$

The rate R (interaction per unit of time for a given initial state) is then given by Eq. (B.88)

$$R = dw_{fi} = \frac{V^{1-n_i}}{(2\pi)^{3n_f}}(2\pi)^4\delta^{(4)}(p_f - p_i)|\mathcal{M}_{fi}|^2 \prod_{i=1}^{n_i} \frac{1}{2E_i} \prod_{f=1}^{n_f} \frac{d^3 p_f}{2E_f}. \quad (2.252)$$

However, in the situation where the momentum of initial state is not definite, but follow a statistical law (as it is the case in the early Universe) one should multiply this rate by the number of quantum states of the initial particles (1 and 2) weighted by their distribution functions f_1 and f_2, respectively:[38]

$$d\langle \tilde{R} \rangle = \prod_{i=1}^{n_i} f_i \frac{V}{(2\pi)^3} d^3 p_i \times R, \quad (2.253)$$

[37]To see it easily, one can use uncertainty principle of Heisenberg, for instance, where the number of states N_f is $\frac{d^3x d^3p}{h^3}$, and $\hbar = 1 \Rightarrow h = 2\pi$, which implies $N_f = \frac{V\, d^3p}{(2\pi)^3}$.

[38]Notice that the dark matter particle is *never* in thermal equilibrium with the plasma during the whole production process.

which gives, combining with Eq. (2.252),

$$\langle \tilde{R} \rangle = V \int \prod_{i=1}^{n_i} f_i \frac{d^3 p_i}{(2\pi)^3 2E_i} (2\pi)^4 \delta^{(4)}(p_f - p_i) |\mathcal{M}_{fi}|^2 \prod_{f=1}^{n_f} \frac{d^3 p_f}{(2\pi)^3 2E_f} \tag{2.254}$$

with $\tilde{R}$ the rate of interaction, per unit of time, in the volume V. To obtain the rate $R(t)$ for the Boltzmann equation (2.244), one just needs to divide $\tilde{R}$ by V, which gives

$$R(t) = \frac{\tilde{R}}{V} = \int \prod_{i=1}^{n_i} f_i \frac{d^3 p_i}{(2\pi)^3 2E_i} (2\pi)^4 \delta^{(4)}(p_f - p_i) |\mathcal{M}_{fi}|^2 \prod_{f=1}^{n_f} \frac{d^3 p_f}{(2\pi)^3 2E_f}. \tag{2.255}$$

A little remark is here: we notice that $\mathcal{T}_{fi}$ has a dimension of $(Energy)^4$, and then $\mathcal{M}_{fi}$ has dimension $(Energy)^{4-n_i-n_f}$, which is obviously a dimensional for $2 \rightarrow 2$ scattering.

In our case of interest ($2 \rightarrow 2$ scattering, $1 + 2 \rightarrow 3 + 4$), using the formula for the 2-body phase space final state Eq. (B.100) gives

$$\frac{d^3 p_3}{(2\pi)^3 2E_3} \frac{d^3 p_4}{(2\pi)^3 2E_4} \delta^{(4)}(p_f - p_i) = \frac{d\Omega_{13}}{512\pi^6} \tag{2.256}$$

with $d\Omega_{13}$ being the solid angle between particles 1 and 3 in the center of mass frame. In the relativistic case for the initial particles,[39] $p_i = E_i$. We can then write

$$\frac{d^3 p_1}{2E_1} \frac{d^3 p_2}{2E_2} = \frac{4\pi \times 2\pi \; d\cos\theta_{12} \; E_1 dE_1 E_2 dE_2}{4}, \tag{2.257}$$

where $\cos\theta_{12}$ is the angle between particles 1 and 2 *in laboratory frame*. Combining with Eq. (2.255), one obtains

$$\boxed{R(t) = \int f_1 f_2 \frac{E_1 E_2 dE_1 dE_2 \; d\cos\theta_{12}}{1024\pi^6} \int |\mathcal{M}_{fi}|^2 d\Omega_{13},} \tag{2.258}$$

with f_1 and f_2 the distribution functions given by Eq. (2.236).

[39] That is the case as the photons and/or Standard Model particles are the 1 and 2 particles, much lighter than the energy at the reheating time, and therefore massless.

We can then suppose $|\mathcal{M}|^2 = \frac{s^{\frac{n+2}{2}}}{\Lambda^{n+2}}$ (corresponding to $\sigma \sim \frac{s^{\frac{n}{2}}}{\Lambda^{n+2}}$). Setting $s = (P_1 + P_2)^2$, P_i being the quadrivector of incoming particle i, and considering massless particles[40] ($m_1 = m_2 = 0$), one can easily integrate Eq. (2.258), which gives

$$R(t) = \frac{4\Gamma_\phi^2 M_P^4 m_\phi^{n-2}}{(n+3)^2(n+4)\pi\,\Lambda^{n+2} t^2}\left[1 - \left(\frac{t_{end}}{t}\right)^{\frac{n+3}{3}}\right]^2 = \frac{\epsilon}{t^2}\left[1 - \left(\frac{t_{end}}{t}\right)^{\frac{n+3}{3}}\right]^2 \tag{2.259}$$

with

$$\epsilon = \frac{4\Gamma_\phi^2 M_P^4 m_\phi^{n-2}}{(n+3)^2(n+4)\pi\,\Lambda^{n+2}}. \tag{2.260}$$

Exercise Compute the production rate in the t and u case ($|\mathcal{M}|^2 \propto t^{\frac{n+2}{2}}$ and $|\mathcal{M}|^2 \propto u^{\frac{n+2}{2}}$).

We can then use the solution Eq. (2.245) to compute $n_\chi(t_{th})$, neglecting $t_{end} \ll t_{th}$

$$n_\chi^{scat}(t_{th}) = \frac{4}{(n+3)^2(n+4)\pi\delta}\,\frac{\Gamma_\phi^{\frac{13}{5}} M_P^{\frac{26}{5}} m_\phi^{n-\frac{14}{5}}}{\Lambda^{n+2}}, \tag{2.261}$$

with δ given by Eq. (2.250). We can then combine Eq. (2.249) with Eq. (2.261) to obtain the ratio $\mathcal{R}$ between the production of dark matter generated by the inflaton decay and the production due to the scattering of particles from the plasma.

[40]There are different reasons to consider massless particles: we are in ultra-high energy regimes, the particles are thus ultra-relativistic, and in the meantime, the electroweak phase transition giving masses to the Standard Model sector has not yet occurred.

We obtain

$$\mathcal{R}(t_{th}) = \frac{n_\chi^{dec}(t_{th})}{n_\chi^{scat}(t_{th})} = \frac{l\ Br_\chi 8\pi^2(n+3)^2(n+4)}{3}\frac{\Lambda^{n+2}}{m_\phi^n M_P^2}$$
$$= \frac{l\ Br_\chi (n+3)^2(n+4)}{1.5\times 10^{-2n-5}}\left(\frac{10^{-5}}{y_\phi}\right)^2\left(\frac{\Lambda}{10^{15}}\right)^{n+2}\left(\frac{10^{13}}{m_\phi}\right)^n, \tag{2.262}$$

2.4 Reheating: Thermal Phase $[10^{-30} - 10^{-28}$ s]

First of all, the first time a graduate student hears about "reheating" he/she should naturally think that the Universe was cold, then hot, and then cold again before having to "re"-heat. That is not exactly the case in fact. This misuse of language is very similar to the one we find when discussing about "re-ionization," "dark age," or "recombination" epochs that we will discuss later on. One reference that the reader can read and is quite accessible to any kind of public is the mini-review made by Lev Kofman [9] for the 60th birthday celebration of Igor Novikov. Lev Kofman collaborated a lot with Andrei Linde and Alexei Starobinsky, two fathers of inflationary models. The term "reheating" is an anachronism as he was used at the time models of inflations supposed a hot and dense phase before an inflationary phase which refresh the Universe, heated again by the inflaton decay: thus the name "re-heating." Contrarily to the previous section, we will consider from now on that the standard model plasma *is thermalized*, which means that the standard model particles are in thermal equilibrium with themselves. In this section we will then neglect all the effects of thermalization in the preheating phase studied before. It has the advantage of ignoring the uncertainties encoded in the x parameter of Eq. (2.237) that was telling us *when* the bath enters in thermal equilibrium. In a sense, in this chapter we will consider that the Universe story began at $t = t_{th}$. This is a generic statement that we can make. The later we consider the *initial condition*, the less dependent we are from any cosmological or early Universe construction. The extreme Weakly Interacting Massive Particle (WIMP) case, as we will see, is even completely independent from the reheating process as we suppose in this case, a thermal equilibrium state between the dark matter and the Standard Model bath.

2.4.1 Understanding the Reheating

We will discuss in detail in Sect. 3.2.4 the general process of thermalization of a dark bath heated by the standard model plasma. In this section, however, we will concentrate on the processes happening *before* this epoch: the thermalization of the standard model bath in itself. In other words, we will describe how a bath of standard model particles produced by the decay of an inflaton reaches a thermal equilibrium. The spectrum of the particles in the plasma at the production can be flat or monochromatic, but the equilibrium is then established in two phases: kinetic equilibrium first where the scattering (elastic or inelastic) or annihilation $2 \to 2$ processes redistribute the temperature to reach a kinetic equilibrium. However, elastic scattering alone cannot bring the decay products of the inflaton into thermal (chemical) equilibrium because the number densities in general will not be correct. Annihilation, which does not change the number of particles, will neither be of any help. Chemical equilibrium is reached after kinetic equilibrium when $2 \to 3$ collisions become operative (see Sect. 3.2 for the difference between kinetic and chemical decoupling and Sect. 3.2.4 for some insights on the thermalization process).

Supposing that the thermal bath is populated by standard model particles f and $\bar{f}$ produced by the inflaton decay, one needs first to approximate at which time (and thus at which temperature) the decay occurs. A first approximation (called "instantaneous reheating") is that the decay of the inflaton ϕ begins to be effective at a time t_d corresponding to a temperature T_d respecting $\Gamma_\phi = H(T_d)$, Γ_ϕ being the width of the inflaton.[41] Indeed, one can interpret the inverse of the Hubble constant $H^{-1}(t)$ as the "doubling time" of the Universe: during the time $t = 1/H(t)$, the radius of the Universe has doubled its size by doubling the scaling factor[42] a. With this interpretation, the condition $\Gamma_\phi = H(T_d)$ corresponds to compute the time t_d at which, for every decay $\phi \to f\,\bar{f}$ the Universe has doubled its size, and the (physical) density of its products f and $\bar{f}$ is not diluted anymore by the expansion: the Universe, after being matter dominated by the massive presence of the inflaton, becomes radiatively dominated by (effectively) massless particles f, $\bar{f}$. This occurs much before the electroweak phase transition and justifies the massless hypothesis.

The Hubble constant in the thermal epoch

Instead of taking the time as the dynamical parameter, once the thermal equilibrium is reached, it is easier (and more physical) to work with the temperature as the dynamical parameter. We let the readers jump to Sect. 3.1.4 to recover the characteristics of number densities and energy densities of a thermal Universe. If one considers a radiation dominated Universe, the evolution of the Hubble parameter as a function of the temperature is easily

(continued)

[41]Or more precisely, the branching ratio of the width into Standard Model particles. Decays into dark or hidden sectors are allowed as we will see in the following section.

[42]$H(t) = \frac{\dot{a}}{a} \Rightarrow da = a\,Hdt$, so during $dt = \frac{1}{H(t)}$, $da = a$: the radius of, the Universe has doubled in size.

derived through the Friedmann equation (2.6),

$$H^2 = \frac{\rho_R(T)}{3M_P^2}, \tag{2.263}$$

with $\rho_R(T) = g_\rho \frac{\pi^2}{30} T^4$ (Eq. 3.35) the radiative energy density of the Universe, g_ρ being the effective degrees of freedom of the plasma (Eq. 3.37). Replacing the expression of $\rho_R(T)$ in Eq. (2.263), we obtain

$$H(T) = \pi \sqrt{\frac{g_\rho}{90}} \frac{T^2}{M_P} \simeq 0.33 g_\rho^{1/2} \frac{T^2}{M_P} = 3.42 \left(\frac{g_\rho}{106.75}\right)^{1/2} \frac{T^2}{M_P}. \tag{2.264}$$

Before that time, particles produced in the bath were diluted by the expansion (see Fig. 2.15 for an illustration of the process). Solving $\Gamma_\phi = H(T_d)$ with the expression (2.264) and approximating at a first step $T_d \simeq T_{\rm RH}$, we obtain

$$\Gamma_\phi = H(T_d) \;\Rightarrow\; T_d \simeq T_{\rm RH} = \left(\frac{90}{g_\rho \pi^2}\right)^{1/4} \sqrt{M_P \Gamma_\phi} \simeq \sqrt{M_P \Gamma_\phi}. \tag{2.265}$$

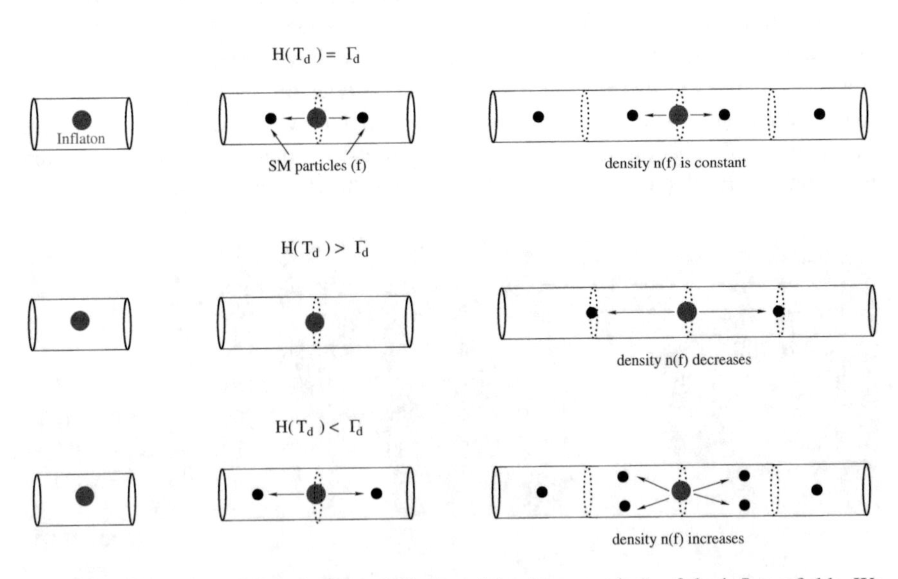

Fig. 2.15 Illustration of the principle of dilution of the decay products of the inflaton fields. We can easily notice that it is when the Universe reaches the temperature obeying $\Gamma_d \simeq H(T_d)$ that the mechanism of population becomes efficient, and the decay products are not diluted anymore by the expansion rate

In the case of Yukawa-type coupling y_ϕ of the inflation to the particles of the bath: $\mathcal{L}_\phi = y\phi \bar{f} f$, we have $\Gamma_\phi = \frac{y_\phi^2}{8\pi} m_\phi$ (Eq. 2.228), which gives finally

$$T_{\rm RH} = \frac{y_\phi}{\pi} \left(\frac{45}{32 g_\rho} \right)^{1/4} \sqrt{M_P m_\phi}. \tag{2.266}$$

The mass of the inflaton being extracted from WMAP/PLANCK data for different kinds of inflationary models (see Sect. 2.2 for more details), a typical mass of inflaton of 3×10^{13} GeV implies a reheating temperature of $T_{\rm RH} = 5\, y_\phi \times 10^{14}$ GeV for $g_\rho = 106.75$, which is the Standard Model effective number of degrees of freedom. This is the typical scale one can expect for a reheating temperature from inflationary model constructions.

Instantaneous reheating and instantaneous thermalization: computing T_{RH}
In the early Universe, the energy density is never totally dominated by one or other species (inflation, radiation, matter, etc.). The processes are smooth, and one usually needs to solve numerically the equations. That is why it is possible to find different definitions of the reheating temperature T_{RH} in the literature. Once the inflaton is settled at its minimum, it decays with a decay rate Γ_ϕ. One usually can consider that the thermalization process in the early Universe is instantaneous. It means that during a doubling time of the Universe, the new "photons" (by photons we mean all relativistic particles in the primordial plasma) produced by the decay of the inflaton enter automatically in equilibrium with the existing plasma [$n_\gamma \langle \sigma v \rangle_{scattering}(T) > H(T)$]. This is called *instantaneous thermalization*. However, one can treat the inflaton as a particle decaying instantaneously (*instantaneous reheating*) or with a time delay. In the instantaneous reheating, we consider that the inflaton decays when $H = \Gamma_\phi$ as we did in this section. The interpretation being that the Universe is dominated by the radiation once the doubling time of the Universe dominates the decay rate, see Fig. 2.15. The other possibility is to consider that the time of reheating t_{RH} can be approximated by Γ_ϕ^{-1}, the inflaton lifetime. This is justified by the fact that the radiation dominates the energy budget of the Universe if the inflaton decays instantaneously at the time Γ_ϕ^{-1}. A last possibility is to consider reheating as the time when the density of radiation reaches the density of the inflaton ($\rho_R = \rho_\phi$). The three definitions of the reheating time differ only by numerical factors of the order one. We can see it by computing the reheating temperature in the second case, where $t_{RH} = \Gamma_\phi^{-1}$. The Friedmann equation (2.6) should be written in a *matter* dominated Universe, as the inflaton is the only source of energy of density $\rho_\phi = n_\phi \times m_\phi$, n_ϕ being the inflaton number density, which, as a

(continued)

matter field, is proportional to a^{-3}:

$$H^2 = \left(\frac{\dot{a}}{a}\right)^2 = \frac{\rho_\phi}{3M_P^2} = \frac{n_\phi\, m_\phi}{3M_P^2} \Rightarrow a(t) = \left(\frac{3}{2}\sqrt{\frac{n_\phi\, m_\phi}{3M_P^2}}\right)^{2/3} t^{2/3}$$

$$\Rightarrow H = \frac{\dot{a}}{a} = \frac{2}{3}\frac{1}{t} \Rightarrow H(t_{RH}) = \frac{2}{3t_{RH}} = \frac{2}{3}\Gamma_\phi. \qquad (2.267)$$

To compute T_{RH}, one supposes that just after t_{RH}, the Universe is dominated by a radiation, of energy density $\rho_R = \frac{g_{RH}\pi^2}{30}T_{RH}^4$, with g_{RH} the relativistic degrees of freedom of the thermal bath at t_{RH}. We then obtain

$$H^2(T_{RH}) = \frac{4}{9}\Gamma_\phi^2 = \frac{\rho_R(T_{RH})}{3M_P^2} = \frac{g_{RH}\,\pi^2}{90M_P^2}T_{RH}^4 \Rightarrow T_{RH}$$

$$= \left(\frac{40}{g_{RH}\pi^2}\right)^{1/4}\sqrt{\Gamma_\phi M_P}. \qquad (2.268)$$

As we claimed above, another definition of the reheating temperature could be when the expansion rate dominates on the decay rate, i.e. $H(T_{RH}) = \Gamma_\phi$. In this case all the results have to be "normalized" to a factor $2/3$ every time the Hubble parameter appears in the expression.

2.4.2 Non-instantaneous Reheating

The reheating temperature discussed above was calculated assuming an instantaneous conversion of the energy density of the inflaton field into radiation when the decay width, Γ_ϕ, is equal to H. The reheating temperature is best regarded as the temperature below which the Universe expands as a radiation dominated Universe, with the scale factor decreasing as $a \propto g_S^{-1/3} T^{-1}$, which can be understood from an entropy conservation argument. In this regard it has a limited meaning. For instance, T_{RH} *should not* be used as the maximum temperature of the Universe during the reheating process. As we will see, the maximum temperature is in fact much larger than T_{RH}. One implication of this is that it is incorrect to assume that the maximum abundance of a massive particle species of mass M, produced after inflation, is suppressed by a factor $e^{-M/T_{RH}}$. We will, show in this section, how to calculate the evolution of the temperature in a hot Universe, in presence of massive states.

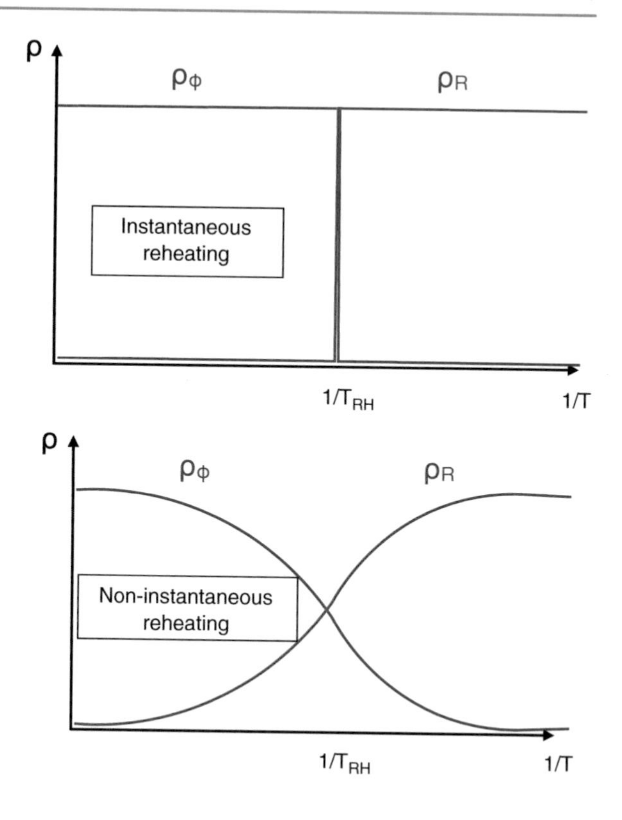

Fig. 2.16 Difference between an instantaneous treatment of the reheating (top) and a treatment taking into account the finite width of the inflaton Γ_ϕ (bottom)

Let us consider first a model Universe with two components: the inflaton field with an energy density ρ_ϕ and the radiation with the energy density ρ_R. We will assume that the decay rate of the inflaton field energy density into light degrees of freedom, from now on referred to as radiation, is $\Gamma_\phi \ll H$. This condition ensures that we place ourselves in a regime where the decay of the inflaton is not yet effective. This regime corresponds to the phase of domination of the inflaton, while it begins to generate the thermal bath. During this first phase, the inflaton can be considered as almost stable because its decay rate is below the expansion rate. We will also assume that the light degrees of freedom are in local thermal equilibrium (instantaneous thermalization.[43]) We show in Fig. 2.16 the difference between the non-instantaneous decay of the inflaton and the instantaneous reheating treatment.

With the above assumption, one needs to describe the evolution of the density of the different components of the Universe. We have 3 equations to deal with: the Boltzmann equations for the inflaton field density and radiation density and the Friedmann equation. We also have 3 parameters: the scale a, the time t (appearing in the definition of Hubble parameter), and the temperature T. As the temperature T is the only measurable quantity of the three, it seems natural to express t and a as

[43]See Sect. 2.3 for the case of non-instantaneous thermalization.

a function of T. This is done by combining the three equations cited above. Let us begin by the inflaton density evolution.

2.4.2.1 Evolution of the Temperature During Reheating

To obtain the evolution of the density of energy and thus of the temperature, we need to solve the set of Eqs. (2.206–2.208) with $\rho_R = g_\rho \frac{\pi^2}{30} T^4$ (Eq. 3.35). With the hypothesis $\Gamma_\phi \ll H$ in the first phase of reheating, the Universe is still dominated by the inflaton, and $N_\phi = \rho_\phi a^3$ is thus almost constant during all the reheating process. To solve the set of equations, one needs to choose first the dynamical parameter. It was naturally the time t when we were analyzing processes in the preheating stage of the Universe, before the formation of a thermal state. The more natural dynamical parameter during the reheating phase, with the instantaneous reheating hypothesis, is the temperature T. The first step is then to eliminate the two other dynamical parameters t and a from the set of equations. Eliminating t is quite straightforward. Indeed, noticing that

$$\frac{d}{dt} = \frac{da}{dt}\frac{d}{da} = \frac{da}{dt}\frac{dT}{da}\frac{d}{dT} = a\,H\,\frac{dT}{da}\frac{d}{dT} = H\,\frac{dT}{d\ln a}\frac{d}{dT},$$

Equation (2.207) then becomes

$$\frac{d\rho_R}{dT} + 4\frac{\rho_R}{a\frac{dT}{da}} = \Gamma_\phi \frac{\rho_\phi}{Ha\frac{dT}{da}}, \tag{2.269}$$

which gives, once we eliminate H through Eq. (2.208),

$$\frac{d\rho_R}{dT} + 4\frac{\rho_R}{a\frac{dT}{da}} = \Gamma_\phi \frac{\sqrt{3\rho_\phi} M_P}{a\frac{dT}{da}}. \tag{2.270}$$

We see appearing little by little the solution. Indeed, considering $\rho_\phi a^3 \simeq$ constant $\Rightarrow \rho_\phi = \rho_\phi^i \left(\frac{a_i}{a}\right)^3$ (ρ_ϕ^i and a_i being the initial value of the density and scale factor, just after the inflationary stage), and using $\rho_R = g_\rho \frac{\pi^2}{30} T^4 = \alpha T^4$, one can solve Eq. (2.270) and obtain

$$4\,T^3\frac{dT}{da} + 4\frac{T^4}{a} = \frac{1}{a^4}\frac{dX^4}{da} = \Gamma_\phi \sqrt{3}\frac{\sqrt{\rho_\phi^i a_i^3}\,M_P}{a^{5/2}\alpha}, \tag{2.271}$$

where we have defined $X = a \times T$. Solving Eq. (2.271) for X, and expressing the solution for T, one obtains

$$T^4 = T_i^4 \left(\frac{a_i}{a}\right)^4 + \frac{12\sqrt{3}}{g_\rho \pi^2} \Gamma_\phi M_P \sqrt{\rho_\phi^i} \left(\frac{a_i}{a}\right)^{3/2} \left[1 - \left(\frac{a_i}{a}\right)^{5/2}\right]. \quad (2.272)$$

Notice that we recover the classical redshift evolution for the temperature if $\Gamma_\phi = 0$. Indeed, this situation corresponds to *no* injection of entropy in the thermal bath. Only adiabatic processes occur, and the thermal system is entropy conserved, $T \times a =$ constant.

Exercise Compute T as a function of a for a generic equation of state, $\rho_\phi \propto a^{-3(1+w)}$.

We show in Fig. 2.17 the evolution of the temperature as a function of the scaling factor (normalized to its initial value, $\frac{a}{a_i}$). We notice the three distinct phases in the evolution of T. After a sharp increase in the temperature, it decreases following a $a^{-3/8}$ law when inflaton density still dominates the Universe and finally a classical entropy conserving a^{-1} law when T reaches T_{RH}. The maximum temperature can be easily extracted from Eq. (2.272), which occurs for values[44]

$$\left(\frac{a}{a_i}\right)_{max} = \left(\frac{8}{3}\right)^{2/5} \simeq 1.5 \quad \text{and} \quad T_{max}^4 = \frac{15\sqrt{3}}{2 g_\rho \pi^2} \Gamma_\phi M_P \sqrt{\rho_\phi^i} \left(\frac{3}{8}\right)^{3/5}. \quad (2.273)$$

The difference of slopes ($-\frac{3}{8}$ compared to -1) between an inflaton dominated Universe and a radiation dominated Universe can be understood from the fact that the inflaton injects some energy continuously in the thermal bath, counterbalancing (but not totally) the redshift due to the expansion. Once the inflaton has completely disappeared (after T_{RH}), the adiabatic relation $T \times a$ holds as there are no any external sources of energy to heat the bath anymore. This effect can also be seen from the expression of the Hubble rate. Whereas it is proportional to $\frac{T^2}{M_P}$ in a radiative Universe, Eq. (2.264), from $H^2 = \frac{\rho_\phi}{3M_P^2}$, one can extract

$$H = \frac{g_\rho \pi^2}{36} \frac{T^4}{\Gamma_\phi M_P^2}, \quad (2.274)$$

[44] We considered $T_i = 0$, in other words, no other thermal sources outside from the inflaton decay.

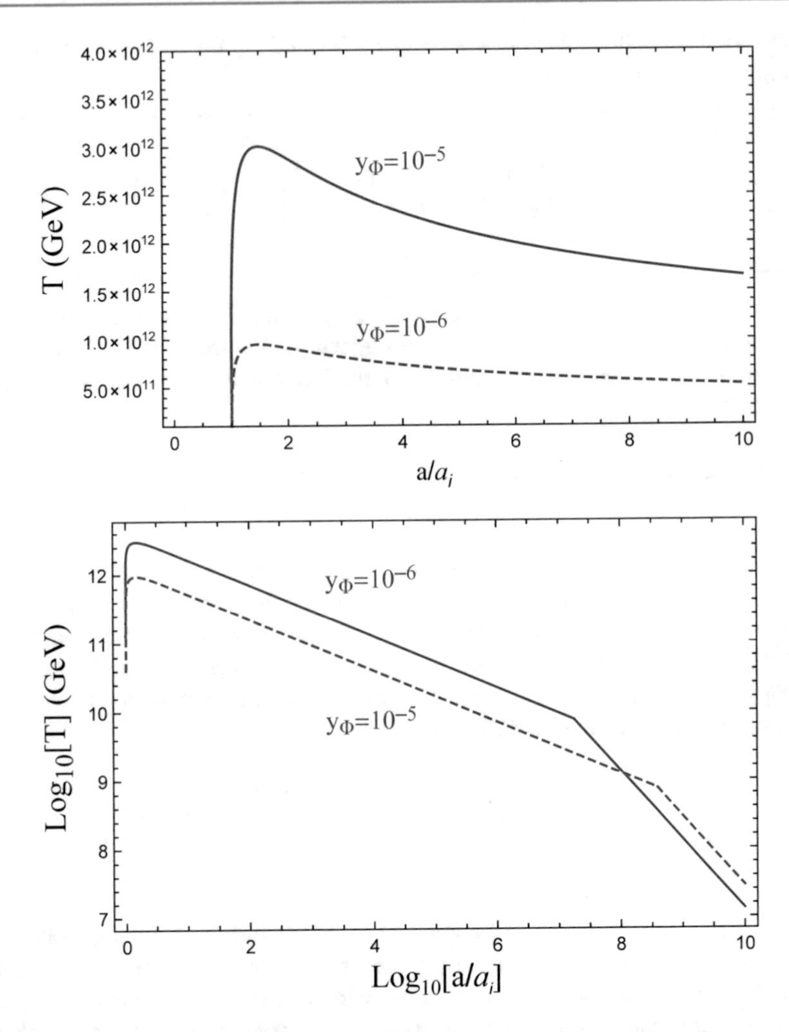

Fig. 2.17 Evolution of the temperature of the Universe as a function of the scaling parameter a/a_i for different values of y_ϕ. One can clearly see the 3 stages of evolution: first, the temperature increases up to a maximum value and then decreases following a $a^{-3/8}$ law when the Universe is still dominated by the inflaton energy density and finally a^{-1} when it becomes purely radiative

which means, naively speaking that "H decreases faster (as a function of T) in a matter dominated Universe than in a radiation dominated Universe." In fact, it is simply coming from the fact that the temperature decreases less in the inflaton dominated Universe because it is a non-adiabatic process from the thermal bath point of view: the latter receives energy injected by the inflaton decay.

Notice that in all the expressions above, we "condensed" by simplicity the initial condition uncertainties into the parameter ρ_ϕ^i. In the literature, it is also common to express the temperature as a function of H_i, the Hubble scale at the end of inflation.

The advantage of using ρ_ϕ^i comes from the possibility of expressing it as a function of the inflaton potential. For instance, if we suppose a potential,

$$V(\phi) = m_\phi^2 \phi^2 \quad \Rightarrow \quad \rho_\phi^i \simeq m_\phi^2 \langle \phi \rangle^2 \simeq m_\phi^2 M_P^2, \tag{2.275}$$

for oscillations with Planck scale amplitudes, expression that can be replaced easily in Eq. (2.272). The fact that the maximum temperature of the Universe can be much larger than the reheating temperature can have dramatic consequences on the production of dark matter during the reheating phase as we will discuss in the following section.

Exercise Recover the expression for T_{max} and a_{max}, and recompute it with a generic equation of state for $\rho_\phi \propto a^{-3(1+w)}$.

2.4.2.2 A Closer Look on the Hubble Constant*

Before studying the production of dark matter in this early phase of the Universe, we want to detail slightly the dependence of H on the temperature in the reheating phase (Eq. 2.274), i.e. when the inflaton still dominates massively the Universe. To be more general, we will do it in the case of an equation of evolution for ρ_ϕ

$$\rho_\phi = \rho_\phi^i \left(\frac{a_i}{a}\right)^m = \eta a^{-m}, \tag{2.276}$$

which is the simplified form of Eq. (2.141). Notice that the problem is in fact more complex because the power m in (2.276) depends on the power of ϕ in the inflationary potential $V(\phi)$, which in his turn can change the Boltzmann equation for ϕ and then the relation (2.276). But we will not enter in such details here. We can then generalize (2.270) to

$$\frac{d\rho_R}{dT} + 4\frac{\rho_R}{a\frac{dT}{da}} = \Gamma_\phi \frac{\sqrt{3\eta} a^{-\frac{m}{2}} M_P}{a\frac{dT}{da}}, \tag{2.277}$$

and using $\rho_R = \alpha T^4$, we obtain

$$4\,T^3\frac{dT}{da} + 4\frac{T^4}{a} = \frac{1}{a^4}\frac{dX^4}{da} = \Gamma_\phi \frac{\sqrt{3\eta} a^{-\frac{m}{2}} M_P}{\alpha a}, \tag{2.278}$$

where as usual we have defined $X = a \times T$ and $\alpha = \frac{g_\rho \pi^2}{30}$. Solving (2.278) for X, we can have the expression of $T = f(a)$:

$$T^4 = \frac{2\Gamma_\phi \sqrt{3\eta} M_P}{\alpha(8-m)} a^{-\frac{m}{2}} \left[1 - \left(\frac{a_i}{a}\right)^{-\frac{m}{2}+4}\right] + T_i^4 \left(\frac{a_i}{a}\right)^4. \tag{2.279}$$

We can then deduce $H(T)$ in the limit $a \gg a_i$ (around $T_R H$, for instance):

$$H(T) = \frac{\sqrt{\rho_\phi}}{\sqrt{3}M_P} = \sqrt{\frac{\eta}{3}\frac{a^{-\frac{m}{2}}}{M_P}} = \frac{8-m}{6}\frac{\alpha T^4}{\Gamma_\phi M_P^2} \tag{2.280}$$

$$H(T) = \frac{\sqrt{\rho_\phi}}{\sqrt{3}M_P} = \sqrt{\frac{\eta}{3}\frac{a^{-\frac{m}{2}}}{M_P}} = \frac{8-m}{6}\frac{\alpha T^4}{\Gamma_\phi M_P^2} \tag{2.281}$$

with $\alpha = \frac{g_\rho \pi^2}{30}$. We recover then Eq. (2.274)

$$H(T) = \frac{5}{6}\frac{\alpha T^4}{\Gamma_\phi M_P^2} \tag{2.282}$$

for a dust-like inflaton ($m = 3$). We then need to be careful when defining the reheating temperature. Indeed, when we used in Eq. (2.267) $H(t_{RH}) = \frac{2}{3t_{RH}} = \frac{2}{3}\Gamma_\phi$ and replaced H by $H(T_{RH})$ with $\frac{\sqrt{\rho_R(T_{RH})}}{\sqrt{3}M_P}$, we used in fact two different definitions of the reheating time without really noticing it. The former equation tells us that the reheating *time* is given by the inverse width of the inflaton field (first definition), whereas the later one tells us that at T_{RH} the Universe is dominated by the radiation. However, it is not exactly true. To be exact, one should use the exact form of $H(T)$ (2.282), evaluate it at T_{RH}, and impose $\rho_\phi(T_{RH}) = \rho_R(T_{RH}) = \alpha T_{RH}^4$. We then obtain[45]

$$\frac{5}{6}\frac{\alpha T_{RH}^4}{\Gamma_\phi M_P^2} = \frac{\sqrt{\alpha T_{RH}^4}}{\sqrt{3}M_P} \Rightarrow T_{RH}^2 = \frac{6}{5\sqrt{3\alpha}}\Gamma_\phi M_P, \tag{2.283}$$

[45]The approximation we need to apply here is that we considered $H(T_{RH}) = \frac{\sqrt{\rho_\phi = \alpha T_{RH}^4}}{\sqrt{3}M_P}$, where we neglected the radiation contribution in ρ because we supposed a Universe still dominated by ρ_ϕ. Using $H(T_{RH}) = \frac{\sqrt{\rho_\phi + \rho_R}}{\sqrt{3}M_P} = \frac{\sqrt{2\rho_R}}{\sqrt{3}M_P}$ will also lead to a misleading factor of $\sqrt{2}$ because the formulae (2.282) could not be applied if the Universe is not anymore dominated by the inflaton field. The exact numerical solution lies between these two approximations.

which corresponds, when plugging it again in $H(T)$ (2.282), to the relation

$$H(T_{RH}) = \frac{2}{3c}\,\Gamma_\phi \tag{2.284}$$

with $c = \frac{5}{3}$, which represents the (tiny) difference between the two definitions of reheating temperature. With this definition of reheating temperature, one can write

$$\rho_\phi = \rho_\phi(T_{RH})\left(\frac{T}{T_{RH}}\right)^8 = \frac{g_\rho \pi^2}{30}\frac{T^8}{T_{RH}^4}. \tag{2.285}$$

How to teach non-instantaneous reheating

I present in this little box a shortcut to derive the evolution of thermal energy density as a function of the scale factor (the way I do it for my students). Instead of writing Equation (2.269) as a function of T, one can directly write it as a function of a:

$$\frac{d\rho_R}{da} + 4\frac{\rho_R}{a} = \frac{\Gamma_\phi}{H}\frac{\rho_\phi}{a}. \tag{2.286}$$

This equation has two advantages. First of all, it shows that it is the ratio $\frac{\Gamma_\phi}{H}$ that governs the process and will determine when the reheating will stop, justifying the approximation that at reheating time $H(T_{RH}) \simeq \Gamma_\phi$. At this temperature, a naive (but interesting) look at Eq. (2.286) gives $\rho_R \simeq \rho_\phi$ for $\Gamma_\phi \sim H$. But, more interesting is the fact that solving this equation is extremely easy if one considers $\rho_\phi \propto a^{-3(1+w)}$ and $H \simeq \frac{\sqrt{\rho_\phi}}{\sqrt{3}M_P}$. Supposing a constant Γ_ϕ, and using the variable $X = \rho_R \times a^4$, we let the reader check that

$$\rho_R = \frac{1}{4+\delta}\frac{\Gamma_\phi}{H}\rho_\phi\left[1 - \left(\frac{a_i}{a}\right)^{4+\delta}\right], \tag{2.287}$$

where $\delta = -\frac{3}{2}(1+w)$, and we supposed $\rho_R(a_i) = 0$. This equation can even be generalized to situations where Γ_ϕ depends also on a (where you have inflaton scattering, for instance). In this case, $\delta = -\frac{3}{2}(1+w) + \delta_\Gamma$, with $\Gamma_\phi \propto a^{\delta_\Gamma}$. This equation is nice to show to students. It explicits the fact that ρ_R is proportional to ρ_ϕ, with a coefficient of proportionality $\sim \frac{\Gamma_\phi}{H}$. We can also write it in different ways, depending what are the initial conditions we

(continued)

are interested in,

$$\rho_R = \frac{\sqrt{3}}{4+\delta}\Gamma_\phi M_P \sqrt{\rho_\phi}\left[1-\left(\frac{a_i}{a}\right)^{4+\delta}\right]$$
$$= \frac{1}{4+\delta}\frac{\Gamma_\phi}{H_i}\rho_\phi^i\left(\frac{a_i}{a}\right)^{\frac{3}{2}(1+w)}\left[1-\left(\frac{a_i}{a}\right)^{4+\delta}\right]\ldots$$

We then recover the condition (2.284), defining the reheating temperature as $\rho_\phi = \rho_R$, $H = \frac{1}{4+\delta}\Gamma_\phi = \frac{2}{5}\Gamma_\phi$ for a dust-like inflaton.

Exercise Compute the factor c of Eq. (2.284), noticing that at T_{RH},

$$\rho_\phi(T_{RH}) = \rho_R(T_{RH}) \Rightarrow H(T_{RH}) = \frac{\sqrt{\rho_R(T_{RH})}}{\sqrt{3}M_P}.$$

Show that taking the definition $H(T_{RH}) = \Gamma_\phi$ as the reheating temperature leads to (for a dust-like inflaton)

$$H(T) = \frac{\alpha T^4}{2\Gamma_\phi M_P^2}, \quad T_{RH}^2 = \frac{2}{\sqrt{5}}\frac{\Gamma_\phi M_P}{\sqrt{\alpha}} \tag{2.288}$$

instead of

$$H(T) = \frac{5}{6}\frac{\alpha T^4}{\Gamma_\phi M_P^4}, \quad T_{RH}^2 = \frac{6}{5\sqrt{3}}\frac{\Gamma_\phi M_P}{\sqrt{\alpha}} \tag{2.289}$$

when supposing $\rho_\phi(T_{RH}) = \alpha T_{RH}^4$.

In summary, one can always use Eq. (2.284) for the definition of the reheating temperature, taking $c = 1$ if one considers $t_{RH} = \Gamma_\phi^{-1}$, $c = \frac{2}{3}$ for $H(T_{RH}) = \Gamma_\phi$, and $c = \frac{5}{3}$ corresponding to the definition $\rho_\phi(T_{RH}) = \alpha T_{RH}^4$. The formulae are easy to switch by a simple "redefinition" of Γ_ϕ.

2.4.3 Producing Dark Matter During the Reheating Phase

2.4.3.1 The Context

We just learned in the previous section how the reheating evolution should be treated with care once we do not suppose instantaneous reheating. It is then natural to ask how the physics of reheating can affect the production of dark matter. In fact, the influence of reheating is especially large for production modes that are heavily dependent on the energy (and thus the temperature). Every model that includes higher dimensional operators or derivative couplings is concerned. It runs from axion-like particles with couplings of the form $aF^{\mu\nu}F_{\mu\nu}$ to longitudinal modes of $U(1)$ gauge groups of the form $\partial_\mu a$ (Stueckelberg notation) or goldstino in the case of the gravitino if SUSY breaking scale is much above the reheating temperature. Indeed, every time the scale of a broken symmetry lies above the mass of the corresponding gauge boson, the only surviving mode is the longitudinal one. That is, the Goldstone boson (goldstino in the case of the gravitino), which is reduced to the derivative of the Higgs phase (or the imaginary part of the Higgs). This degree of freedom, which is "eaten" in the standard Higgs mechanism, stays as a physical degree of freedom in the low energy theory and in fact as the only polarization mode of the massive gauge boson, appearing under the form of the derivative of the phase of the Higgs and thus highly temperature dependent.

Those were just some examples where one should be cautious when looking at the production of dark matter in the early Universe. Two effects should then be treated separately:

- The effect of a *non-instantaneous reheating*. In this case, the inflaton does not decay at a given time, but the bath produced while it still dominates the energy budget of the Universe can already begin to produce dark matter (through annihilation) before the Universe is dominated by the radiation.
- The effect of a *non-instantaneous thermalization*. This happens if the standard model particles do not have time to thermalize before the production of dark matter from the annihilation of standard particles just produced by the inflaton decay (of energy $m_\phi/2$, see Sect. 2.3).

Indeed, the Universe in the presence of the inflaton is particular as we already discussed in Sect. 2.4.2. To be complete, one should solve the set of three equations, the combined ones of the radiation, the inflaton (conservation of energy), and the dark matter (Boltzmann equation). We can then rewrite the set of Eqs. (2.206–2.208) adding the possibility for dark matter production

$$\frac{d\rho_\phi}{dt} + 3H\rho_\phi = -\Gamma_\phi\rho_\phi$$
$$\frac{d\rho_R}{dt} + 4H\rho_R = \Gamma_\phi\rho_\phi$$
$$H^2 = \frac{\rho_\phi}{3M_P^2} + \frac{\rho_R}{3M_P^2}$$
$$\frac{dn_\chi}{dt} + 3H\, n_\chi = R(T), \tag{2.290}$$

where n_χ is the dark matter density. Several remarks should be done at this point. First of all, the Boltzmann equation (the last one) is linked to the three other ones through the dependence of t on T. Indeed, if you remember what we told in the previous section, to express all the equations as a function of one variable (we have chosen T because it corresponds to what is measured nowadays by CMB experiments), one needs to eliminate the scale factor a and the time t by subtle combinations of the Friedmann equation and the conservation of energy equations. We will not obtain the simple relation $T \propto a^{-1}$ as we have in a pure radiation domination. That can be understood as in those cases, one can treat the thermal bath as an isolated system whose entropy is conserved. In the case of a massive particle (inflaton of density ρ_ϕ), which decays into radiation of density ρ_R, we are in the presence of two open thermodynamical systems in contact. The energy of the inflaton is diffused to the thermal bath. The total entropy is conserved, but not inside the thermal bath that receives entropy from an external system, proportional to the decay width of the inflaton as it is clear from Eq. (2.290).

In this section, we will suppose a non-instantaneous reheating, which means we suppose that the decay of the inflaton "takes its times" in a sense. In other words, $\Gamma_\phi \ll H(T)$. This can be considered as a "long-lived" inflaton. That means that, during a time, the Universe will be dominated by the inflaton, while the thermal bath will begin to form. Dark matter can then be produced during this period of time, where the particles are very energetic because the temperature is very high, around the inflaton mass. In the first step, we will consider instant thermalization. That means that two photons (I call indifferently photon any relativistic particle of the thermal bath) have time to scatter to a photon newly produced by the inflaton to bring it to the thermal bath before the expansion forbids it. From Eq. (2.272) one can write for $a \gg a_i$

$$T = \beta \left(\frac{a_i}{a}\right)^{3/8}, \quad \text{with } \beta = \left[\frac{12\sqrt{3}}{g_\rho \pi^2}\Gamma_\phi M_P \sqrt{\rho_\phi^i}\right]^{1/4}. \tag{2.291}$$

Once we know the dependence of T on a, one can define $\frac{d}{dt}$ as a function of $\frac{d}{dT}$:

$$\frac{d}{dt} = \frac{d}{da}\frac{da}{dt} = \frac{d}{da}aH = \frac{d}{dT}\frac{dT}{da}aH = -\frac{3}{8}HT\frac{d}{dT}, \tag{2.292}$$

from where we can express the Boltzmann equation of Eq. (2.290) as a function of the temperature

$$\frac{dn_\chi}{dT} - 8\frac{n_\chi}{T} = -\frac{8R(T)}{3H\,T}. \tag{2.293}$$

We can then define a "covariant" Yield $Y = \frac{n}{T^8}$, permitting to "absorb" the Hubble expansion in the Boltzmann equation, giving

$$\frac{dY}{dT} = -\frac{8}{3}\frac{R(T)}{H\,T^9}. \tag{2.294}$$

It is *always* possible to define a covariant Yield Y. $Y = \frac{n}{T^3}$ in entropy conserved system, $\frac{n}{T^8}$ in mixed matter–radiation dominated Universe... The power of T in the definition of Y depends only on the dependence of T on a and thus on how the temperature evolves with the scale of the Universe. The fact of having $T \propto a^{-3/8}$ in the inflaton-type model means that the temperature decreases slower than in typical radiation dominated Universe where $T \propto a^{-1}$ (which is logical as matter as tendency to slow down expansion). We then need higher power of temperatures to keep covariant Yields.

2.4.3.2 Production from Inflaton Decay

We can then easily compute the relic abundance of a dark matter χ obtained by the inflaton decay if we allow a fraction B_R of the inflaton decay into dark matter. One then needs to solve the Boltzmann equation (2.294) between T_{max} and T_{RH}, defining the production rate (the number of dark matter particles produced per unit of time and volume) as

$$\begin{aligned} R_{decay}(T) &= B_R\,\Gamma_\phi \frac{\rho_\phi(T)}{m_\phi} = B_R\,\Gamma_\phi \frac{\rho_\phi^i}{m_\phi}\left(\frac{a_i}{a}\right)^3 \\ &= B_R \Gamma_\phi \frac{\rho_\phi^i}{m_\phi}\left(\frac{T}{\beta}\right)^8 = B_R \times \frac{g_\rho^2 \pi^4}{432}\frac{T^8}{\Gamma_\phi m_\phi M_P^2}. \end{aligned} \tag{2.295}$$

We can then implement Eq. (2.274) and Eq. (2.295) into Eq. (2.294) to write

$$\frac{dY_\chi^{dec}}{dT} = -\frac{2}{9}\,B_R \times \frac{g_\rho \pi^2}{m_\phi T^5}, \tag{2.296}$$

which gives, after integration between T_{max} and T_{RH},

$$\begin{aligned} Y_\chi^{dec}(T_{RH}) &\simeq \frac{1}{18}\,B_R \times \frac{g_\rho \pi^2}{m_\phi T_{RH}^4} \\ \Rightarrow\ n_\chi^{dec}(T_{RH}) &= B_R \times \frac{g_\rho \pi^2}{18\, m_\phi} T_{RH}^4 \end{aligned} \tag{2.297}$$

with $Y_\chi(T) = \frac{n_\chi(T)}{T^8}$. Notice that, if we would *not* have taken into account the non-instantaneous effect, Eq. (2.297) would have been written simply as

$$n_\chi^{dec}(T_{RH}) = B_R \times n_\phi = B_R \times \frac{\rho_\phi}{m_\phi} = \frac{g_\rho \pi^2}{30\, m_\phi} T_{RH}^4, \tag{2.298}$$

i.e. $\frac{3}{5}$ times less.

The density of a dark matter of mass m_χ measured today can be written using Eq. (E.1)

$$\Omega = \frac{\rho}{\rho_c} = \frac{n_\chi(T_0) \times m_\chi}{10^{-5} h^2 \text{ GeV.cm}^{-3}} \tag{2.299}$$

$$\Rightarrow \Omega h^2|_{\text{decay}} = 10^5 \left(\frac{g_s^0}{g_s^{RH}}\right) \left[\frac{n_\chi^{dec}(T_{RH})}{\text{cm}^{-3}}\right] \left(\frac{T_0}{T_{RH}}\right)^3 \left(\frac{m_\chi}{1 \text{ GeV}}\right)$$

$$\simeq \frac{3}{2} \times 10^8 \left(\frac{g_s^0}{g_s^{RH}}\right) \left(\frac{n_\chi^{dec}(T_{RH})}{T_{RH}^3}\right) \left(\frac{m_\chi}{1 \text{ GeV}}\right),$$

where we used[46] $T_0 = 3.3 \times 10^{-4}$ eV, and $1 \text{ cm}^{-3} \simeq 8 \times 10^{-42} \text{ GeV}^3$. We can then implement the expression (2.297) into Eq. (2.299) to obtain

$$\Omega h^2|_{\text{decay}} = \frac{g_\rho \pi^2}{12} \left(\frac{B_R}{10^{-8}}\right) \left(\frac{T_{RH}}{m_\phi}\right) \left(\frac{m_\chi}{1 \text{ GeV}}\right)$$

$$\simeq 0.1 \left(\frac{B_R}{10^{-8}}\right) \left(\frac{10^{13}\text{GeV}}{m_\phi}\right) \left(\frac{T_{RH}}{10^{10}\text{GeV}}\right) \left(\frac{m_\chi}{1 \text{ GeV}}\right). \tag{2.300}$$

2.4.3.3 Production by Scattering

If one defines the scattering rate of production of dark matter from the thermal bath as[47] $R(T) = \frac{T^{n+6}}{\Lambda^{n+2}}$. As we did for the inflaton decay, we can combine Eq. (2.294) and Eq. (2.274) to write

$$\frac{dY_\chi^{scat}}{dT} = -\frac{96}{g_\rho \pi^2} \frac{T^{n-7}}{\Lambda^{n+2}} \Gamma_\phi M_P^2, \tag{2.301}$$

[46] And approximated 1.7 by $\frac{3}{2}$ to have simplified expressions.

[47] The power of T has been chosen so that n corresponds to the dependence of the *cross section* in the temperatures: $R(T) \propto n_{SM}^2 \langle \sigma v \rangle \propto T^6 \langle \sigma v \rangle$.

which gives after integration between T_{max} and T_{RH} depending on the value of n:

$$\begin{aligned}
\text{n} < 6: \;\; n(T_{RH}) &= \frac{96}{g_\rho \pi^2 (6-n)} \Gamma_\phi M_P^2 \left(\frac{T_{RH}}{\Lambda}\right)^{n+2} \\
\text{n} = 6: \;\; n(T_{RH}) &= \frac{96}{g_\rho \pi^2} \Gamma_\phi M_P^2 \left(\frac{T_{RH}}{\Lambda}\right)^{8} \ln\left(\frac{T_{max}}{T_{RH}}\right) \\
\text{n} > 6: \;\; n(T_{RH}) &= \frac{96}{g_\rho \pi^2 (n-6)} \Gamma_\phi M_P^2 \frac{T_{RH}^8 T_{max}^{n-6}}{\Lambda^{n+2}},
\end{aligned} \tag{2.302}$$

which gives, for the relic abundance, using Eq. (2.299),

$$\begin{aligned}
\text{n} < 6: \;\; \Omega h^2|_{\text{scat}} &= 0.1 \times \frac{48\, c \times 10^9}{\pi \sqrt{10 g_\rho}(6-n)} \left(\frac{M_P}{T_{RH}}\right)\left(\frac{T_{RH}}{\Lambda}\right)^{n+2} \left(\frac{m_\chi}{1\text{ GeV}}\right) \\
\text{n} = 6: \;\; \Omega h^2|_{\text{scat}} &= 0.1 \times \frac{48\, c \times 10^9}{\pi \sqrt{10 g_\rho}} \left(\frac{M_P}{T_{RH}}\right)\left(\frac{T_{RH}}{\Lambda}\right)^{8} \left(\frac{m_\chi}{1\text{ GeV}}\right) \ln\left(\frac{T_{max}}{T_{RH}}\right) \\
\text{n} > 6: \;\; \Omega h^2|_{\text{scat}} &= 0.1 \times \frac{48\, c \times 10^9}{\pi \sqrt{10 g_\rho}(n-6)} \left(\frac{T_{RH}^7 T_{max}^{n-6} M_P}{\Lambda^{n+2}}\right) \left(\frac{m_\chi}{1\text{ GeV}}\right),
\end{aligned}$$

where we used

$$\Gamma_\phi = c H(T_{RH}) = c \sqrt{\frac{g_\rho \pi^2}{90}} \frac{T_{RH}^2}{M_P} \tag{2.303}$$

with c of the order one represents the freedom in defining the reheating temperature, see Eq. (2.268). It then becomes interesting to combine the production of dark matter by the thermal bath, Eq. (2.303) with the direct product of the inflaton decay Eq. (2.300),

$$\Omega h^2|_{\text{tot}} = \Omega h^2|_{\text{scat}} + \Omega h^2|_{\text{decay}},$$

we obtain

$$\boxed{\begin{aligned}
\Omega h^2|_{\text{n}<6} = 0.1 \times \Bigg[& \frac{48\, c \times 10^9}{\pi \sqrt{10 g_\rho}(6-n)} \left(\frac{M_P}{T_{RH}}\right)\left(\frac{T_{RH}}{\Lambda}\right)^{n+2} \\
& + \frac{g_\rho \pi^2}{12} \left(\frac{B_R}{10^{-9}}\right)\left(\frac{T_{RH}}{m_\phi}\right)\Bigg] \left(\frac{m_\chi}{1\text{ GeV}}\right)
\end{aligned}}$$

$$\Omega h^2|_{n<=6} = 0.1 \times \left[\frac{48\, c \times 10^9}{\pi \sqrt{10 g_\rho}} \left(\frac{M_P}{T_{RH}}\right)\left(\frac{T_{RH}}{\Lambda}\right)^8 \ln\left(\frac{T_{max}}{T_{RH}}\right) + \frac{g_\rho \pi^2}{12}\left(\frac{B_R}{10^{-9}}\right)\left(\frac{T_{RH}}{m_\phi}\right)\right]\left(\frac{m_\chi}{1\text{ GeV}}\right)$$

$$\Omega h^2|_{n>6} = 0.1 \times \left[\frac{48\, c \times 10^9}{\pi \sqrt{10 g_\rho}(n-6)} \left(\frac{T_{RH}^7 T_{max}^{n-6} M_P}{\Lambda^{n+2}}\right) + \frac{g_\rho \pi^2}{12}\left(\frac{B_R}{10^{-9}}\right)\left(\frac{T_{RH}}{m_\phi}\right)\right]\left(\frac{m_\chi}{1\text{ GeV}}\right). \tag{2.304}$$

We show in Fig. 2.18 the result on the scan of the parameter space in the case $n = 4, 6$, and 8, respecting the relic density constraint $\Omega h^2 = 0.11$ for a large range of dark matter mass for $T_{RH} = 10^{11}$ GeV and $T_{max} = 10^{13}$ GeV. We notice that

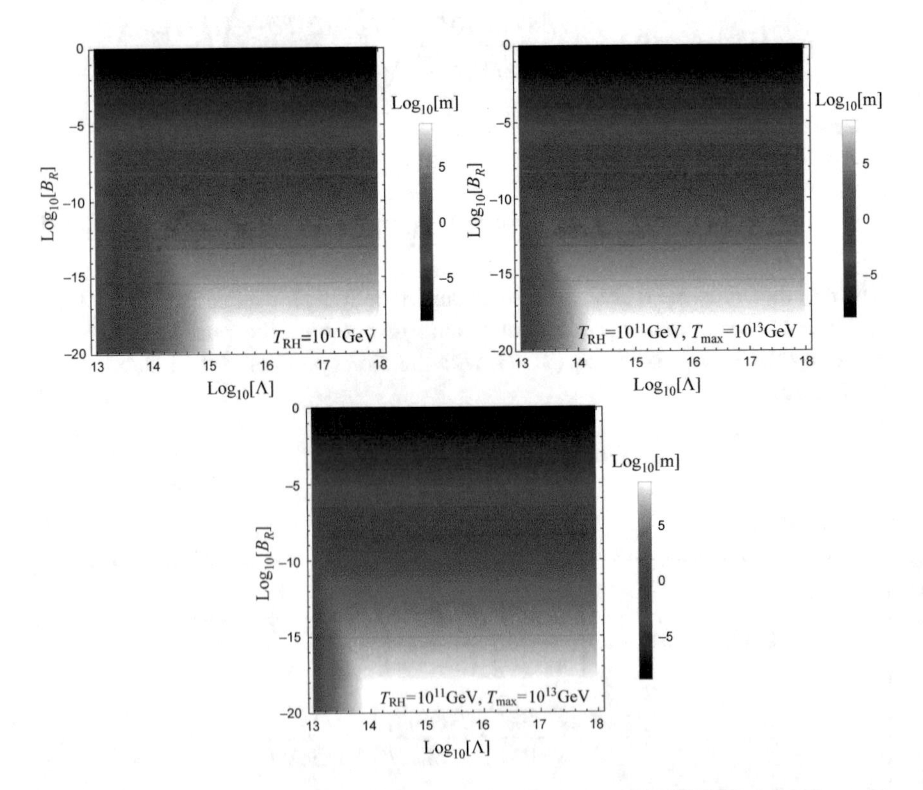

Fig. 2.18 Parameter space in the plane (Λ, B_R) of points respecting PLANCK constraints, with the corresponding dark matter mass m in the case $n = 4$, 8, and 8, respectively, see Eq. (2.304)

the range of dark matter mass allowed is very large (from keV to Eev scale), that is, a direct consequence of the strong power dependence of the relic abundance on the scale Λ. We also remark that a branching ratio of $\simeq 1$ is possible to obtain for dark matter mass of keV, whereas PeV scale dark matter requires very tiny branching fraction of the order of $B_R \simeq 10^{-18}$ to avoid an overclosure of the Universe.

We can also imagine the limit, in the plane (Λ, m) where the production of dark matter from the inflaton decay dominates on the production from annihilation. From Eq. (2.304), we obtain the conditions

$$B_R^{min}|_{n<6} = \frac{576\sqrt{10}}{(6-n)(g_\rho \pi^2)^{3/2}} \frac{T_{RH}^n M_P m_\phi}{\Lambda^{n+2}}$$

$$B_R^{min}|_{n<6} = \frac{576\sqrt{10}}{(g_\rho \pi^2)^{3/2}} \frac{T_{RH}^6 M_P m_\phi}{\Lambda^8} \ln\left(\frac{T_{max}}{T_{RH}}\right)$$

$$B_R^{min}|_{n>6} = \frac{576\sqrt{10}}{(n-6)(g_\rho \pi^2)^{3/2}} \frac{T_{RH}^6 T_{max}^{n-6} M_P m_\phi}{\Lambda^{n+2}}.$$

2.5 The Thermal Era $[10^{-28} - m_\chi]$

In this last section, we will consider the instantaneous reheating *and* instantaneous thermalization phases. For dark matter production processes that do not exhibit strong dependence on the energy and thus on the temperature, this treatment is much more simpler and usually gives reasonably good results compared to the one obtained numerically. We will in a sense repeat the previous exercise, considering that the inflation decayed instantaneously, which means, the history of the Universe begins at T_{RH}. The advantage is that we do not have to make any assumptions on the inflationary models, the thermalization process, or the inflaton width. In this case, T_{RH} is the only initial condition one needs.

2.5.1 Instantaneous Reheating and Instantaneous Thermalization

The Boltzmann equation in a Universe where the dark matter is not in thermal equilibrium in the bath, because of its small density n_χ, forbids the back reaction $n_\chi\, n_\chi \to n_{SM}\, n_{SM}$ and can be written (Eq. 2.244):

$$\frac{dn_\chi}{dt} + 3Hn_\chi = R(T) \tag{2.305}$$

with (see Eq. 2.258)

$$R(T) = \int f_1 f_2 \frac{E_1 E_2 dE_1 dE_2\, d\cos\theta_{12}}{1024\pi^6} \int |M_{fi}|^2 d\Omega_{13}, \tag{2.306}$$

for a process $1 + 2 \to 3 + 4$ with 1 and 2 the standard model particles of the bath and 3 and 4 the dark matter candidates, with f_1 and f_2 being the distribution function of the incoming particles 1 and 2, and $d\Omega_{13}$ is the solid angle between the particles 1 and 3 in the center of mass frame.

To solve the Boltzmann equation, the strategy is always the same: eliminating the time t to express it as a function of the temperature T (because rates and cross sections depend on T) using the (dimension) scale factor a as an intermediate state. Indeed, the Hubble parameter H connects a and t through $H = \frac{\dot{a}}{a}$ and also a and T through the Friedmann equation $H^2(T) = \frac{8\pi\rho_{tot}}{3M_{Pl}^2} = \frac{\rho_{tot}}{3M_P^2}$, M_P being the reduced planck mass $M_P = \frac{1}{\sqrt{8\pi}}M_{Pl} = 2.4 \times 10^{18}$ GeV. More concretely, one can write

$$\frac{d}{dt} = H \times a\frac{d}{da} \quad \text{and} \quad \frac{d}{da} = \frac{dT}{da}\frac{d}{dT} \quad \Rightarrow \quad \frac{d}{dt} = H\frac{dT}{d\ln a}\frac{d}{dT}. \tag{2.307}$$

We see that, knowing $H(T)$ from the Friedmann equation, if one can obtain the dependence, $a = f(T)$, the Boltzmann equation (2.305) only as a function of the temperature T. Without loss of generalities, we can generalize the dependence of T on a as $T = a^{-m}$. In this case, (2.305) becomes

$$-HTm\frac{dn}{dT} + 3Hn = R(T) \quad \Rightarrow \quad \frac{dn}{dT} - 3\frac{n}{mT} = -\frac{R(T)}{mHT}. \tag{2.308}$$

It can seem strange to "complexify" the Boltzmann equation introducing the m dependence. In fact, it comes from the fact that in a lot of dark matter scenarios, the physics occurs at temperatures well below the reheating temperature T_{RH}, in a regime where the Universe is totally dominated by radiation. In this case, $m = 1$ because the temperature is just redshifted by the expansion. That is also a (hidden) consequence of the conservation of entropy: $s \times a^3 = \text{cst} \Rightarrow T \propto a^{-1}$. However, in a Universe where entropy is injected in the thermal bath from the decay of the inflaton, for instance, non-instantaneous reheating, the entropy is not conserved (locally): the bath is continuously receiving energy from an outer system. In this case, the relation $s \times a^3$ does not hold anymore, and one needs to generalize it. It is even more true when one considers non-instantaneous thermalization, where even the concept of entropy and temperature does not make sense (or at least, not in the usual understanding).

Let us now define Yields $Y(T)$ such that the Hubble term of Eq. (2.308), $-3\frac{n}{mT}$, cancels. That is equivalent, in the radiation dominated era (or any entropic conservation framework), to the definition $Y \propto \frac{n}{T^3} \propto n \times a^3$, which permits to work in a comoving frame,[48] absorbing the effect of space dilatation in Y, which is in fact

[48] In the literature, it is very common to define $Y = \frac{n}{s}$ and common to define $Y = \frac{n}{n_\gamma}$. In any case, the dependence in both cases is of the form $Y = \text{cte} \times \frac{n}{T^3}$. As we will deal with epochs where (locally) the entropy will not be conserved (and thus the definition of s is more difficult), we prefer to define $Y = \frac{n}{T^n}$ in this section, with $n = 3$ in a Universe with local entropy conservation.

proportional to the *total number* of particles and not the density. If one generalizes $Y = \frac{n}{T^k}$, imposing $m \times k = 3$, Eq. (2.308) becomes

$$\frac{dY}{dT} = \frac{-R(T)}{mHT^{\frac{3}{m}+1}}. \tag{2.309}$$

Notice that we recover for $m = 1$, the classic Boltzmann equation in a radiation dominated Universe $\frac{dY}{dT} = -\frac{R(T)}{HTT^3}$, or if one defines $Y = \frac{n}{s}$, $\frac{dY}{dT} = -\frac{R(T)}{HTs}$. Let us have a look at the solution of (2.309) in 2 different scenarios: radiation dominated Universe and matter dominated Universe. The case of a Universe dominated by a decaying matter (as an inflaton) has been treated in the previous section.

2.5.1.1 Radiation Dominated Universe

To obtain the dependence of the Hubble rate on the temperature $H(T)$, one needs to solve the Friedmann equation (2.51) neglecting the curvature

$$H^2 = \frac{8\pi}{3} G\rho_R = \frac{\rho_R}{3M_P^2} = \frac{\alpha T^4}{3M_P^2} \Rightarrow H = \sqrt{\frac{\alpha}{3}} \frac{T^2}{M_P}, \tag{2.310}$$

M_P being the reduced Planck mass $M_P = \frac{M_{Pl}}{\sqrt{8\pi}}$ and $\alpha = \frac{g_\rho \pi^2}{30}$ [$g_\rho = 106.75$ in the Standard Model, see Eq. (3.38)]. To know m, one needs to write the law of conservation of energy

$$\begin{aligned} &\frac{d\rho_R}{dt} + 4H\rho_R = 0 \Rightarrow Ha\frac{d\rho_R}{da} + 4H\rho_R = 0 \\ &\Rightarrow \frac{d\ln\rho_R}{d\ln a} = -4 \Rightarrow \rho_R \propto a^{-4} \Rightarrow T^4 \propto a^{-4}, \end{aligned} \tag{2.311}$$

which gives $m = 1$. We can then write

$$\frac{dY}{dT} = -M_P \sqrt{\frac{3}{\alpha}} \frac{R(T)}{T^6}. \tag{2.312}$$

If one defines $R(T) = \frac{T^{n+6}}{\Lambda^{n+2}}$, Λ representing the BSM scale, one obtains

$$Y(T) = \int_T^{T_{RH}} dY \Rightarrow Y(T) = \sqrt{\frac{3}{\alpha}} \frac{M_P T_{RH}^{n+1}}{(n+1)\Lambda^{n+2}}$$

$$\Rightarrow n(T) = T^3 \times Y = \sqrt{\frac{3}{\alpha}} \frac{M_P T_{RH}^{n+1}}{(n+1)\Lambda^{n+2}} T^3. \tag{2.313}$$

We integrated from T up to T_{RH} because before T_{RH} the Universe is not anymore radiation dominated and one should then be careful in the treatment of the energy conservation equations, which deals with matter + radiation components as we will see below. From Eq. (2.313), one can deduce the relic abundance of a dark matter candidates of mass m_χ nowadays ($T = T_0$):

$$\Omega = \frac{n(T_0) \times m_\chi}{\rho_c^0} \Rightarrow \Omega h^2 = 1.6 \times 10^8 \frac{g_0}{g_{RH}} \sqrt{\frac{3}{\alpha_{RH}}} \frac{M_P}{(n+1)} \frac{T_{RH}^{n+1}}{\Lambda^{n+2}} \left(\frac{m_\chi}{1\text{ GeV}}\right), \tag{2.314}$$

where $g_0 = 3.91$ and $g_{RH} = 106.75$ in the Standard Model, which gives

$$\Omega h^2 \simeq 4 \times 10^{-3} \left(\frac{m_\chi}{(n+1)\text{ GeV}}\right) \left(\frac{T_{RH}}{10^{10}\text{ GeV}}\right)^{n+1} \left(\frac{10^{13}\text{ GeV}}{\Lambda}\right)^{n+2} \times 10^{11-3n}. \tag{2.315}$$

As we defined n from $R(T) = \frac{T^{6+n}}{\Lambda^{n+2}}$, we can approximate $\langle \sigma v \rangle \propto \frac{T^n}{\Lambda^{n+2}}$. In other words, the exchange of a massive virtual field corresponds to $n = 2$, equivalent to the exchange of a massless field with two (mass) reduced couplings. $n = 4$ corresponds to one reduced coupling and a massive exchanged fields, whereas $n = 6$ is the case of the exchange of a massive field and two reduced couplings, like in the gravitino case in high-scale SUSY. Indeed, the couplings of the gravitino is Planck-mass suppressed and the cross section is suppressed by the exchange of very massive supersymmetric particles. For $n = 6$, for instance, we obtain

$$\Omega h^2 = 0.112 \left(\frac{m_\chi}{7 \times 10^7\text{ GeV}}\right) \left(\frac{T_{RH}}{10^{10}\text{ GeV}}\right)^7 \left(\frac{10^{13}}{\Lambda}\right)^8. \tag{2.316}$$

It is interesting to note several points before developing further on the reheating process. First of all, the dependence on the reheating temperature (to the power 7) is very important. It means that a small change in the temperature can affect drastically the abundance. On the other way, it also means a weak dependence on the dark matter mass, as it can be easily corrected by the reheating temperature. For instance, a change of 10^7 in the dark matter mass corresponds "only" to one order of magnitude in the reheating temperature. That is also an interesting remark:

natural values of reheating temperatures of the order of 10^{10} lead to heavy dark matter (around the PeV scale). That can also be understood easily by the fact that producing dark matter in such circumstances is not so easy. The feeble amount of matter is thus compensated by heavier candidates.

2.5.1.2 Matter Dominated Universe

When the Universe is matter dominated, by any massive field, the entropy is still conserved (locally and globally). However, the dependence of H on T is different, due to the Friedmann equation

$$H^2 = \frac{\rho}{3M_P^2} = \frac{\rho_m}{3M_P^2}, \quad \text{with } \rho_m = \frac{N \times M}{a^3}. \tag{2.317}$$

In Eq. (2.317), ρ_m is the dominant density of the Universe, given by N particles of mass M. That is of course an interpretation: replacing $N \times M$ by any mass scale M does not change anything in the following. To obtain the dependence of ρ_m on a, we just solved the equation of conservation of energy:

$$\frac{d}{dt}\rho_m + 3H\rho_m = 0 \ \Rightarrow \dot{a}\frac{d}{da}\rho_m + 3\frac{\dot{a}}{a}\rho_m = 0 \ \Rightarrow \rho_m \propto a^{-3}. \tag{2.318}$$

The dependence of $a = f(T)$ is the same as in the radiation dominated case. Indeed, we can see the system massive particle + radiation as two independent systems, which do not discuss together, isolated. The entropy is then conserved individually in the two sectors (which was not the case in the previous section when dealing with a Universe with an energy budget dominated by a decaying massive inflaton). We can say that the total entropy S is the sum of a "matter" entropy S_m and a "radiation" entropy S_R, $S = S_m + S_R$. The entropy of a massive particle is zero, so $S = S_R = T^3 \times a^3 = \text{constant} \ \Rightarrow \ a \propto T^{-1}$. An equivalent way is to solve the energy conservation for the radiation

$$\frac{d\rho_R}{dt} + 4H\rho_R = 0 \ \Rightarrow \ Ha\frac{d\rho_R}{da} + 4H\rho_R = 0 \ \Rightarrow \ \rho_R \propto a^{-4} \ \Rightarrow \ T \propto a^{-1},$$

as we did in the radiation dominated Universe. The main difference between a matter dominated and radiation dominated Universe is then only the liked between the time t and the scale factor a through H in the Friedmann equation but not between the temperature T and the scale factor a. We then obtain for the Yield Y

$$\frac{dY}{dT} = -\frac{R(T)}{3HT^4} = -\frac{R(T)M_P a^{3/2}}{\sqrt{3N \times M}T^4} \propto -\frac{M_P}{\sqrt{3N \times M}}\frac{R(T)}{T^{11/2}}. \tag{2.319}$$

Considering as above,[49] $R(T) = \frac{T^{n+6}}{\Lambda^{n+2}}$, we can solve the preceding equation, which gives

$$Y(T) = \frac{M_P}{\sqrt{3N \times M}(n+3/2)} \frac{T_{RH}^{n+3/2}}{\Lambda^{n+2}} \Rightarrow \tag{2.320}$$

$$\Omega h^2 = 0.1\sqrt{\frac{5 \times 10^{13}\ \text{GeV}}{N \times M}} \left(\frac{m_\chi}{(n+3/2)\ \text{GeV}}\right)\left(\frac{T_{RH}}{10^{10}\ \text{GeV}}\right)^{n+3/2} \times \left(\frac{10^{13}\ \text{GeV}}{\Lambda}\right)^{n+2},$$

which gives, in the case $n = 6$,

$$\Omega h^2 = 0.1\sqrt{\frac{5 \times 10^{13}\ \text{GeV}}{N \times M}} \left(\frac{m_\chi}{7.5\ \text{TeV}}\right)\left(\frac{T_{RH}}{10^{10}\ \text{GeV}}\right)^{15/2} \left(\frac{10^{13}\ \text{GeV}}{\Lambda}\right)^{8}. \tag{2.321}$$

However, in the case of the inflaton, the Universe is not simply matter dominated. It is dominated by a particle (the inflaton), which decays into relativistic SM particles (the radiation). One should then deal with coupled energy conservation equations (2.207) to obtain the value of m and of $H(T)$.

References

1. R. d'Inverno, *Introducing Einstein's Relativity* (Clarendon, Oxford, 1992), 383 p.
2. J.B. Hartle, *Gravity: An Introduction to Einstein's General Relativity* (Pearson, London, 2002), 616 p.
3. S. Weinberg, *Gravitation and Cosmology: Principles and Applications of the General Theory of Relativity* (John Wiley and Sons, Inc., Hoboken, 1972), 657 p.
4. E. Komatsu et al. [WMAP Collaboration], Astrophys. J. Suppl. **192**, 18 (2011). [arXiv:1001.4538 [astro-ph.CO]]
5. I.L. Shapiro, [arXiv:1611.02263 [gr-qc]]
6. D.S. Gorbunov, V.A. Rubakov, Introduction to the theory of the early universe: Cosmological perturbations and inflationary theory. https://doi.org/.1142/7874
7. L. Kofman, A.D. Linde, A.A. Starobinsky, Phys. Rev. D **56**, 3258–3295 (1997). https://doi.org/10.1103/PhysRevD.56.3258 [arXiv:hep-ph/9704452 [hep-ph]]
8. V. Mukhanov, *Physical Foundations of Cosmology* (Cambridge University Press).
9. L.A. Kofman, astro-ph/9605155

[49] And integrating the coefficient of proportionality in the definition of Λ by simplicity.

A Thermal Universe $[T_{RH} \rightarrow T_{CMB}]$

3

Abstract

Once the Universe has terminated its inflationary phase, and produced a thermal bath through the Inflaton decay, its evolution follows the evolution of a classical plasma in an expanding Universe. The law undergoes (relativistic) statistical laws and one can apply our knowledge of this field to the production and decoupling of elements, from neutrino to dark matter. We propose in this chapter to review in details the thermal evolution of the primordial plasma, and the possibility to produce weakly interacting massive particles (WIMP) from it.

3.1 Thermodynamics

3.1.1 A Brief Thermal History of the Universe in Some Dates and Numbers

The history of the Universe is rich and complex and depends strongly on its temperature. In this section, we will give a quick snapshot of the thermal history of the Universe that will be developed through the chapter. Each phase will be studied in detail later on. Let us begin by recalling some temperatures, possibly realized in the hot Universe, the related cosmic times, and the connection with microscopic physics at the corresponding energies:

- $T \simeq 0.1\text{eV}$ [$t \simeq 10^{13}$ s $\simeq$ 380,000 years]
  Light nuclei and electrons form neutral atoms and the Universe becomes transparent to photons. They decouple from the plasma (see Sect. 3.3.4) and are observable today as cosmic microwave background (CMB).

Y. Mambrini, *Particles in the Dark Universe*,
https://doi.org/10.1007/978-3-030-78139-2_3

- $T \simeq 0.1 - 10\,\text{MeV}$ [$t \simeq 10^2 - 10^{-2}\,\text{s}$]
  Light nuclei are formed from protons and neutrons (primordial nucleosynthesis, BBN) and neutrinos decouple from the plasma (see Sect. 3.3.1).
- $T \simeq 10\,\text{GeV}$ [$t \simeq 10^{-8}\,\text{s}$]
  Weakly interacting massive particles (WIMPs), the most popular dark matter candidates, decouple from the plasma (see Sect. 3.5).
- $T \simeq 100\,\text{GeV}$ [$t \simeq 10^{-10}\,\text{s}$]
  The Higgs vacuum expectation value forms, and all Standard Model particles become massive. Baryon and lepton number changing "sphaleron process" are no longer in thermal equilibrium.
- $T \simeq 10^8 - 10^{11}\,\text{GeV}$ [$t \simeq 10^{-22} - 10^{-28}\,\text{s}$]
  Baryogenesis via leptogenesis takes place and gravitino dark matter or particles coupling via GUT interactions to the Standard Model (see Sect. 3.6) can be thermally produced.
- $T \simeq 10^{12}\,\text{GeV}$ [$t \simeq 10^{-30}\,\text{s}$]
  This corresponds to the reheating process, where the thermal bath is created as inflaton decay modes, perturbatively or through parametric resonance (see Sect. 2.4).

This little thermal history is summarized in Figs. 3.1 and 3.2. Nowadays, the known Universe is mainly composed by baryonic matter, photons, and neutrinos, while the unknown components are dark matter and dark energy, which involve physics beyond the Standard Model (BSM). The content of the Universe in particles

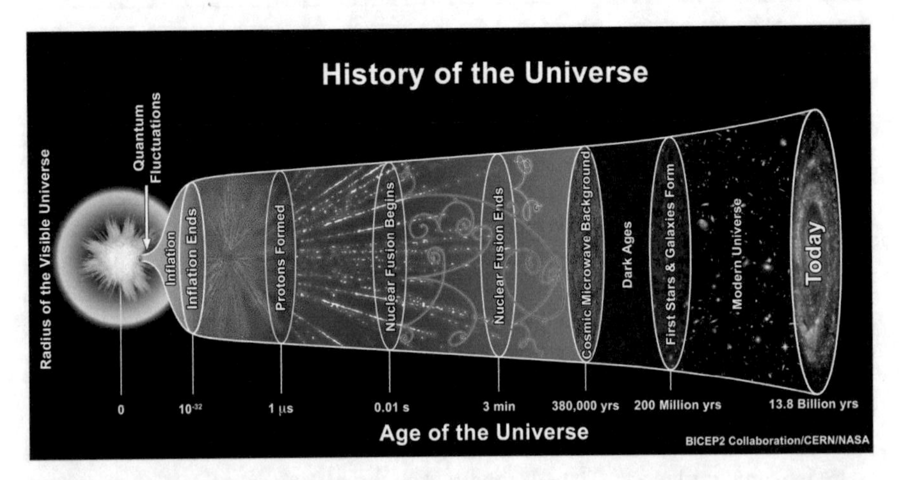

Fig. 3.1 Epochs of the hot early Universe with their cosmic times scale (The Astronomy Bum, under the Creative Commons CC0 1.0 Universal Public Domain Dedication)

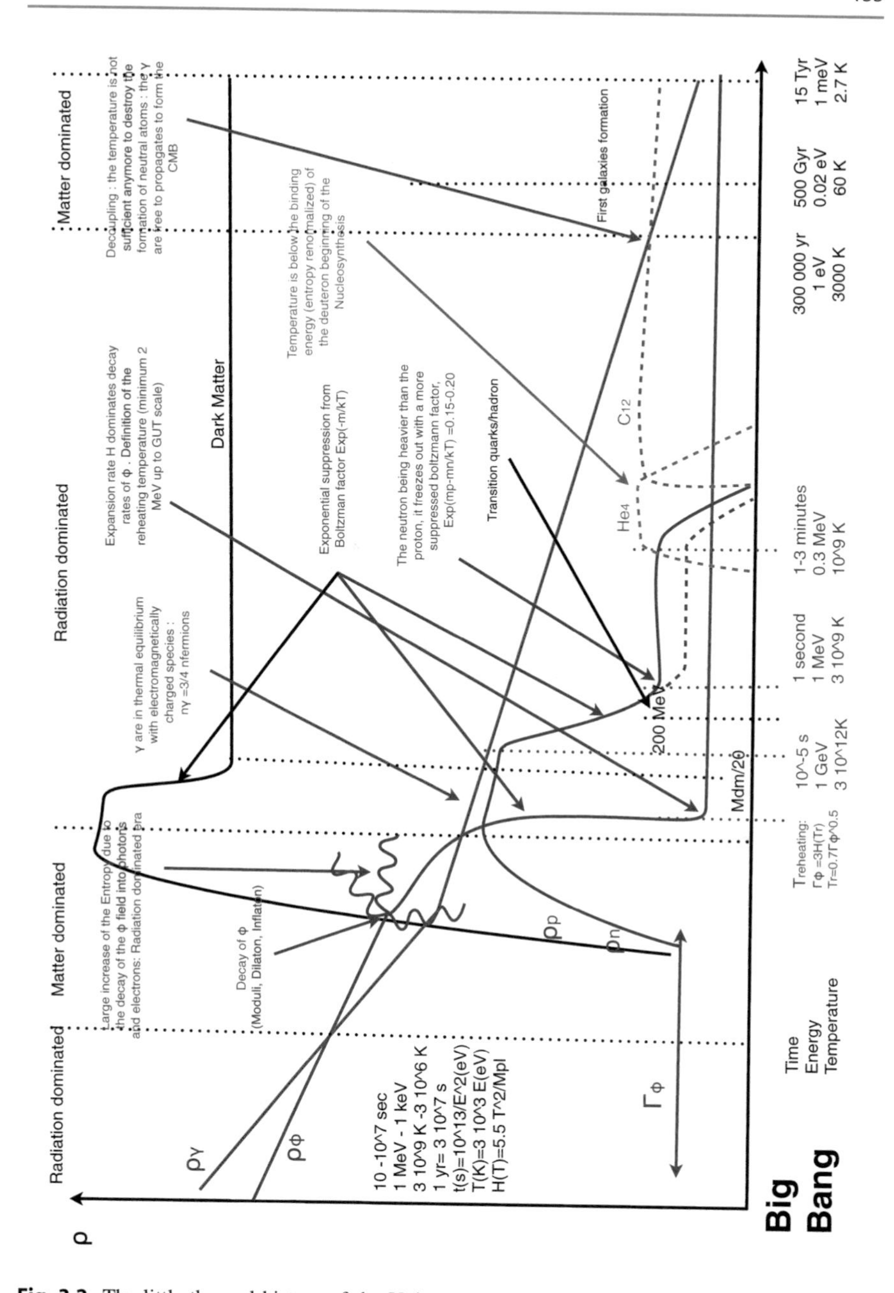

Fig. 3.2 The little thermal history of the Universe with each process described in the following sections

of the Standard Model is known thanks to a set of observables ranging from Cosmic Microwave Background (CMB) measurements to interstellar gas data. The baryon density, ρ_b, is obtained from a combination of CMB data and primordial deuterium abundance extracted from the absorption spectrum of high-redshift quasars. The mean value is

$$\Omega_b h^2 \simeq 0.022 = \frac{\rho_b}{\rho_c^0} h^2 \tag{3.1}$$
$$\Rightarrow \rho_b \simeq 2.3 \times 10^{-7}\ \mathrm{GeV/cm^3} = 2.46 \times 10^{-7} m_p/\mathrm{cm}^3.$$

m_p being the proton mass and where we used $\rho_c^0 = 1.05 \times 10^{-5} h^2\ \mathrm{GeV/cm^3}$, see Eq. (2.56). This implies a baryonic density

$$n_b = 2.46 \times 10^{-7}\ \mathrm{cm}^{-3}, \tag{3.2}$$

which corresponds, roughly to 0.25 baryons per m^3, or 10^{11} solar mass per Mpc^3. As we will see, whereas baryons dominate around the Earth, it is really subdominant at larger scale. The most abundant species in interstellar space is the photon. As we have seen, its density n_γ is intimately related to the temperature of the thermal bath.[1] This will be explicitly calculated later, see Eq. (3.27):

$$n_\gamma = \frac{2\zeta(3)}{\pi^2} T_\gamma^3 = 411\ \mathrm{cm}^{-3}, \tag{3.3}$$

where we took $T = 2.725\ K = 2.39 \times 10^{-4}$ eV. Combining (3.2) and (3.3) we obtain

$$\boxed{\eta = \frac{n_b}{n_\gamma} = 6 \times 10^{-10},} \tag{3.4}$$

very near from the measured value of η given by PLANCK collaboration,

$$\eta_{PLANCK} = 6.12 \times 10^{-10}. \tag{3.5}$$

[1] By default when one talks about temperature, T represents the photon temperature T_γ, or more generically, the temperature of the thermal bath.

This means that for each baryon, we currently find on average 2×10^9 photons in the Universe. The other Standard Model particle which coexists with the photons and baryons in the interstellar medium is the neutrino. Its density follows the same law than the photon, but with a different temperature (see Sect. 3.3.1):

$$T_\nu = \left(\frac{4}{11}\right)^{1/3} T_\gamma = 1.96\,\text{K}\,(1.7 \times 10^{-4}\,\text{eV}) \;\Rightarrow\; n_\nu = \frac{\frac{3}{2}}{2}\frac{4}{11} n_\gamma \simeq 112\,\text{cm}^{-3},$$

where the first ratio corresponds to the difference between the degrees of freedom[2] of a Majorana (fermionic) neutrino ($\frac{3}{4} \times 2$) and a (bosonic) photon (2). This corresponds roughly to 1/3 neutrino per photon. One could also multiply by 3 to take into account the three neutrino flavors.

The last component is the dark matter, with a global density measured by Planck

$$\Omega_{dm} h^2 \simeq 0.1 \;\Rightarrow\; n_{DM} \simeq 10^{-6} \left(\frac{1\,\text{GeV}}{M_{dm}}\right) \text{cm}^{-3}, \tag{3.6}$$

which is globally 5 times more than the baryonic one (3.2), and a local one measured by astrophysical observations

$$n_{DM} \simeq 0.3 \left(\frac{1\,\text{GeV}}{M_{dm}}\right) \text{cm}^{-3}. \tag{3.7}$$

Let us go into the details to understand how the Standard Model of cosmology combined with the Standard Model of particle physics to explain these numbers.

3.1.2 Statistics of Gas, Pressure, and Radiation: The Classic Case

Most of the matter in the Universe exists in gaseous form. As we just discussed, a fraction ($\sim$20%) of it is baryonic matter, like what we are familiar with in everyday experience, and a larger fraction ($\sim$80%) is non-baryonic dark matter, the nature of which we try to identify in this chapter. We infer the presence of dark matter through its gravitational interactions, all visible matter are therefore still baryonic. The baryonic matter of the Universe consists mainly of Hydrogen (about three quarter of the mass) and Helium (about a quarter). There is a small fraction of heavier elements which, collectively, are referred to as metals. The abundance of metals in the solar neighborhood is about 2% by mass, while the Hydrogen abundance is 71%. Table 3.1 shows the mass fraction of several elements in the solar neighborhood. As we saw in the previous chapter, a keypoint to understand the dynamic and the evolution of a gas is to know the dependency of its essential parameters, i.e. pressure

[2]The computation of the degrees of freedom will be detailed in Sect. 3.1.5.

Table 3.1 Mass fraction of different elements in the Solar neighborhood

H	0.71
He	0.28
C	0.34×10^{-2}
N	0.99×10^{-3}
O	0.96×10^{-2}
Ne	0.18×10^{-2}
Na	0.35×10^{-4}
Mg	0.66×10^{-3}
Al	0.56×10^{-4}
Si	0.70×10^{-3}
S	0.3×10^{-3}
Cl	0.47×10^{-5}
Ar	0.11×10^{-3}
Ca	0.65×10^{-4}
Cr	0.18×10^{-4}
Fe	0.13×10^{-2}
Co	0.36×10^{-5}
Ni	0.73×10^{-4}

and distribution function of the energy density in the system. We will first see how to calculate these quantities in a classical system.

The equation of state of a gas is the relation connecting its variables like pressure, density, temperature, or internal energy. We will first concentrate on the case where the energy of mutual interactions between particles are negligible. This type of gas is composed by particles whose kinetic distribution follows random motion, a kinetic energy E_k being associated with a momentum $p(E_k)$, the total energy being written $E = E_0 + E_k$ with $E_0 = mc^2$ the rest energy. We remind the reader that the pressure is defined as the rate of momentum transfer in a given direction through a unit of area per unit time, and since the direction of the momentum is randomly distributed in three dimensions,[3] the pressure P is given by

$$P = \frac{1}{3} \int_0^\infty f(E)p(E)v(E)dE, \tag{3.8}$$

where $f(E)dE$ is the number of particles with kinetic energy between E and $E + dE$ in a gas unit volume, and $v(E)$ is the velocity associated with the energy E. To simplify, it will be assumed that the particles in the gas do not have internal degrees of freedom (like spin or charges), and that the kinetic energy is entirely due to the

[3] For a more technical explanation of the factor $\frac{1}{3}$ originated from an integration of $\cos^2\theta$ (θ being the scattering angle), see Eq. (2.203).

random translational motion.[4] If the random motion in the gas is non-relativistic, then we can write $v = p/m$ and $E_k = p^2/2m$, resulting in

$$P = \frac{2}{3}\int_0^\infty f(E)E_k dE = \frac{2}{3}u, \tag{3.9}$$

where

$$u = \int_0^\infty f(E)E_k dE \tag{3.10}$$

is the internal energy density. This corresponds to the total kinetic energy of all the particles per unit volume of the gas. If the particles are relativistic, $v = c$ and $E_k = pc$, giving

$$P = \frac{1}{3}\int_0^\infty f(E)E_k dE = \frac{1}{3}u. \tag{3.11}$$

Notice that the two relations $P = \frac{2}{3}u$ and $P = \frac{1}{3}u$ for the two limits are very general and do not depend on the details of the distribution function $f(E)$.

Exercise Using $E^2 = (mc^2)^2 + (pc)^2$, recover the values for E_k in the two extreme cases (relativistic and non-relativistic). Compare then the $P = f(\rho)$ relations obtained classically above with those calculated in the treatment of the inflaton in Eq. (2.139). Which value of k corresponds to a relativistic inflaton? A non-relativistic inflaton? Are the relations identical in both cases? If not, why?

The coefficient of proportionality between P and u has a physical meaning: it is equal to $\gamma - 1$, where γ, called the "adiabatic index" (Eq. A.115), corresponding to the ratio of specific heats for a thermal gas. Reversible adiabatic processes yield PV^γ =constant,[5] (Eq. A.114) where $V \propto R^3$ is the volume of the gas. As a consequence one finds that for a specific adiabatic expansion, the total energy $U = uV$ of a gas being proportional to $\frac{1}{V^w}$ drops as R^{-2} for a non-relativistic gas and as R^{-1} for a relativistic gas, where $V \propto R^3$. Another way to obtain the same result is to note that the energy of a relativistic species is $U_{rel} = pc$ and scales as R^{-1}, whereas the energy of non-relativistic species is $U_{non-rel} = \frac{1}{2}mv^2$ and scales as R^{-2}.

[4]This assumption is obviously not valid in the case of the Standard Model, since all its particles have internal degrees of freedom.

[5]See Sect. A.5 for the demonstration of this law called Laplace's law.

At this point, let us recall what we mean by a "thermal gas." For a classical gas, this means that all energy levels of the gas, both discrete and continuous, are occupied according to the Boltzmann distribution:

$$N(E) \propto g(E)e^{-E/kT},$$

where $g(E)$ is the so-called density of states. In case of quantum statistics, the corresponding distribution is

$$N(E) \propto \frac{g(E)}{e^{(E-\mu)/kT} \pm 1}$$

μ being the chemical potential and where the positive sign in the denominator corresponds to a Fermi gas and the negative sign to a Bose gas. Quantum statistics comes into play only when the number of particles per phase space cell of volume h^3 is of order unity. For dilute gases, as encountered in most astrophysical situations, classical description is quite adequate, whereas in the case of the dense primordial plasma in the early Universe, a complete detailed quantum analysis is necessary.

3.1.3 Statistics of Gas, Pressure, and Radiation: The Quantum Case

3.1.3.1 Distribution Functions and Thermodynamics Quantities

From elementary quantum mechanics we can remember the "particle in the box" interpretation of the phase space occupation number. Let suppose a cubic box of length L (volume $V = L^3$) with periodic boundary conditions. Solving the Schrodinger equation to determine the energy and momentum eigenstates, we obtain a discrete set of momentum eigenvalues

$$\mathbf{p} = \frac{h}{L}(n_x\mathbf{i} + n_y\mathbf{j} + n_z\mathbf{k}), \quad n_i = 0 \pm 1, \pm 2, \ldots, \tag{3.12}$$

where $h = 4.14 \times 10^{-24}$ GeV is the Planck's constant. The density of states in momentum space $\mathbf{p}$ (i.e. the number of states per $\Delta p_x \Delta p_y \Delta p_z$) is thus[6]

$$\frac{L^3}{h^3} = \frac{V}{h^3},$$

[6]Another way to understand it is to imagine a square box made of sides corresponding to $n_i = 0$ and $n_i = 1$. Each side of the box has a size of h/L, so the box has a volume of h^3/L^3. However, each of the eight corners of the box corresponds to a state, pertaining to 8 boxes surrounding it, so counting as 1/8 of states par box. Then, a box of 8 corners contains $8 \times 1/8 = 1$ state in its h^3/L^3 volume. In other words, $\Delta p_x \Delta p_y \Delta p_z/(h^3/L^3)$ represents the number of states [number of boxes] having momentum $\mathbf{p}$ in the range $(\Delta p_x, \Delta p_y, \Delta p_z)$.

and the state density in phase space $\mathbf{x}, \mathbf{p}$ is $1/h^3$. If the particle has g internal degrees of freedom (e.g. spin), then the density of states becomes

$$\frac{g}{h^3} = \frac{g}{(2\pi)^3},$$

where in the second equality we used natural units with $\hbar = h/(2\pi) = 1$. To obtain the number density n of a gas of particles, we need to know how the momentum eigenstates are distributed. This information is contained in the phase space distribution function $f(\mathbf{x}, \mathbf{p}, t)$. Because of homogeneity, the distribution function should, in fact, be independent of the position $\mathbf{x}$. Moreover, at such early times isotropy requires that the momentum dependance is only in terms of the magnitude of the momentum $p = |\mathbf{p}|$. We will typically leave the time dependance implicit: it will manifest itself in terms of the temperature dependance of the distribution function. The particle density in phase space is then the density of states multiplied by the distribution function:

$$\frac{g}{(2\pi)^3} \times f(\mathbf{p}). \tag{3.13}$$

The *number density* of particles (in real space) is found by integrating (3.13) over momentum

$$n = \frac{g}{(2\pi)^3} \int d^3p f(p). \tag{3.14}$$

To obtain the energy density of the gas of particles, we have to weight each momentum eigenstates by its energy. To a good approximation, the particles in the early Universe were *weakly interacting*. This allows us to ignore the interaction energies between the particles and write the energy of a particle of mass m and momentum p simply as

$$E(p) = \sqrt{m^2c^4 + p^2c^2}. \tag{3.15}$$

Integrating the product of (3.13) and (3.15) over momentum then gives the *energy density*

$$\rho = \frac{g}{(2\pi)^3} \int d^3p f(p) E(p). \tag{3.16}$$

And, just like we did in Sect. 3.1.2, we can define the pressure in a quantum gas as

$$P = \frac{g}{(2\pi)^3} \int d^3p f(p) \frac{p^2}{3E}, \tag{3.17}$$

where we rewrote Eq. (3.8) using the relation $v_i = \frac{p_i}{E}$. Notice that we recover the expressions (3.9) and (3.11) for $E = p$ and $E = p^2/2m$, respectively.

3.1.4 In the Primordial Plasma

The original gas, denoted as "*Ylem*" by Gamow himself, exists under the form of a complete ionized plasma of elementary particles. As far as their kinetic energy is dominant on their potential energy, we can suppose that the primordial gas is perfect.[7] Moreover, the interaction rates between all these particles are often much larger than the expansion rate of the Universe driven by the Hubble constant H. The situation is then very different from the previous chapter, where at times around the inflationary period, the expansion played an important role on the distribution functions and dark matter production. After the reheating, the Universe entered in a phase of thermodynamic equilibrium, thermal *and* chemical equilibrium. We dedicate a complete Sect. 3.2 on the subject of the equilibriums coexisting in a hot gas, but we want in this chapter to give the main ideas. From now on, to simplify the presentation of the results, one will consider $c = 1$.

The thermodynamical evolution of a complex system, composed of multiple particles, is expressed in terms of the chemical potentials μ_i for each of the particles i. If one considers the reactions between particles A_i and B_i

$$A_1 + A_2 + \ldots + A_n \leftrightarrow B_1 + B_1 + \ldots + B_m, \tag{3.18}$$

then the *chemical* equilibrium condition is written

$$\mu_{A_1} + \mu_{A_2} + \ldots + \mu_{A_n} = \mu_{B_1} + \mu_{B_2} + \ldots + \mu_{B_m}. \tag{3.19}$$

This relation reflects the fact that the rate of production of particles "B" is the same as the rate of destruction of these particles. Therefore, the concentration of the species A and B does not vary over time. A system in *thermal* equilibrium is a system where the temperature within the system is spatially uniform and kept temporally constant thanks to efficient scatterings of particles interacting in the plasma. Processes involved in thermal equilibrium do not change the nature or the number of species. For instance:

- The plasma is in *thermal* equilibrium under collisional effect of the type:

$$e + \gamma \to e + \gamma. \tag{3.20}$$

[7] It is important to note that this hypothesis will no longer be valid when it comes to the quark-hadrons phase transition, where the strong interactions confine quarks inside protons, neutrons, or pions.

- The *chemical* equilibrium is realized through reactions like

$$3\gamma \leftrightarrow e^+e^- \leftrightarrow 2\gamma. \tag{3.21}$$

From the last reaction, one can immediately deduce that the chemical potential of photon is null ($3\mu_\gamma = 2\mu_\gamma$) and that the chemical potential of positron and electrons are opposite ($\mu_{e^-} = -\mu_{e^+}$). Moreover, from the very tiny ratio of the baryon to photon ratio of today ($n_B/n_\gamma \simeq 6\times 10^{-10}$), one deduces that $\mu_{e^-} = \mu_{e^+}$. The two previous hypotheses imply that $\mu_{e^-} = \mu_{e^+} = 0$ and one can describe the primordial plasma as a group of bosonic and fermionic population at temperature T with *null* chemical potential. In this case, for a particle A, in a more general context its statistic (homogeneous) distribution is given by

$$f_A(\mathbf{p}) = \frac{g_A}{e^{(E-\mu_A)/kT} \pm 1} \tag{3.22}$$

with $E^2 = m^2c^4 + p^2c^2$ and ± 1 correspond to fermionic (+) or bosonic (−) statistic. g_A is the internal (spin, helicity, polarization, color factors. . .) degree of freedom of the particle A (2 for a fermion, 3 for a massive vector, 2 for a photon. . .). The number density $n_A(T)$ and energy density $\rho_A(T)$ of a species A is then given by

$$\begin{aligned} n_A(T) &= \frac{g_A}{(2\pi)^3}\int f_A(\mathbf{p})d^3\mathbf{p} = \frac{g_A}{2\pi^2}\int_{m_A}^{\infty}\frac{(E^2-m_A^2)^{1/2}}{e^{(E-\mu_A)/kT}\pm 1}EdE \\ &= \frac{g_A}{2\pi^2}\int_{m_A}^{\infty}\frac{p^2dp}{e^{\left(\sqrt{p^2+m_A^2}-\mu_A\right)/kT}\pm 1} \end{aligned} \tag{3.23}$$

$$\begin{aligned} \rho_A(T) &= \frac{g_A}{(2\pi)^3}\int E(\mathbf{p})f(\mathbf{p})d^3\mathbf{p} = \frac{g_A}{2\pi^2}\int_{m_A}^{\infty}\frac{(E^2-m_A^2)^{1/2}}{e^{(E-\mu_A)/kT}\pm 1}E^2dE \\ &= \frac{g_A}{2\pi^2}\int_{m_A}^{\infty}\frac{\left(\sqrt{p^2+m_A^2}\right)p^2dp}{e^{\left(\sqrt{p^2+m_A^2}-\mu_A\right)/kT}\pm 1}. \end{aligned} \tag{3.24}$$

In the relativistic limit ($T \gg m$) and for negligible chemical potential ($T \gg \mu_A$) one can extract an analytical expression of 3.24 using the relation

$$\int_0^{\infty} dx\left(\frac{x^n}{e^x-\delta}\right) = \Gamma(n+1)\zeta(n+1)y(\delta) \tag{3.25}$$

with $y(\delta) = 1$ if $\delta = 1$ and $\left(1 - \frac{1}{2^n}\right)$ if $\delta = -1$. Γ is the Euler function ($\Gamma(z+1) = z\Gamma(z)$) and $\zeta(s)$ is the zeta function defined by

$$\zeta(s) = \sum_{n=1}^{\infty} \frac{1}{n^s}, \quad [\zeta(0) = -1/2; \zeta(1) = \infty; \zeta(2) = \pi^2/6; \zeta(3) \simeq 1.2; \zeta(4) = \pi^4/90; ..].$$

We give a list of very useful integrals in the Appendix A.6.4.

Integrating Eq. (3.24) gives

$$\rho_A(T)_{T \gg m_A, \mu} = \frac{\pi^2}{30} g_A T^4 \quad \text{(bosons)}$$

$$\rho_A(T)_{T \gg m_A, \mu} = \frac{7}{8} \frac{\pi^2}{30} g_A T^4 \quad \text{(fermions)} \tag{3.26}$$

$$n_A(T)_{T \gg m_A, \mu} = \frac{\zeta(3)}{\pi^2} g_A T^3 \quad \text{(bosons)}$$

$$n_A(T)_{T \gg m_A, \mu} = \frac{3}{4} \frac{\zeta(3)}{\pi^2} g_A T^3 \quad \text{(fermions).} \tag{3.27}$$

For non-relativistic species (when the temperature of the plasma approaches the mass of a particle A), the Boltzmann factor dominates the denominator in Eq. (3.24), making bosonic and fermionic distribution identical. Developing (3.23) for $p^2/m_A^2 \ll 1$ and using $\int_0^\infty x^2 e^{-ax^2} = \frac{1}{4}\sqrt{\frac{\pi}{a^3}}$ we obtain

$$n_A(T)_{T \ll m_A} = g_A \left(\frac{m_A T}{2\pi}\right)^{3/2} e^{-(m_A - \mu_A)/T} \tag{3.28}$$

$$\rho_A(T)_{T \ll m_A} = g_A m_A \left(\frac{m_A T}{2\pi}\right)^{3/2} e^{-(m_A - \mu_A)/T}. \tag{3.29}$$

One can then compute the mean energy *per particle* for non-degenerate relativistic species, $\langle E \rangle = \frac{\rho}{n}$:

$$\langle E \rangle_{T \gg m_A}^{per\ particle} = \frac{\pi^4}{30\zeta(3)} T \simeq 2.7\, T \quad \text{(bosons)}$$

$$\langle E \rangle_{T \gg m_A}^{per\ particle} = \frac{7\pi^4}{180\zeta(3)} T \simeq 3.2\, T \quad \text{(fermions).} \tag{3.30}$$

Taking concrete values, we can estimate the energy density of a gas composed of relativistic particles at a temperature T. For that we need to reintroduce the fundamental numbers c and h and multiply the expression (3.26) by $\frac{k^4}{\hbar^3 c^3} \simeq 1.15 \times 10^{-21}\text{J.cm}^{-3}.K^{-4}$ for the energy density and $(k/\hbar c)^3$ for the number density (3.27). Remembering that $g_\gamma = 2$, we obtain

$$\rho_\gamma(T) \simeq 9 \times 10^{40}\ \text{GeV.cm}^{-3} \left(\frac{T}{\text{GeV}}\right)^4 ;$$

$$n_\gamma(T) \simeq 20 \left(\frac{T}{\text{K}}\right)^3 \simeq 3.3 \times 10^{40} \left(\frac{T}{\text{GeV}}\right)^3 .$$

When the particle is non-relativistic, but still having some kinetic energy, it is an intermediate regime. If we develop $E = \sqrt{m^2 + p^2} \simeq m + p^2/2m$ in the non-relativistic limit, integrating on $\mathbf{p}$ and using $\int_0^\infty dx x^4 e^{-ax^2} = \frac{3}{2^3 a^2}\sqrt{\frac{\pi}{a}}$, we can obtain the kinetic energy of a gas of non-relativistic particles:

$$\begin{aligned} \langle E_c \rangle_{T \ll m_A} &= \frac{3 g_A}{16} \left(\frac{2 m_A T}{\pi}\right)^{3/2} T\, e^{-m_A/T} \\ \Rightarrow \langle E_c \rangle_{T \ll m_A}^{\text{per particle}} &= \frac{\langle E_c \rangle}{n_A(T)_{T \ll m_A}} = \frac{3}{2} T. \end{aligned} \tag{3.31}$$

Gathering the mass and kinetic energy we finally obtain

$$\langle E \rangle_{T \ll m_A}^{per\ particle} = m + \frac{3}{2} T, \tag{3.32}$$

which is obviously the same for a boson or a fermion because the exponential suppression is the dominant contributor to the density and it washes up the spin-statistics differences. In other words, one can say that, in average, a relativistic particle has only kinetic energy (by definition) of 2.7 T if it is a boson, or 3.2 T if it is a fermion. On the other hand, if the particle is non-relativistic, its average kinetic energy is 1.5 T (fermion or boson).

Another possibility (especially if one needs analytical solutions) is to use the Boltzmann distribution instead of the Fermi-Dirac or Bose-Einstein one. Indeed, in regimes where $(T \ll E, \mu_A)$, $f_A(\mathbf{p}) \simeq g_A e^{-\frac{E}{T}} = f_{\text{Boltzmann}}$. In the Boltzmann approximation ($e^{\frac{E}{T}} \gg 1$), one can have an analytical solution of n_{eq} of a population of particles with g_A internal degrees of freedom

$$n_{eq} \sim g_A \int \frac{d^3p}{(2\pi)^3} e^{-\frac{E}{T}} = \frac{g_A}{2\pi^2} \int |p| E dE e^{-\frac{E}{T}} \tag{3.33}$$

$$= \frac{g_A\, m_A^3}{2\pi^2} \int \sqrt{(E/m_A)^2 - 1} \left(\frac{E}{m_A}\right) d\left(\frac{E}{m_A}\right) e^{-\frac{E}{m_A}\frac{m_A}{T}}$$

$$= \frac{g_A\, m_A^3}{2\pi^2} \int \sqrt{z^2 - 1}\, z dz e^{-zx} = \frac{g_A\, m_A^2\, T}{2\pi^2} K_2\left(\frac{m_A}{T}\right) \tag{3.34}$$

$K_2(x)$ being the modified Bessel function of second kind described in Appendix A.6.3. In fact, we can show that this expression is still valid at the order of 20 to 25% for $T \gg E$. Indeed, comparing Eqs. (3.27) and (3.34), a 100 GeV Majorana particle ($g_A = 2$) at a temperature of $T = 10^8$ GeV has a density of 1.8×10^{23} GeV^3, whereas the Boltzmann approximation (3.34) gives $2. \times 10^{23}$ GeV^3.

3.1.5 Degrees of Freedom

3.1.5.1 Computation of $g_\rho(T)$

So far we have considered plasma of specific species. The primordial plasma is a mix between different kind of particles, some are relativistic, some are not. However, since the energy density of relativistic species is much greater than that of no-relativistic ones (Boltzmann suppressed), it suffices to include the relativistic species only.[8] The energy density can thus can be written

$$\rho(T) = \frac{\pi^2}{30} T^4 \sum_{i=all\ species} \frac{30}{\pi^2} \left(\frac{T_i}{T}\right)^4 \frac{g_i}{2\pi^2} \int_{x_i}^{\infty} \frac{(\epsilon^2 - x_i^2)^{1/2} \epsilon^2 d\epsilon}{e^{\epsilon - \mu_i/T} \pm 1} = g_\rho(T) \frac{\pi^2}{30} T^4 \tag{3.35}$$

with $x_i = m_i/T$ and

$$g_\rho(T) = \frac{30}{\pi^2} \sum_{i=all\ species} \frac{g_i}{2\pi^2} \left(\int_{x_i}^{\infty} \frac{(\epsilon^2 - x_i^2)^{1/2} \epsilon^2 d\epsilon}{e^{\epsilon - \mu_i/T} \pm 1} \right) \left(\frac{T_i}{T}\right)^4 \tag{3.36}$$

[8]This is true in the early Universe, but not at a later time when eventually the rest masses of the particles left over from annihilation begin to dominate and we enter a matter dominated era, see Sect. 3.1.6.

which gives with a pretty good approximation, approximating the Boltzmann factor as a step function:

$$g_\rho(T) \simeq \sum_{b=bosons} g_b \left(\frac{T_b}{T}\right)^4 + \frac{7}{8} \sum_{f=fermions} g_f \left(\frac{T_f}{T}\right)^4. \tag{3.37}$$

The relative factor of $\frac{7}{8}$ accounts for the difference in Fermi and Bose statistics. The individual temperatures T_b and T_f are equal to the photon temperature of the bath ($T_\gamma = T$) as long as they are relativistic in equilibrium in the bath. One can interpret g_ρ as being the effective internal degree of freedom of a boson which composes the gas. The degrees of freedom of all the real relativistic species are integrated in g_ρ.

As an example we can compute the total degrees of freedom of a gas in the $SU(3)_c \times SU(2)_L \times U(1)_Y$ Standard Model. The contents are: 6 Dirac (4) quarks colored (3) [$6 \times 4 \times 3$] plus 3 Dirac (4) leptons [3×4] plus 3 Majorana (2) neutrino [3×2] plus 3 massive (3) vector fields Z^0 $W^\pm$ [3×3], 9 non-massive vector fields (2) (1 photon and 8 gluon) [9×2] and finally one real scalar (1) Higgs field [1] which gives

$$g_\rho^{SM} = (3 \times 3 + 9 \times 2 + 1) + \frac{7}{8}(6 \times 4 \times 3 + 3 \times 4 + 3 \times 2) = \frac{427}{4} = 106.75. \tag{3.38}$$

Exercise Show that g_ρ^{SUSY} for the minimal supersymmetric Standard Model (MSSM), where each boson has a fermionic partner and each fermion has a bosonic partner,[9] with the addition of a Higgs doublet and its fermionic part, is

$$g_\rho^{SUSY} = 122 + 122 \times \frac{7}{8} = \frac{915}{4} = 227.75. \tag{3.39}$$

Note that for the Higgs field, one can also do the calculation above the electroweak breaking scale and count 4 degrees of freedom for the Higgs, and only two for each gauge boson (which have not yet "eaten" the degrees of freedom of the Higgs). If one adds a Dirac dark matter candidate, $g_\rho^{SM+DM} = g_\rho^{SM} + \frac{7}{8} \times 4 = 110.25$. We represent in Fig. 3.3 the evolution of the degree of freedom $g_\rho = g_{eff}$ as function of the temperature. We observed that the effective number of relativistic species decreases especially around 200 MeV where the quark/hadrons transition absorbs the up-type, down-type, and gluonic degrees of freedom to form the first

[9]Notice that the 2 degrees of freedom of the gravitino $h_{\mu\nu}$ and the 2 degrees of freedom of its partner the gravitino are not counted here because they do not participate to the thermal bath.

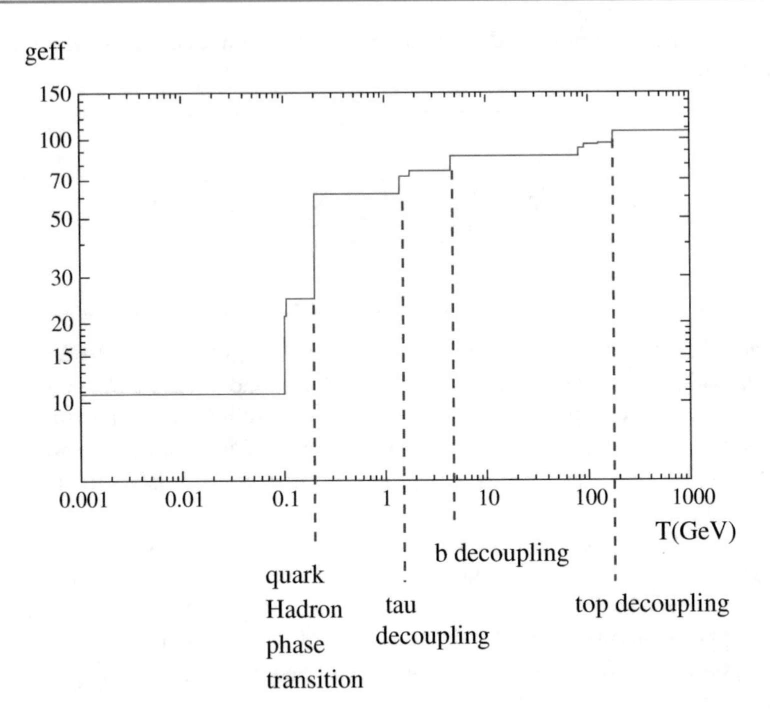

Fig. 3.3 Effective degree of freedom of the primordial plasma as function of the temperature

nucleus (see the Big Bang Nucleosynthesis section for more details) going from 61.75 to 10.75. Below 511 keV, we enter in a region where the neutrino decoupling plays a role.

We can compute the value of g_ρ nowadays, noticing that, a priori the 2 relativistic species still present in the bath are the photons and the neutrinos, even if the neutrinos have a lower temperature T_ν than the photons due to the entropy transferred from the electrons to the photons when they decoupled (see Sect. 3.3.1), $T_\nu = (4/11)^{1/3} T_\gamma = (4/11)^{1/3} \times 2.725 \approx 1.95\,\text{K}$ (1.7×10^{-4} eV). What is left is thus 1 family of neutrino (fermions) with 2 degrees of freedom (SU(2)) and a temperature[10] $T_\nu = 1.95\,\text{K}$ and a photon (boson) with 2 degrees of freedom and a temperature $T_\gamma = T_0 = 2.725\,\text{K}$. We can then extract from Eq. (3.37),

$$g_\rho^{\text{today}} = 2 + \frac{7}{8} \times 2 \times \left(\frac{4}{11}\right)^{4/3} = 2.45 \tag{3.40}$$

[10] At least two neutrino flavors being nowadays non-relativistic from the oscillations measurements, see the text in the box below Eq. (3.90) for a discussion on the subject.

($g_\rho = 3.36$ for 3 neutrino flavors) which implies

$$\rho_{\rm rel} = \frac{\pi^2}{30} g_\rho T_0^4 = 5.9 \times 10^{-34} \text{ g cm}^{-3} \tag{3.41}$$

and

$$n_\gamma^{\rm today} = \frac{2\zeta(3)}{\pi^2} T_0^3 = 411 \text{ cm}^{-3}; \quad n_\nu^{today} = 3 \times \frac{\frac{3}{2}}{2} \frac{4}{11} n_\gamma \simeq 336 \text{ cm}^{-3}. \tag{3.42}$$

3.1.5.2 QCD (Quark–Hadrons) Phase Transition

Before the strange quark had time to annihilate, something else happens: matter undergoes the *QCD phase transition* (also called quark-hadron transition). This takes place at $T \sim 150$ MeV. While quarks are *asymptotically free* (i.e. weakly interacting) at high energies, below 150 MeV, the strong interaction between the quarks and the gluons becomes important. The quarks and gluons then form bound three-quark systems, called *baryons*, and quark-antiquark pairs, called *mesons*, and below this temperature we should treat these as our new degrees of freedom.[11]

The lightest baryons are the proton and the neutron. The lightest mesons are the pions $\pi^\pm$, π^0. Baryons are fermions mesons are bosons. There are many different species of baryons and mesons, but all except the pions are non-relativistic below the temperature of the QCD phase transition. Thus, the only particle species left in large number are the pions, electrons, muons, neutrinos, and photons. The three pions (spin 0) correspond to $g = 3$ internal degrees of freedom. We therefore get $g_\rho = 2 + 3 + \frac{7}{8} \times (4 + 4 + 6) = 17.25$. Soon after the QCD phase transition, the pions and muons annihilate and $g_\rho = 17.25 - 3 - \frac{7}{8} \times 4 = 10.75$. Next electrons and positrons annihilate (see Sect. 3.3.1).

3.1.5.3 A Little History of $g_\rho(T)$: Summary

We show in the Table 3.2 the different values of $g_\rho(T)$ after the annihilation is over assuming the next annihilation would not have begun yet. In reality they overlap in many cases. The temperature on the left is the approximate mass of the particle in question and indicates roughly when the annihilation begins. The temperature is much smaller when the annihilation ends. Therefore top annihilation is placed after the electroweak transition. The top quark receives its mass after the electroweak transition, so annihilation only begins after the transition.

---

[11]The assumption of a weakly interacting gas of particles still holds for the baryons and the mesons, but not for the individual quarks and gluons.

Table 3.2 History of $g_\rho(T)$

T	Particles	$g_\rho(T)$
200 GeV	All present	106.75
100 GeV	Electroweak transition	(No effect)
<170 GeV	Top annihilation	96.25
<80 GeV	$W^\pm$, Z^0, H^0	86.25
<4 GeV	Bottom	75.75
<1 GeV	charm, τ^-	61.75
<150 MeV	QCD transition	17.25
<100 MeV	$\pi^\pm, \pi^0, \mu^-$	10.75
<500 keV	e^- annihilation	(7.25) $2 + 5.25(4/11)^{4/3} = 3.36$

3.1.6 Time and Temperature

From the relations we have just obtained and thanks to the current data, we can extract the evolution of temperature, number density, and energy density of the Universe as a function of time. Indeed, from the measurement of the CMB background temperature T_0 we can deduce the present energy density of the photons with Eq. (3.26),

$$\rho_\gamma^0 = \frac{\pi^2}{30} \times 2 \times T_0^4 = 2.62 \times 10^{-10}\ \text{GeV/cm}^3. \tag{3.43}$$

Moreover, we saw that the neutrino temperature is 1.91 K in the absence of any kind of "non-standard" phenomena that could increase the temperature or the effective degrees of freedom of the neutrino (see Sect. 3.3.3 for some exceptions). This gives us for the energy density of the neutrinos,

$$\rho_\nu = \frac{\pi^2}{30} \times \frac{7}{8} \times 2 \times 3 \times \left(T_0^\nu\right)^4 = 1.78 \times 10^{-10}\ \text{GeV/cm}^3, \tag{3.44}$$

where we considered 3 types of neutrinos. This expression is therefore valid until two types of neutrinos become non-relativistic. If we note by m_1, m_2, and m_3 their mass eigenstate, the measurements of the mass differences ($|m_3 - m_1|$ and $|m_2 - m_1|$) give

$$|m_3 - m_1| \simeq 0.049\,\text{eV} > |m_2 - m_1| \simeq 0.008\,\text{eV} > T_0^\nu = 1.7 \times 10^{-4}\,\text{eV} \tag{3.45}$$

which would mean that at least two neutrinos are non-relativistic. So there are two types of neutrinos, currently, that participate in the "hot" component of dark matter. This will be discussed in more detail in Sect. 3.3.1. We also want to add that the factor "3" we used in Eq. (3.44) considers an instantaneous decoupling of the neutrinos. The ratio $\frac{T^\nu}{T_\gamma}$ is not exactly the same when we take into account non-instantaneous processes and the neutrino oscillation effects, which effectively corresponds to a neutrino number of $N_{eff} \simeq 3.045$ where N_{eff} is defined by

$$\rho_\nu^{CMB} = \frac{\pi^2}{30} \times \frac{7}{8} \times 2 \times N_{eff} \times \left(T_0^\nu\right)^4 . \tag{3.46}$$

Finally, the present day radiation is thus

$$\rho_R = \rho_\gamma + \rho_\nu = 2.4 \times 10^{-10} + 5.7 \times 10^{-11} \simeq 3 \times 10^{-10}\,\text{GeV/cm}^3, \tag{3.47}$$

where we took one family of neutrino.[12] For the matter content, we measured around 10^{11} solar mass per megaparsec which implies a matter density

$$\rho_{\text{matter}} = 10^{11}\text{ solar masses/Mpc}^3 \simeq 4.1 \times 10^{-6}\,\text{GeV/cm}^3 \tag{3.48}$$

in relatively good accordance with WMAP data which gives

$$\rho_{\text{baryons+dark matter}} \simeq 1.6 \times 10^{-6}\,\text{GeV/cm}^3;$$

$$\rho_{\text{baryons+dark matter+dark energy}} \simeq \rho_c^0 = 5.3 \times 10^{-6}\,\text{GeV/cm}^3, \tag{3.49}$$

from which we can deduce $\rho_{\text{dark energy}} \simeq 3.7 \times 10^{-6}\,\text{GeV/cm}^3$. Using Eq. (2.53), we can also write

$$\Omega^0_{matter} h^2 = \frac{\rho_{matter}}{\rho_c^0} h^2 \simeq 0.15 \qquad \Omega^0_{radiation} h^2 = \frac{\rho_{radiation}}{\rho_c^0} h^2 \simeq 2.9 \times 10^{-5}. \tag{3.50}$$

Matter dominated the Universe during almost 9.8 billion years after the CMB. Since then, the dark energy is the main contributor to the density. As we can see, presently Universe density of matter is 5000 times greater than radiation density.[13] However, it was not always the case. In the early Universe, all Standard Model particles were relativistic and contributed to the radiation. The evolution over time of a system dominated by "dust" (matter) is indeed very different from the evolution of a Universe dominated by radiation.

---

[12] Keeping in mind the possibility that all the three families of neutrino are massive and thus do not contribute at all to the radiation density.

[13] Be careful to not be confused with the ratio $\eta_b = n_b/n_\gamma \simeq 6 \times 10^{-10}$. Indeed, a photon energy is around 10^{-4} eV at present time, whereas a baryon mass is 1 GeV. $\rho_b/\rho_\gamma \simeq 1000$ gives you $n_b/n_\gamma \simeq 10^{-10}$. The number density of the photon is much larger than the number density of the baryons nowadays, this is the baryogenesis concept.

The density of matter changes because the volume V of the Universe changes, while no new matter is created: $\rho_m \sim 1/V \propto 1/a^3$ where a is the scaling factor defined from the current radius of the Universe R_0: $R(t) = R_0 \times a(t)$ $(0 < a(t) < 1)$. For radiation, it is easier to think in terms of photons. Its *number* density evolves as $\sim 1/V$ as well, however, each photon wavelength is redshifted, so energy of individual photons also changes as $E_\gamma \sim 1/a$. Therefore the energy density of radiation evolves as $\frac{E_\gamma}{V} \propto 1/a^4$. We can then deduce from Eq. (3.26) that the scale factor evolves as $a \propto 1/T$ and thus the matter density evolves as $\rho_m \propto T^3$. From the actual values obtained by WMAP we then have

$$\rho_m = 1.6 \times 10^{-6} \left[\frac{T}{T_0}\right]^3 \text{GeV/cm}^3; \quad \rho_R = 3 \times 10^{-10} \left[\frac{T}{T_0}\right]^4 \text{GeV/cm}^3 \tag{3.51}$$

with T_0 is the present temperature of the Universe $T_0 = 2.34 \times 10^{-13}$ GeV. The result is shown in Fig. 3.4 where we can see that the matter has begun to dominate at a temperature of around 1 eV, corresponding to the decoupling time, see Sect. 3.3.4.

If we want to calculate the expression of energy densities as a function of time rather than temperature, we must first calculate the Hubble constant H which represents the relation between time and the scaling factor a. As a first approximation, we can write $H \equiv \dot{a}/a = \frac{1}{2t}$ with $\dot{a} = \frac{da}{dt}$ (see discussions around Eq. (2.216) for details). Combining Eqs. (2.263) and (3.26) we then deduce the time

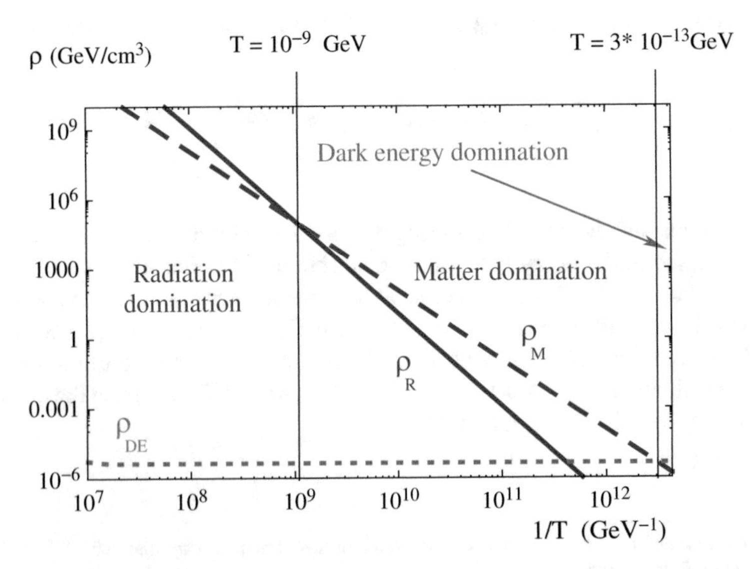

Fig. 3.4 Evolution of the different components of the Universe density as function of the temperature of the photons computed from Eq. (3.51)

t_R in the radiation dominated era:

$$H = \frac{\dot{a}}{a} = \frac{1}{2t_R} = \sqrt{\frac{\rho_R}{3M_P^2}} = \frac{\pi}{3}\sqrt{\frac{g_\rho}{10}}\frac{T^2}{M_P} \tag{3.52}$$

$$\Rightarrow t_R \simeq 2.4 \times 10^{-6}\frac{1}{\sqrt{g_\rho}}\left(\frac{\text{GeV}}{T}\right)^2 \text{ s [Radiation dominated era : T > 1 eV]}.$$

Later on, when the matter begins to dominate the Universe, the density is dominated by ρ_m given in Eq. (3.51) and the time dependance now becomes t_m:

$$H = \frac{2}{3t_m} = \sqrt{\frac{\rho_m}{3M_P^2}} \tag{3.53}$$

$$\Rightarrow \ t_m \simeq 5.9 \times 10^{-2} \times \left(\frac{1\ \text{GeV}}{T}\right)^{3/2} \text{ s [Matter dominated era : T < 1 eV]}.$$

We can then inverse the previous relations to obtain the evolution of the temperature of the Universe as function of time in the radiation era $[T_R(t)]$ and the matter era $[T_m(t)]$.

$$T_R(t) \simeq 1.5 \times 10^{-3} \left(\frac{1\ \text{s}}{t}\right)^{1/2} \text{GeV}$$
$$T_m(t) \simeq 0.15 \times \left(\frac{1\ \text{s}}{t}\right)^{2/3} \text{GeV}. \tag{3.54}$$

We represent the evolution of the temperature as function of the time in Fig. 3.5 where we can notice the "slowdown" of the decrease at $t \simeq 100$ s, corresponding to a temperature $T \simeq 0.2$ MeV. This is the BBN epoch when the quarks/colors degrees of freedom are converted to heat the photon of the plasma following the entropy conservation. There is a second slowdown at around $t \simeq 10^{13}$ s corresponding to the recombination epoch (last scattering surface) when the Universe begins to be matter dominated: the slop of $T = f(t)$ is thus changing from $-1/2$ to $-2/3$ as it is clear from Eq. (3.54).

We can compute z_{EQ}, the redshift at the matter density energy begins to dominate over the radiation density energy. For that, one needs first to compute the present radiation density, which content neutrino and photons. We will suppose that the neutrino was relativistic all the time, even if we know that at least two have decoupled from the recombination time. That is a relatively good approximation. The neutrino temperature being $T_\nu = \left(\frac{4}{11}\right)^{1/3} T_\gamma$, with T_γ the photons temperature, and having also 2 degrees of freedom (because only left-handed neutrino has been

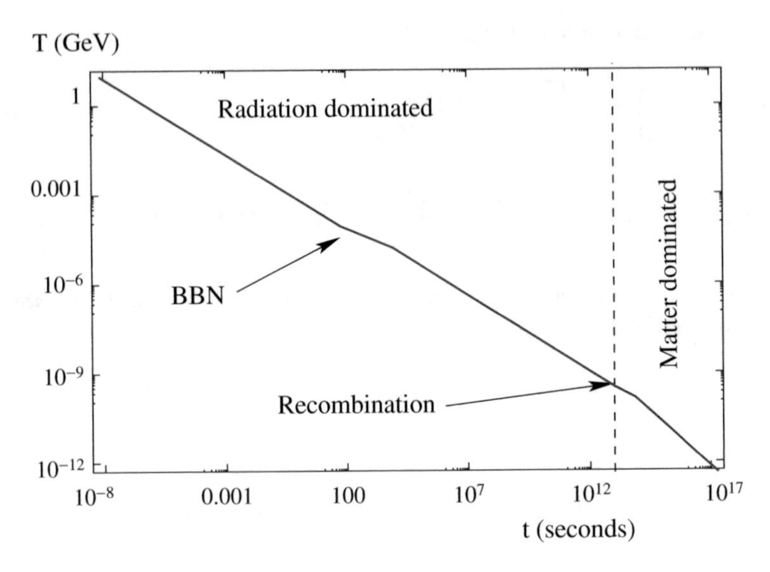

Fig. 3.5 Evolution of the temperature as function of the time computed from Eq. (3.54)

detected so far), we can write using Eq. (3.26)

$$\rho_\nu = 3 \times \frac{7}{8} \times \left(\frac{4}{11}\right)^{4/3} \times 2 \times \frac{\pi^2}{30} T_\gamma^4 = 3 \times \frac{7}{8} \left(\frac{4}{11}\right)^{4/3} \rho_\gamma = 0.68\, \rho_\gamma \tag{3.55}$$

implying

$$\rho_R = \rho_\nu + \rho_\gamma = 1.68\, \rho_\gamma, \tag{3.56}$$

where ρ_γ is the (measured) background radiation. Asking for $\Omega_\gamma(z_{EQ}) = \Omega_m(z_{EQ})$, we obtain

$$\frac{\Omega_R}{\Omega_m} = \frac{1.68 \times \Omega_\gamma}{\Omega_m} = \frac{1.68\, \Omega_\gamma^0 (1 + z_{EQ})^4}{\Omega_m^0 (1 + z_{EQ})^3} = 1$$

$$\Rightarrow z_{EQ} = \frac{0.311}{1.68 \times 5.5 \times 10^{-5}} = 3366 \simeq 3400, \tag{3.57}$$

corresponding to a time $t_{EQ} \simeq$ 40,000 years from the Big Bang.

Exercise Taking the dark energy density and matter density to be, respectively, $\Omega_\Lambda = 0.689$ and $\Omega_m = 0.311$ show that the redshift z_Λ corresponding to the dark energy domination, $\Omega_\Lambda(z_\Lambda) = \Omega_m(z_\Lambda)$, is $z_\Lambda = 0.304$. Show that it happens ∼10

Gyrs after the Big Bang. Find then the temperature/time relation in the phase where the Universe is dominated by dark energy, 10 Gyrs after the Big Bang.
Hints: use the relation in a $\Lambda-$ dominated Universe, $1 + z = e^{H_0 t}$.

The cosmological constant problem

The introduction of the cosmological constant has proved to be a real challenge in the history of the construction of theoretical models, as we have seen in our discussion at the end of the Sect. 2.1.6. However, a model dominated by a cosmological constant is the most natural if we want a homogeneous and isotropic Universe as Albert Einstein imagined it. It is indeed more difficult for the density of matter to be homogeneously and isotropically distributed when it is dominated by unstable dynamic forces, like the gravity forces. Moreover, this hypothesis, known as the cosmological principle, was unnatural, especially in view of the night sky before Edwin Hubble's measurements. The discovery of Hubble's law was above all the confirmation of a homogeneous and isotropic Universe. The crucial point is, if we assume the existence of this cosmological constant, what would be its most natural value today? That is a reasonable question, leading to what is commonly called "the cosmological constant problem," a complete review being accessible at [1]. The main idea is that combining Eq. (2.30) with Eq. (2.89), for the Higgs Lagrangian (B.229), in the electroweak breaking phase, we can write an "effective" electroweak cosmological constant Λ_{EW}

$$V(H = v) = -\frac{1}{8}M_H^2 v^2 \;\Rightarrow\; \Lambda_{EW} = -\frac{1}{8}\frac{M_H^2}{M_P^2}v^2 \simeq -2 \times 10^{-29}\ \mathrm{GeV}^2,$$

where we used Eq. (B.232) and $v = 246\,\mathrm{GeV}$. From Eq. (2.54), this value for Λ_{EW} corresponds to a density $\rho_\Lambda = M_P^2 \Lambda_{EW} = -1.2 \times 10^8\ \mathrm{GeV}^4$. This number is huge and has to be compared with the values of the cosmological constant measured today, $\rho_\Lambda^0 = \Omega_\Lambda \times \rho_c^0 = 2.9 \times 10^{-47}\ \mathrm{GeV}^4$, 55 orders of magnitude less that what we should expect. We can always tune the parameters, adding an uplifting term in the Higgs potential to counterbalance exactly the $V(H = v)$ term and obtain the cosmological constant measured today. This is clearly not satisfactory and would imply that the vacuum energy density was huge prior to the electroweak phase transition, not to mention the fact that such fine tuning would not resist loop corrections. This is what is called the cosmological constant problem.

3.1.7 The Entropy

To compute the entropy of the Universe, we need the second law of thermodynamics. The first law was used to compute the relation between pressure and volume in the Appendix A.5 in the case of adiabatic transformations. The second principle of thermodynamics establishes the link between entropy, pressure, temperature, and internal energy. It can be written as

$$TdS = dU + PdV = \delta Q, \tag{3.58}$$

where dS is the variation of entropy at a temperature T under a pressure P, corresponding to an exchange of heat δQ. If one defines ρ as the energy density ($U = \rho V$) and s the density of entropy ($S = sV$), one obtains

$$d\rho - Tds = (Ts - \rho - P)\frac{dV}{V}. \tag{3.59}$$

In equilibrium, the entropy density, energy density, and pressure are intensive quantities that can be written as functions only of the temperature $\rho(T)$, $s(T)$, $P(T)$, such that $d\rho - Tds \propto dT$. The coefficients in front of dT and dV are then independent and must vanish separately because one is intensive (independent of volume) and the other one is extensive (depends on the size of the system). This relates the entropy density to the energy density and pressure

$$s = \frac{\rho + P}{T} \tag{3.60}$$

using (3.11) and (3.26) one has

$$s = \frac{2\pi^2}{45} g_s T^3 \tag{3.61}$$

$$g_s = \sum_{b=bosons} g_b \left(\frac{T_b}{T}\right)^3 + \frac{7}{8} \sum_{f=fermions} g_f \left(\frac{T_f}{T}\right)^3. \tag{3.62}$$

Except in the case where particles transfer their entropy to photons and not to already decoupled particles (such as neutrinos), all particles have the same temperature and we can reasonably approximate $g_s \simeq g_\rho$. One can also notice that $s_\gamma \simeq 3.6\, n_\gamma$, and constant entropy imposes $s \propto 1/R^3$. Then, working with comoving frame densities ($Y_i = n_i/s \propto n_i \times R^3$) is equivalent to working with particle/photon ratio density $Y_i \simeq n_i/n_\gamma$. It is also important to notice that, like

$n(T), s(T)$ evolves as T^3. However, the degrees of freedom (especially the statistical factors are 7/8 and not 3/4 as in $n(T)$) are the ones of a statistic in T^4, as the entropy is proportional to $\rho(T)/T$ and not $n(T)$. Conservation of entropy is then directly linked to the conservation of energy: $\rho(T_i)/T_i = \rho(T_f)/T_f$.

As in the case of degrees of freedom for energy density, we can calculate the effective degrees of freedom for entropy density s today, g_s^{today}, which is different from g_ρ^{today} because the decoupling of the neutrino has created a difference between the neutrino and photon temperature (see Sect. 3.3.1 for details). One obtains[14]

$$g_s^{\text{today}} = 2 + \frac{7}{8} \times 2 \times 3 \times \frac{4}{11} = 3.91. \tag{3.63}$$

The fact that we considered 3 types of neutrinos (and not only one relativistic) is subtle and explained in the box below Eq. (3.90). From that number one can deduce the present entropy s_0, which will be very useful to compute relic abundance of stable species

$$s_0 = \frac{2\pi^2}{45} g_s^{\text{today}} T_0^3 = 34.71 \text{ K}^3 = 2.2 \times 10^{-38} \text{ GeV}^3 = 2909 \text{ cm}^{-3}. \tag{3.64}$$

Asking for constant entropy of the Universe, $S = sR^3$, has several consequences. First, from Eq. (3.61) the temperature *of the photons and other relativistic particles* follows a law $T \propto R^{-1}$. The other consequence is that, every time a particle decouples from the thermal bath, this decoupling happens at constant entropy. It means that this particle species "gives" its entropy to the relativistic particles to which it is coupled and still present in the bath, before leaving it. This information is in fact encoded in the degrees of freedom g^s_{eff}: after the decoupling of a particle species i, the effective degrees of freedom in the bath decrease: $g^s_{eff} \rightarrow g_s - g_i$. the entropy being constant (and so its density as the Universe do not have time to evaluate during this process considered adiabatic) $g_s T^3$=cst implies that the decoupling of a specie increases the temperature of the bath (i.e. of the photons) following $T_\gamma^{\text{after}} = T_\gamma^{\text{before}} \times (g_s^{\text{before}}/g_s^{\text{after}})^{1/3}$. After this heating of the bath, the temperature of the plasma follows the R^{-1} law.

On the other hand, the particle which has decoupled from the plasma follows different laws if it is a massless or massive particle. Indeed, after decoupling the energy of each massless particle is redshifted by the expansion of the Universe

[14]The fact of taking into account three species of neutrino, even if it is known from the measurement of Δm_{atm} and Δm_{sol} that some are non-relativistic nowadays, comes from the fact that, once decoupled, the massive neutrinos still follow a classical relativistic distribution function (see Sect. 3.2 for more details).

$E(t) = E(t_{dec})R(t_{dec})/R(t) = E_{dec}R_{dec}/R = p(t)$ in the relativistic case. As $n(t)$ decreases proportionally to R^{-3} because the density is "frozen" (no any way of producing it thermally nor destroying it after decoupling time) the distribution function $f(\mathbf{p}) = d^3n/d^3\mathbf{p}$ is constant during the expansion. This implies $e^{E/T} = e^{E_{dec}/T_{dec}} = e^{ER/R_{dec}T_{dec}} \Rightarrow T = T_{dec}R_{dec}/R \propto R^{-1}$. The relativistic species thus follows the R^{-1} evolution after they decoupled from the plasma but are not affected by the $(g_s)^{1/3}$ enhancement that affects the photons and relativistic species still living in the bath.

The massive non-relativistic species ($m \gg T_{dec}$) follows another law after being decoupled. Their kinetic momentum follows the classical redshift $p = p_{dec}R_{dec}/R$ from which it follows that the kinetic energy ($\propto \frac{p^2}{2m}$) of each particle red shifts as R^{-2}: $E_c = E_c^{dec}R_{dec}^2/R^2$. Conservation of the density function thus implies $e^{p_{dec}^2/2mT_{dec}} = e^{p^2/2mT} \Rightarrow T = T_{dec}R_{dec}^2/R^2 \propto R^{-2}$. As a conclusion, after the decoupling of a species "i"

$$T_i \propto R^{-1} \text{ if relativistic} \quad T_i \propto R^{-2} \text{ if non-relativistic.} \tag{3.65}$$

We have summarized this effect in Fig. 3.6

Exercise Taking the radius of the Universe found in Eq. (2.65) show that the total entropy in the Universe is $S_{tot} = 9.3 \times 10^{89}$, the total number of photons $N_\gamma = 1.3 \times 10^{89}$, and the total number of baryons $N_b = 7.9 \times 10^{79}$. Compute the same numbers for a radius of 46.3×10^9 lyrs.

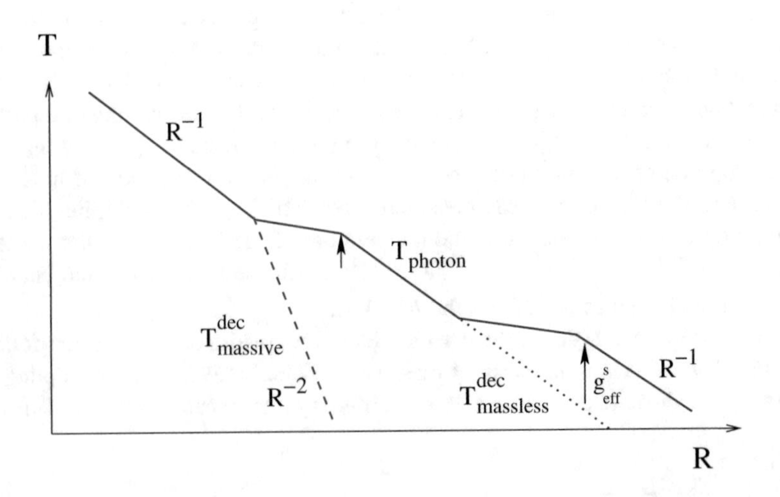

Fig. 3.6 Evolution of the temperature of different species (relativistic/massless and non-relativistic/massive) after their decoupling from the thermal bath. It can be, for instance, the dark matter decoupling followed by the neutrino decoupling

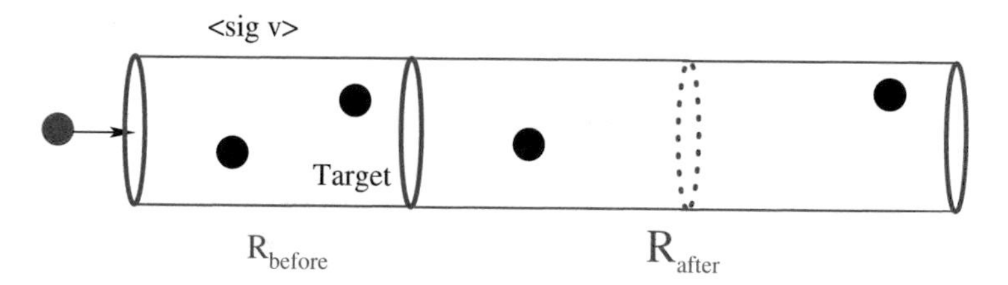

Fig. 3.7 Illustrative example of the decoupling epoch when the number of interactions is divided by 2 during a time Δt due to the dilution of the target. The volume necessary to have 2 collisions (R_{before}) is now just sufficient to give one collision (R_{after})

3.1.8 The Meaning of Decoupling

In the previous paragraphs, we have often referred to the notion of decoupling of primordial plasma, without really giving a definition of what is meant by "decoupling." Suppose a particle A interacting in the bath with a rate per particle $\Gamma = n\langle\sigma v\rangle$, n being the density of the target particle, and $\langle\sigma v\rangle$ the average cross section times the relative velocity. $\Delta t = 1/\Gamma$ represents the mean time between two collisions. During this time Δt the Universe has expanded by a factor ΔR such as $\Delta R/R = H\Delta t = H/\Gamma$. In another word, when $H \simeq \Gamma$, the size of the Universe has doubled and the density n of the target has been divided by 8, as the interaction rate Γ which is proportional to n. In another word, the time (or the temperature) of the Universe when $H \simeq \Gamma$ is the epoch where the particles decouple from the bath and their interaction rates with the plasma decreases exponentially. We illustrate it in Fig. 3.7. The exact way to treat the decoupling problem is to solve the Boltzmann equation. However, the approximation $H \simeq \Gamma$ to obtain the decoupling time of particles is usually quite accurate. We will give two specific examples to understand how the nature of the interaction can change drastically the temperature of decoupling of species. We will consider two distinct cases:

- (i) interactions mediated by a massless gauge boson (like the photon)
- (ii) interactions mediated by a massive gauge boson (Z or Z').

The exchange of a massless gauge field between two particles S and $\tilde{S}$ can be parameterized in a case of bosonic particles[15] by a Lagrangian of the form $\mathcal{L} = (D_\mu S)(D^\mu S)^\dagger + (D_\nu \tilde{S})(D^\nu \tilde{S})^\dagger$ which includes the terms of interactions $ig\ p^\mu A_\mu S S^\dagger + ig\ \tilde{p}^\nu A_\nu \tilde{S}\tilde{S}^\dagger$, A_μ being the massless vectorial field (photon, for instance) and g its coupling to the particles in the bath. One can then compute the amplitude of the interaction (see Appendix C)

$$\mathcal{M} = g^2 p_\mu \eta^{\mu\nu} \tilde{p}_\nu / p^2 = g^2 p.\tilde{p}/p^2 \simeq E\tilde{E}(1-\cos\theta)/E^2,$$

[15]The analysis does not depend on the nature –fermionic/bosonic– of the particles we consider.

$\cos\theta$ being the diffused angle between S and $\tilde{S}$ which implies

$$|\mathcal{M}|^2 \simeq g^4 E\tilde{E}(1 - \cos\theta)/E^2.$$

Using Eq. (B.110), $\sigma \simeq |\mathcal{M}|^2/64\pi^2 s$, considering that the particles are relativistic in the plasma and using Eq. (3.30) ($m_{S,\tilde{S}} \ll T \Rightarrow E \simeq \tilde{E} \simeq T$) one can deduce $\sigma \simeq g^4/T^2$ implying when combined with Eq. (3.23) $\Gamma = n_S\langle\sigma v\rangle = n_S\langle\sigma c\rangle \simeq g^4 T$. The particle S will thus be decoupled from the primordial plasma when

$$\frac{\Gamma}{H} \lesssim 1 \Rightarrow \frac{g^4 M_P}{T} \lesssim 1 \Rightarrow T \gtrsim g^4 M_p \simeq 10^{14}\text{GeV}, \tag{3.66}$$

where we took $g = 0.1$ as illustration. It is important to notice that T in this case is an upper bound, which means that as soon a $T \lesssim 10^{14}$ GeV the long-range force mediated by the massless boson (as it is the case for electroweak interactions) will always be sufficient to maintain charged relativistic particles in equilibrium in the bath: the decoupling will appear only when the temperature will reach m_S, where the density will be exponentially suppressed by the Boltzmann factor (the term n_S in the expression of Γ).

In the case of the exchange of a massive gauge boson Z' ($M_{Z'} \gtrsim T$) the amplitude of the reaction can be written $\mathcal{M} \simeq g^2E^2/M_{Z'}^2 \simeq g^2T^2/M_{Z'}^2 \Rightarrow \Gamma \simeq g^4T^5/M_{Z'}^4$. The decoupling temperature will then be given by the usual condition $\Gamma/H \lesssim 1$:

$$\frac{\Gamma}{H} \lesssim 1 \Rightarrow \frac{g^4 M_P T^3}{M_{Z'}^4} \lesssim 1 \Rightarrow T \lesssim \left(\frac{M_{Z'}}{g}\right)^{4/3} M_p^{-1/3} \simeq 0.1 \left(\frac{M_{Z'}}{1\ \text{TeV}}\right)^{4/3} \text{MeV}. \tag{3.67}$$

In this case we clearly see that the decoupling of particles charged only under $SU(2)$ (Z boson exchange) will decouple quite late in the history of the Universe (around 1 MeV) which is precisely the case of the neutrino, studied in detail in Sect. 3.3.1. We represent in Fig. 3.8 the evolution of $H(T)$ and $\Gamma(T)$ in different cases (taking $g = 0.1$). The following relations help to understand the diagram.

$$\log H \simeq -2\log 1/T - 19\,;$$

$$\log \Gamma_\gamma \simeq -\log 1/T - 4\,;$$

$$\log \Gamma_{M_{Z'}} \simeq -5\log 1/T - 4\log M_{Z'} - 4.$$

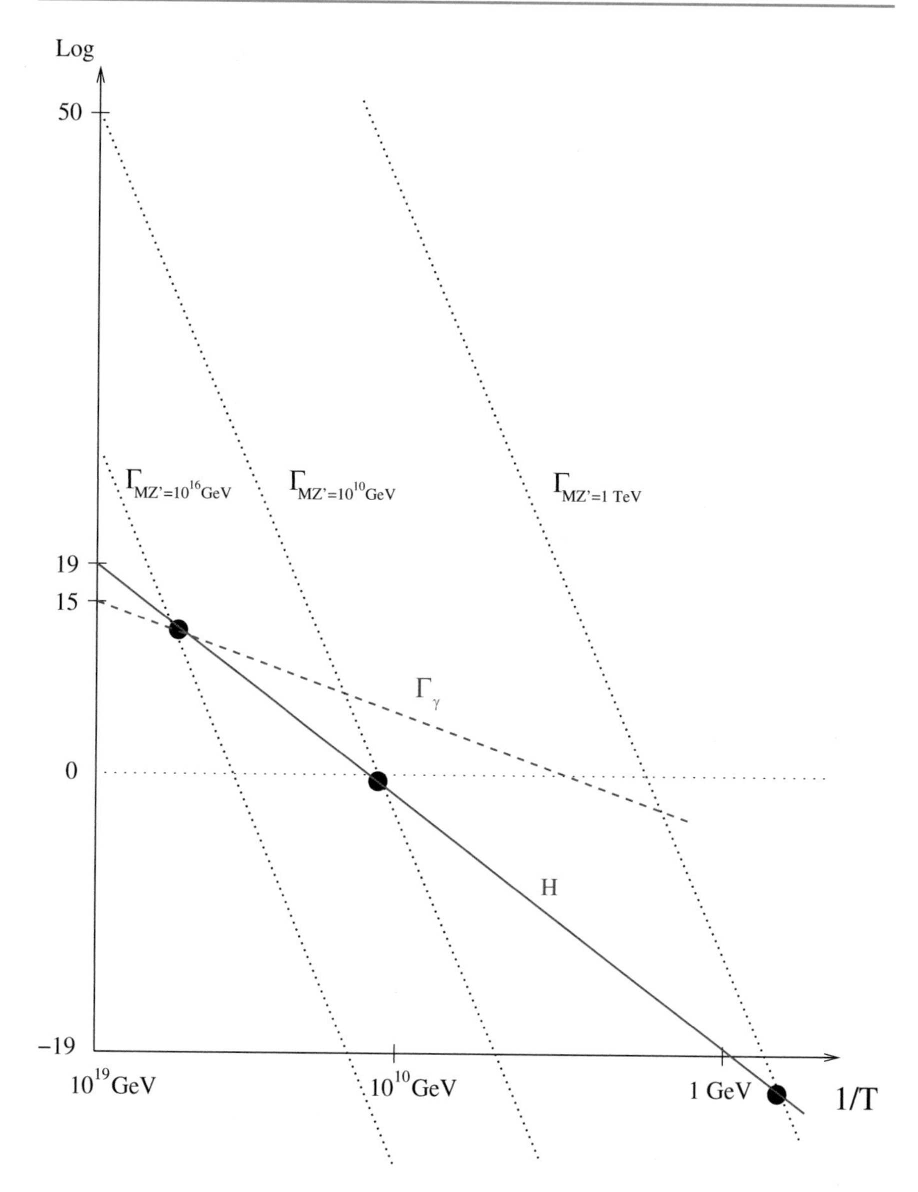

Fig. 3.8 Evolution of $H(T)$, $\Gamma_{\gamma}(T)$, and $\Gamma_{M_{Z'}}(T)$ for different masses of Z' and relativistic species. The black dots show the temperature when the decoupling occurs ($\Gamma/H \simeq 1$)

3.2 Chemical Decoupling or Kinetic/Thermal Decoupling?

3.2.1 The Main Idea

Because we will need to analyze the decoupling of a dark matter from the thermal bath, it is important to point out that all the discussions above concerned mainly the *chemical* decoupling, i.e. the temperature for which the production rate of a species (Standard Model particles or dark matter) is too small to maintain it in equilibrium with the thermal bath. In other words the thermal bath cannot change anymore the *number* of dark species, so the name "chemical" decoupling in analogy with chemical reactions. It is very important to realize that even though dark matter is a chemically distinct particle species after this decoupling, it is still a constituent of the local hot bath. Indeed, the relevant target density for elastic scattering processes to maintain *thermal* equilibrium is provided by the number density of relativistic Standard Model particles and thus decreases only at a rate proportional to T^3, and not exponentially as the "self" target, the dark matter itself. Eventually, at a temperature T_k, the elastic scattering (and not annihilating) rate[16] $\Gamma_{scat} = n_{SM}^{eq}\langle\sigma_{scat}\rangle$, with n_{SM}^{eq} the equilibrium density, cannot compete with that of the expansion of the Universe, and the dark matter particles start to decouple from kinetic equilibrium. Elastic scattering processes cease. In fact, to be precise, even before the last scattering occurs, the temperature of the dark matter has no time to be relaxed because of the Hubble expansion. It is thus the relaxation time which will determine the kinetic decoupling of the dark matter, i.e. the time when the temperature of the dark matter is not maintained anymore to the plasma one through scattering and drops quickly with the scale factor a as a^{-2} instead of a^{-1}, as we saw in Eq. (3.65). The temperatures of the dark matter particles χ and the radiation background are then approximately related by $T_\chi \simeq \langle|\mathbf{p}|^2/2m_\chi\rangle \sim T^2/T_k$, $T_k \simeq m_\chi$ being the temperature of the kinetic decoupling.

3.2.2 Approximate Solution*

Let us estimate the relaxation time for a WIMP of mass m_χ living in a thermal bath of temperature T. Let suppose that after each shock, the Standard Model particles transfer a momentum $\Delta\mathbf{p}$ (tri-vector), roughly given by its energy $|\Delta\mathbf{p}| \sim T$ to the dark matter corresponding to a velocity $\Delta\mathbf{v}$. After the number of collisions N_{coll}, the velocity of the dark matter is $\mathbf{v}_{N_{coll}} = \mathbf{v}_{N_{coll}-1} + \Delta\mathbf{v}$, implying

$$|\mathbf{v}_{N_{coll}}|^2 = |\mathbf{v}_{N_{coll}-1}|^2 + 2|\mathbf{v}_{N_{coll}-1}||\Delta\mathbf{v}|\cos\theta + |\Delta\mathbf{v}|^2,$$

[16] Notice that $\sigma_{scat} v_{SM} = \sigma_{scat}$ as the SM particles are still largely relativistic in the thermal bath : $v_{SM} = c$.

θ being the angle between the two colliding (dark matter and Standard Model) particles. We can thus deduce $\langle v^2_{N_{coll}} \rangle = \langle v^2_0 \rangle + N_{coll}(\Delta v)^2$. In other words, after N_{coll} collisions, the velocity (and thus momentum) gained by the dark matter (if we suppose it at rest, for instance, at the beginning, $v_0 = 0$) is $\Delta p_{N_{coll}} = N^{1/2}_{coll} \Delta p$ which should be equal to $(2m_\chi E_c)^{1/2} \sim (m_\chi T)^{1/2}$ for a dark matter particle at the temperature T. Replacing $\Delta p \sim T$ we then can compute the number of collisions needed to keep the dark matter in kinetic equilibrium:

$$N_{coll} \sim \frac{m_\chi}{T} \gg 1. \tag{3.68}$$

The relaxation time (time needed to still keep the kinetic equilibrium, i.e. to obtain N_{coll} collisions) with a scattering Γ_{scat} is

$$\tau_r \sim \frac{N_{coll}}{\Gamma_{scat}} = \frac{m_\chi}{T} \frac{1}{n^{eq}_{SM} \sigma_{scat}} \sim \frac{m_\chi}{T^4 \sigma_{scat}} \sim \frac{10^9 \; m_\chi}{T^4 \left(\frac{\sigma_{scat}}{10^{-9}} \right)}. \tag{3.69}$$

Then, kinetic decoupling occurs approximately at the temperature T_k, for which $\tau_r(T_k) = H^{-1}(T_k)$ which gives with $H = \frac{\pi}{3} \sqrt{\frac{g_\rho}{10}} \frac{T^2}{M_P}$

$$T_k \sim 10^{-4} g_\rho^{1/4} \left(\frac{\sigma_{scat}}{10^{-9}\ \text{GeV}^{-2}} \right)^{1/2} \left(\frac{m_\chi}{100\ \text{GeV}} \right)^{-1/2} \text{GeV} \tag{3.70}$$

corresponding roughly to the MeV scale for a 100 GeV dark matter and an electroweak-like scattering cross section ($\sigma_{scat} \simeq \sigma_{EW} \sim 10^{-9}$ GeV^{-2}). When the temperature of the plasma drops below T_k, the number of collisions needed to still keep the dark matter in kinetic equilibrium is such that the time to reach it becomes larger that the Hubble expansion time: the dark matter reaches the kinetic decoupling. This effect can play an important role when the velocity appears in processes like Sommerfeld enhancement, for instance. A more detailed analysis of the thermalization in the dark bath is explained in Sect. 3.2.4.

3.2.3 What Is Happening After the Decoupling?

Once Γ (scattering or decay) falls below the expansion rate H, the particles decouple from the plasma and propagate freely along geodesics of the space-time. **The form of the distribution function $f(p)$ is conserved**, while the momentum redshifts as $p(t) \propto 1/a$, implying $p(t) = p(t_{\text{dec}})a(t_{\text{dec}})/a(t)$. The form of the distribution function is indeed conserved as long as the particle is not in thermal bath with another dark sector (which can also have decoupled from the primordial plasma before) because of the conservation of energy/momentum of the decoupled system: there is no source (thermal bath or collisions) that can modify the distribution in energy of the decoupled gas. It follows that the distribution function for any $t > t_{\rm dec}$ is given by

$$f(p, t > t_{\rm dec}) = f(\frac{a(t)}{a(t_{\rm dec})} p, t_{\rm dec}). \quad (3.71)$$

We see that the distribution function of the decoupled particles is simply a rescaled version of the distribution function at decoupling time $t_{\rm dec}$.

- *Decoupling while relativistic*: if a particle of mass m decouples when relativistic, $T_{\rm dec} \gg m$, then the distribution function (3.71) takes the form

$$f(p, t > t_{\rm dec}) = \frac{1}{e^{E/\tilde{T}(t)} \pm 1}, \quad \text{with } \tilde{T}(t) = T_{\rm dec} \frac{a(t_{\rm dec})}{a(t)} . \quad (3.72)$$

  So the temperature of a decoupled relativistic species falls strictly as a^{-1}. If the particle later becomes non-relativistic, we have $E \sim m$ but the distribution function keeps the form (3.72). This is an important point, especially when computing the relativistic degrees of freedom for the present entropy, for instance. What is important is that the particle was relativistic during the decoupling time. Once it is decoupled, it is as if the thermal bath gave them a punch, and they continue their way in space without any interaction at the same velocity, just redshifted by the scale factor. The density number distribution is a relativistic one, respecting $n(\tilde{T}) \propto \tilde{T}^3$ (3.27) even if $T < m$ (no exponential suppression). In other words, all particles that were relativistic during their decoupling time count as relativistic degrees of freedom in the Universe nowadays. Notice that the conservation of the entropy, in the radiation dominated era, imposes $S = sR^3 = sa^3R_0^3$=cte, using (3.61), one deduces $T \propto a^{-1}$. This means that even after decoupling, a gas of relativistic particle keep a "virtual" temperature $\tilde{T}$ following the one of the thermal bath.
- *Decoupling while non-relativistic*: if, on the other hand, a particle is non-relativistic at the time of decoupling, $T_{\rm dec}$, then $E \simeq m + p^2/2m$ and the distribution function is given by

$$f(p, t > t_{\rm dec}) = e^{-m/T_{\rm dec}} e^{-p^2/2m\tilde{T}}, \quad \tilde{T}(t) = T_{\rm dec} \left(\frac{a(t_{\rm dec})}{a(t)} \right)^2 . \quad (3.73)$$

  So the temperature of decoupled non-relativistic species falls as a^{-2}, which means much faster than the relativistic ones. Their number density thus decreases *not exponentially* but as T^2 in a radiation dominated Universe.

It is important to emphasize that, despite the different scaling behavior of the temperature for relativistic and non-relativistic species after decoupling, in both cases the equilibrium distribution is maintained. The particles are *not* anymore in

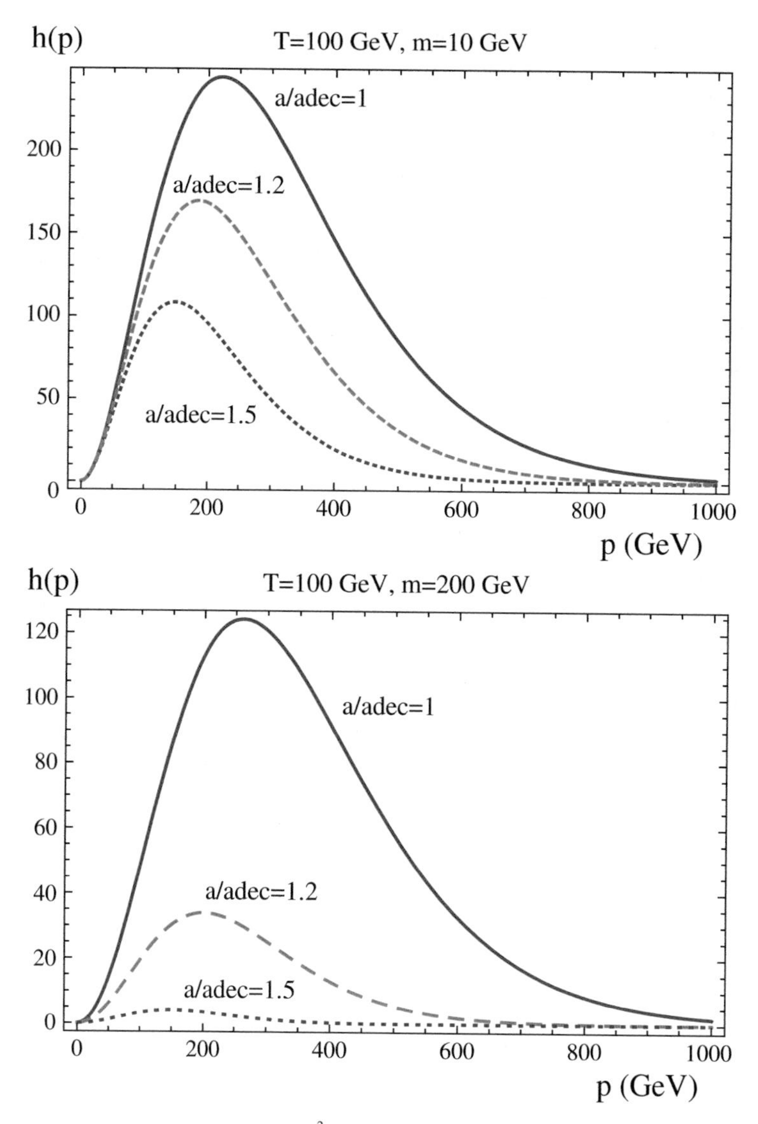

Fig. 3.9 Distribution function $h(p) = \frac{p^2}{2\pi^2} \times f(p)$ for $T = 100\,\text{GeV}$ for different values of the scale factor $a(t)$ in the relativistic fermion (top) and non-relativistic fermion (bottom)

thermal equilibrium but the *shape* of the distribution is maintained around a *virtual* temperature $\tilde{T}$. We illustrate the behavior of the two cases (relativistic and non-relativistic) in the Fig. 3.9 for $T = 100\,\text{GeV}$ and two masses of particles: 10 and 200 GeV for the relativistic and non-relativistic case, respectively.

3.2.4 Transfer of Energy and Thermalization

3.2.4.1 Generalities

There exists a second Boltzmann equation which concerns the transfer of energy. If a set of particles i is all produced by the same mechanism and the scattering is more efficient than the expansion of the Universe, they naturally reach a common thermal temperature (the one of the photons), $T = T_i$. However, in scenario where the dark sector is not produced with the Standard Model ones or very feebly coupled with the visible sector, the temperature of the dark sector increases slowly due to the transfer of the energy from the thermal bath to the dark bath of energy density ρ' and pression P'. If the density of dark particles is very small at the reheating time, the Boltzmann equation for the transfer from the thermal bath energy density ρ to ρ' from $1 + 2 \rightarrow 3 + 4$ reaction can be written

$$\frac{d\rho'}{dt} + 3H(\rho' + P') = n_{EQ}^2 \langle \sigma v (E_1 + E_2) \rangle \tag{3.74}$$

with n_{EQ} being the density of particles in the bath. The phenomenon is illustrated in Fig. 3.10. The right-hand side of Eq. (3.74) should also content terms proportional to n_{dark}^2 for the inverse transfer process $\rho'(3+4) \rightarrow \rho(1+2)$ but as we considered the dark bath composed of particles of mass M with feeble interactions with the thermal bath, this process can be neglected. It is a situation similar to FIMP (Freeze in Massive Particle) or heavy mediators which will be discussed is Sect. 3.6. With the hypothesis that the thermal bath is in equilibrium, combining Eqs. (3.74) and (3.23) and the fact that in the relativistic case[17] $P' = \rho'/3$ and $\frac{d}{dt} = -HT\frac{d}{dT}$

$$\rho \frac{d\,(\rho'/\rho)}{dT} = \frac{d\rho'}{dT} - 4H\rho' \;\Rightarrow\; \frac{d(\rho'/\rho)}{dT} = -\frac{n_{EQ}^2}{\rho H T} \langle \sigma v (E_1 + E_2) \rangle \tag{3.75}$$

with

$$\langle \sigma v (E_1 + E_2) \rangle = \frac{1}{n_{EQ}^2} \int f_1(E_1) f_2(E_2) \frac{d^3 p_1}{(2\pi)^3} \frac{d^3 p_2}{(2\pi)^3} \frac{|\bar{\mathcal{M}}|^2}{128\pi^2 E_1 E_2} (E_1 + E_2) d\Omega \tag{3.76}$$

[17]Notice that we made the supposition in this section that the dark matter bath is in thermal equilibrium with itself (with a temperature T') or from the self-scattering of the dark matter on itself or from the scattering of other particles of the dark sector on themselves, i.e. $\Gamma_{\text{scatter}}(T) > H(T)$. For a more detailed description of the mechanism of thermalization, see the next section.

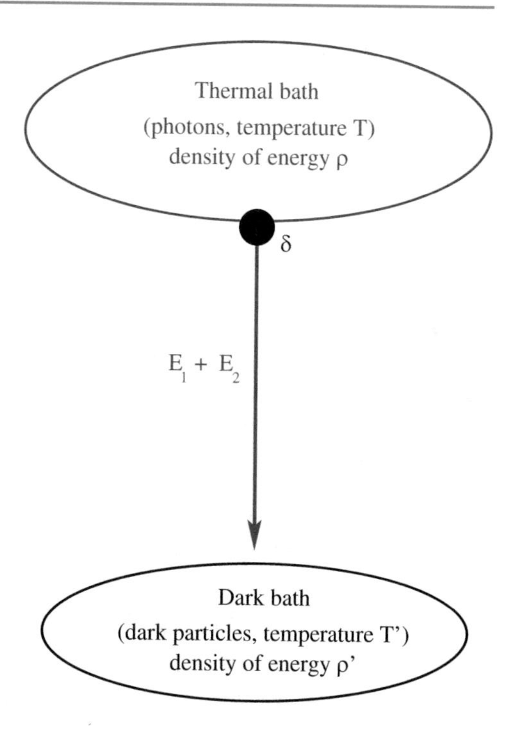

Fig. 3.10 Illustration of the Boltzmann equation describing the transfer of energy from the Standard Model bath to the dark bath

$\bar{\mathcal{M}}$ being the mean of the amplitude squared on the spin of the initial particles and f_i their statistical distribution. Following a classical procedure of integration that will be more detailed when we will treat the WIMP case, see Eq. (3.180), we can write

$$n_{EQ}^2\langle\sigma v(E_1+E_2)\rangle$$
$$=\frac{g_1g_2}{32(2\pi)^6}\int_{4M^2}^{\infty}\sqrt{s-4M^2}|\bar{\mathcal{M}}|^2 s\,ds d\Omega\int_1^{\infty}t\sqrt{t^2-1}e^{-t\frac{\sqrt{s}}{T}}dt$$

implying

$$\frac{d(\rho'/\rho)}{dT}=-45\frac{g_1g_2\sqrt{10}\,M_P}{512\pi^9T^3g_\rho^{3/2}}\int_{2\frac{M}{T}}^{\infty}|\bar{\mathcal{M}}|^2x^2\sqrt{x^2T^2-4M^2}\,K_2(x)\,dx d\Omega \tag{3.77}$$

g_1 and g_2 being the internal degrees of freedom of particles 1 and 2. We can then divide the integral (3.77) in two regimes. In the first one the particle can be considered as relativistic ($M \ll T$), and the second one when the temperature drops below the mass of the dark matter ($T \ll M$) where the integral in Eq. (3.77) tends to 0:

$$\frac{d(\rho'/\rho)}{dT} = -45\frac{g_1 g_2\sqrt{10}\,M_P}{512\pi^9 T^2 g_\rho^{3/2}}\int_0^\infty |\bar{\mathcal{M}}|^2 x^3\, K_2(x)\, dx d\Omega \qquad [\mathrm{M} \ll \mathrm{T}]$$

$$\frac{d(\rho'/\rho)}{dT} = 0 \qquad [\mathrm{T} \ll \mathrm{M}] \qquad (3.78)$$

3.2.4.2 A Specific Case: Exchanged of a Massless Gauge Boson*

To illustrate the general result obtained in Eq. (3.77) we will take a simple toy model, based on mirror dark matter, where the dark matter, a Dirac fermion of mass M_χ, is charged under the electromagnetic charge with a coupling δ. This happens, for instance, in the case of the presence of a kinetic mixing between an extra U(1)[dark photon] and photon. This is also called "millicharged dark matter" as it is equivalent to suppose that the dark matter is very feebly coupled to the electromagnetic fields (coupling of the order $\delta \ll 1$). The amplitude can then be written

$$\mathcal{M} = \delta\bar{\chi}\gamma^\mu\chi\frac{\eta_{\mu\nu}}{(p_1+p_2)^2}q_f e\bar{f}\gamma^\nu f \qquad (3.79)$$

p_i being the quadrivector of the incoming particles, and q_f the electromagnetic charge of f. We than can compute

$$|\bar{\mathcal{M}}|^2 = 16\, q_f^2 e^2\delta^2\left[(1+\cos^2\theta) + \frac{M_\chi^2}{s}(1-\cos^2\theta)\right]$$

$$\Rightarrow \int |\bar{\mathcal{M}}|^2\, d\Omega = 128\pi\, q_f^2 e^2\delta^2\left[\frac{2}{3}+\frac{1}{3}\frac{M_\chi^2}{s}\right]$$

$$\Rightarrow \frac{d(\rho'/\rho)}{dT} = -\frac{15}{4}\sqrt{10}\frac{g_1 g_2 M_P q_f^2 e^2\delta^2}{\pi^8 T^3 g_\rho^{3/2}}\int_{2\frac{M_\chi}{T}}^\infty\left(2+\frac{M_\chi^2}{T^2}\right)$$

$$\times\, x^2\sqrt{x^2T^2 - 4M_\chi^2}\; K_2(x)\, dx.$$

Noticing that $\int x^3 K_2(x)dx = 8$, after integration on T we obtain, in the regime $T \gg M_\chi$ for Dirac annihilating particles ($g_1 = g_2 = 4$) and $q_f = 1$ as an example,

$$\frac{d(\rho'/\rho)}{dT} = -15360\sqrt{10}\frac{\alpha\alpha_D M_P}{\pi^6 T^2 g_\rho^{3/2}} \;\Rightarrow\; \left(\frac{\rho'}{\rho}\right) = 2560\sqrt{\frac{45}{\pi}}\frac{\alpha\alpha_D M_P}{\pi^6 T g_\rho^{3/2}} \qquad (3.80)$$

with $\alpha = e^2/4\pi$ and $\alpha_D = \delta^2/4\pi$. We made the hypothesis that $\rho' \simeq 0$ at the beginning of the transfers and integrated from the reheating temperature T_{RH} until T considering $T_{RH} \gg T$. Notice that we neglected in Eq. (3.76) a second term in the right-hand side corresponding to the inverse transfer from the dark bath to the

Standard Model (photons) bath. This comes from the fact we supposed that the dark system has not been populated by dark particles from the beginning of the thermal history (no coupling to the inflaton, for instance).

Once the density of dark particles reaches (through the Boltzmann equation describing the evolution of the thermal density of dark particles) an equilibrium appears between ρ' and ρ similar to the one for the Yields Y_i: $d(\rho'/\rho)/dT \simeq 0 \Rightarrow \rho' \propto \rho$. As the we supposed the dark bath is in thermal equilibrium ($\Gamma_{\text{scatter}}(T) > H(T)$), we can use Eq. (3.23) and write $\frac{\rho'}{\rho} = \left(\frac{T'}{T}\right)^4$. Eq. (3.80) then becomes

$$T' = (\alpha\alpha_D)^{1/4}(15360)^{1/4}10^{1/8}\frac{M_P^{1/4}T^{3/4}}{\pi^{3/2}g_\rho^{3/8}} \Rightarrow \left(\frac{T'}{1\text{ GeV}}\right)$$
$$\simeq 3000\sqrt{\delta}\left(\frac{T}{1\text{ GeV}}\right)^{3/4}. \tag{3.81}$$

The numerical result is shown in Fig. 3.11 where we plotted $\frac{T'}{T}$ as a function of T for $M_\chi = 100\,\text{GeV}$ and different values of δ from 10^{-6} down to 10^{-9}. We clearly see the two regimes $T \ll M_\chi$ and $T \gg M_\chi$ with the two behaviors given by Eqs. (3.78) and (3.81), respectively. We also remark that the value $\delta \simeq 10^{-6}$ is the limit value for which the dark bath does not enter in thermal equilibrium with the thermal bath.

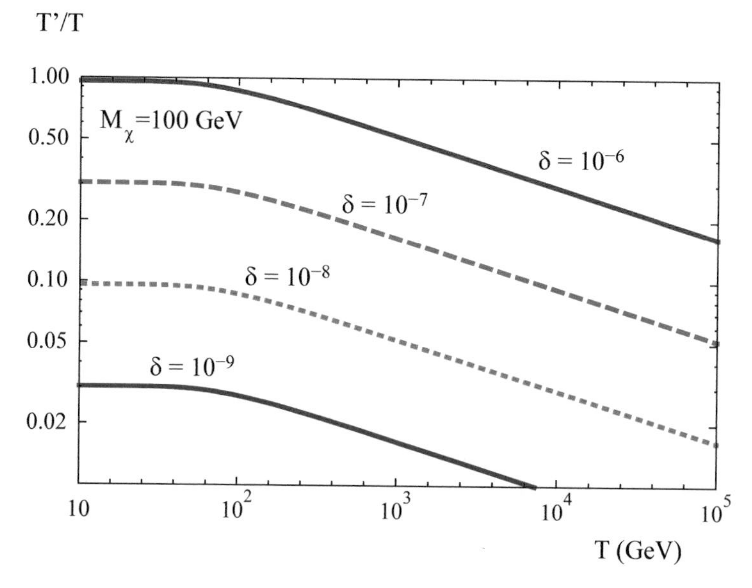

Fig. 3.11 Evolution of T', temperature in the dark bath as function of T for different values of the coupling δ and for $M_\chi = 100\,\text{GeV}$

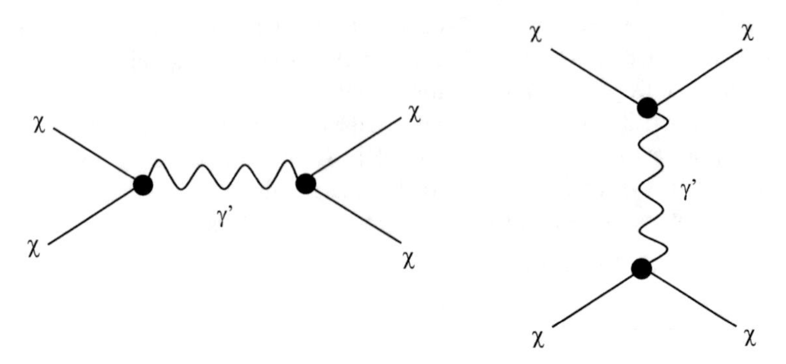

Fig. 3.12 Feynman diagrams of the $\chi\chi \to \chi\chi$ scattering which contributes to the thermalization in the dark sector of temperature T'

Indeed, for any values of δ larger than 10^{-6}, the temperature T' (and thus the density of energy ρ') will reach T (respectively, ρ) and then form a common thermal bath.

Exercise Do the same analysis with the Higgs-portal model.

3.2.4.3 Thermalization

In the previous sections, we always made the supposition (explicitly or implicitly) that the system(s) we were studying were in thermal equilibrium. We can check when this condition is realized and, if not realized, how the dark bath reaches equilibrium through a (somewhat) complex thermalization process. The condition for the dark bath to be in thermal equilibrium is $\Gamma_{scatter/dark \leftrightarrow dark}(T) > H(T)$. This translates into, before the Universe had time to double in size, there was (at least) one scattering. We show the different scattering processes in Fig. 3.12. The scattering is present to keep the distribution function of the systems of particles (Boltzmann, Fermi-Dirac, or Bose-Einstein) in equilibrium state.

These processes depicted are equivalent to the Bhabha scattering, and to a good approximation, one can keep the s-channel exchange diagram and apply Eq. (3.181)

$$|\bar{\mathcal{M}}|^2 \simeq g_D^4(1+\cos^2\theta)$$

$$\Rightarrow \langle\sigma v\rangle n_\chi^2 = \frac{T' g_\chi^2}{32(2\pi)^6}\int \sqrt{s}K_1\left(\frac{\sqrt{s}}{T'}\right)|\bar{\mathcal{M}}|^2 ds d\Omega = \frac{g_D^4 T'^4 g_\chi^2}{3(2\pi)^6}, \quad (3.82)$$

where we made the hypothesis that the dark matter is in equilibrium at a temperature T'. We want to know for which temperature T the condition of equilibrium $\Gamma_{scat}(T) > H(T)$. We then obtain at equilibrium using Eq. (3.27), $n_\chi = \frac{3}{4}\frac{\zeta(3)}{\pi^2} g_\chi T'^3$:

$$\Gamma_{scat}(T') = n_\chi \langle\sigma v\rangle_{scat} \simeq \frac{g_D^4 T' g_\chi}{9\zeta(3)(2\pi)^4}. \quad (3.83)$$

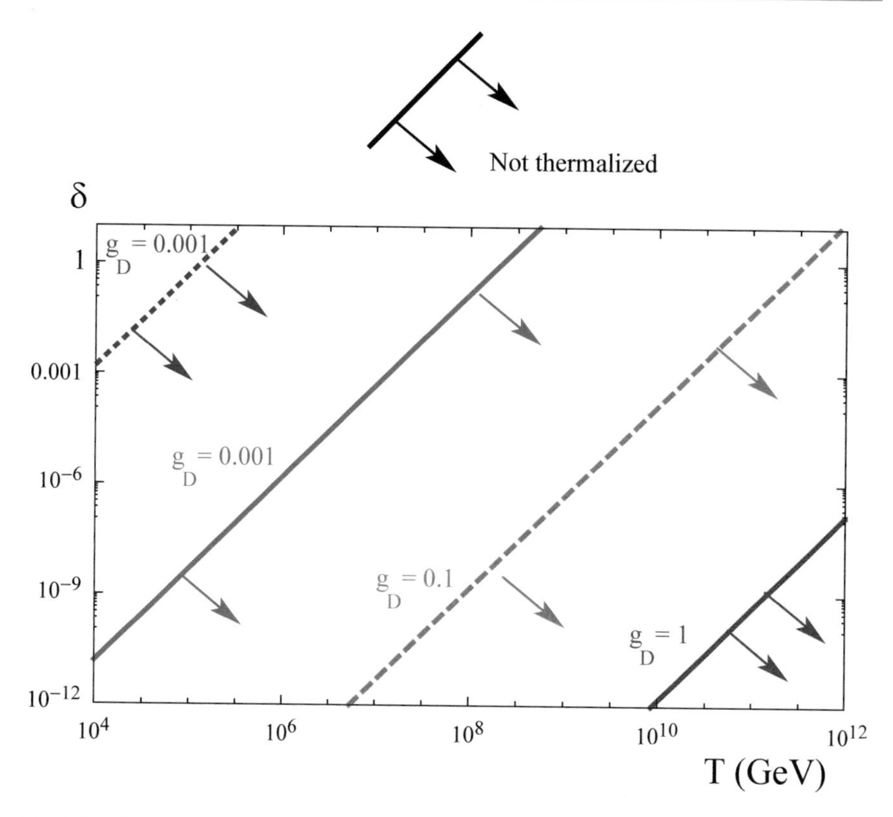

Fig. 3.13 Region of the parameter space (T, δ), T being the temperature of the Standard Model bath, where we can consider that the dark bath of temperature T' is in equilibrium with itself through scattering

Using the expression of $H(T)$ given by Eq. (2.264) and the relation between T' and T we obtained[18] in Eq. (3.81)

$$\frac{\Gamma_{scat}(T')}{H(T)} = \frac{2\, g_D^4 g_\chi}{3\zeta(3)(2\pi)^5}\sqrt{10}\frac{M_P}{T}\left(\frac{T'}{T}\right) \simeq 2.6\times 10^{14}\left(\frac{g_D}{0.1}\right)^4 \sqrt{\delta}\left(\frac{1\text{ GeV}}{T}\right)^{5/4}. \tag{3.84}$$

We show in Fig. 3.13 the region of the parameter space where thermal equilibrium is satisfied. We notice that the temperature T of the Standard Model bath from which the thermal equilibrium of the dark bath of temperature T' is attained is smaller for small values of δ and of g_D. The reasons are different for both cases. Indeed, small δ implies a few energy transfer between the two systems, and thus a low temperature T', and as a consequence, a low density of dark particles: the scattering is thus not sufficiently efficient to bring the new created dark matter from

[18] We took Dirac fermions for the dark matter and particles 1 and 2: $g_1 = g_2 = g_\chi = 4$.

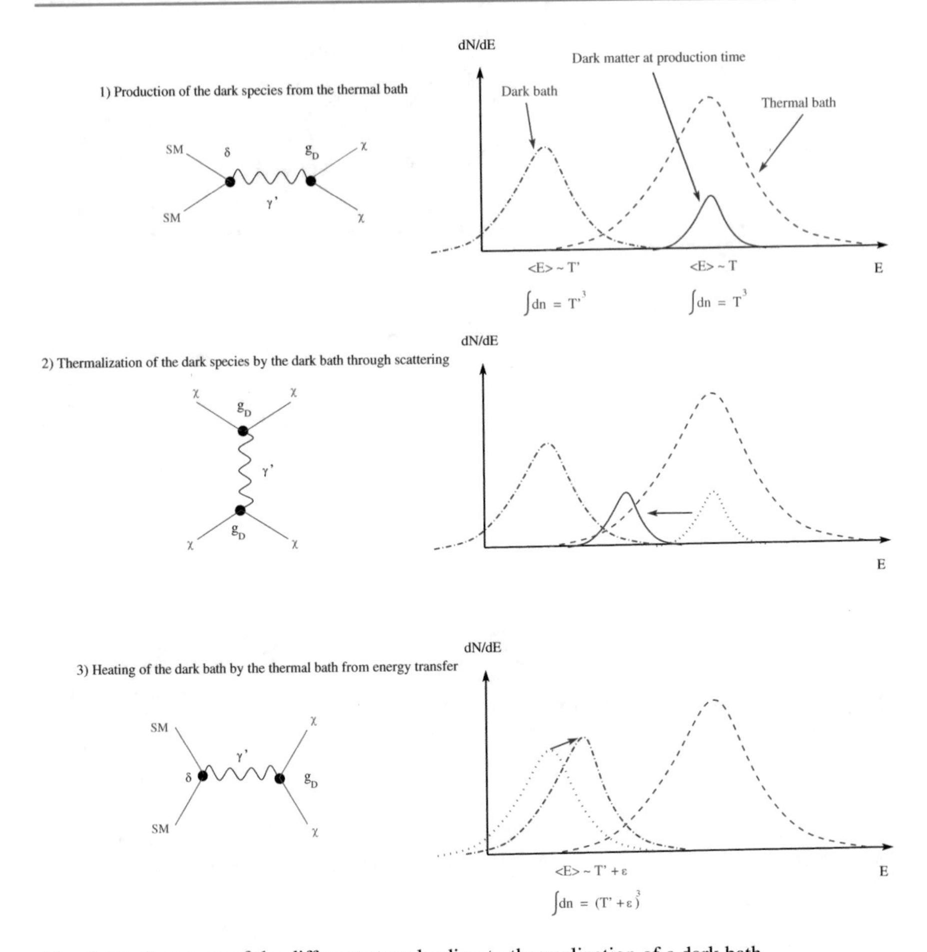

Fig. 3.14 Summary of the different steps leading to thermalization of a dark bath

annihilation of the Standard Model in the thermal bath into a new dark equilibrium state. A small value of g_D renders the scattering less efficient also but due from the too weak coupling of the self interaction of the dark matter on itself. We can also remark that the condition of thermal equilibrium we supposed to obtain Fig. 3.11 is valid.

We summarize in Fig. 3.14 the process of thermalization. The dark mater χ is produced from the thermal (Standard Model) bath with a distribution of energy dn_χ/dE_γ which follows the distribution of energy of the photons from the thermal bath, by conservation of energy: $E_\gamma = E_\chi$, so $(T_\chi)_{\text{after production}} \simeq T_\gamma = T$. However, the interaction between the dark matter and the photon is very weak (proportional to δ) and thus, the scattering of the photons on the dark matter is not sufficient to maintain the thermal equilibrium of χ. However, other χ particles already produced before and in dark equilibrium at a temperature T' between themselves can scatter on the newly produced χ to bring it into the dark bath and to follow the new distribution of energies dn_χ/dE_χ which defines the temperature T'. But in the meantime, there was a small transfer of energy between the thermal bath and the dark bath which has increased a little bit $T' \rightarrow T' + \epsilon$. Then the process loops until T' reach T and then the dark matter enters in equilibrium with the thermal bath, or until the temperature of the thermal bath T drops below the dark matter mass M_χ: the photons are then not sufficiently energetic anymore to heat the dark bath.

3.2.4.4 γ' Entering in the Dance

After some time the γ' will become sufficiently abundant to enter in thermal, and chemical equilibrium with the dark matter χ. When it was only the mediator of the self-scattering process, its density was not of real interest for the evolution of T' or n_χ. However, once it enters in the dark bath ($\Gamma_{\chi\chi\leftrightarrow\gamma'\gamma'}(T) > H(T)$), its effect on T' will be equivalent to the inverse process of the electron decoupling (see Sect. 3.3.1). Whereas electron gives its degrees of freedom to heat the bath of photons when it decouples, in this case, from conservation of entropy, the inclusion of the γ' in the thermal bath will *decrease* the temperature T' from entropy conservation in the dark system, S':

$$S'_{\text{before}} \propto g_\chi (T'_{\text{before}})^3; \quad S'_{\text{after}} \propto (g_\chi + g_{\gamma'})(T'_{\text{after}})^3$$

$$\Rightarrow \quad T'_{\text{after}} = T'_{\text{before}} \times \left(\frac{g_\chi}{g_\chi + g_{\gamma'}} \right)^{1/3} . \tag{3.85}$$

This result can also be observed from the Boltzmann equation where a new term expressing the effect of γ' on the abundance of n_χ from its annihilation enters in the game.

Exercise Recover this result by solving the Boltzmann equation.

We understood in this section how important is the process of thermalization, and how complex can be the mechanism if it involved several species of particles. Three kinds of equilibrium can co-exist in the primordial plasma: kinetic equilibrium, where all the particles share the same distribution of energy; chemical equilibrium, where the number density of particles are maintained constant from annihilation/production processes; and thermal equilibrium which is the state of kinetic + chemical equilibrium. We will now apply these techniques to the case of the neutrino decoupling and the recombination time at the CMB epoch.

3.3 The Case of Light Species

3.3.1 The Neutrino Decoupling

Before the last scattering (CMB) period, the neutrino decoupled from the plasma. Indeed, the neutrino is the only particle of the Standard Model which interacts only weakly and not electromagnetically nor strongly. Because of that, the interaction rate $\Gamma_\nu = n_\nu \sigma v$ from the scattering reaction $e\nu \to e\nu$ or annihilation $\nu\nu \to ee$ becomes smaller than the Hubble expansion rate $H(T)$ much earlier than for the other particles of the Standard Model in the plasma. We can easily compute this rate combining Eqs. (3.30) and (B.139)

$$\Gamma_\nu \simeq n_\nu \frac{G_F^2}{8\pi} 9.93 \times T^2, \text{ with } G_F = 1.1664 \times 10^{-5} \text{GeV}^{-2} \text{ and } n_\nu = \frac{3}{4}\frac{\zeta(3)}{\pi^2} g_\nu T^3.$$

At such earlier time, the Universe is radiation dominated, which means $H(T) = 0.33 g_\rho^{1/2} T^2 / M_P$, see Eq. (2.263). The condition $\Gamma_\nu(T_{\nu dec}) \lesssim H(T_{\nu dec})$ gives

$$\frac{3}{4}\frac{\zeta(3)}{\pi^2} 2 \frac{9.93}{8\pi}\left(1.16 \times 10^{-5}\right)^2 T_{\nu dec}^5 \lesssim 0.33(10.75)^{1/2} \frac{T_{\nu dec}^2}{M_P} \Rightarrow T_{\nu dec} \lesssim 3.58 \text{ MeV}. \tag{3.86}$$

From 3.6 MeV, the neutrino density is frozen and will not increase even when $T \lesssim m_e = 511$ keV because the process $ee \to \nu\nu$ is too weak compare to the interaction with the only light particles left in the plasma at such low temperature, the photons. The entropy in $e^\pm$ pair is transferred to the photon through the stronger electromagnetic interaction (Thomson scattering), but not to the neutrino through a too weak interaction.

When the temperature is below 3 MeV, the neutrinos are decoupled, which means they are "invisible" in the thermal plasma where only electrons, positrons, and photons survive. However, even if decoupled, their temperature followed the classical law $T_\nu^3 R^3 = $ cte. This is due to the fact that the effective number of neutrinos is fixed because they escaped from the plasma. The photons follow the same law, which means than even after the neutrino decoupling, $T_\nu = T_\gamma$: there is no "transfer" of the neutrino degrees of freedom to the temperature of the plasma from entropy conservation, as the neutrino still participates to the entropy of the Universe. They decoupled from the photon but still keep the same temperature and evolution. The system can be seen as the composition of two independent boxes, hermetic to each other. In one of the boxes the density of photons, electrons, and positrons follows the thermal distribution law with a greater number of degrees of freedom than in the "neutrino" box, whose neutrino energy follows a classical redshift law. It is an abuse of language to talk about "temperature" in the neutrino system, since there is no longer thermal equilibrium. However, as they are still relativistic, this

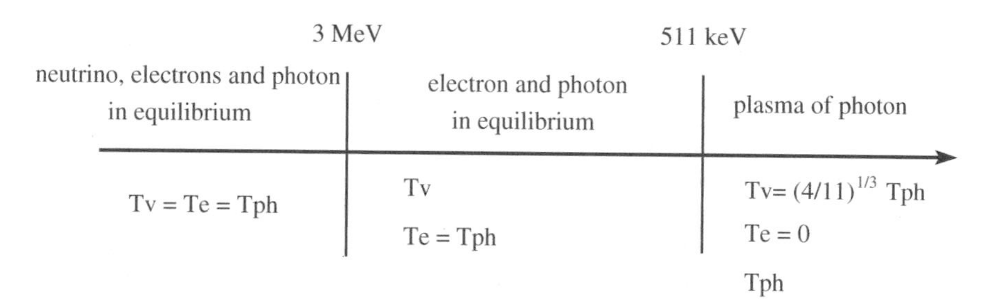

Fig. 3.15 A brief history of the neutrino

abuse of language remains coherent. Each system keeps its own entropy, without sharing it (see the box below for a more detailed explanation).

When the temperature drops below 511 keV, the electrons and positrons decouple from the plasma. They transfer their part of entropy to the photons (and not to the neutrinos which are already decoupled) and increase the photons temperature relative to the neutrinos temperature by this way. The entropy in the system (photons, electrons, positrons) being constant, we can write

$$g_s^{e+e-\gamma}(T_\gamma^{e+e-\gamma})^3 R^3 = g_s^\gamma (T_\gamma^\gamma)^3 R^3$$

with $g_s^{e+e-\gamma} = 2 + 4 \times \frac{7}{8} = \frac{11}{2}$ and $g_s^\gamma = 2$ (see Eq. (3.62), T_γ^γ being the photons temperature after the electrons decoupling and $T_\gamma^{e+e-\gamma}$ the plasma temperature before the electrons decoupling[19]). We then obtain

$$T_\gamma^\gamma = \left(\frac{11}{4}\right)^{\frac{1}{3}} T_\gamma^{e+e-\gamma} = \left(\frac{11}{4}\right)^{\frac{1}{3}} T_\nu. \tag{3.87}$$

The present photons temperature being 2.725 K on deduce that the neutrinos temperature today is

$$T_\nu = \left(\frac{4}{11}\right)^{\frac{1}{3}} \times 2.725 = 1.95\ \text{K} = 1.68 \times 10^{-4}\ \text{eV}. \tag{3.88}$$

The processes are summarized in Fig. 3.15.

As a final note we want to stress that there are finally 2 ways to obtain a freeze out condition: $n_i(T) \times \langle \sigma v \rangle < H(T)$. Or the density decreases due to the exponential suppression of the Boltzmann factor $e^{-E_i/T}$ (*kinetic* freeze out) when $T \lesssim m_i$ (case of the quarks, gauge bosons or dark matter candidate as we will see later on), or the particle is light and still relativistic, but the expansion rate $H(T)$ becomes

[19] Two degrees of freedom for the photon, 2 fermionic states for the electron, plus 2 for the positrons.

greater than the annihilation cross section $\langle\sigma v\rangle$ because of weak couplings (like in the case of the neutrino, *chemical* freeze out). In the case of the recombination temperature (decoupling of the photon) we will see that it occurred not at $T =$ 13.6 eV which is the binding energy of the atom of hydrogen but a little bit later, at 0.3 eV as we can see in Fig. 3.16, because what is important is not only that the photons have a temperature below the binding energy B_H but then $n_\gamma\langle\sigma v\rangle$ is below the rate to destroy and forbid the hydrogen atom formation. In other words, even at $T = T_\gamma = 1$ eV, there are still sufficient photons present in the queue of the distribution with $T = 13.6$ eV, able to destroy the structure of the hydrogen atom.

Another interesting point is the measurement of the relic abundance of neutrinos nowadays. From $T_\nu = 1.95$ K ($\simeq 1.7 \times 10^{-4}$ eV) one can compute from Eq. (3.27) the neutrino density par family to be $n_\nu \simeq 112$ cm^{-3} implying[20] $\rho_\nu^0 = m_\nu n_\nu^0$

$$\Omega_\nu = \frac{\rho_\nu^0}{\rho_c^0} = \frac{m_\nu\ 112\ \text{cm}^{-3}}{1.05 \times 10^{-5} h^2\ \text{GeV cm}^{-3}} \approx \frac{m_\nu}{94\ h^2\ \text{eV}} \quad \Rightarrow \quad \Omega_\nu h^2 \approx \frac{m_\nu}{94\ \text{eV}}. \tag{3.89}$$

Asking for the neutrino abundance to be less than the CMB bound on $\Omega_M \sim 0.3$ and taking $h \simeq 0.7$ we can derive the limit on the mass of a stable neutrino:

$$\sum_\nu m_\nu \lesssim 13.8\ \text{eV}. \tag{3.90}$$

This cosmological bound to the mass of a stable, light neutrino species is often referred to as the Cowsik-McClelland bound.

How many degrees of freedom are for the neutrinos nowadays?

When proposing lectures to master students, often appears the question "why should we take 3 degrees of freedom for the calculation of the entropy nowadays, whereas we know that at least 2 neutrino flavors have masses above T_ν and are then non-relativistic, out of equilibrium from their own bath." The question is very legitimate and arises when I present the computation of the degrees of freedom for the entropy, Eq. (3.63) which takes into account 3 generations of neutrino to obtain $g_s^{\text{today}} = 3.91$. First of all, even once a neutrino becomes non-relativistic (we remind that the relative

(continued)

[20] Notice that the mass density of a massive neutrino does not follow the Boltzmann suppressed evolution of classical massive particles as neutrino have already decoupled from the thermal bath, and there density thus always follows the T^3 evolution by number of particle conservation ($n_\nu(t) \times a(t) \propto n(t) \times T^{-3} = $ cst), $a(t)$ being the scale factor of the Universe.

mass measurements extracted from the observations of the oscillation gives $\Delta m^2_{21} \simeq 7.4 \times 10^{-5}$ eV^2 and $\Delta m^2_{31} \simeq 2.6 \times 10^{-3}$ eV^2 [3] implying at minima (massless lightest neutrino), $m_2 \gtrsim 8.6 \times 10^{-3}$ eV and $m_3 \gtrsim 5.1 \times 10^{-2}$ eV, both masses being larger than $T_\nu = 1.95$ K = 1.7×10^{-4} eV), it does not *quit* its thermal bath as does the dark matter, or even the electrons or any Standard Model particles. Indeed, after their decoupling, neutrinos are neither in thermal equilibrium with the photons nor with themselves. The three flavors are out of equilibrium *but* with a thermal distribution: their number density n_ν, for instance, is proportional to T_ν^3, as it is the case for the photons. However, T_ν should not be considered as a "temperature" as there is no thermal equilibrium between neutrinos. It should more be considered as "the parameter in e^{-E_ν/T_ν} in the distribution function." That is usually where the confusion is coming from. So, when computing the entropy, it is indeed the three neutrino species that should be taken into account. By conservation of the entropy $S = s \times a^3$ nothing happens when T_ν reaches values below m_2 or m_3 because T_ν has no thermal meaning. In the case of the electrons, when T_{e^-} reaches m_{e^-}, their degrees of freedom are transferred to the photon through the thermal bath, conserving that way the entropy. In the case of neutrinos when $T_\nu \lesssim m_3$, nothing happens as T_ν has no thermal meaning. To be more precise, if one separates $S = S_\gamma + S_\nu$ into its radiation and neutrino components, the neutrino part $S_\nu = (s_\nu^1 + s_\nu^2 + s_\nu^3) \times a^3 \propto \left(\frac{T_\nu^1}{T_\gamma}\right)^3 + \left(\frac{T_\nu^2}{T_\gamma}\right)^3 + \left(\frac{T_\nu^3}{T_\gamma}\right)^3$ should be invariant, independently of the value $\frac{T_\nu^i}{m_i}$. One should then really count the 3 flavors of neutrino in the computation of the present entropy.

However, another subtlety appears when one needs to compute the pressure of the neutrino gas today, and its contribution to the radiative energy density ρ_{rel} (3.40). In this case, one should only consider the relativistic species, i.e. the lightest neutrino and the photon. This explains why in the computation of g_ρ^{today} we considered only one neutrino degrees of freedom and not three. Remark that there is also the possibility that *none* of the neutrinos is relativistic today, as we do not constraint their masses but their mass differences. The same argument holds to compute the pressure of the neutrino gas π_ν. The neutrino gas should be pressureless (equation of state with $w_\nu = 0$) as $\frac{\pi_\nu}{\rho_\nu} \ll 1$. This last relation can be computed exactly from the distribution function of the massive neutrinos today but can also be understood qualitatively from the kinetic energy of a non-relativistic gas with a momentum distribution $f_\nu(p_\nu)$, $E_c^\nu = \langle \frac{p_\nu^2}{2m_\nu} \rangle \Rightarrow \frac{\pi_\nu}{\rho_\nu} \propto \frac{T_\nu}{m_\nu} \ll 1$.

(continued)

Exercises Recover the expression $\frac{\pi_\nu}{\rho_\nu}$ for a non-relativistic gas.

Show that a neutrino of 0.05 eV becomes non-relativistic at a redshift of 297, concluding that neutrinos are almost certainly relativistic at all epochs where the radiation content in the Universe is significant.

Considering by simplification that 3 species of neutrino are relativistic, show that the redshift where the radiation equal the matter is $z_{EQ} \simeq 3400$.

3.3.2 The Tremaine-Gunn Bound

The Cowsik-McClelland limit discussed above is often counterbalanced by another constraint imposed by Tremaine and Gunn in 1979. Supposing a massive neutrino as the dark matter (which was one of the first valid candidate) they noticed that the upper limit on the density in phase space of one spin state of one kind of neutrino is

$$f_\nu = \frac{1}{e^{\frac{E}{T}} + 1} \lesssim \frac{1}{2}$$

where we defined

$$dn_\nu = f_\nu(p) \frac{d^3 p}{(2\pi)^3}.$$

Notice that this upper limit on f_ν is only valid because ν is a neutrino. There does not exist such a bound for bosons which tend to form a condensate by gathering into a single state. We then obtain

$$n_\nu^{max} \lesssim \int_0^{p_{max}} \frac{1}{2} \frac{d^3 p}{(2\pi)^2} = \frac{1}{4\pi^2} \frac{|p|^3_{max}}{3}.$$

Supposing a galaxy of mass M and radius R, conservation of energy gives us

$$|p|^2_{max} = \frac{2m_\nu^2 GM}{R} \quad \Rightarrow \quad n_\nu \lesssim n_\nu^{max} = \frac{m_\nu^3}{12\pi^2} \left(\frac{2GM}{R} \right)^{\frac{3}{2}}, \tag{3.91}$$

or

$$M = \frac{4}{3} \pi R^3 n_\nu m_\nu \lesssim \frac{8}{81} \frac{M^3 G^3 R^3}{\pi} m_\nu^8. \tag{3.92}$$

Noticing that (virial theorem) $v^2 \simeq \frac{GM}{R}$ we can rewrite the limit

$$m_\nu^4 \gtrsim \sqrt{\frac{81}{8}} \frac{\pi}{R^2 v_\nu G} \tag{3.93}$$

which is called the *Tremaine-Gunn* bound. If we consider typical velocities $v_\nu \simeq 300$ km/s and $R \simeq 20$ kpc, we obtain

$$m_\nu \gtrsim 15 \text{ eV}, \tag{3.94}$$

and even

$$m_\nu \gtrsim 280 \text{ keV} \tag{3.95}$$

for a dwarf galaxy with $v_\nu \simeq 100$ km/s and $R \simeq 10$ pc. Notice that the above limits are incompatible with the Cowsik-McClelland bound (3.90). This is one of the reasons why we usually exclude hot particles as possible candidates for dark matter.

In general, light dark matter enters easily in conflict with the physics of structure formations, destroying a large amount of substructure during the streaming time between their decoupling from the bath and the time they become non-relativistic, where they begin to be the web structuring the future galaxies and clusters of galaxies. The machinery to compute the Jean mass generated during a phase where the Universe is filled by a relativistic dark constituent is technically very complex and necessitates months of simulations and statistical analysis for a trustworthy result. However, the physics leading to the minimal halo mass obtained after the smoothing passage of a dark component of mass[21] m_ν can be understood with simple orders of magnitude reasoning.[22]

Between its decoupling time and a time t, a neutrino streaming about with a velocity $v(t)$ would smooth primeval departures from a homogeneous neutrino distribution on the length scale $\lambda_s \simeq v(t) \times t$, generating a smoothing mass

$$M_s = \frac{4\pi}{3}\rho(t)\lambda_s^3 = \frac{\pi\sqrt{3}}{2}\frac{M_P^3}{\sqrt{\rho}}v^3, \tag{3.96}$$

where we used $t = \frac{1}{2H}$ and $H = \frac{\sqrt{\rho}}{\sqrt{3}M_P}$, see Eq. (3.52). Primeval mass density fluctuations would be dissipated on scales smaller than M_s. Until the matter-domination era, at t_{EQ}, the neutrino is relativistic, and then $v = c$. A quick look

[21]We keep in this paragraph the notation m_ν for the dark matter mass because it was the first candidate of that sort proposed in the literature and still can be under the form of a sterile neutrino, for instance.

[22]For a more realistic and detailed study, the reader can have a look at Sect. 5.12.1 and Eq. (5.184).

at Eq. (3.96) shows that the mass M_s increases with time, proportionally to $a^{3/2}$, a being the scale factor. On the other hand, once υ becomes non-relativistic, $\upsilon \propto a^{-1}$ and $M_s \propto a^{-3/2}$. This decreasing effect illustrates the fact that the expansion freezes the proto-structure, and no smoothing is sufficient enough to counterbalance the expansion effect. There exists then a maximum mass M_s^{max} below which the mass distribution, at the start of the structure formation (that will grow M_s even further), had been smoothed. This mass is attained at $t = t_\nu$ defined by $T_\nu(t_\nu) = m_\nu$, just before the time where the neutrino becomes relativistic. From (3.96) we obtain

$$M_s^{max} = \frac{3\sqrt{5}}{2}\frac{M_P^3}{m_\nu^2} \simeq 5 \times 10^{14} M_\odot \left(\frac{10\,\text{eV}}{m_\nu}\right)^2. \tag{3.97}$$

This means that, at $T_\nu = (\frac{4}{11})^{1/3} T_\gamma \simeq m_\nu$, all structures having masses below $\sim 10^{15}\ M_\odot$ would have been smoothed out, forbidding the presence of structures ("seeds") at the level of galaxies and dwarf galaxies, which is clearly not compatible with the observations. To ensure the preservation of structures of the order of $M_s \simeq 5 \times 10^8\ M_\odot$, one needs to add the constraint

$$m_\nu \gtrsim 10\,\text{keV}, \tag{3.98}$$

which is a limit of the same order of magnitude than (3.95). These kinds of constraints, resulting from structure formation or stability, tend to exclude hot dark matter candidates, but still allowing a window for warm dark matter, i.e. above the $\sim$keV scale, see Fig. 5.24 for a more detailed analysis.

3.3.3 Dark Radiation

3.3.3.1 Generalities

A lot of work have been published recently focusing on increasing the effective number of neutrino species in the Universe N_{eff} defined in Eq. (3.46), trying to solve the "issue" observed by spectrum of WMAP and certain telescopes. Indeed, the relativistic degrees of freedom in have been measured $N_{eff} = 3.7^{+0.8}_{-0.7}(2\sigma)$ from the inferred abundance of primordial ^{4}He. The Atacama Cosmology Telescope (ACT) and the South Pole Telescope have measured a damping tail of the CMB spectrum and, in combination with data from WMAP, Baryon acoustic oscillation (BAO), and the Hubble constant H_0 also find a preference for a value of N_{eff} greater than three, obtaining $N_{eff} = 4.6 \pm 0.8\ (1\sigma)$ and $N_{eff} = 3.9 \pm 0.4\ (1\sigma)$, respectively.

The simplest solution is to introduce new particles, which decouple at a temperature above the neutrino decoupling. The cosmic radiation content being usually expressed in term of the effective number of thermally excited neutrino species, N_{eff}, the standard value being $N_{eff} = 3.045 \simeq 3$, the number of relativistic degrees of freedom of a plasma containing the standard relativistic particles (photon γ and neutrino ν) with a new candidate $\tilde{\nu}$ will then be written with the help of Eq. (3.26):

$$\rho_R = \rho_\gamma + \rho_\nu + \rho_{\tilde{\nu}} = \frac{\pi^2}{30}\left(g_\gamma T_\gamma^4 + 3 \times \frac{7}{8} g_\nu T_\nu^4 + N_{\tilde{\nu}} \times \frac{7}{8} g_{\tilde{\nu}} T_{\tilde{\nu}}^4\right) \quad (3.99)$$

$$= \frac{\pi^2}{30}\left(g_\gamma T_\gamma^4 + \frac{7}{8} g_\nu T_\nu^4 \left[3 + N_{\tilde{\nu}} \frac{g_{\tilde{\nu}}}{g_\nu}\left(\frac{T_{\tilde{\nu}}}{T_\nu}\right)^4\right]\right) = \frac{\pi^2}{30}\left(g_\gamma T_\gamma^4 + \frac{7}{8} g_\nu T_\nu^4 N_{eff}\right)$$

with g_i the internal degree of freedom of particle i (2 for photons and neutrino) and $N_{\tilde{\nu}}$ the number of types (generations) of particles $\tilde{\nu}$ and

$$N_{eff} = 3 + N_{\tilde{\nu}} \frac{g_{\tilde{\nu}}}{g_\nu}\left(\frac{T_{\tilde{\nu}}}{T_\nu}\right)^4 = 3 + \Delta N_{eff}. \quad (3.100)$$

Following the procedure described in the Sect. 3.3.1 for the decoupling of the standard neutrino, we need to impose the conservation of entropy $g_s^i T_i^3 = \text{constant}$:

$$g_s^{\gamma\nu\tilde{\nu}} T_{\tilde{\nu}}^3 = g_s^{\gamma\nu} T_\nu^3, \quad \Rightarrow \quad \frac{T_{\tilde{\nu}}}{T_\nu} = \left(\frac{g_s^{\gamma\nu}}{g_s^{\gamma\nu\tilde{\nu}}}\right)^{1/3} \quad \Rightarrow \quad \Delta N_{eff} = N_{\tilde{\nu}} \frac{g_{\tilde{\nu}}}{g_\nu}\left(\frac{g_s^{\gamma\nu}}{g_s^{\gamma\nu\tilde{\nu}}}\right)^{4/3} \quad (3.101)$$

with the notation $g_s^{\gamma\nu\tilde{\nu}}$ being the number of degrees of freedom while γ, ν, and $\tilde{\nu}$ are still coupled while $g_s^{\gamma\nu}$ corresponds to the number of degrees of freedom when only γ and ν are coupled. This is equivalent to the phenomena of the electron decoupling transferring its degrees of freedom to the temperature of the photon. Indeed the $\tilde{\nu}$ particle is massive and couples to the neutrino and photons (through the exchange of a Z', for instance) and, as it is massive, when it decouples, $\tilde{\nu}$ disappears and thus heats the plasma. As an example we can imagine right-handed neutrinos ($N_{\tilde{\nu}} = 3$, $g_{\tilde{\nu}} = 2$) decoupling at a temperature of 3 GeV. The number of degrees of freedom at this temperature is $g_s^{\gamma\nu\tilde{\nu}}(3 \text{ GeV}) = 81$, where we used Tab.(3.2). Remembering that[23] $g_s^{\gamma\nu}(3 \text{ MeV}) = \frac{43}{4}$ we obtain $\Delta N_{eff} = 3 \times (\frac{43}{324})^{\frac{4}{3}} \simeq 0.2$.

3.3.3.2 One Example to Increase N_{eff}*

Another possibility to increase N_{eff} is to introduce a new particle χ which can play for the neutrino the same role than the electron for the photon. In other words, if a χ of mass $m_\chi \sim 10\,\text{MeV}$ couples *only* with neutrino, it decouples from the plasma *after* the neutrino decoupling temperature of $\sim$3 MeV (at $T_d \simeq m_\chi/26 \sim 0.4\,\text{MeV}$, see Eq. 3.150) it will reheat only the neutrino temperature without modifying the photon one, thus increasing N_{eff} through the increase of T_ν. The entropy can thus be decomposed, at the temperature of the neutrino system decoupling T_d into 2

[23]Which corresponds to 3 active neutrinos, $e^\pm$ and photons.

different entropies S^γ and S^ν. Each entropy is thus constant after T_d

$$S = S^\gamma + S^\nu; \quad S^\gamma = g_s^\gamma T_\gamma^3 a^3; \quad S^\nu = g_s^\nu T_\nu^3 a^3 \Rightarrow \frac{T_\nu^3}{T_\gamma^3} = \frac{S^\nu}{S^\gamma}\frac{g_s^\gamma}{g_s^\nu} \tag{3.102}$$

we can determine $\frac{S^\nu}{S^\gamma}$ noticing that at $T = T_d$, $T_\gamma = T_\nu$ we then can write

$$\frac{T_\nu^3}{T_\gamma^3} = \frac{g_s^\nu|_{T_d}}{g_s^\gamma|_{T_d}}\frac{g_s^\gamma}{g_s^\nu} \quad \Rightarrow \quad \frac{T_\nu}{T_\gamma} = \left(\frac{g_s^\nu|_{T_d}}{g_s^\gamma|_{T_d}}\frac{g_s^\gamma}{g_s^\nu}\right)^{1/3}. \tag{3.103}$$

We can notice indeed that if the neutrino is alone in its own plasma, $g_s^\nu|_{T_d} = g_s^\nu$ as neutrino of course cannot decouple from itself. We recover then the result $\frac{T_\nu}{T_\gamma} = (\frac{4}{11})^{\frac{1}{3}}$. Looking at Eq. (3.103) it is obvious that if neutrino is in equilibrium in a system isolated from the thermal plasma, if one particle decouples from the neutrino plasma it will transfer its entropy to the neutrino, increasing its temperature T_ν relative to T_γ. As a consequence, following Eq. (3.46) it is equivalent to increase the effective degree of freedom N_{eff}.

3.3.3.3 Another Example: The Case of the Mirror Dark Matter*

The mirror dark matter arises in models where you suppose a completely "mirrored" world from the Standard Model like q' quarks or ν' neutrinos, even mirrored hydrogen H' or oxygen O'. The interaction between the mirror world and our visible Universe is only due to gravity (except if one allows for some kinetic mixing between the photon γ and γ'). The two worlds are supposed to be independently in thermal equilibrium, but not in equilibrium with each other. One can then define two different temperatures (and entropies) T and S for the visible thermal bath, and T', S' for the mirror bath. Both entropy being constant, and the volume being obviously the same, we can deduce that

$$\frac{S'}{S} = \frac{a^3 s'}{a^3 s} = \frac{s'}{s} = \frac{g_s(T')T'^3}{g_s(T)T^3} \equiv x^3 = \text{constant} \Rightarrow \quad \frac{T'}{T} = x \times \left(\frac{g_s(T)}{g_s(T')}\right)^{1/3}. \tag{3.104}$$

One can then deduce the total radiation density $\bar{\rho}_R$ including the mirror bath

$$\bar{\rho}_R = \frac{\pi^2}{30}(g_\rho T^4 + g'_\rho T'^4) = \frac{\pi^2}{30}T^4 g_\rho\left[1 + \frac{g'_\rho(T')}{g_\rho(T)}\left(\frac{g_s(T)}{g'_s(T')}\right)^{4/3} x^4\right]$$
$$= \frac{\pi}{30}g_\rho T^4(1 + bx^4).$$

At the temperature we consider, above 1 MeV, $b \simeq 1$. The deviation from g_ρ can be interpreted as a new neutrino effective degree of freedom. g_ρ with 3 neutrinos (2 degrees of freedom each), 1 photon (2 degrees of freedom), and 1 electron (4 degrees of freedom) gives

$$g_\rho = 3 \times 2 \times \frac{7}{8} + 2 + 4 \times \frac{7}{8} = \frac{43}{4} = 10.75.$$

The difference between $\bar{g}_\rho = g_\rho(1 + x^4)$ and g_ρ can be written as $\Delta N_{eff} \times 2 \times \frac{7}{8} = 1.75\ \Delta N_{eff}$. Thus we have

$$\Delta N_{eff} = 6.14\, x^4. \tag{3.105}$$

Imposing the recent constraints, one obtains

$$\Delta N_{eff} < 1 \quad \Rightarrow \quad x < 0.70. \tag{3.106}$$

This means that the mirror matters should have an equilibrium temperature lower than the observable one in our Universe. It is interesting to note that the mirror dark matter is just a motivated framework. In any model where two different thermal baths are present, the constraints from N_{eff} on its temperature T' (3.106) are always valid.

3.3.4 The Recombination: Decoupling of the Photons

This period of the Universe happens quite late (∼380,000 years after the Big Bang) and is at the origin of the cosmic microwave background (CMB). Indeed, at this time, 2 phenomena took place: the temperature of the plasma in equilibrium decreased such that the photon living in the plasma did not have anymore sufficient kinetic energy to destroy the forming atom of hydrogen. The would-be-free electrons begin to form bound state (the binding energy of the hydrogen is 13.6 eV) with the proton from the plasma to compose an electromagnetic neutral atom of hydrogen. This is the *recombination era*. As a consequence of this process, the population of free electrons drastically decreases opening the way to the photons which do not scatter anymore on them and begin their very long journey toward us to create the CMB, decoupling era as it is depicted in the illustration (2.3). This corresponds to the *last scattering surface*. In this section, we will rederivate "a la Gamow" the temperature of these two processes and show that they happened in a very short period of time. As soon as the first atoms of hydrogen were formed, the photons were liberated from the plasma.

3.3.4.1 The Recombination

To simplify the picture, we will suppose that the plasma is made of electrons (e), protons (p), photons (γ), and hydrogen atoms (H) in thermal equilibrium:

$$p + e \leftrightarrow H + \gamma. \tag{3.107}$$

The Universe being electrically neutral, the electrons and protons density are equal: $n_e = n_p$. The density of each species can thus be written

$$n_i = g_i \left(\frac{m_i T}{2\pi}\right)^{3/2} e^{\frac{\mu_i - m_i}{T}} \tag{3.108}$$

and, knowing from Eq. (3.21) that $\mu_\gamma = 0$, the condition of chemical equilibrium imposes $\mu_e + \mu_p = \mu_H$ implying

$$n_H = \frac{g_H}{g_p g_e} n_p n_e \left(\frac{2\pi}{m_e T}\right)^{3/2} e^{\frac{B_H}{T}} \tag{3.109}$$

B_H being the binding energy of the hydrogen, $B_H = m_e + m_p - m_H$ and were we supposed $m_p \simeq m_H$. For $T \gg B_H$, one has $n_H \propto n_e \times n_p$ which can be understood as the probability of forming a hydrogen is proportional to the probability of presence of an electron multiplied by the probability of presence of a proton. Once the temperature approaches the hydrogen atom binding energy $T \sim B_H$, the exponential factor dominates the hydrogen density: the photons of the backgrounds are not able to split the atom of hydrogen, and its formation is exponential. If one defines the ionization fraction

$$X_e = \frac{n_p}{n_b} = \frac{n_p}{n_p + n_H} = \frac{n_e}{n_b}. \tag{3.110}$$

Using $\eta = \frac{n_b}{n\gamma} = 6 \times 10^{-10}$, Eq. (3.4), and $n_\gamma = 2\frac{\zeta(3)}{\pi^2} T^3$, Eq. (3.27), we can rewrite Eq. (3.109)[24]

$$1 - X_e = \eta\, n_\gamma\, X_e^2 \left(\frac{2\pi}{m_e T}\right)^{\frac{3}{2}} e^{\frac{B_H}{T}} = X_e^2 \sqrt{\frac{2}{\pi}} \times 2.4 \times 10^{-9} \zeta(3) \left(\frac{T}{m_e}\right)^{\frac{3}{2}} e^{\frac{B_H}{T}} \tag{3.111}$$

which is a second order equation called Saha equation that we can solve straightforwardly. The result is shown in Fig. 3.16 where we plotted the ionization fraction as function of the temperature. The exponential nature of the formation of the atom of Hydrogen is clearly visible for $T \simeq 0.3$ eV. If one defines the recombination temperature as the temperature where only 10% of the electrons are still free. Solving $X_e(T_{rec}) = 0.1$ we obtained $T_{rec} = 0.297$ eV.

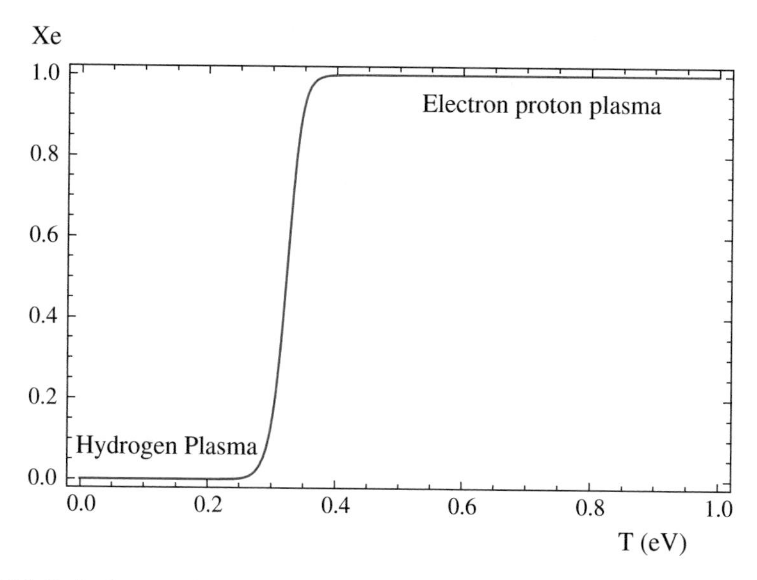

Fig. 3.16 Ionization rate X_e as function of the temperature (Eq. 3.111). The Hydrogen begins to be formed for $T \simeq 0.3$ eV. The recombination time is defined when only 10% of the electrons are still free

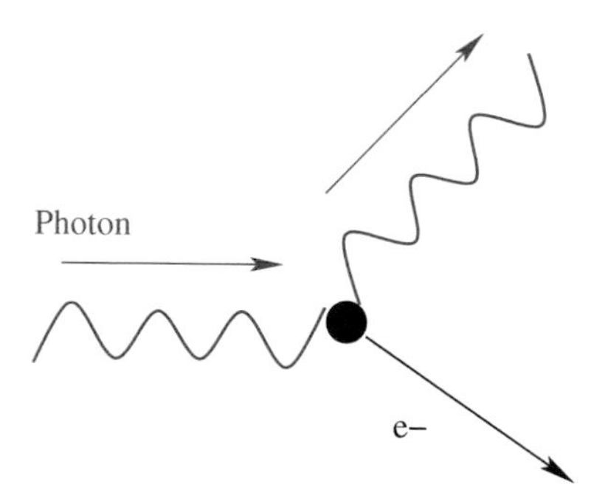

Fig. 3.17 Thomson scattering of a photon off an electron

3.3.4.2 The Last Scattering Surface

After the formation of the atoms of hydrogen, electromagnetically neutral, the mean free path of the photons increases dramatically as there are very few electrons left in the plasma to interact. We compute in this section the temperature for which the free path of the photons becomes larger than the horizon of the Universe (Hubble distance), which is named the *decoupling temperature*. Photons interact with electrons through the Thomson scattering process showed in Fig. 3.17. The

[24] with $g_e = g_p = 2$ (spin 1/2) and $g_H = 4$ (combination of two spins 1/2).

interaction rate can be written (see the Sect. 5.4.4 for more details)

$$\Gamma_\gamma = \sigma_T c n_e \;[\mathrm{s}^{-1}] \tag{3.112}$$

$$\text{with } \sigma_T = \frac{8\pi}{3}\left(\frac{q^2}{4\pi\epsilon_0 m_e c^2}\right)^2 \simeq 6.65\times 10^{-25}\mathrm{cm}^2 = 0.665\ \mathrm{barn}$$

$$\simeq 2500\ [\mathrm{GeV}^{-2}]$$

Γ_γ^{-1} corresponds to the mean time between two collisions, thus $c\Gamma_\gamma^{-1}$ is the distance that a photon will go through before colliding an electron. A photon will decouple at the temperature for which its mean path equals the Hubble horizon $H(T)$, as we depicted in Fig. 3.7. In a matter dominated Universe (which is already the state of the Universe 70,000 years after the Big Bang due to the presence of dark matter) we have from Eq. (2.58)

$$H(T) = H_0\sqrt{\Omega_M^0}\left(\frac{T}{2.35\times 10^{-4}}\right)^{3/2}$$

$$= 2.13\times 10^{-42} h\sqrt{\Omega_M^0}\left(\frac{T}{2.35\times 10^{-4}}\right)^{3/2}\ \mathrm{GeV}, \tag{3.113}$$

where T is expressed in eV. Asking for $c\Gamma_\gamma^{-1} = H^{-1}(T)$ one obtains $H(T) = n_e\sigma_T$ which can be written

$$2500\, n_e(T_{dec}) = 2.13\times 10^{-42}\sqrt{\Omega_M^0 h^2}\left(\frac{T_{dec}}{2.35\times 10^{-4}\ \mathrm{eV}}\right)^{3/2}. \tag{3.114}$$

Combining $n_e(T) = X_e(T)n_b$ with Eqs. (3.4) and (3.111) one can solve numerically the condition (3.114). We obtained $T_{dec} = 0.234$ eV, which means that *just after* the recombination, the photons become free. This results from the exponential formation of the atoms of Hydrogen, implying that very quickly n_e decreases and no electron are available anymore for any Thomson scattering.[25]

3.3.5 The Dark Ages, or Re-Ionization

There is also another epoch called "dark ages" or "re-ionization time" which runs from the decoupling/recombination time (380,000 years, see Sect. 3.3.4.1) till 100

[25]The Thomson scattering is the low energy limit of the Compton scattering. Both effects are a diffusion on the electromagnetic field by a charged particle. In the CMB, the electron being non-relativistic, the Thompson limit is a valid one.

million or one billion years. At the recombination time, the rate of recombination of electrons and protons to form neutral hydrogen was higher than the re-ionization rate. Universe was opaque due to the large rate of scattering of photons on electrons, and then became transparent once the first atom of hydrogen formed. The dark ages of the Universe started at that point, because there was no light sources other than the gradually redshifting cosmic background radiation. The second phase change occurred once objects started to condense in the early Universe that were energetic enough to re-ionize neutral hydrogen. As these objects formed and radiated energy, the Universe reverted from being neutral, to once again being an ionized plasma. This occurred between 150 million and one billion years after the Big Bang. At that time, however, matter had been diffused by the expansion of the Universe, and the scattering interactions of photons and electrons were much less frequent than before electron-proton recombination. Thus, a Universe full of low density ionized hydrogen will remain transparent, as is the case today.

3.4 The Big Bang Nucleosynthesis

3.4.1 The Context

The very early Universe was hot and dense, resulting in particle interactions occurring much more frequently than today. For example, we saw that while photon can today travel across the entire Universe without interacting with the interstellar medium (deflection or capture), resulting in a mean-free path greater than 10^{28} cm (the horizon), the mean-free path of a photon when the Universe was 1 s old was about the size of an atom. This resulted in a large number of interactions which kept the interacting constituents of the Universe in equilibrium. As the Universe expanded, the mean-free path of particles increased (thus decreasing the rates of interactions) to the point where equilibrium conditions could no longer be maintained. Different constituents of the Universe decoupled (fell out of equilibrium with the rest of the Universe) at different times, which determined their abundance.

We have previously seen the history of the decoupling of the neutrino and the photon. In this chapter, we will concentrate on the decoupling of the baryons. This decoupling is a little bit more complicated (in a sense) than that of neutrinos. The main reason is that even if it occurs around the same time (1 MeV), the neutrino is a fundamental "free" particle, unable to form bound states with particles surrounding it, which is not the case of the quarks and gluons. They are *partons*, the main constituents of baryons, and form bound states beginning by protons and neutrons. Then protons and neutrons begin to form the first nucleus (deuterium, helium, and lithium), but why at $T = 1$ MeV?

Whereas $T = 3$ MeV corresponds to the decoupling of the neutrinos which cannot "talk" anymore with the thermal bath due to the weakness of their interactions, 511 keV is the decoupling scale for the electrons due to their underproduction from a thermal bath with $T \lesssim 511$ keV as shown in Fig. 3.15, 1 MeV corresponds to the energy at which the photons are not able anymore to forbid nuclear bound state through its scattering processes. The MeV scale is indeed the scale of nuclear boundary energies. A nice summary of the mechanisms in play in these different processes can be found in [2]. As we just described in the previous chapters, when the temperature of the Universe reaches 1 MeV, the cosmic plasma consists of:

- **Relativistic particles in equilibrium: photons, electrons, and positrons.** These interact among themselves via electromagnetic interaction $e^+e^- \leftrightarrow \gamma\gamma$. The abundances of these constituents are given by Fermi-Dirac and Bose-Einstein statistics.
- **Decoupled relativistic particles: neutrinos.** They decoupled at around 3 MeV due to their weak coupling with the photons and electrons (which couple together quite "strongly" through the electromagnetic interaction). Even if decoupled from the photon bath, they continue to be present in the Universe, contributing to its entropy as we saw in Sect. 3.3.1.
- **Non-relativistic parcels: the baryons.** The baryon is present thanks to the baryonic asymmetry in the Universe. Indeed, if there was as many baryons (b) as antibaryon ($\bar{b}$), after being frozen out they should annihilate $b + \bar{b} \to \gamma\gamma$ (through a t-channel b exchange, for instance) which is clearly not the case as we are made of baryons.

What happened is that there was an initial asymmetry between baryons and anti-baryons

$$\frac{n_b - n_{\bar{b}}}{s} \simeq 10^{-10}, \tag{3.115}$$

throughout the early history of the Universe, until the anti-baryons were annihilated away at about $T \simeq 1$ MeV. We can express the ratio baryon density to photon density η_b at 1 MeV as function of the present ratio (n_b^0/n_γ^0) which gives

$$\eta_b \equiv \frac{n_b}{n_\gamma}(1\text{ MeV}) = \frac{n_b^0}{n_\gamma^0} = \frac{\frac{\rho_b^0}{m}}{n_\gamma} = \frac{\frac{\rho_c \Omega_b}{m}}{n_\gamma} = \Omega_b \left(\frac{1\text{ GeV}}{m}\right) \frac{1.05 \times 10^{-5} h^2\text{ cm}^{-3}}{411\text{ cm}^{-3}}$$
$$= 2.4 \times 10^{-8} \Omega_b h^2 \left(\frac{1\text{ GeV}}{m}\right) = 5.8 \times 10^{-10} \left(\frac{1\text{ GeV}}{m}\right)\left(\frac{\Omega_b h^2}{0.02}\right), \tag{3.116}$$

where we used the observed values given in [3]. By a simple argument one can understand why the ratio n_b/n_γ is constant with respect to the time after the decoupling from the thermal plasma. Just from the Boltzmann equation $\frac{dn_b}{dt} = -3Hn_b$, with $H = \frac{\dot{R}}{R}$. Imposing a constant entropy,

$$S = R^3 s \propto R^3 T^3, \; \frac{dS}{dt} = 0 \Rightarrow \frac{\dot{R}}{R} = -\frac{1}{T}\frac{dT}{dt}$$

implying

$$\frac{dn_b}{dt} = -\frac{3}{T}\frac{dT}{dt}n_b \Rightarrow \frac{n_b}{T^3} \propto \frac{n_b}{n_\gamma} = \text{constant},$$

where we used the formulae (3.27) and (3.61) for n_γ and s, respectively.

Therefore, there are orders of magnitude more relativistic particles than baryons at about $T \simeq 1$ MeV (and nowadays of course). The keypoint of this observation is that baryons were frozen out at 1 MeV ($t \sim 10^{-4}$ s), the number of photons and baryons/antibaryon pairs must have been nearly equal, since they were constantly exchanging (via pair production/annihilation). Indeed, the matter-anti-matter pairs must also have been equal, i.e. the Universe was equal parts matter and anti-matter. At freeze out, there must have been a slight excess of matter over anti-matter (1 part in 10^9), so the present Universe is dominated by matter and not anti-matter.

The aim of this chapter is to understand how the baryons combine themselves from 1 MeV town to keV scale to form the first light nucleus. This process bears the name "Big Bang Nucleosynthesis" (or BBN) in contrast with the Solar Nucleosynthesis developed by Hoyle and Fowler some years after. The first idea of BBN genesis was the Gamow work in 1948 [4], see [5] for historical details. If the baryons were in thermal equilibrium throughout the history of the Universe, they will naturally produce lead and iron which are some of the more stable nuclear element, with the higher binding energy. However, it is far to be the case as the number of interactions between baryons decreases as the Universe evolve (as soon as $H(T) \gtrsim \langle\sigma v\rangle_{equilibrium}$) letting baryon living with themselves but out of any equilibrium. At that moment, the binding energy plays a role. Indeed, baryons will tend naturally to form binding states, which lower their ground energetic state as soon as the photons, scattering around, are not able to interfere in the binding process by destroying newly formed bounded state.

Historical aspect of the BBN

When teaching the rudiments of the Big Bang Nucleosynthesis, I usually begin my lecture by the method originally employed by Gamow and Alpher in 1946–1948. I will not go into the historical details that can be found in the excellent book [6] but give the main line of reasoning. Gamow was aware that when the energy of the plasma dropped below the binding energy of the deuterium (2.5 MeV) its dissociation by photons scattering stopped, and the process of helium production through the deuterium interactions is open. Even if not explicitly stated in his papers, we can suppose he knew the Saha equation (3.109) and did not take the temperature to be $T_d \simeq 2 \times 10^{10}$ K, corresponding to the deuterium binding energy, but $T_d = 10^9\,\text{K} \simeq 10^{-4}$ GeV, which is the temperature obtained by solving the Saha equation. Physically, this means that even at a temperature 10 times lower, there are still sufficient

(continued)

photons to dissociate deuterium. This temperature corresponds to a time (supposing a radiation dominated Universe and $H \sim \frac{T^2}{M_P}$),

$$\Delta t_{BBN} = 10^{27} \text{ GeV}^{-1} = 667 \text{ s.} \sim 10^3 \text{ s.} \tag{3.117}$$

The baryonic density to initiate the nucleosynthesis can be found asking for at least one production:

$$n_b \times \sigma v \times \Delta t_{BBN} \gtrsim 1 \;\Rightarrow\; n_b \gtrsim 10^{17} \text{ cm}^{-3}, \tag{3.118}$$

where Gamow took for σv the Bethe value, $\sigma v \simeq 10^{-20}$ cm^{3}s^{-1}. In the 40's, the density of the Universe was evaluated by astrophysical observations to be $\rho(T_0) = 10^{-30}$ kg/cm$^3 \simeq 10^{-6}$ proton per cm^3, which gives for the present temperature T_0

$$\frac{n_b(T_d)}{n_b(T_0)} = \left(\frac{T_d}{T_0}\right)^3 \;\Rightarrow\; T_0 \simeq 20 \text{ K}, \tag{3.119}$$

which is already quite an impressive prediction considering all the approximations made, and the very few data available at this time. Notice that in their 1948 paper, Alpher and Hermann took a more precise value $T_d = 0.6 \times 10^9$ K, which gives (good exercise) $T_0 = 5.5$ K (!!).

3.4.2 Overview

To visualize the generation of the light nuclei, especially helium and lithium, one needs to imagine the system of neutrons, protons, electrons, and neutrinos in thermal equilibrium as long as the reaction

$$n + \nu \leftrightarrow p + e^- \tag{3.120}$$

respects

$$\Gamma_p = n_\nu \langle \sigma v \rangle > H(T). \tag{3.121}$$

Notice that there are more reactions than (3.120) involved in the equilibrium processes, like the neutron decay. The set of equations needs to be solved numerically, but we want to give an idea of the primordial nucleosynthesis in this section. The binding energy of deuterium or helium being around the MeV scale, the protons and neutrons are highly non-relativistic, and their density number follows Eq. (3.28), which means

$$\frac{n_n}{n_p} \simeq e^{-\frac{Q}{T}} \tag{3.122}$$

with $Q = m_n - m_p$=1.293 MeV. To determine the decoupling temperature given by (3.121), one needs to compute the cross section of the scattering process $n + \nu \to p + e^-$, via the t-channel exchange of a W boson. Using the couplings given in Appendix B.2.3 we can write an effective 4-fermions interaction “a la Fermi” for $T \ll M_W$

$$\mathcal{L}_{eff} = \frac{g}{2\sqrt{2}}\bar{\nu}\gamma^{\mu}(1-\gamma^5)e\,\frac{i\eta_{\mu\nu}}{M_W^2}\frac{g}{2\sqrt{2}}\bar{n}\gamma^{\nu}(c_V - c_A\gamma^5)p. \tag{3.123}$$

Notice that the couplings to the protons and neutrons take a more complex form than for the leptons, because they are bound states of quarks and cannot be determined only on theoretical basis but should be measured. The most recent analysis gives $c_V \simeq 1$, $c_V \simeq 1.26$. To be more precise, this correction accounts for the possibility that gluons inside the nucleon split into quark-antiquark pairs. The amplitude squared is then given by

$$|\mathcal{M}|^2 = \frac{g^4}{2M_W^2}(1+3c_A^2)(P_n.P_\nu)(P_p.P_e) \simeq 16G_F^2(1+3c_A^2)\; m_n\; E_\nu \times m_p\; E_e, \tag{3.124}$$

where the factor 3 comes from the spin structure on the nucleons, as we can see from Eq. (4.46) and we supposed the neutron and protons almost at rest at the MeV scale. Using the same method presented in the Appendix B.4.4.6 we obtain from Eq. B.110

$$\sigma = \frac{|\mathcal{M}|^2}{16\pi s} \simeq \frac{G_F^2}{\pi}(1+3c_A^2)\;(3.2T)\left(\frac{3}{2}T\right), \tag{3.125}$$

where we approximate $s = (P_p + P_e)^2 \simeq m_p^2$ and make use of Eqs. (3.30) and (3.31) for E_ν and E_ϵ, where we supposed E_e almost not relativistic. With the help of Eq. (3.27) we can compute the decoupling temperature T_d satisfying Eq. (3.121) with the help of Eq. (2.264):

$$\begin{aligned} &\Gamma_p = n_\nu\langle\sigma v\rangle \simeq 1.6\, G_F^2 T^5, \\ &\Gamma_p \lesssim H(T) = \frac{\pi}{3}\sqrt{\frac{10.75}{10}}\frac{T^2}{M_P} \quad \Rightarrow \quad T \lesssim 0.86\,\text{MeV} = T_d, \end{aligned} \tag{3.126}$$

where we supposed the radiation dominated Universe with $g_\rho = 10.75$ as we computed in Table 3.2. Implementing (3.126) in (3.122), we obtain at the freeze out temperature T_d

$$\left(\frac{n_n}{n_p}\right)_{FO} \simeq \frac{1}{5}. \tag{3.127}$$

In fact the exact numerical value, solving the complete set of reactions like the neutron decay, or electron-neutron scattering gives

$$\left(\frac{n_n}{n_p}\right)_{FO}^{num.} \simeq \frac{1}{7}. \tag{3.128}$$

It is, however, surprising how a simple analytical analysis gives a result not so far from the numerical one.

Once the protons and neutrons are frozen out, they will quickly bound to ^{4}He, the most tightly bound of all the light nuclei. 1 neutron being present for 7 protons (2 neutron per 14 protons), the mass ratio of the Helium ^{4}He (composed of 2 neutrons and 2 protons) on the total baryonic mass is then

$$Y_p = \frac{n_4}{baryons} = \frac{4}{2+14} = \frac{1}{4} = 25\%. \tag{3.129}$$

This prediction was one of the more outstanding proofs of the Standard Model of Cosmology, especially the superiority of the Big Bang Model above other alternatives like the Steady State Universe of Gold, Bondi, and Hoyle. Notice how lucky we (the Universe) are that the decoupling temperature T_d of the couple (proton, neutron) is very close to their mass difference. Indeed, if we take a closer look at Eq. (3.122), it would be enough for the temperature to be just 10 times smaller so that, at the moment of freeze out, there is not enough neutron in the thermal bath to be able to form helium in large enough quantities.

Exercise Repeat the previous analysis, explicitly calculating the average interaction rate:

$$\Gamma_p = \frac{1}{(2\pi)^5} \int \frac{d^3p_E}{2E_e} \frac{d^3p_\nu}{2E_\nu} \frac{d^3p_n}{2E_n} \delta^{(4)}(P_p + P_e - P_\nu - P_n)$$
$$\times f_e(E_e)(1 - f_\nu(E_\nu))|\mathcal{M}|^2 \tag{3.130}$$

with $|\mathcal{M}|^2$ given by the equation (3.124). Then show that

$$\Gamma_p = 1.636 \frac{G_F^2}{2\pi^3}(1+3c_A^2) m_e^5 \left(\frac{T}{m_e}\right)^3 e^{-\frac{Q}{T}} \quad [T \ll Q, m_e]$$

$$\Gamma_p = \frac{7\pi}{60}(1+3c_A^2) G_F^2 T^5 \qquad [T \gg Q, m_e], \qquad (3.131)$$

and find the decoupling temperature of protons and neutrons.

3.4.3 The Deuterium Formation

Even if the method we developed in the previous paragraph gives sufficiently accurate results to have an idea of the mechanisms in play during the Big Bang Nucleosynthesis era, we show in this section how to be a little bit more precise. Indeed, the formation of Helium is not instantaneous and passes through the deuterium bottleneck phase. Indeed, the first reaction is the binding of a neutron and a proton into a nucleus of deuterium (D):

$$n + p \rightarrow D + \gamma. \qquad (3.132)$$

The binding energy of deuterium is $B_D = 2.2\,\text{MeV}$ (see Appendix E.2), which means that $m_D = m_p + m_n - B_D = 1877.62\,\text{MeV}$. As soon as the temperature of the Universe falls below 2.2 MeV, the photons from the thermal bath are not sufficiently energetic to destroy a "molecule" of deuterium. So, naively, one would expect that the formation process of the first deuterium appears at around 1–2 MeV. It happens in fact much later due to the very weak concentration of baryons (protons and neutrons) at this temperature, reduced by the measured baryon to photon ratio (around 10^{-10}) from the particle/antiparticle asymmetry (see above). Let us check it quantitatively. At $T \simeq 1\,\text{MeV}$, we are clearly in the regime $m_i \gg T$ and can apply the formulae (3.29) and compute, not forgetting that deuterium has $g_D = 3$ because of 3 spin states of D and $g_p = g_n = 2$:

$$\frac{n_D}{n_n n_p} = \frac{g_D \left(\frac{m_D T}{2\pi}\right)^{3/2} e^{-m_D/T}}{g_n \left(\frac{m_n T}{2\pi}\right)^{3/2} e^{-m_n/T} g_p \left(\frac{m_p T}{2\pi}\right)^{3/2} e^{-m_p/T}} \qquad (3.133)$$

$$= \frac{g_D}{g_n g_p} \left(\frac{T}{2\pi}\right)^{-3/2} \left(\frac{m_D}{m_n m_p}\right)^{3/2} e^{-(m_D - m_n - m_p)/T}$$

$$= \frac{3}{4} \left(\frac{2\pi m_D}{m_n m_p T}\right)^{3/2} e^{B_D/T}$$

because $B_D = m_n + m_p - m_D$. With reasonable approximations one can write

$$\frac{n_D}{n_n n_p} \simeq \frac{3}{4}\left(\frac{4\pi}{m_p T}\right)^{3/2} e^{B_D/T}. \tag{3.134}$$

Noticing that at 1 MeV $n_p \simeq n_n \simeq n_b/2$ (the only baryons at 1 MeV are mainly protons and neutron), we can rewrite Eq. (3.134)

$$\frac{n_D}{n_b} \simeq \eta_b n_\gamma \frac{3}{4}\left(\frac{4\pi}{m_p T}\right)^{3/2} e^{B_D/T} = \frac{12\zeta(3)}{\sqrt{\pi}}\eta_b\left(\frac{T}{m_p}\right)^{3/2} e^{B_D/T}, \tag{3.135}$$

where we have used $n_\gamma = \frac{2\zeta(3)}{\pi^2}T^3$ from Eq. (3.27). Having a look to this result, we understand easily that as $\eta_b \ll 1$ and $m_p \gg 1$ MeV, the temperature at which the density of the deuterium is of the order of magnitude of the proton one is below 1 MeV. A simple solution of Eq. (3.135) gives

$$\log\left(\frac{n_D}{n_b}\right) \simeq -23 + \frac{B_D}{T} \Rightarrow \left(\frac{B_D}{T}\right)_{n_D \simeq n_b} \simeq 23. \tag{3.136}$$

We then expect the beginning of the BBN to occur at around 50 keV which is around the minute scale if we look at the relation (3.52). Small baryon to photon ratio thus inhibits nuclei production until the temperature drops well beneath the binding energy ($T \ll B_D$). This is why at temperatures $T > 0.1$ MeV virtually all baryons are in the form of neutrons and protons. Around this temperature, the production of deuterium and helium starts, but the reaction rates are too low to produce heavier elements. The heavier elements are formed in stars (triple alpha process):

$${}^4He +^4 He +^4 He \to^{12} C, \tag{3.137}$$

but that is only much later. The early Universe is not sufficiently dense for this reaction to take place, i.e. for three helium nuclei to find another on relevant timescales. The three alpha processes have been found to take place in stars by Burbidge, Burbidge, Fowler, and Hoyle in 1957 [7].

3.4.4 The Lithium Problem

Standard Big Bang Nucleosynthesis (SBBN) comprises the set of first-order Boltzmann equation on the abundances of the different species

$$\frac{dY_i}{dt} = -H(T)T\frac{dY_i}{dT} = \sum(\Gamma_{ij}Y_j + \Gamma_{ikl}Y_k Y_l + \ldots). \tag{3.138}$$

Assuming thermal distribution and the expression of $H(T)$ given in Eq. (2.263) one can solve the system of equations (3.138). This system is quite complex and generally involves numerical solutions. It should be noted that this happens in a Universe where the number of remaining baryons is very small ($\eta_b = 6 \times 10^{-10}$) and apart from the production of Helium 4 which can be seen in a fairly good approximation as a mechanism for absorbing free neutrons before they decay, the other mechanisms, as we have seen for deuterium formation, require nuclear reactions whose rates are obviously proportional to η_b, as is explicit in Eq. (3.135).

Concerning the Lithium, it is formed through a reaction between Helium 4 and Helium 3 following:

$$^{4}He +^{3} He \rightarrow^{7} Be + \gamma \rightarrow^{7} Li \tag{3.139}$$

the last reaction resulting from the capture of an ambient electron by the beryllium nucleus. This trace of Lithium, although extremely weak compared to helium-4 or even deuterium, remains an important source of data for the study of physics beyond the Standard Model. The most recent predictions give ^{7}Li/H ratio of the order of 5×10^{-10}, while observations are around 1.5×10^{-10}:

$$\left(\frac{^7Li}{H}\right)_{th} \simeq 5 \times 10^{-10}; \quad \left(\frac{^7Li}{H}\right)_{exp} \simeq 1.5 \times 10^{-10}. \tag{3.140}$$

It is not yet clear where this discrepancy comes from. It could come from the physics of the formation of the first stars that could absorb some of the lithium produced during the BBN phase. Research is still very active in this area. Another application of the study of primordial nucleosynthesis is the study of metastable relic particles.

Indeed, if a particle disintegrates during the BBN cycle or a little later, this will cause a change in the abundance ratios of the various primordial elements. One of the most effective constraints is to require that the lifetime of metastable particles respects $\tau \lesssim 100$ s. This is, for example, the case in models such as supersymmetry with the partner of the graviton (the gravitino) as dark matter candidate. In this case, the just slightly more massive supermetrics particle (usually the stau $\tilde{\tau}$, the supersymmetric partner of the tau lepton) will have a relatively long lifetime due to reduced gravitational coupling. The constraint $\tau_{\tilde{\tau}} < 100$ s is very often applied in this case. More generally, the constraints from BBN measurements applied also for the searches of unstable relics. Indeed, very generally, the BBN bounds require that unstable relics previously present in the Universe, decay with lifetime smaller than 100 s, unless the abundance of these particles is very small or only a tiny fraction of these particles decay with energetic hadrons in the final state. Among the MSSM particles, the latter condition is satisfied by the lightest neutrino or also the lightest stau.

Before concluding we want to point out that several papers have studied the Big Bang Nucleosynthesis, as well as from a model independent point of view as for more specific models. A nice and clear introduction of the subject can be

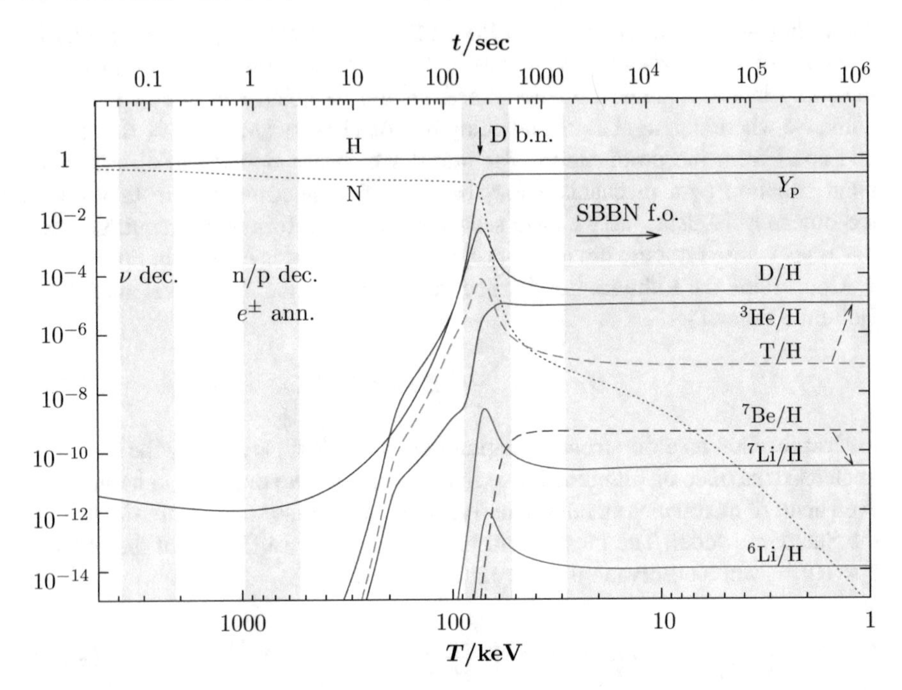

Fig. 3.18 Time and temperature evolution of all standard Big Bang nucleosynthesis relevant nuclear abundances. The vertical arrow indicates the moment at $T \simeq 8.5 \times 10^8$ K at which most of the helium nuclei are synthetized. The gray vertical bands indicate main BBN stages. From left to right: neutrino decoupling, electron-positron annihilation and n/p freeze out, D bottleneck, and freeze out of all nuclear reactions. Proton (H) and neutrons (N) are given relative to n_b, whereas Y_p denotes the ^{4}He mass fraction [2]

found in [2]. Obviously, historically speaking, the idea and first calculation of relic abundance of nuclei were done by Alpher, Herman, and Gamow in [4]. Figure 3.18 is extracted from [2].

3.5 Producing Dark Matter in Thermal Equilibrium

Before BBN occurred, the dark matter had already decoupled from the thermal bath. If it was relativistic during its decoupling the mechanism is exactly the same as in the case of the neutrino described in the Sect. 3.3.1. However, if, as in the case of a Weakly Interacting Massive Particles (WIMPs), its decoupling is done when it becomes non-relativistic, the treatment is different. This is what we will see in detail in this chapter.

3.5.1 The Boltzmann Equation

As we already discussed, the Boltzmann equation formalizes the interactions of a particle with the photons (and other relativistic species) in the thermal bath before they decoupled from it. It can be written as:

Evolution of particle A per unit of volume = particle A created by annihilation of particles in the bath B minus self-annihilation of A.

Putting this in equation, we obtain

$$\frac{1}{a^3}\frac{d\left(n_A a^3\right)}{dt} = \langle\sigma v\rangle_{B\to A} n_B^2 - \langle\sigma v\rangle_{A\to B} n_A^2 \tag{3.141}$$

$$\Rightarrow \quad \dot{n}_A + 3Hn_A = \langle\sigma v\rangle_{B\to A} n_B^2 - \langle\sigma v\rangle_{A\to B} n_A^2. \tag{3.142}$$

Using $H = \frac{1}{a}\frac{da}{dt}$ and knowing that by definition, when particles A are in equilibrium, its number per comoving frame is constant[26] and writing the evolution as function of the temperature from the relations $T^3 \times a^3$ =cte (constant entropy) $\Rightarrow da/a = -dT/T$ and $H = \frac{da/dt}{a}$ one can deduce $\frac{d}{dt} = \frac{dT}{dt}\frac{d}{dT} = -HT\frac{d}{dT}$ and write (forgetting from now on the index A)

$$\frac{dn}{dT} - 3\frac{n}{T} = \frac{\langle\sigma v\rangle}{HT}\left(n^2 - n_{eq}^2\right). \tag{3.143}$$

We can distinguish 3 regimes before giving an explicit solution of the equation.

(A) When the temperature of the plasma is large compared to the mass of the particle A ($T \gg m_A = m$) and the Hubble expansion is negligible, the particle A is relativistic and stays in thermal equilibrium with its other relativistic companions, like the photons, and their number density is given by Eq. (3.27): the Boltzmann equation (3.143) can be simplified as $n \simeq n_{eq} \propto T^3$.

(B) When the temperature of the plasma reaches the value $T \lesssim m$, then the value of $n_{eq} \propto e^{-m/T}$ decreases very steeply, almost like a step function Eq. (3.29) and the production rate of Eq. (3.143) proportional to n_{eq}^2 is almost completely suppressed. We can interpret it as if the photons in the plasma do not possess sufficient kinetic energy ($\propto T$) to produce particles A onshell. We are left with a bath of particles A which does not "discuss" anymore with the bath of the photons[27] and can only annihilate. Their density then

[26] $\frac{d[n_A \times a^3]}{dt} = 0$ when $n_A = n_A^{eq}$.

[27] More precisely, the particles A talk to the photons but cannot hear them.

decreases, but not as steeply as $e^{-m/T}$ as they have decoupled from the bath: their evolution is dictated by their annihilation rate $n^2\langle\sigma v\rangle$ whose temperature behavior depends strongly on the model and on σv. Up to now, we are still not in the "frozen" regime but in a semi-decoupled one (the Hubble parameter H is still negligible compared to $n\langle\sigma v\rangle$). The Eq. (3.143) can then be approximated by $\frac{dn}{dT} \simeq \frac{n^2\langle\sigma v\rangle}{HT}$.

(C) After a while, the Hubble parameter H dominates on the annihilation rate $\Gamma = n\langle\sigma v\rangle$ and we recover the decoupling limit ("freeze out") we discussed previously $\Gamma/H \lesssim 1$. In this regime, the number of particles A per comoving frame is constant and frozen and the Boltzmann equation (3.143) can be simply approximated by $dn/dT \simeq 3n/T \Rightarrow n \propto T^3$. The law has the same behavior than a relativistic particle in thermal bath but for a different reason (expansion rate in this case). In fact, fundamentally speaking, the reason is not so different. The keypoint being the conservation of the entropy which, at the same time, is responsible for the thermal equilibrium *and* for the evolution of the scale factor as function of the temperature: asking for a constant number of particle per comoving frame gives similar evolution for a relativistic particle in thermal equilibrium with the photons than for a particle out of equilibrium which density evolve solely due to the expansion. These 3 regimes (A), (B), and (C) are summarized in Fig. 3.19.

3.5.2 Overview

Neglecting the evolution of $g_s(T)$ we can write (3.143) in a more workable way. Indeed, introducing $Y = n/s$, s being the entropy density, and supposing adiabatic processes, we can deduce $sa^3 = S = \text{cste} \Rightarrow \dot{s}/s = -3\dot{a}/a = -3H \Rightarrow \dot{Y}/Y = \dot{n}/n - \dot{s}/s = \dot{n}/n - 3H$. We then deduce

$$\dot{n} + 3Hn = s\dot{Y} = -H\, s\, T \times \frac{dY}{dT}, \tag{3.144}$$

where we have used as previously $\frac{d}{dt} = \frac{dT}{dt}\frac{d}{dT} = -HT\frac{d}{dT}$. We can then write

$$\begin{aligned} \frac{dY}{dx} &= \frac{\langle\sigma v\rangle s}{Hx}\left(Y_{eq}^2 - Y^2\right) \\ &\simeq \frac{6.6 \times 10^9}{x^2}\frac{g_s}{g_\rho^{1/2}}\left(\frac{m}{\text{GeV}}\right)\left(\frac{\langle\sigma v\rangle}{3 \times 10^{-26}\text{cm}^3\text{s}^{-1}}\right)\left(Y_{eq}^2 - Y^2\right) \end{aligned} \tag{3.145}$$

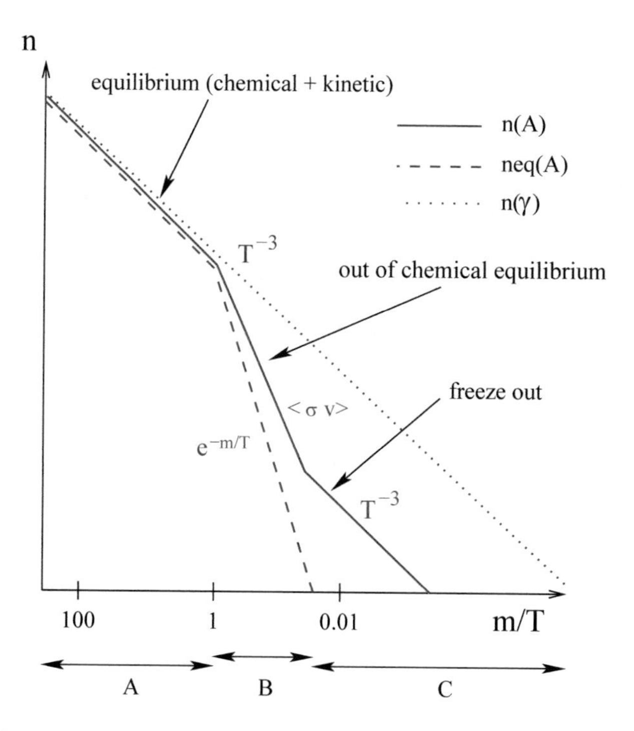

Fig. 3.19 Schematically evolution of a massive particle A as a function of the temperature following the Boltzmann equation (red-full) in comparison with the photon thermal bath density (brown-dotted) and the equilibrium state (blue-dashed) i the regimes A, B, and C (see the text for details). Kinetic equilibrium is equivalent to thermal equilibrium

with (see Sect. 3.1.3 for more details)

$$x = \frac{m}{T}, \quad s = \frac{2\pi^2}{45} g_s \frac{m^3}{x^3},$$

$$Y_{eq} = \frac{n_{eq}}{s} = \frac{45}{2\pi^4}\left(\frac{\pi}{8}\right)^{1/2} \frac{g_A}{g_s} x^{3/2} e^{-x} \tag{3.146}$$

$$\simeq 0.14 \frac{g_A}{g_s} x^{3/2} e^{-x} \quad \text{(for } x \gg 3)$$

$$\simeq 0.278 \frac{g_A}{g_s} \left[\frac{3}{4}\right] \quad \text{(for } x \ll 3) \quad \text{if } A \text{ is a bosons [fermion],} \tag{3.147}$$

g_A being the internal degrees of freedom of the particle A, and H is given by Eq. (2.264):

$$H(x) = \frac{\pi}{3}\sqrt{\frac{g_\rho}{10}} \frac{m^2}{x^2 M_P}. \tag{3.148}$$

To quantify the above argument concerning the transition between the thermal equilibrium and the freeze out time (the region B). Defining the deviation from the equilibrium Δ such as $Y = (1 + \Delta)Y_{eq}$, after neglecting Δ^2, $d\Delta/dx \ll \Delta$ Eq. (3.145) becomes

$$\Delta \simeq \frac{H}{2\Gamma_{eq}}\left(x - \frac{3}{2}\right) \simeq 5.5 \times 10^{10} \frac{g_s}{g} \frac{(x - 3/2)}{x^{1/2}} e^x \left(\frac{1\ \text{GeV}}{m}\right)\left(\frac{3 \times 10^{-26}\ \text{cm}^3\text{s}^{-1}}{\langle\sigma v\rangle}\right). \tag{3.149}$$

We can easily understand from Eq. (3.149) that the yield Y will departure from Y_{eq} when $\Gamma_{eq} = n_{eq}\langle\sigma v\rangle$ begins to be small. For $m = 100$ GeV and $\langle\sigma v\rangle = 3 \times 10^{-26}$ cm^3 s^{-1} one obtains

$$\Delta \simeq 1 \text{ (decoupling condition)} \Rightarrow x = \frac{m}{T} \simeq 26, \tag{3.150}$$

which means that the decoupling does not begin when $T \lesssim m$ but much later, when $T \lesssim m/26$. At this temperature, $Y \simeq 2Y_{eq} > Y_{eq}$.

Another way to understand the decoupling phenomenon is by writing Eq. (3.145) as

$$\frac{dY}{dx} = \frac{\Gamma_{eq}}{H} \frac{1}{xY_{eq}} \left(Y_{eq}^2 - Y^2\right). \tag{3.151}$$

It is then easy to understand that in early Universe, as long as $\Gamma_{eq} \gg H\ Y_{eq}$, the number density Y tries to track the equilibrium value Y_{eq}. This is because Y wants to change to match Y_{eq}. However, Γ_{eq} is decreasing. Eventually $\Gamma_{eq} \simeq H$ at some "time" x_f. From that point on, dY/dx becomes small and Y does not want to change anymore. We are left with $Y(x) \simeq Y(x_f)$, so that the number of particles per comoving volume has frozen out. For neutrinos this occurs while the species are still relativistic (see above), whereas for WIMPs, this occurs when the particles are already non-relativistic (see below).

That is also a fundamental point to underline. We made the hypothesis, in the neutrino case as in the WIMP case, that the dark matter, or the neutrino, is in thermal equilibrium with the bath before decoupling from it. In one case (neutrino) the decoupling comes from the fact that its coupling to the bath is too weak to counterbalance the expansion, whereas in the dark matter case, it comes from the density, reduced by the Boltzmann suppression factor $e^{-m/T}$. Their exist a third possibility, which is the production from the thermal bath while *not being* in equilibrium with it. In this case, all the particles are relativistic, the initial ones (bath) as the final ones (dark matter) and the treatment above is not valid anymore as we will see in Sect. 3.6.

Computing the freeze out temperature

I present here, a simplified way to recover quickly the relation (3.150) between the decoupling temperature and the dark matter mass. As we explained in Sect. 3.1.8, the freeze out occurs when the Hubble expansion rate begins to dominate the annihilation rate of the process $dm\ dm \rightarrow SM\ SM$. In other words, when

$$n_{dm}\langle\sigma v\rangle_{dm\rightarrow sm} \lesssim H(T) \simeq \frac{T^2}{M_P}. \tag{3.152}$$

Using the non-relativistic expression for the dark matter density n_{dm} (3.29), we can write for the decoupling temperature T_d:

$$(mT_d)^{3/2}e^{-\frac{m}{T_d}}\langle\sigma v\rangle \simeq \frac{T_d^2}{M_P}$$

$$\Rightarrow \quad \frac{m}{T_d} \simeq \ln M_P + \frac{3}{2}\ln m - \frac{5}{2}\ln T_d + 4\ln g_{dm} \simeq \ln M_P - \ln m + 4\ln g_{dm}$$

$$\Rightarrow \quad \frac{m}{T_d} \simeq \ln\frac{g_{dm}^4 M_P}{m},$$

where we supposed $\langle\sigma v\rangle \simeq \frac{g_{dm}^4}{T^2}$, and approximated $T_d \simeq m$ on the right-hand side of the equation. This gives $\frac{m}{T} = 28.5$ for $m = 100\,\text{GeV}$ and $g_{dm} = 10^{-1}$, which is quite in good agreement with (3.150). In fact, this ratio is not really sensitive to the dark matter mass and varies from 33 to 19 for dark matter masses between 1 GeV and 1 PeV.

3.5.3 Solving the Equation

Before solving the equation (3.145), let us write it in a convenient way

$$\frac{dY}{dx} = \frac{2g_s\pi}{15}\sqrt{\frac{10}{g_\rho}}\frac{m\ M_P}{x^2}\langle\sigma v\rangle\left(Y_{eq}^2 - Y^2\right) = \frac{\lambda}{x^2}\left(Y_{eq}^2 - Y^2\right) \tag{3.153}$$

with

$$\lambda = \frac{2g_s\pi}{15}\sqrt{\frac{10}{g_\rho}}\,mM_P\langle\sigma v\rangle.$$

We can also consider $g_\rho = g_s$, which is the case at such temperature as non-massive degrees of freedom like neutrino has not decoupled yet. Combining Eq. (B.112) and (B.137), a typical electroweak cross section can be written for a fermionic dark matter

$$\langle \sigma_{EW} v \rangle \simeq \frac{|\mathcal{M}|^2}{2 \times 2 \times 8\pi s} \simeq \frac{2G_F^2 s^2}{2 \times 2 \times 8\pi s} \simeq 10^{-9} \frac{s}{(2 \times 10)^2} \text{GeV}^{-2}, \tag{3.154}$$

where we normalized to a 10 GeV dark matter. We then obtain $\lambda \simeq \langle \sigma_{EW} v \rangle m M_P \simeq 10^{11} \gg 1$ for a 100 GeV particle ($m = 100$ GeV). As a consequence, such a large value of λ depletes very quickly the density of the particle (even if much slower than the equilibrium density which is Boltzmann suppressed).

3.5.3.1 s-wave

The equation (3.153) is a type of Riccati equation with no analytic solution. Despite the fact that it is not exactly solvable, we can still see through by invoking some physics intuition. We know that all happen around $x \sim 1$. In this region we can see that the left-hand side of (3.153) is $O(Y)$ while the right-hand side is $O(\lambda Y)$. We just understood that $\lambda \gg 1$, so the right-hand side must have a cancelation in the $Y^2 - Y_{eq}^2$ term.

After freeze out, Y_{eq} will continue to decrease according to the thermal suppression $e^{-m/T}$, so that $Y \gg Y_{eq}$. This happens at late times $x \gg 1$ where the Boltzmann equation reduces to

$$\frac{dY}{dx} \approx -\frac{\lambda(x)}{x^2} Y^2 \tag{3.155}$$

s-wave annihilation is characterized by $\langle \sigma v \rangle$ = cst. In this case, the resolution of the Eq. (3.153), integrating between the freeze out time x_f and the present time x_0, one obtains (noticing $x_0 \gg x_f$ and $Y_0 \ll Y_f$)

$$\frac{1}{Y_0} - \frac{1}{Y_f} = \lambda \left(\frac{1}{x_f} - \frac{1}{x_0} \right) \Rightarrow Y_0 \simeq \frac{x_f}{\lambda} \tag{3.156}$$

with x_f, time of freeze out obtained as a first approximation by solving

$$n_{eq}(x_f) \langle \sigma v \rangle = H(x_f) \tag{3.157}$$

which, for a weakly cross section, gives $x_f \simeq 20$ (a more precise solution will be given in the following section). From Y_0 one can then deduce the relic abundance of the particle A of mass m:

$$\Omega_A = \frac{\rho_A}{\rho_c^0} = \frac{m n_A}{\rho_c^0} = \frac{m Y_0 s_0}{\rho_c^0} = \frac{m\, x_f}{\lambda} \frac{2\pi^2 g_s^{\text{today}} T_0^3}{45 \rho_c^0} \tag{3.158}$$

with $\rho_c^0 = 3H_0^2/8\pi G = 1.88h^2$ g cm^{-3} $= 1.05 \times 10^{-5}h^2$ GeV cm^{-3} as we computed in Eq. (2.53) and the value of $g_s^{\text{today}} = 3.91$ as we found in Eq. (3.63). With these numbers, we obtain

$$\Omega_A = \frac{x_f}{\frac{\langle\sigma v\rangle}{(10^{-9}\text{GeV}^{-2})}} \frac{8.8 \times 10^{-7}}{\sqrt{g_\rho}\rho_c^0} \text{ GeV cm}^{-3}. \tag{3.159}$$

Taking $g_\rho \simeq 100$ (see Fig. 3.3), $x_f \sim 20$ (Eq. (3.170)) and the value of ρ_c^0 of Eq. (2.53) we can write[28]

$$\Omega_A h^2 \simeq \frac{0.17}{\frac{\langle\sigma v\rangle}{(1.2\times 10^{-26}\text{ cm}^3\text{ s}^{-1})}}. \tag{3.160}$$

This is often called "WIMP miracle." Indeed, we see that for a typical electroweak cross section the relic abundance Ω_A reaches $0.17/h^2 \simeq 0.3$ which is the measured value of the matter content in the Universe. Some corrections have to be taken into account: the velocity at decoupling time is not c, the value of x_f should be computed iteratively (see next section for a more complete calculation) and the dependence on the effective degree of freedom or mass of dark matter should be looked carefully. However, this approximation is surprisingly quite accurate in any models with s-wave dominated annihilation process.

3.5.3.2 General Solution

Now that we understood how to compute the relic abundance in a specific case, we can apply the same method in the generic case, developing $\sigma v = a + bv^2$, v being the relative velocity between the two annihilating particles.[29] Notice that in the Boltzmann equation, it is not σv which enters in the definition of λ in Eq. (3.153) but the thermal averaged cross section $\langle\sigma v\rangle$. At the temperature of interest at freeze out ($x_f = m/T_f \approx 20$ as we will compute more in detail later on) we can consider that the annihilating particles "1" and "2" are non-relativistic and thus their distributions (3.22) can be approximated by a Boltzmann distribution $f_i \simeq e^{-E_i/T} \simeq e^{-(m+p_i^2/2m)/T}$. One thus can write

$$\langle\sigma v\rangle = \frac{\int_1\int_2 d^3p_1 d^3p_2 (a+bv^2) e^{-E_1/T} e^{-E_2/T}}{\int_1\int_2 d^3p_1 d^3p_2 e^{-E_1/T} e^{-E_2/T}}. \tag{3.161}$$

[28] $\langle\sigma v\rangle$ has been normalized to a typical electroweak cross section for a 100 GeV particle: 10^{-9} GeV^{-2} $= 1.2 \times 10^{-26}$cm^3 s^{-1}, Eq. (3.154).

[29] We define the relative velocity between two particles i and j by $v_{ij} = \frac{\sqrt{(p_i\cdot p_j)^2 - m_i^2 m_j^2}}{E_i E_j}$, with p_i and E_i being four-momentum and energy of particle i.

We can then develop

$$\langle a + bv^2 \rangle = a + b\langle v^2 \rangle$$

with, Eq. (5.71)

$$v = \frac{|\mathbf{p}_2 - \mathbf{p}_1|}{\gamma m} \simeq \frac{|\mathbf{p}_2 - \mathbf{p}_1|}{m}.$$

We then have

$$v^2 = (|\mathbf{p}_1|^2 + |\mathbf{p}_2|^2 - 2\mathbf{p}_1\mathbf{p}_2 \cos\theta)/m^2,$$

θ being the angle between the two colliding particles. Noticing by symmetry that $\langle \cos\theta \rangle = 0$ and $\langle |\mathbf{p}_1|^2 \rangle = \langle |\mathbf{p}_2|^2 \rangle$, using Eq. (A.122) we then can write[30]

$$\langle v^2 \rangle = 2\frac{\int_1 p_1^2 dp_1 (p_1^2/m^2) e^{-p_1^2/2mT}}{\int_1 p_1^2 dp_1 e^{-p_1^2/2mT}} = \frac{2}{m^2}\frac{3/8\sqrt{\pi}(2mT)^{5/2}}{1/4\sqrt{\pi}(2mT)^{3/2}} = 6\frac{T}{m} = 6x^{-1} \tag{3.162}$$

giving

$$\langle \sigma v \rangle = a + \frac{6b}{x}. \tag{3.163}$$

We can now solve the Boltzmann equation (3.153) in the regime $x \gg x_f$ to obtain $Y_0 = Y(x_0)$ nowadays. Defining $\Delta = Y - Y_{eq}$, we can neglect Y_{eq} and dY_{eq}/dx which are negligible due to the Boltzmann suppression at such late times. Expressing

$$\lambda = \langle \sigma v \rangle \lambda_0, \quad \text{with } \lambda_0 = \frac{2g_s\pi}{15}\sqrt{\frac{10}{g_\rho}}\, m M_P. \tag{3.164}$$

Equation (3.153) can be rewritten

$$\frac{d\Delta}{dx} \simeq -\frac{\left(a + \frac{6b}{x}\right)\lambda_0}{x^2}\Delta^2. \tag{3.165}$$

[30]See the Sect. 3.5.5.1 for another way to lead the integration for the mean $\langle \sigma v \rangle$.

After integration from $x_f \to x_0$ and noticing $\Delta_f = \Delta(x_f) \ll \Delta(x_0) = \Delta_0$ due to the larger value of λ_0, using $(g_\rho(x_f) = g_s(x_f))$

$$\lambda_0 = \frac{2g_s\pi}{15}\sqrt{\frac{10}{g_\rho}}\, mM_P, \quad s_0 = \frac{2\pi^2}{45} g_s^0 T_0^3, \quad \text{and} \quad \rho_c^0 = 3H_0^2 M_P^2 \tag{3.166}$$

one obtains

$$Y_0 \simeq \Delta_0 = \frac{x_f}{\lambda_0}\frac{1}{\left(a + 3\frac{b}{x_f}\right)} \quad \Rightarrow \quad \Omega_0 = \frac{mY_0 s_0}{\rho_c^0} \tag{3.167}$$

$$= \frac{\pi}{9}\sqrt{\frac{g_\rho}{10}}\frac{T_0^3}{H_0^2 M_P^3}\left(\frac{x_f}{a + 3\frac{b}{x_f}}\right) \quad \approx \quad \frac{4\times 10^{-2}}{h^2\sqrt{g_\rho(x_f)}}\frac{x_f}{\frac{(a+3b/x_f)}{3\times 10^{-26}\ \mathrm{cm^3 s^{-1}}}}.$$

To complete the solution we need to compute a more precise value of x_f. For that we can "define" x_f such that the temperature when the deviation of the density from the equilibrium value is of the order of the equilibrium density itself. Putting in equation, one needs to solve $\Delta(x_f) = cY_{eq}(x_f)$, c being of the order of unity and can be calculated exactly numerically. We will keep it as a free parameter during all our calculation. However, in this regime, we cannot neglect anymore Y_{eq} and dY_{eq}/dx as around x_f the evolutions of the densities are very large and supposing the evolution of $\Delta(x)$ negligible with respect to the evolution of Y_{eq} ($d\Delta/dx \ll dY_{eq}/dx$). We can then write (3.153)

$$\frac{d\Delta}{dx} = -\frac{dY_{eq}}{dx} - \frac{\lambda_0\left(a + \frac{6b}{x}\right)}{x^2}\Delta\left(2Y_{eq} + \Delta\right) \tag{3.168}$$

with, Eq. (3.146)

$$Y_{eq} = \frac{45}{2\pi^4}\left(\frac{\pi}{8}\right)^{1/2}\frac{g_A}{g_s}x^{3/2}e^{-x} = \alpha x^{3/2}e^{-x}$$

which gives (considering $x_f \gg 1$)

$$e^{x_f} = \frac{\lambda_0\alpha}{\sqrt{x_f}}\left(a + \frac{6b}{x_f}\right)c(c+2) \tag{3.169}$$

which gives using (3.166)

$$x_f = \ln\left[c(c+2)\sqrt{\frac{45}{8}}\frac{1}{2\pi^3}\frac{g_A M_P m(a+6b/x_f)}{\sqrt{x_f}\sqrt{g_s(x_f)}}\right]$$
$$\approx \ln\left[2c(c+2)\times 10^8 \frac{m\, g_A}{\sqrt{g_s(x_f)}}\frac{(a+6b/x_f)}{3\times 10^{-26}\mathrm{cm}^3\mathrm{s}^{-1}}\right] \quad (3.170)$$

g_A being the internal degree of freedom of the annihilating particle, and c an order unity parameter which is normally determined numerically from the solution of the Boltzmann equation. It is usually set equal to 0.5 for analytical approximation. Equation (3.170) gives $x_f \simeq 23$ for a 100 GeV mass particle which confirms our approximation $x_f \gg 1$.

3.5.3.3 Hot Dark Matter

It can be interesting to see what is happening with a particle which decoupled while still relativistic (a neutrino-like particle, but decoupling at a freeze out time x_f). In this case,

$$Y_{eq}(x) = \frac{n_{eq}}{s} = \frac{45}{2\pi^4}\frac{g_{eff}}{g_s(x)}\zeta(3) \quad (3.171)$$

with $g_{eff} = g_A$ $(\frac{3}{4}g_A)$ for a bosonic (fermionic) dark matter. Applying the same reasoning than above, we can easily show that as $dY_{eq}/dx = 0$ around the freeze out time, contrarily to the cold/massive case where $dY_{eq}/dx \propto Y_{eq}$. As a consequence, $\Delta(x) \approx 0$ and so Y follows Y_{eq} and stays constant after the freeze out:

$$Y_0 = Y_{eq}(x_f) = \frac{45}{2\pi^4}\frac{g_{eff}}{g_s(x_f)}\zeta(3)$$
$$\Rightarrow \Omega_0 = \frac{g_{eff}\, g_s^0}{g_s(x_f)}\frac{8m(T_0)^3\zeta(3)}{3\pi H_0^2 M_P^2} \approx \frac{9.6\times 10^{-2}}{h^2}\frac{g_{eff}}{g_s(x_f)}\left(\frac{m}{\mathrm{eV}}\right).$$

From this result, we can obtain an upper limit for a hot dark matter (right-handed neutrino like, for instance, with $g_{eff} = 2\times 3/4$ and $g_s(x_f) \simeq 10$) from the condition $\Omega h^2 \lesssim 0.1$ we obtain $m \lesssim 7$ eV.

WIMP in brief

When I teach the Boltzmann equation or the WIMP paradigm, I like to recall the historical way it was done in 1977. It allows for a reasonably good WIMP-relic abundance calculation, without the need of the heavy Boltzmann equation machinery. Several great physicists (Weinberg, Lee, Hut, Zeldovich, Dolgov…) got the same idea at the same time. Using arguments similar to the one used by Gamow to compute the present CMB temperature (3.117), they noticed (see above) that for a "hot" species, the relic abundance can be written (I keep "ν" for historical reason, a heavy neutrino being, at this time, the more popular dark matter candidate):

$$\Omega_\nu h^2 \simeq 0.1 \frac{m_\nu}{10\,\text{eV}}. \tag{3.172}$$

They remarked that a 3 GeV candidate would need a suppression of 3×10^8 in its density to still be compatible with the cosmological observations. Considering the Boltzmann suppression factor of $e^{\frac{m_\nu}{T_\nu}}$, this corresponds to $T_\nu \simeq \frac{m_\nu}{20}$, or, considering a radiation dominated Universe,

$$t_\nu \simeq 6.67 \times 10^{-4} \left(\frac{1\text{ GeV}}{m_\nu}\right)^2 \text{ s}. \tag{3.173}$$

Asking for the annihilation to stop at t_ν (freeze out time, time when you have less than one annihilation)

$$n_\nu \times \sigma v \times m_\nu \lesssim 1 \;\Rightarrow\; n \lesssim 10^{30}\text{ cm}^{-3}, \tag{3.174}$$

where we used the electroweak cross section $\sigma v = 3 \times 10^{-27}$ cm^{3}s^{-1}, corresponding to a present density

$$n(T_0) = \left(\frac{T_0}{T_\nu}\right)^3 n_\nu \;\Rightarrow\; \Omega h^2 = \frac{n(T_0)m_\nu}{\rho_c/h^2} \simeq \left(\frac{1\text{ GeV}}{m_\nu}\right)^2. \tag{3.175}$$

We recover in this simple manner a cosmological limit $m_\nu \gtrsim 3$ GeV, called the Lee-Weinberg bound.

3.5.3.4 Another Approach to Average the Annihilation Cross Section

We just saw that the thermal average of the annihilation cross section times velocity can be done by expanding the cross section at low relative velocity, $\langle \sigma v \rangle = a + \frac{6b}{x}$ from $\sigma v = a + bv^2$, v being the relative velocity between the two colliding particles. However, there are cases where the approximation is not valid anymore due to some divergences. An integral formulation of the solution becomes then useful.

Let us consider the case of two dark matter particles with energy E_1 and E_2 and momenta $\mathbf{p_1}$ and $\mathbf{p_2}$ colliding with an angle $\cos\theta$. It is important to note that in this case, one should make the computation in the "gas rest frame" and not in the mass frame of the two colliding particles, because one should take into account a statistical population of colliding particles with a statistical distribution of momenta and angles. First of all, one should compute the cross section of two particles "1" and "2" of masses M_1 and M_2 into two particles "3" and "4" of masses m_3 and m_4 representing the relativistic particles in the plasma. From Eq. (B.110), taking into account that $\sqrt{(P_1.P_2)^2 - M_1^2 M_2^2} = E_1 E_2 v_{12}$, v_{12} being the relative velocity between the particle 1 colliding the particle 2 (from now denoted simply by v) we deduce

$$\frac{d\sigma}{d\Omega} = \frac{|\bar{\mathcal{M}}|^2}{128\pi^2 s} \frac{\sqrt{s^2 - 2m_3^2 s - 2m_4 s + (m_3 - m_4)^2}}{E_1 E_2 v} \tag{3.176}$$

$|\bar{\mathcal{M}}|^2$ being the squared mean of the amplitude over the initial spin states. As a good approximation, for the temperature of interest, we can relatively safely consider that the particles living in the bath in thermal equilibrium with the massive particles 1 and 2 are massless (they are electrons, neutrinos, quarks, and gauge bosons mainly). The massive ones decoupled from the bath (Boltzmann suppression) very quickly and do not "discuss" anymore with the other particles in thermal equilibrium. Moreover, we will concentrate on annihilation of dark matter (the generalization to coannihilation is straightforward) and we will thus suppose $M_1 = M_2 = M$. Note that during all the demonstration, we will consider annihilation of dark matter $1 + 2 \to 3 + 4$, the computation would be the same if one considers the production process $3 + 4 \to 1 + 2$.

The average of the annihilations cross section per solid angle[31] can be written (considering $m_2 = m_3 \ll s$):

$$\langle v\, d\sigma \rangle = \frac{\int f_1(E_1) f_2(E_2) \frac{d^3 p_1}{(2\pi)^3} \frac{d^3 p_2}{(2\pi)^3} \frac{|\bar{\mathcal{M}}|^2}{128\pi^2 E_1 E_2} d\Omega}{\int f_1(E_1) f_2(E_2) \frac{d^3 p_1}{(2\pi)^3} \frac{d^3 p_2}{(2\pi)^3}} = \frac{A}{n_{eq}^2} \tag{3.177}$$

[31] We must be careful that the solid angle $d\Omega$ is the one between the outgoing particles m_3 and m_4 in the center of mass of the colliding particles, to not confuse with the solid angle of the colliding particles $\cos\theta$ in which we perform the statistical average.

n_{eq} being the density of the colliding particles at equilibrium (the dark matter candidate in our case). The phase space $d^3p_1d^3p_2$ can be rewritten as

$$d^3p_1d^3p_2 = 4\pi|p_1|E_1dE_1 4\pi|p_2|E_2dE_2\frac{1}{2}d\cos\theta$$

giving

$$A = \int f_1 f_2 |p_1| dE_1 |p_2| dE_2 d\cos\theta \frac{|\bar{\mathcal{M}}|^2}{(2\pi)^4 64\pi^2} d\Omega. \tag{3.178}$$

As $\mathcal{M}$ usually is expressed as function of the center of mass energy it would be easier to make a benefit change of variables:

$$\begin{aligned} E_+ &= E_1 + E_2 \\ E_- &= E_2 - E_1 \\ s &= (P_1 + P_2)^2 = 2M^2 + 2E_1E_2 - 2|p_1||p_2|\cos\theta. \end{aligned} \tag{3.179}$$

Using the Jacobian and the new limit on integration one obtains

$$A = \int_{s=4M^2}^{\infty} \int_{E_+=\sqrt{s}}^{\infty} \int_{E_-=-\sqrt{E_+^2-s}\sqrt{1-4M^2/s}}^{E_-=+\sqrt{E_+^2-s}\sqrt{1-4M^2/s}} \frac{f_1 f_2}{(2\pi)^4} \frac{|\bar{\mathcal{M}}|^2}{256\pi^2} d\Omega dE_+ dE_- ds. \tag{3.180}$$

For the regime where $T \simeq M$ (in other words when the dark matter candidate is not relativistic at the freeze out temperature) one can approximate the Fermi-Dirac or Bose-Einstein statistical distribution f_1 and f_2 by the Boltzmann one = $f \simeq e^{-E_1/T}$; $f_2 \simeq e^{-E_2/T}$. We let the reader have a look to the Sect. 3.6 to see how this change in the case of non-thermal production, when the approximation $T \lesssim M$ is not valid anymore. In the present case, $f_1 f_2 = e^{E_+/T}$ and one can integrate over E_- and E_+ directly using the definition of modified Bessel function $K_1(z)$ of the Appendix A.6.3:

$$\begin{aligned} A &= \int_{s=4M^2}^{\infty} \int_{E_+=\sqrt{s}}^{\infty} \frac{e^{-\frac{E_+}{T}}\sqrt{s-4M^2}\sqrt{E_+^2-s}}{(2\pi)^6 32\sqrt{s}} dE_+ ds d\Omega \\ &= \frac{T}{32(2\pi)^6} \int_{4M^2}^{\infty} \sqrt{s-4M^2} K_1\left(\frac{\sqrt{s}}{T}\right) |\bar{\mathcal{M}}|^2 ds d\Omega = n_{eq}^2 \langle v d\sigma \rangle. \end{aligned} \tag{3.181}$$

This result is one of the more important of the chapter and will be very useful to solve the Boltzmann equation analytically. One can also compute n_{eq} as it is done in Eq. (3.34)

$$\langle v d\sigma \rangle = \frac{\frac{T}{32(2\pi)^6} \int_{4M^2}^{\infty} \sqrt{s - 4M^2} |\bar{\mathcal{M}}|^2 K_1 \left(\frac{\sqrt{s}}{T}\right) ds d\Omega}{\left[\frac{m^2 T}{2\pi^2} K_2 \left(\frac{m}{T}\right)\right]^2}$$

$$= \frac{1}{256\pi^2 T M^4 K_2^2(m/T)} \int_{4M^2}^{\infty} \sqrt{s - 4M^2} |\bar{\mathcal{M}}|^2 K_1 \left(\frac{\sqrt{s}}{T}\right) ds d\Omega. \quad (3.182)$$

It is also interesting to notice that if we would have looked at the production process $3 + 4 \rightarrow 1 + 2$ the result would have been the same. Indeed, setting $m_3 = m_4 = M$ in the Eq. (3.176) (the final state being in this case the dark matter), one would have obtain the same factor $\sqrt{s - 4M^2}$ of the the $1 + 2 \rightarrow 3 + 4$ case, generated this time from the integration on E_-, Eq. (3.181). This equation is thus valid for the annihilation as for the production process.

3.5.4 The Lee-Weinberg Bound

For any kind of cold dark matter, we can apply Eq. (3.160) and find the minimal mass for a typical WIMP (heavy neutrino) still in accordance with WMAP limits. Indeed, in the case of a heavy neutrino ν of mass m_ν, the Lagrangian needed to compute the annihilation cross section is

$$\mathcal{L}_\nu = Z_\mu \left[\frac{g}{2\cos\theta_W} \bar{\nu}\gamma^\mu \left(c_V^\nu - c_A^\nu \gamma^5\right) \nu + \frac{g}{2\cos\theta_W} \bar{f}\gamma^\mu \left(c_V^f - c_A^f \gamma^5\right) f \right]. \quad (3.183)$$

In the regime $M_Z \gg m_\nu$ we can write the Z_μ propagator as

$$i \frac{\eta_{\mu\nu} + P_\mu^Z P_\nu^Z / M_Z^2}{P_Z^2 - M_Z^2} \approx \frac{i\eta_{\mu\nu}}{M_Z^2}$$

which gives

$$\mathcal{M} = i \frac{g^2}{4\cos^2\theta_W M_Z^2} \bar{v}(p_{\bar{\nu}})\gamma^\mu \left(c_V^\nu - c_A^\nu \gamma^5\right) u(p_\nu)\, \bar{u}(p_f)\gamma_\mu \left(c_V^f - c_A^f \gamma^5\right) v(p_{\bar{f}}) \quad (3.184)$$

with $c_V^f = T_{3L}^f - 2q_f \sin^2\theta_W$ and $c_A^f = T_{3L}^f$, T_{3L}^f being the isospin of the particle f. We keep the c_i^ν as free parameter for our calculation. Computing $|\mathcal{M}|^2$

$$|\mathcal{M}|^2 = 32\left(\frac{g^2}{4\cos^2\theta_W M_Z^2}\right)^2 \sum_f c_f \left[\left(c_V^f\right)^2 + \left(c_A^f\right)^2\right] \tag{3.185}$$
$$\times\left[\left[(c_V^\nu)^2 + (c_A^\nu)^2\right]\left(p_f{\cdot}p_\nu\, p_{\bar f}{\cdot}p_{\bar\nu} + p_f{\cdot}p_{\bar\nu}\, p_{\bar f}{\cdot}p_\nu\right)\right.$$
$$\left.+\,2m_\nu^2\left[(c_V^\nu)^2 - (c_A^\nu)^2\right]\, p_f{\cdot}p_{\bar f}\right]$$

with c_f being the color factor of $SU(3)$ charged particles. At the limit of zero velocity, $E_\nu \simeq m_\nu$ and Eq. (3.185) can be written as

$$|\mathcal{M}|^2 = \frac{8g^4 m_\nu^4}{\cos^4\theta_W M_Z^4}\,(c_V^\nu)^2 \sum_f c_f \left[\left(c_V^f\right)^2 + \left(c_A^f\right)^2\right]$$
$$\Rightarrow \sigma v \simeq \frac{1}{4}\frac{|\mathcal{M}|^2}{32\pi^2}\frac{4\pi}{s} = \frac{g^4 m_\nu^2}{16\pi\cos^4\theta_W M_Z^4}\,(c_V^\nu)^2 \sum_f c_f \left[\left(c_V^f\right)^2 + \left(c_A^f\right)^2\right]$$

For light masses, $\sum_f c_f\ [(c_V^f)^2 + (c_A^f)^2] \simeq 6.2$. If we impose $\Omega h^2 \lesssim 0.1$, Eq. (3.160) gives $\sigma v \gtrsim 2\times 10^{-26}\text{cm}^3\ \text{s}^{-1} = 1.7\times 10^{-9}\text{GeV}^{-2}$ we then obtain

$$m_\nu \gtrsim 4.1\ \text{GeV}\ \ \text{for}\ \ \Omega_\nu h^2 \lesssim 0.1. \tag{3.186}$$

Although it is often called the Lee–Weinberg bound [8], it was discovered independently by a number of people.[32] A viable way of building light dark matter models avoiding the Lee–Weinberg bound is by postulating new light bosons. This increases the annihilation cross section and reduces the coupling of dark matter particles to the Standard Model making them consistent with accelerator experiments.

This limit was used during a long time in several publication as it justified relatively heavy (~100 GeV, or electroweak scale) dark matter. However, since the nineties, several lighter candidates appeared and when one looks in more details the solution to the Boltzmann equation, we find several exceptions. The main point is that the expression (3.184) is somewhat too simplistic as (1) we cannot decouple the Z boson so easily every time (especially near its pole mass), and (2) maybe the particle exchanged is not the Z boson, but a Z' or a supersymmetric particle. In any case, it is important to keep in mind that the Lee-Weinberg bound is only valid for a dark matter candidate coupled only to the Z boson, far from its pole and with

[32] For instance, by Pete Hut [9].

electroweak like strength coupling. We will review some exceptions to this bound in the following discussion.

3.5.5 Two Exceptions to the Boltzmann Equation*

The two exceptions we will discuss in this section are the pole region and the kinematic threshold. You can find a detailed analysis of such regime in [10].

3.5.5.1 The Pole Region

Computing the annihilation cross section of a particle χ through an intermediate particle exchanged, with a mass M_{ex} gives

$$\sigma v = \frac{\alpha_D \alpha_V s}{(s - M_{ex}^2)^2 + M_{ex}^2 \Gamma_{ex}^2}, \tag{3.187}$$

where $\alpha_D, \alpha_V \sim 0.01$ corresponds to the coupling of the exchanged particle to the dark sector (D) and the visible one (V). Γ_{ex} is the total width of the exchanged particle and[33] $s = \frac{4m_\chi^2}{1-v^2/4}$. If one defines

$$r = \left(\frac{2m_\chi}{M_{ex}}\right)^2 \text{ and } \epsilon = \left(\frac{\Gamma_{ex}}{M_{ex}}\right)^2$$

one obtains

$$\sigma v = \frac{\alpha_D \alpha_V}{M_{ex}^2} \frac{\frac{r}{1-v^2/4}}{\left[1 - \frac{r}{1-v^2/4}\right]^2 + \epsilon}. \tag{3.188}$$

In the case of a Z exchange, $\epsilon = (2.5/91.2)^2 \simeq 7.5 \times 10^{-4}$, and one can understand easily why the velocity in Eq. (3.188) should be treated with care. At zero relative velocity, the pole occurs at $r = 1$ while in general it occurs at $r = 1 - v^2/4$. The value of the cross section is then

$$\sigma v_{pole} = \frac{\alpha_D \alpha_V}{M_{ex}^2 \epsilon}. \tag{3.189}$$

However, as we saw in the previous section, in relic abundance calculation, the cross section must be thermal averaged. The standard method was to Taylor expand it to first order in v^2 and then substitutes $\langle v^2 \rangle = 6\frac{T}{m_\chi} = \frac{6}{x}$ (see Eq. (3.162) and below).

[33] See Eq. (B.111) and discussion below for details and the expression (B.152) for a concrete example of a Z' exchanged.

However, for small values of ϵ the expansion in v^2 breaks down near the pole and the standard method gives extremely poor results, even allowing the cross section to become negative!

Two methods have usually been used to treat the pole region, with quantitatively bad results. A common approach has been to factorize the "pole factor" $P(v^2) = \left(\left[1-\frac{r}{1-v^2/4}\right]^2+\epsilon\right)^{-1}$ before making the Taylor expansion and then multiply the result by the pole factor at zero velocity which would give

$$\langle\sigma v\rangle_1 = \frac{\alpha_D\alpha_V r}{M_{ex}^2}\left[1+\frac{3}{2x}\right]P(0) = \frac{\alpha_D\alpha_V r}{M_{ex}^2}\left[1+\frac{3}{2x}\right]\frac{1}{(1-r)^2+\epsilon}. \tag{3.190}$$

Another possibility is to approximate $\langle\sigma v\rangle$ by just substituting $v^2 \to \frac{6}{x}$ in σv

$$\langle\sigma v\rangle_2 = \sigma v\left(v^2=\frac{6}{x}\right) = \frac{\alpha_D\alpha_V r}{M_{ex}^2}\frac{\frac{1}{1-3/2x}}{\left(1-\frac{r}{1-3/2x}\right)^2+\epsilon}. \tag{3.191}$$

Both approximations can be valid far from the pole, but a real numerical treatment should be used when approaching it. The numerical thermalized cross section $\langle\sigma v\rangle$ is found from the calculation of the mean of σv in the center of mass frame ($\mathbf{p}_1 = -\mathbf{p}_2$)

$$\langle\sigma v\rangle = \frac{\int_1\int_2 d^3p_1 d^3p_2\delta^3(\mathbf{p}_1+\mathbf{p}_2)(\sigma v)e^{-E_1/T}e^{-E_2/T}}{\int_1\int_2 d^3p_1 d^3p_2\delta^3(\mathbf{p}_1+\mathbf{p}_2)e^{-E_1/T}e^{-E_2/T}} \tag{3.192}$$

in the non-relativistic limit, one can approximate

$$E_1 \simeq E_2 \simeq m + \frac{p^2}{2m} \quad \text{with } |p_1| = |p_2| = |p| = \gamma(v_1)mv_1 = \gamma(v/2)mv/2,$$

v being the relative velocity between particle 1 and particle 2 (see discussions after Eq. (B.111) for this point) and $\gamma(v_1) = 1/\sqrt{1-v_1^2}$. We then obtain with $x = \frac{m}{T}$ and the help of the functions (A.122)

$$\langle\sigma v\rangle = \langle\sigma v\rangle_{num} = \frac{\int p^2(\sigma v)e^{-p^2/mT}dp}{\int p^2 e^{-p^2/mT}dp} = \frac{x^{3/2}}{2\sqrt{\pi}}\int_0^\infty dv v^2(\sigma v)e^{-xv^2/4}. \tag{3.193}$$

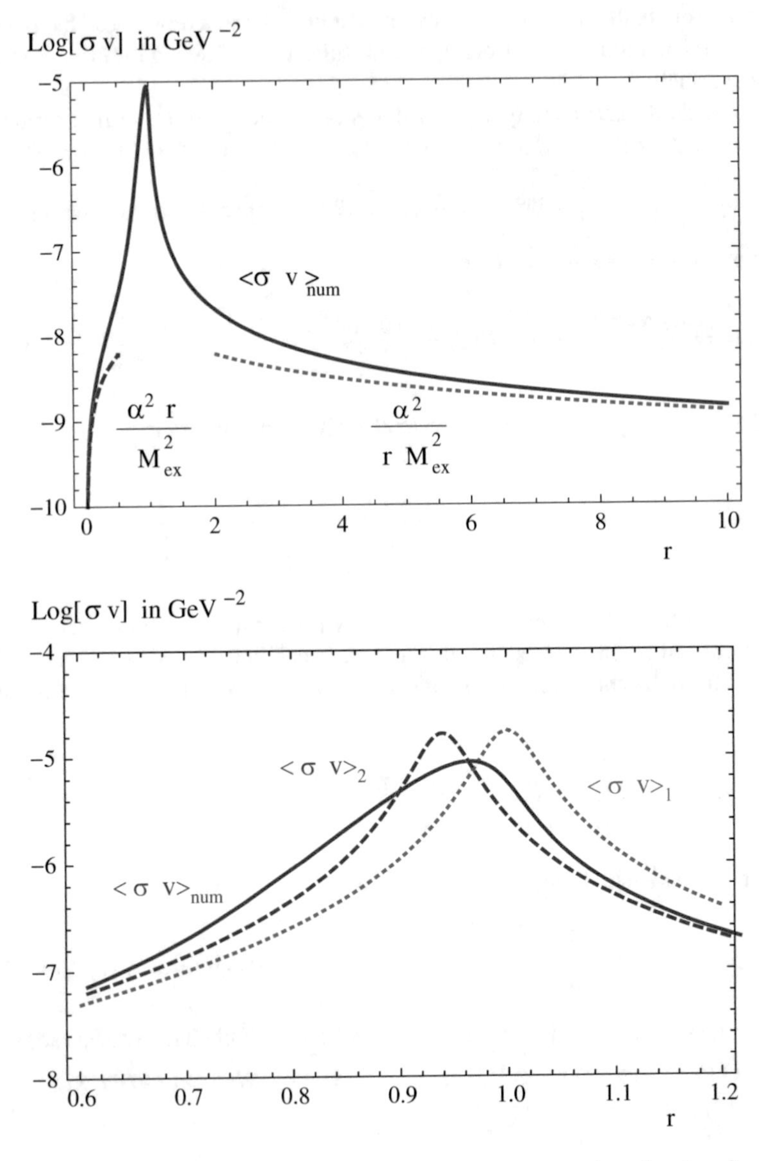

Fig. 3.20 Numerical thermal averaged cross section $\langle \sigma v \rangle_{num}$ near a pole as function of $r = \frac{2m_\chi}{M_{ex}}$ in comparison with the two asymptotic behaviors (top) and with the two approximations (called "1" and "2" in the text, bottom)

The results are shown in Fig. 3.20 where we plotted $\langle \sigma v \rangle$ as function of r for $x = 25$ and $M_{ex} = 91$ GeV, $\alpha_D = \alpha_V = 0.01$. As expected, in the first method, the pole appears at $r = 1$, whereas in the second one appears at $r = 1 - \frac{3}{2x} = 0.94$. We also show the behavior of the two asymptotic values $\sigma v = \frac{\alpha^2 r}{M_{ex}^2}$ and $\sigma v =$

$\frac{\alpha^2}{r M_{ex}^2}$ corresponding to the average cross section far from the pole where there is no dependence on the velocity, $\langle \sigma v \rangle = \sigma v$ in this case.

3.5.5.2 Kinematic Threshold

Suppose a channel where $2m_\chi \lesssim m_1 + m_2$, m_1 and m_2 being the mass of the two Standard Model-like particles. The annihilation channel $\chi\ \chi \to 1\ 2$ should then be forbidden, except in the early Universe where the temperature at the decoupling time $T \simeq \frac{m_\chi}{25}$ can be sufficient for s to reach the threshold $m_1 + m_2$. Since the χ particles are Boltzmann suppressed, the annihilation takes place at a certain rate. If the masses of the annihilation products are not too much greater, this kind of annihilation can dominate the cross section. This possibility as well as the coannihilation processes are very well described in [10].

3.6 Non-thermal Production of Dark Matter

3.6.1 The Idea

The previous discussion was based on a fundamental hypothesis: all the particles, including the dark matter, are thermal, which means they are produced, at the reheating temperature in equilibrium with the photons. There is another possibility that we should take into account. If some particles, including the dark matter, are not produced after the reheating time with the thermal bath, their evolution follows a completely different way. In this case, one should neglect in Eq. (3.153) the second term corresponding to the annihilation of the dark matter, because its number density Y is too small, and only consider the production rate from the particles living in the bath. The evolution of the number density Y is then given by

$$\frac{dY}{dx} \simeq \frac{\lambda}{x^2} Y_{eq}^2 = \frac{2 g_s \pi}{15} \sqrt{\frac{10}{g_\rho}}\, m M_P \frac{\langle \sigma v \rangle}{x^2} Y_{eq}^2 = \frac{2 g_s \pi}{15} \sqrt{\frac{10}{g_\rho}}\, m M_P \frac{\langle \sigma v \rangle}{\tilde{s}^2 x^2} n_{eq}^2$$

$\tilde{s}$ being the density of entropy of the system (3.61) at the temperature T and $x = \frac{m}{T}$. Using Eq. (3.181) ($A = \langle \sigma v \rangle n_{eq}^2$) we obtain

$$\frac{dY}{dx} = \frac{2 g_s \pi}{15} \sqrt{\frac{10}{g_\rho}} \frac{m M_P}{x^2} \frac{T^4}{16 (2\pi)^6 \tilde{s}^2} \int_{\frac{2m}{T}}^{\infty} z \sqrt{z^2 - 4x^2} K_1(z) |\mathcal{M}|^2 dz d\Omega \tag{3.194}$$

with $z = \frac{\sqrt{s}}{T}$. Developing $\tilde{s} = \frac{2\pi^2}{45} g_s T^3$ we obtain

$$\frac{dY}{dx} = \frac{270\sqrt{10}}{\pi} \frac{M_P}{m} \frac{1}{g_s \sqrt{g_\rho} 16 (2\pi)^8} \int_{2x}^{\infty} z \sqrt{z^2 - 4x^2} K_1(z) |\mathcal{M}|^2 dz d\Omega. \tag{3.195}$$

We can find solutions to this equation for canonical regimes. Before studying a microscopic defined model, let us first suppose a simplified model with a mediator H coupling to the SM and dark matter with a strength α'. The amplitude squared can then be approximated (to some numerical factors of the order of unity) by

$$|\mathcal{M}|^2 \simeq \frac{(\alpha')^2 s^2}{(s - M_H^2)^2} \quad \text{with } \alpha' = \frac{(g')^2}{4\pi}. \tag{3.196}$$

At the temperature of interest (around the reheating time) we can safely neglect the mass of the dark matter compared to the center of mass energy s: $m \ll \sqrt{s}$. We then have to consider 2 cases: very heavy mediators, with $M_H \gg T_{RH}$ and thus $M_H \gg \sqrt{s}$ or light mediator, $M_H \ll \sqrt{s}$. We will distinguish the two cases in the following.

3.6.1.1 Case A: Heavy Mediator: $M_H \gg T_{RH}$

In this case, $|\mathcal{M}|^2$ can be approximated by $|\mathcal{M}|^2 \approx (\alpha')^2 s^2 / M_H^4$. At the temperature of interest, we can consider that $g_\rho = g_s$. The Eq. (3.195) then becomes

$$\frac{dY}{dx} = \frac{48(\alpha')^2}{(2\pi)^7} \frac{270\sqrt{10}}{g_s^{\frac{3}{2}}\pi} \frac{M_P m^3}{x^4 M_H^4} \Rightarrow Y(T) = \frac{4}{(2\pi)^9} \frac{270\sqrt{10}}{g_s^{\frac{3}{2}}\pi} \frac{M_P}{(M_H/g')^4} \left[T_{RH}^3 - T^3\right] \tag{3.197}$$

T_{RH} being the reheating temperature. From Eq. (3.197) we can see that all the dark matter are produced around the reheating temperature. Very quickly Y reach its maximum value $Y_\infty = Y(T \to 0)$. Physically speaking, it means that only the photon and relativistic particles which possess large energy (around T_{RH}) can contribute to the production of the dark matter candidate. As soon as the temperature decreases below T_{RH} the production process is too weak compared to the expansion rate. From Eq. (3.197) one can also deduce the relation between the heavy particle mediator and the temperature of reheating to produce sufficiently dark matter to respect WMAP bound, $\Omega h^2 = 0.1$. We can indeed write

$$Y_\infty = \frac{\Omega \rho_c^0}{m\, s_0} = \left(\frac{\Omega}{0.1}\right)\left(\frac{100\text{ GeV}}{m}\right)\frac{135 \times 10^{-3} H_0^2 M_P^2}{2\pi^2 g_s^0 T_0^3}\text{ GeV}^{-1} \Rightarrow \tag{3.198}$$

$$M_H^4 = (g')^4 \frac{3\sqrt{10}}{4\pi} \frac{g_s(T_0)}{g_s^{3/2}(T_{RH})} \left(\frac{16 \times 10^3}{3(2\pi)^7}\right)\left(\frac{T_{RH}^3 T_0^3}{H_0^2 M_P}\right)\left(\frac{0.1}{\Omega}\right)\left(\frac{m}{100\text{ GeV}}\right)\text{ GeV}$$

$$\Rightarrow \quad M_H \simeq 4.6 \times 10^5\, g' \left(\frac{m}{100\text{ GeV}}\right)^{1/4}\left(\frac{0.1}{\Omega h^2}\right)^{1/4} T_{RH}^{3/4}\text{ GeV}^{1/4}. \tag{3.199}$$

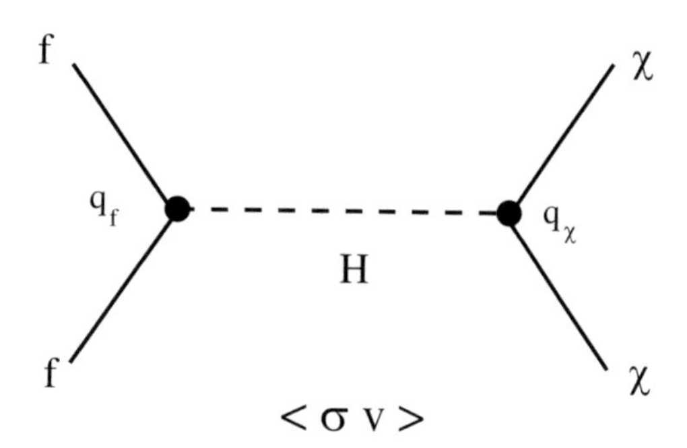

Fig. 3.21 Feynman diagram representing the process of production of dark matter from annihilation of the Standard Model particles living in the bath through the exchange of a mediator H

The last expression[34] gives us the minimum value required by M_H for each reheating temperature to avoid the overclosure of the Universe ($\Omega h^2 < 0.1$):

$$M_H^{\min} = 4.6 \times 10^5 \, g' \left(\frac{m}{100 \text{ GeV}}\right)^{1/4} T_{RH}^{3/4} \text{ GeV}^{1/4}. \tag{3.200}$$

The results are shown in Fig. 3.21.

It is important to notice several keypoints at this stage. First of all, the approximation concerning the distribution (Maxwell-Boltzmann approximation) we used to obtain Eq. (3.195) is not valid anymore for large temperature. Strictly speaking we should take the Fermi-Dirac expression of the distributions f_1 and f_2. A complete numerical analysis gave us a 44% of differences between the two treatments. The reader can have a look at Sect. 2.4.3 where we treat this case in detail. Secondly, α' is in fact the interaction strength multiplied by the charges of the interacting particles, so factor of unities can also enter in the results: $\alpha' = q_{DM}^2 q_{SM}^2 \alpha'_{real}$. Finally our condition is valid in the approximation $M_H > T_{RH}$ (which was our hypothesis) which corresponds to, from Eq. (3.199)

$$g' > 2.1 \times 10^{-6} \, T_{RH}^{1/4} \left(\frac{100 \text{ GeV}}{m}\right)^{1/4} \left(\frac{\Omega h^2}{0.1}\right)^{1/4} \text{ GeV}^{-1/4}, \tag{3.201}$$

lower bound which stay perturbative in any cases. There is, however, other scenario possible if the mediator is light (let suppose massless for a good approximation). We will study them in the next section.

3.6.1.2 Case B: Light Mediator: $M_H \ll s$, Weak-Like coupling: $\alpha' \simeq \alpha_{EW}$

When the particle exchanged between the dark sector and the visible world is very light or even massless, the thermal history is quite different. Depending on the value of the coupling α', the dark matter can reach the thermal equilibrium before that the temperature of the plasma drops below the mass of the dark matter candidate

[34] As an indication, $Y_\infty \simeq 3.3 \times 10^{-12} \left(\frac{100 \text{ GeV}}{m}\right)\left(\frac{\Omega h^2}{0.1}\right)$.

m, or later. The former case corresponds to relatively large value of α' (of the order of electroweak coupling) and we recover the classical WIMP-like scenario, even if the dark matter was not produced with the SM ones at the reheating time, but progressively. In the former case, however, the coupling is so feeble that the dark matter abundance reaches the critical value before reaching the thermal bath equilibrium. Approximating $M_H \ll \sqrt{s}$, we can approximate $|\mathcal{M}|^2 \approx (\alpha')^2$. The equation (3.195) becomes

$$\frac{dY}{dx} = \frac{270\sqrt{10}}{\pi} \frac{(\alpha')^2}{g_s\sqrt{g_\rho}4(2\pi)^7} \frac{M_P}{m} \tag{3.202}$$

$$\Rightarrow \; Y(T) = \frac{270\sqrt{10}}{\pi} \frac{(\alpha')^2}{g_s\sqrt{g_\rho}4(2\pi)^7} \left(\frac{M_P}{T}\right) \simeq (\alpha')^2 \, \frac{4.2 \times 10^{11} \text{ GeV}}{T}$$

contrarily to the heavy mediator case, the evolution is very fast with decreasing temperature. For a relativistic particle, one can compute the temperature for which the dark matter number density reach its equilibrium number:

$$Y_{eq} = \frac{n_{eq}(T)}{s(T)} = \frac{45\zeta(3)g_{eff}}{2\pi^4 g_s(T)} \approx 2.6 \times 10^{-3} \, g_{eff} \tag{3.203}$$

giving $Y_{eq} \simeq 7.8 \times 10^{-3}$ for a Dirac fermion. If we suppose a classical coupling ($\alpha' \sim 10^{-3}$) and using Eq. (3.202) one can see that the equilibrium is reached for $T \simeq 5 \times 10^7$ GeV so very quickly after the reheating. Once $Y = Y_{eq}$, $\frac{dY}{dx} \simeq 0$ from Eq. (3.153), and we recover the classical WIMP thermal history with freeze out scenario. In fact, for weakly like coupling dark matter, there is no real difference if the candidate is produced at the reheating time at the same time that the Standard Model particles (through the decay of the inflaton, for instance) or if it is produced by Standard Model particles annihilation because in the later case it reaches its thermal value vary fast. The process is depicted in Fig. 3.22.

3.6.1.3 Case C: Light Mediator: $M_H \ll s$, Feebly Like Coupling: $\alpha' \ll \alpha_{EW}$

In this scenario, the coupling is so feeble that the dark matter particle is not sufficiently produced by SM annihilation: when the temperature of the thermal bath reaches $T \simeq m$, the production is completely frozen "in." The particles produced by this mechanism are called FIMP as Freeze In (or Feebly Interacting) Massive Particle. The evolution follows the Eq. (3.202) but with very tiny values of α'. The value of the coupling needed to respect WMAP can be obtained asking for the number density to reach the relic abundance density just before the decoupling temperature ($T_f \simeq \frac{m}{23}$, Eq. 3.170) where the particles of the plasma do not have sufficient energy to produce the dark matter species (exponential suppression). Strictly speaking, we cannot really talk about "decoupling temperature" or "decoupling time" as the particle was never coupled to the plasma. This happens when $Y(T_f \approx \frac{m}{23}) = Y_\infty$

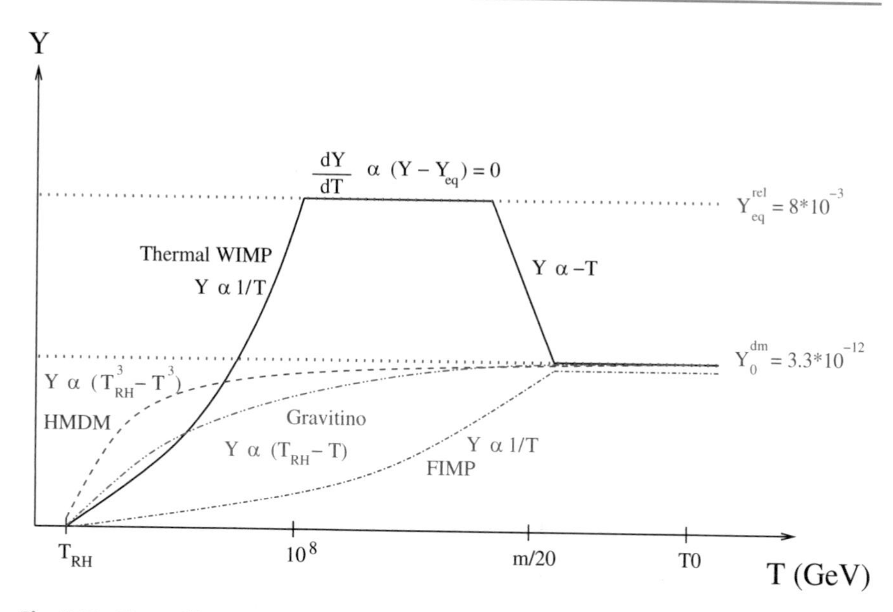

Fig. 3.22 Three different ways to produce dark matter from the Standard Model bath and obtain the correct critical relic abundance $Y_\infty \simeq 3.3 \times 10^{-12} \left(\frac{100\ \text{GeV}}{m}\right)\left(\frac{\Omega h^2}{0.1}\right)$: Thermal WIMP, Heavy Mediated Dark Matter (HMDM), and Feebly Interacting Dark Matter (FIMP)

which gives

$$Y\left(\frac{m}{23}\right) = Y_\infty \;\Rightarrow\; (\alpha')^2 = \left(\frac{\Omega}{0.1}\right)\frac{g_s^2}{\sqrt{g_\rho}}\, 3\times 10^{-1}(2\pi)^4 \frac{M_P H_0^2}{T_0^3} \simeq 10^{-22}$$
$$\Rightarrow \alpha' = \alpha_{FIMP} \approx 10^{-11} \tag{3.204}$$

this scenario is summarized in Fig. 3.22.

3.6.2 Axion as a Dark Matter Candidate*

Several dark matter candidates are analyzed in this book. We want to detail in this section the calculation of the relic abundance for one of the first that appears in the literature: the axion dark matter proposed by Pierre Sikivie and Laurence Abbott [11].

3.6.2.1 The Thermal Production

Axion particles have different origins. It was introduced in 1977 to solve the Strong-CP problem (see Appendix B.7 for details). One property of an axion ϕ_a is its coupling to the Standard Model:

$$\mathcal{L}_a = \frac{\phi_a}{f_a} F_{\mu\nu} \tilde{F}^{\mu\nu}, \tag{3.205}$$

where $F_{\mu\nu}$ is the field strength of a gauge boson A_μ, $\tilde{F}_{\mu\nu} = \frac{1}{2}\epsilon_{\mu\nu\rho\sigma}F^{\rho\sigma}$ and f_a a typical breaking scale of the theory. Historically, the coupling involved only the gluonic fields, but it is now common to find in the literature this kind of candidate under the name ALP for Axion Like Particle. One possibility of production is from the thermal bath. As we will see, processes of the type

$$A + A \to A \to A + \phi_a \tag{3.206}$$

can produce ϕ_a in a sufficiently large amount.

Exercise Is this process possible for Abelian gauge group?

Suppose that at the earliest stages of the Universe the axions were in thermal equilibrium. This is possible if the cross section of the reactions (3.206) respects[35]

$$n\langle\sigma v\rangle_{AA\leftrightarrow A\phi_a} \gtrsim H(T) \simeq \frac{T^2}{M_P}, \tag{3.207}$$

where we supposed a Universe dominated by the radiation (2.264), with a number density n. Approximating the amplitude of the process $\mathcal{M} \sim g_3 \frac{\sqrt{s}}{f_a}$, where A is a gluon field and $\sqrt{s}$ the center of mass energy, we deduce

$$n\langle\sigma v\rangle \sim T^3 \frac{|\mathcal{M}|^2}{8\pi s} \sim \frac{\alpha_3}{2} \frac{T^3}{f_a^2} > \frac{T^2}{M_P} \quad \Rightarrow \quad T > T_d = \left(\frac{f_a}{10^{16}\ \text{GeV}}\right)^2 \frac{10^{14}}{\alpha_3}\ \text{GeV}, \tag{3.208}$$

with $\alpha_3 = \frac{g_3^2}{4\pi}$ and T_d represents the decoupling temperature. At T_d, we can write $n_a(T_d) = \frac{\zeta(3)}{\pi^2} T_d^3$ (Eq. 3.27) and implement it in Eq. (E.1) to write

$$\Omega_a h^2 = 1.6 \times 10^8 \left(\frac{g_0}{g_d}\right) \frac{n(T_d)}{T_d^3} \frac{m_a}{1\ \text{GeV}}, \tag{3.209}$$

[35] See the Sect. 3.1.8 for a deeper understanding of this decoupling condition.

where g_d is the effective degree of freedom at the decoupling time that we will take $g_d = 106.75$ (3.38), whereas the present one is $g_0 = 3.91$. We then obtain for the thermal component of a

$$\frac{\Omega_a^{th} h^2}{0.1} \simeq \frac{m_a}{140\ \text{eV}}, \tag{3.210}$$

which is in tension with constraints from free-streaming (see Sect. 5.12.1). One also needs to add a condition on f_a from the fact that its lifetime $\tau_a = \Gamma_a^{-1}$ should exceed $X-$rays constraints at this mass range, which means $\Gamma_a \lesssim 10^{-52}$ GeV, or

$$\Gamma_{a \to A\,A} \simeq \frac{m_a^3}{4\pi f_a^2} < 10^{-52}\ \text{GeV} \ \Rightarrow \ f_a > 1.5 \times 10^{15} \left(\frac{m_a}{140\ \text{eV}}\right)\ \text{GeV}. \tag{3.211}$$

This lower bound on f_a implemented in (3.208) means that $T_d \gtrsim 10^{12}$ GeV, which is in tension with the maximum temperature of the thermal bath if one considers that the primordial plasma is produced from the inflaton decay. We understand that the thermal production of axion is not really convincing, but another mechanism can populate the Universe of ultra-light axions, without being excluded by free-streaming constraints or the temperature during the reheating. This is called the misalignment mechanism.

3.6.2.2 The Misalignment Mechanism

Another possibility is indeed to consider the axion field as a background field, i.e. an homogeneous field varying with time, in a very similar way the inflaton does. Before applying it directly to the QCD-axion, let us look in detail the calculation of the relic abundance for a generic background field $\phi_a(t)$. From its action

$$S_a = \int d^4x \sqrt{-g}\mathcal{L}_a = \int d^4x \sqrt{-g}\left[\frac{1}{2} g_{\mu\nu}\partial^\mu \phi_a \partial^\nu \phi_a - \frac{m_a^2}{2}\phi_a^2\right], \tag{3.212}$$

the equation of motion

$$\frac{\partial}{\partial \phi_a}\sqrt{-g}\mathcal{L}_a - \partial^\mu \frac{\partial}{\partial \partial^\mu \phi_a}\sqrt{-g}\mathcal{L}_a = 0 \tag{3.213}$$

for a field $\phi_a = \phi_a(t)$ reduces to

$$\ddot{\phi}_a + 3H\dot{\phi}_a + m_a^2 \phi_a = 0. \tag{3.214}$$

Exercise Recover the previous equation.

A solution of the form $\phi_a(t) = f(t)\cos m_a t$ can be extracted from this equation, with

$$\dot{f} = -\frac{3}{4t}f \quad \Rightarrow f(t) \propto t^{-\frac{3}{4}} \propto a^{-\frac{3}{2}}, \tag{3.215}$$

where we supposed a radiation dominated Universe. We then obtain

$$\phi_a(t) = \phi_a^i \cos m_a t \left(\frac{a_i}{a}\right)^{\frac{3}{2}}, \tag{3.216}$$

where a_i is the scale factor at the time where $m_a \simeq H$, in other words when the oscillation modes begin to dominate the energy density. Before this time, for scale factors below a_i where $H \gg m_a$, the equation (3.214) can be reduced to ϕ_a =constant$\equiv \phi_a^i$ if one takes $\dot{\phi}_a^i = 0$.

The energy density ρ_a is then (2.95)

$$\rho_a = \frac{1}{2}\dot{\phi}_a^2 + \frac{m_a^2}{2}\phi_a^2 = \frac{m_a}{2}(\phi_a^i)^2\left(\frac{a_i}{a}\right)^3. \tag{3.217}$$

Notice that ρ_a behaves like a dust, whatever is the value of m_a. From the perturbation point of view, and the structure formation constraints, we do not have the free-streaming tensions we had for the neutrino. Indeed, the background field $\phi_a(t)$ can be viewed as a sum of oscillators *at rest* around which structures can begin to form in the early thermal stages, respecting conditions (5.200) well before the recombination time. ϕ_a is indeed a *cold dark matter candidate*. To compute its relic abundance, we use Eq. (E.1) with

$$n_a(T) = \frac{\rho_a}{m_a} = \frac{m_a}{2}\left(\phi_a^i\right)^2\left(\frac{a_i}{a}\right)^3 \tag{3.218}$$

which gives

$$\Omega_a h^2 \sim 0.8 \times 10^8 \frac{m_a(\phi_a^i)^2}{(m_a M_P)^{3/2}}\left(\frac{g_0}{g_i}\right)\frac{m_a}{1\ \text{GeV}}, \tag{3.219}$$

or

$$\boxed{\frac{\Omega_a^{mis} h^2}{0.1} = \left(\frac{m_a}{0.4\ \text{eV}}\right)^{\frac{1}{2}}\left(\frac{\phi_a^i}{10^{12}\ \text{GeV}}\right)^2 = \left(\frac{m_a}{0.4\ \text{eV}}\right)^{\frac{1}{2}}\left(\frac{\theta_i f_a}{10^{12}\ \text{GeV}}\right)^2} \tag{3.220}$$

where $\theta_i = \frac{\phi_a^i}{f_a}$ is introduced because such and axionic field appears often as a pseudo-Nambu-Goldstone boson of a broken chiral symmetry described by a field

$\Phi = A(x)\, e^{i\frac{\phi_a}{f_a}} \equiv A(x)\, e^{i\theta}$. The misalignment terminology comes from the fact that, at the beginning of the domination of the oscillation modes, the angle $\theta_i \neq 0$.

3.6.2.3 QCD-Axion Dark Matter

In the case of the *QCD-axion*, two additional features should be taken into account. Firstly, the fact that the axion mass is generated through quantum corrections and therefore is dependent on f_a. Secondly, the mass being a radiative product, it also depends on the temperature and the time, rendering the equation of motion (3.214) a little bit more complex. Let first find the equation for $\rho_a(t)$ in the case of a varying mass $m_a(t)$. Multiplying Eq. (3.214) by $\dot\phi_a$, we obtain

$$\frac{1}{2}\frac{d}{dt}\dot\phi_a^2 + 3H\dot\phi_a^2 + \frac{m_a^2}{2}\frac{d}{dt}\phi_a^2 = 0, \tag{3.221}$$

or

$$\frac{d}{dt}\left[\frac{1}{2}\dot\phi_a^2 + \frac{m_a^2}{2}\phi_a^2\right] - m_a\dot m_a\phi_a^2 + 3H\dot\phi_a^2 = 0. \tag{3.222}$$

Taking the mean of the previous relation, and noticing that for $m_a \gg H$, $\langle\dot\phi_a^2\rangle = m_a(t)\langle\phi_a^2\rangle$, where the averages are over the oscillation period, we obtain

$$\dot\rho_a - \frac{\dot m_a}{m_a}\rho_a + 3H\rho_a = 0 \;\Rightarrow\; \rho_a \propto \frac{m_a}{a^3}. \tag{3.223}$$

It is remarkable that even if the mass is time-dependent, the density still follows the law of a dust-like component.

To find the temperature T_i for which the oscillation mode begins to dominate, one needs to solve

$$m_a(T_i) \gtrsim H(T_i) \;\Rightarrow\; T_i^6 \lesssim m_a M_P \Lambda_{QCD}^4 \sqrt{\frac{10}{g_i}\frac{3}{\pi}} \tag{3.224}$$

where we supposed

$$m_a(T \gg \Lambda_{QCD}) = m_a(\Lambda_{QCD})\left(\frac{\Lambda_{QCD}}{T}\right)^4 = m_a\left(\frac{\Lambda_{QCD}}{T}\right)^4. \tag{3.225}$$

The evolution of the mass for the *QCD*-axion needs a highly non-trivial computation, the previous relation being a reasonable approximation. For $T \lesssim \Lambda_{QCD}$, $m_a \sim$ constant. Applying once more Eq. (E.1) with

$$n_a(T_i) = \frac{m_a(T_i)}{2}(\phi_a^i)^2 \;\Rightarrow\; \frac{n_a(T_i)}{T_i^3} = \frac{(\phi_a^i)^2}{2m_a^{1/6}M_P^{7/6}\Lambda_{QCD}^{2/3}}\left(\frac{g_i}{10}\right)^{\frac{7}{12}}\left(\frac{\pi}{3}\right)^{\frac{7}{6}}, \tag{3.226}$$

and taking $\phi_a^i = f_a\theta_i$ and $f_a = \frac{m_\pi f_\pi}{2m_a}$ from Eq. (B.225), we obtain

$$\Omega_a^{qcd}h^2 = 2\times 10^7 \left(\frac{g_0}{g_i}\right)\left(\frac{g_i}{10}\right)^{\frac{7}{12}}\left(\frac{\pi}{3}\right)^{\frac{7}{6}} \frac{(m_\pi f_\pi)^2}{M_P^{7/6}\Lambda_{QCD}^{2/3}}\theta_i^2\frac{1}{m_a^{7/6}}$$

$$\Rightarrow \quad \frac{\Omega_a^{qcd}h^2}{0.1} \simeq \theta_i^2\left(\frac{1.5\times 10^{-6}\,\text{eV}}{m_a}\right)^{\frac{7}{6}}, \tag{3.227}$$

where we took $\Lambda_{QCD} = 200\,\text{MeV}$.

3.6.3 The Special Case of the Gravitino**

3.6.3.1 What Is a Gravitino

It is obviously very far from the objective of the book to describe the supersymmetry theory of fields, let alone to discuss the supergravity foundations. However, we will try in this section to give the main clues that should help the reader to appreciate the motivations and the gravitino-SM coupling with a minimum of hypothesis. The literature is full of very good textbook treating the supergravity models. For a complete introduction, we suggest [12] and [13].

In 1972, Volkov and Akulov noticed that the Nature possesses a massless boson, the photon, in which mass is protected by the gauge invariance of the electromagnetic interaction. At this epoch, another particle seemed also massless, the neutrino. Knowing that any broken symmetry generates a massless goldstone boson, they proposed in [14] to extend the idea to spinors, introducing a “Goldstone fermion,” what we presently call a “goldstino.” Enlarging the Poincaré group to spinors transformation, we can consider an upgraded space-time (ψ, X^μ), ψ being a spinor. If we suppose that the neutrino, or any other spinor field ψ is massless, it should respect the condition

$$i\gamma^\mu\partial_\mu\psi = 0. \tag{3.228}$$

This equation is invariant under transformations of the Poincaré group and the chiral transformations as well as under a new kind of translation of the type

$$\psi \to \psi + \zeta\ ; \quad X_\mu \to x_\mu = X_\mu, \tag{3.229}$$

where ζ is a constant spinor, anticommuting with ψ. Replacing the transformation (3.229) by

$$\psi \to \psi + \zeta\,; \quad \bar{\psi} \to \bar{\psi} + \bar{\zeta}$$
$$X^\mu \to x^\mu = X^\mu + \frac{i}{2\Lambda^4}\left[\bar{\zeta}\gamma^\mu\psi - \bar{\psi}\gamma^\mu\zeta\right],$$

we create a group structure with ten commuting and four anticommuting parameters. This is typical from a supersymmetric transformation. Looking at an infinitesimal transformation $\zeta = d\psi$

$$dx^\mu = dX^\mu + \frac{i}{2\Lambda^4}\left[d\bar{\psi}\gamma^\mu\psi - \bar{\psi}\gamma^\mu d\psi\right], \tag{3.230}$$

we can define the vierbein

$$e^a_\mu = \frac{\partial X^a}{\partial x^\mu} = \delta^a_\mu + \frac{i}{2\Lambda^4}\left[\bar{\psi}\gamma^a\partial_\mu\psi - \partial_\mu\bar{\psi}\gamma^a\psi\right] \tag{3.231}$$

such that from the definition of the metric

$$ds^2 = \eta_{ab}dX^a dX^b = g_{\mu\nu}dx^\mu dx^\nu = \eta_{ab}\frac{\partial X^a}{\partial x^\mu}\frac{\partial X^b}{\partial x^\nu}dx^\mu dx^\nu$$

we deduce

$$g_{\mu\nu} = \eta_{ab}\frac{\partial X^a}{\partial x^\mu}\frac{\partial X^b}{\partial x^\nu} = \eta_{ab}e^a_\mu e^b_\nu = \eta_{\mu\nu} + \frac{i}{\Lambda^4}\left[\bar{\psi}\gamma_\mu\partial_\nu\psi - \partial_\nu\bar{\psi}\gamma_\mu\psi\right]$$
$$= \eta_{\mu\nu} + \delta g_{\mu\nu}.$$

The coupling to the Standard Model fields can then be calculated from

$$\mathcal{L}_\psi = \frac{1}{2}\delta g_{\mu\nu}T^{\mu\nu}_{SM} \supset \frac{i}{2\Lambda^4}\left[\bar{\psi}\gamma_\mu\partial_\nu\psi - \partial_\nu\bar{\psi}\gamma_\mu\psi\right]T^{\mu\nu}_{SM} \tag{3.232}$$

$T^{\mu\nu}_{SM}$ being the stress-energy tensor of the Standard Model. This coupling appears when the scale of the supersymmetry, represented by[36] Λ, is much above the energy we consider for our calculation (inflaton mass in the case of reheating processes). ψ can be considered as the goldstino, or the longitudinal mode of the gravitino. In the case of the Volkov-Akulov paper, they considered ψ to be the neutrino.

For energies above the supersymetry breaking scale, one needs to look at the effect of other particles in the spectrum, especially the supersymmetric partners of

[36]To be more precise, $\Lambda \sim \sqrt{M_P m_{3/2}}$, $m_{3/2}$ being the gravitino mass.

the particles present in the thermal bath. The more natural coupling one can write, implies gravitino, gluon and its partner, the gluino:

$$\mathcal{L}_{3/2} = \frac{1}{4M_P} \bar{\Psi}^{\mu}_{3/2} \gamma_{\mu} \left[\gamma^{\alpha}, \gamma^{\beta}\right] \tilde{G}\, G_{\alpha\beta}, \tag{3.233}$$

where $G^a_{\mu\nu} = \partial_\mu A^a_\nu - \partial_\nu A^a_\mu + g_3 f_{abc} A^b_\mu A^c_\nu$, A^a_μ being the gluon field. Considering the longitudinal mode, reminding the classical longitudinal mode for the gauge boson ($\partial_\mu \theta$, θ being the parameter of the transformation), we can write $\psi^{\mu}_{3/2} = \partial^\mu \psi$, implying

$$\mathcal{L}_{3/2} = -\frac{i}{4M_P} \frac{\partial_\mu \bar{\psi}}{m_{3/2}} \gamma^{\mu} \left[\gamma^{\alpha}, \gamma^{\beta}\right] \tilde{G}\, G_{\alpha\beta}, \tag{3.234}$$

which gives after integration by part and applying the Dirac equation on the gluino,

$$\mathcal{L}_{3/2} = \frac{M_3}{4M_P m_{3/2}} \bar{\psi} \left[\gamma^{\alpha}, \gamma^{\beta}\right] \tilde{G}\, G_{\alpha\beta} \tag{3.235}$$

with M_3 the (onshell) gluino mass.

3.6.3.2 $M_{SUSY} < T_{RH}$

In supersymmetric models, the gravitino is the partner of the graviton. If the supersymmetric spectrum lies below the gravitino mass, the gravitino is produced also from the annihilation of Standard Model particles from the bath, mainly the ones charged under $SU(3)$. The more common process is the annihilation of gluons g into gluino[37] $\tilde{g}$ of mass M_3 and gravitino $\tilde{G}$ of mass $m_{3/2}$ through the exchange of a gluon: $g + g \to g \to \tilde{g} + \tilde{G}$. The sum of all the scattering process has been calculated and gives, with a good approximation:

$$\begin{aligned}
\langle \sigma v \rangle_{3/2} n^2_{eq} &\simeq \frac{4 g_3^2 T^6}{\pi^2 M_P^2} \left(1 + \frac{M_3^2}{3 m_{3/2}^2}\right) \\
\Rightarrow \frac{dY}{dx} &= \frac{1}{4 g_s \sqrt{g_\rho}} \frac{270\sqrt{10}}{\pi} \frac{m M_P}{x^2 T^6} \langle \sigma v \rangle_{3/2} n^2_{eq} \\
&\simeq \frac{270\sqrt{10}}{g_s^{\frac{3}{2}} \pi} \frac{m g_3^2}{x^2 \pi^2 M_P} \left(1 + \frac{M_3^2}{3 m_{3/2}^2}\right)
\end{aligned} \tag{3.236}$$

[37] Supersymmetric partner of the gluon.

which implies after integrating on x

$$Y(T) = \frac{270\sqrt{10}}{g_s^{\frac{3}{2}}\pi}\frac{g_3^2}{M_P\pi^2}\left(1+\frac{M_3^2}{3m_{3/2}^2}\right)\left[T_{RH} - T\right]. \tag{3.237}$$

The evolution of the population of gravitino is thus proportional to the temperature and its rate of production is softer than in the case of the heavy mediator we discussed in the previous section. Moreover, the ratio of the gluino to gravitino mass becomes important in this case. In specific scenarios where $m_{3/2} \gg M_3$ (anomaly mediation models, for instance) one obtains, with $g_3^2 \sim 0.5 \sim (g_3^{GUT})^2$ at such high temperatures,

$$Y_\infty \simeq 2 \times 10^{-12}\left(\frac{T_{RH}}{10^{10}\ \text{GeV}}\right) \tag{3.238}$$

giving a reheating temperature of $T_{RH} \simeq 10^8$ GeV for a 10 TeV gravitino.[38] We have represented the evolution of the gravitino production mechanism in Fig. 3.22.

3.6.4 Non-thermal Production Through Decays

3.6.4.1 Generalities

Another possibility is that heavy particles like scalars, Z', or moduli fields can decay into the dark matter candidate to populate the Universe. Let us consider a particle A in thermal equilibrium in the bath, and decaying into 2 dark matter candidates χ of momentum p_1 and p_2: $2m_\chi \lesssim M_A$. Supposing that initially the dark matter is not produced in the thermal bath: it can happen if the dark matter is not coupled with the inflaton (through symmetries, for instance). In this case, the number of particles produced per second from the decay of the parent A is $dn_\chi/dt = \Gamma_A n_A$, with Γ_A being the width of A. We can then rewrite Eqs. (3.142) and (3.143) neglecting the annihilation rates (as the annihilation process is an order of coupling weaker than the decay) (Fig. 3.23)

$$\frac{dn_\chi}{dt} + 3Hn_\chi = g_A n_A \langle\Gamma_A\rangle, \quad \Rightarrow \frac{dY_\chi}{dT} = -g_A\frac{\langle\Gamma_A\rangle}{HT}Y_A \quad \Rightarrow \frac{dY_\chi}{dx}$$
$$= g_A\frac{\langle\Gamma_A\rangle}{Hx}Y_A \tag{3.239}$$

[38] After using $Y_\infty \simeq 3.3 \times 10^{-12}\left(\frac{100\ \text{GeV}}{m}\right)\left(\frac{\Omega h^2}{0.1}\right)$ from Eq. (3.198).

Fig. 3.23 Feynman diagram of a particle A in thermal bath decaying into 2 dark matter candidates χ

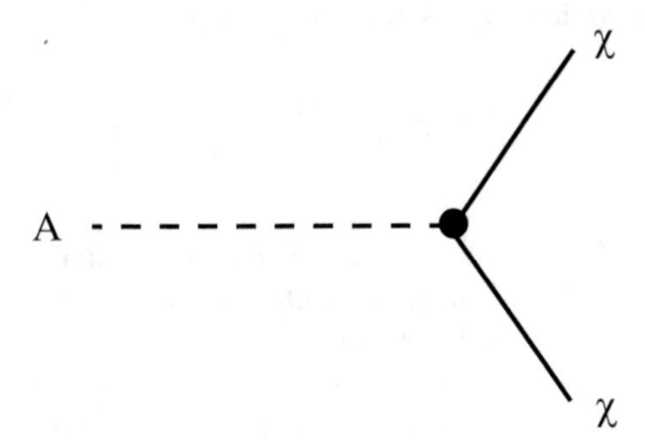

with $x = m_\chi/T$, g_A the number of degrees of freedom of A and $Y_i = n_i/s$, s being the entropy density given by Eq. (3.61), $s = 2\pi^2/45 g_s T^3$. In fact, in this specific case, each parent A decays into *two* dark matter candidates. To include all the possibilities, like the decay into 3 or 4 particles, we have included this multiplicity in the definition of g_A. It is important to notice that $\langle\Gamma_A\rangle$ is the *averaged* width of the parent A and not the *pure* width. We need to stop some time on this issue. Indeed, there is a little subtlety in the Boltzmann equation that is not obvious at the first glance.[39] The real Boltzmann equation (3.239) should in fact be written as[40]

$$\frac{dn_\chi}{dt} + 3Hn_\chi$$
$$= g_A \int \frac{d^3p_A}{(2\pi)^3 2E_A} f_A \int\int \frac{d^3p_1}{(2\pi)^3 2E_1}\frac{d^3p_2}{(2\pi)^3 2E_2}(2\pi)^4\delta^4(p_A - p_1 - p_2)|\mathcal{M}|^2$$
$$= g_A \int \frac{d^3p_A}{(2\pi)^3} f_A \frac{M_A}{E_A}\Gamma_A = g_A n_A\langle\Gamma_A\rangle = S\frac{dY}{dt} = -HTs\frac{dY}{dT} = Hxs\frac{dY}{dx} \tag{3.240}$$

f_A being the statistical distribution function of A in the thermal bath and $x = m_A/T$. Comparing with the general formula Eq. (B.157) we note the presence of the Lorentz invariant phase space $\frac{d^3p_A}{(2\pi)^3 2E_A}$ in the right hand side of (3.240). This is somewhat similar to the annihilation process, where the "$2E_A$" factor necessary to satisfy Lorentz invariance in Eq. (3.240) is absorbed in the "$2M_A/2E_A$" factor, the "$2E_1 2E_2$" factor in annihilating case is absorbed in the "v" of σv from the formula

$$4\sqrt{(P_1.P_2)^2 - M_1^2 M_2^2} = 2E_1 2E_2 \times v \tag{3.241}$$

[39]This remark is also valid in the case where the dominant process is the annihilation one.

[40]We have incorporated in g_A the multiplicity of the decay rate as A decays into 2 dark matter particles.

that one finds in Eq. (B.106). Another way to see it is that the factor $\frac{M_A}{E_A} = \frac{1}{\gamma}$ corresponds to the time contraction factor. Some particles A being in the thermal bath with some high velocities are a smaller effective width (a longer lifetime). Γ_A is the width of A at rest.

We can then solve step by step the evolution equation for Y_χ supposing that A is in thermal equilibrium and that all the physics is happening for temperatures of the plasma T much above M_A, the thermal distribution can be approximated by the Maxwellian one from $T = M_A$ to the reheating temperature T_{RH}, $f_A \simeq e^{-E_A/T}$ we then can write

$$\begin{aligned} Hxs\frac{dY}{dx} &= M_A\Gamma_A \int_{M_A}^{T_{RH}} e^{-E_A/T} p_A \frac{dE_A}{(2\pi)^3} 4\pi \\ &= \frac{M_A\Gamma_A}{2\pi^2}\int_{M_A}^{\infty} e^{-E_A/T}\sqrt{E_A^2 - M_A^2}\, dE_A \\ &= \frac{T M_A^2 \Gamma_A}{2\pi^2} K_1\left(\frac{M_A}{T}\right) \Rightarrow \frac{dY}{dx} = \frac{M_A^3\Gamma_A}{Hx^2 s 2\pi^2} K_1(x). \end{aligned} \tag{3.242}$$

K_1 being the first modified Bessel function of the second kind that the reader can find in Sect. A.6.3. Using the common relations (3.27) for $s(x)$ and (3.148) for $H(x)$

$$s(x) = g_s \frac{2\pi^2}{45}\frac{M_A^3}{x^3}, \quad H(x) = \frac{\pi}{3}\sqrt{\frac{g_\rho}{10}}\frac{M_A^2}{x^2 M_P}$$

we finally obtain the equation

$$Y_0 = Y(x \to \infty) = \frac{270\sqrt{10}}{\pi}\frac{1}{g_s\sqrt{g_\rho}}\frac{M_P\Gamma_A g_A}{8\pi^4 M_A^2}\int_{\frac{M_A}{T_{RH}}}^{\infty} x^3 K_1(x)dx. \tag{3.243}$$

To have an idea of the order of magnitude of the yield of the dark matter produced from the decay of a massive particle of mass M_A, we can approximate $\Gamma_A \simeq g_D^2 M_A/(16\pi)$, g_D being a gauge coupling in the case of a vectorial particle A, a Yukawa-type coupling in the case of a scalar one.[41] Noticing that $g_S \simeq g_\rho$ at the energy of interest, we can write

$$Y_0 \simeq \frac{270\sqrt{10}}{\pi}\frac{g_D^2 g_A M_P}{128\pi^5 M_A}\int_{\frac{M_A}{T_{RH}}}^{\infty} x^3 K_1(x)dx. \tag{3.244}$$

We then need to distinguish two cases: 1) $M_A < T_{RH}$, A is in thermal equilibrium with the bath and $\int_{\frac{M_A}{T_{RH}}}^{\infty} x^3 K_1(x)dx \simeq \int_0^{\infty} x^3 K_1(x)dx = 4.7$ and 2) $M_A > T_{RH}$.

[41] See Eq. (B.182) and (B.185) for the exact expressions.

In this case, using the equations (A.122) and the approximation of $K_1(x)$ given in Sect. A.6.3 we have

$$\int_X^\infty x^3 K_1(x)dx \simeq \sqrt{2\pi}\int_{\sqrt{\frac{M_A}{T_{RH}}}}^\infty t^6 e^{-t^2}dt \simeq \sqrt{\frac{\pi}{2}}e^{-\frac{M_A}{T_{RH}}}\left(\frac{M_A}{T_{RH}}\right)^{5/2}. \tag{3.245}$$

Applying then Eq. (3.167)

$$\Omega_0 = \frac{mY_0 s_0}{\rho_c^0} = m_\chi Y_0 \frac{g_s^0 2\pi^2 (T_0)^3}{135 M_P^2 (H_0)^2} \Rightarrow \Omega_0 h^2 \simeq 2.8 \times 10^8\, m_\chi\, Y_0 \tag{3.246}$$

which gives

$$\Omega_0 h^2 \simeq 2 \times 10^{22}\, g_D^2\, \frac{m_\chi}{M_A} \qquad [M_A < T_{RH}],$$

$$\Omega_0 h^2 \simeq 5 \times 10^{21}\, g_D^2\, \frac{m_\chi}{M_A} e^{-\frac{M_A}{T_{RH}}}\left(\frac{M_A}{T_{RH}}\right)^{3/2} \qquad [M_A > T_{RH}]. \tag{3.247}$$

To respect WMAP constraints one thus needs $g_D \simeq 10^{-11}$ if A and the dark matter are at TeV scale in the plasma (first case) or $M_A \simeq 4 \times 10^{10}$, for $T_{RH} = 10^9$ GeV and a 100 GeV weakly interacting particle ($g_D \simeq 10^{-1}$) in the second case.

3.6.4.2 An Example: Decay of the Gravitino to Populate Dark Matter**

In several minimal supersymmetry extension of the Standard Model, the dark matter candidate is a *wino*, fermionic partner of the W^a gauge vector. The wino being charged under SU(2), the coupling to the Standard Model bath (added to the coannihilation process with its charged supersymmetric partner which have the same mass at tree level) underproduced dark matter. Typical relic abundance for a wino dark matter is around $\Omega h^2 \simeq 10^{-3}$ There are different ways to overcome this issue. One solution is to populate the Universe through the decay of the gravitino $\tilde{G} \to \chi\chi$ of mass $m_{3/2}$. As we discussed earlier on, the gravitino is super-weakly interacting and decouples from the plasma at an early stage when the plasma has temperature T_d, around the reheating temperature T_{RH}, $T_d \simeq T_{RH}$. Its density is then fixed by T_{RH} by Eq. (3.238) $Y_{\tilde{G}} \simeq 2 \times 10^{-12}(T_{RH}/10^{10}\text{ GeV})$. The equilibrium condition can be written $\Gamma_{\tilde{G}} \lesssim H(T)$.

To understand physically speaking what is happening in the plasma, it is useful to express Eq. (3.242) as function of $\Gamma_{\tilde{G}}$ and H only:

$$\frac{dY_\chi}{dx} \propto \frac{\Gamma_{\tilde{G}}}{H} x K_1(x). \tag{3.248}$$

As we did when we tried to understand the Boltzmann equation and the decoupling effect for annihilating dark matter, we see that a similar phenomenon exists for a decaying particle into dark matter. Indeed, when the temperature of the plasma reaches $\Gamma_{\tilde{G}} \lesssim H(T = T_{3/2})$, $\frac{dY_\chi}{dx} \simeq 0$. In other words, the number of dark matter particle per co-volume (Y_χ) resulting from the decay of the gravitino is constant. Physically speaking one can understand it as during the doubling Hubble time $t = 1/H$, the decay of $\tilde{G}$ populates the Universe with 2 particles: the density is thus constant. We need to impose that this decay occurs before the Big Bang Nucleosynthesis (BBN) time to avoid any disturbance in the nucleosynthesis process ($T_{BBN} \simeq 1$ MeV). Indeed, at least one possible channel of decay must include either a photon, a charged lepton or a meson, each of which would be energetic enough to destroy a nucleus if it strikes one. One can show that enough such energetic particles will be created in the decay as to destroy almost all the nuclei created in the era of nucleosynthesis, in contrast with observations. In fact, in such a case the Universe would have been made of hydrogen alone, and star formation would probably be impossible.

From the expression of $\Gamma_{\tilde{G}}$ [15]

$$\Gamma_{\tilde{G}} = \frac{1}{4}\left(n_v + \frac{n_m}{12}\right)\frac{m_{3/2}^3}{M_{Pl}^2} = 24\left(\frac{10\ \text{TeV}}{m_{3/2}}\right)^{3/2} \text{sec}, \tag{3.249}$$

where $n_v = 12$ and $n_m = 49$ are the number of vector and chiral matter multiplets, respectively, and $M_{Pl} = \sqrt{8\pi} M_P$. The condition $H(T_{3/2}) = \Gamma_{\tilde{G}}$ can then be written in the radiation dominated era

$$\begin{aligned} T_{3/2} &= \left(\frac{45}{4 g_\rho \pi^3}\right)^{1/4} \sqrt{M_{Pl}\Gamma_{\tilde{G}}} = \left(\frac{45}{4 g_\rho \pi^3 M_{Pl}^2}\right)^{1/4} \left(\frac{n_v + \frac{n_m}{12}}{32\pi}\right)^{1/2} m_{3/2}^{3/2} \\ &\simeq 0.24 \left(\frac{10.75}{g_\rho(T_{3/2})}\right)^{1/4} \left(\frac{m_{3/2}}{10\ \text{TeV}}\right)^{3/2} \text{MeV}. \end{aligned} \tag{3.250}$$

We thus can observe that for a gravitino lighter than 10 TeV, its decay happens at a temperature below T_{BBN}. This is what is called as the gravitino problem. As the result of the late time decays of the gravitino, the wino (or any supersymmetric particle which can play the role of the dark matter) is non-thermally produced at around $T_{3/2}$ which is lower than the freeze out temperature of the wino. After the decay of all the gravitino, the resultant relic density for χ is given by the gravitino yield, Eq. (3.238) applied to the formula (3.247) which gives

$$\Omega_\chi^{decay} h^2 \simeq 0.168 \times \left(\frac{m_\chi}{300\ \text{GeV}}\right)\left(\frac{T_{RH}}{10^{10}\ \text{GeV}}\right). \tag{3.251}$$

Altogether, the relic abundance of a wino (or any non-thermally produced dark matter from a decay) is the sum of the thermal production plus the non-thermal one

$$\Omega_\chi h^2 = \Omega_\chi^{thermal} h^2 + \Omega_\chi^{decay} h^2. \tag{3.252}$$

As a conclusion, the thermal history of the Universe in the presence of a metastable gravitino begins by the decoupling of the gravitino itself at a temperature T_d around T_{RH} when $H(T_d) \simeq \langle\sigma v\rangle_{SM\ SM\to\tilde{G}\ \tilde{G}}$. Then the dark matter candidate χ decouples (freezes out) at a temperature around $T_\chi = m_\chi/20$ which can give $\Omega_\chi h^2 \ll 0.1$, corresponding to $H(T_\chi) = \langle\sigma v\rangle_{SM\ SM\to\chi\ \chi}$.

The population of dark matter is, however, still populated by the gravitino decay. Finally the end of the decay (kind of freeze out too) of the gravitino $\tilde{G} \to \chi\chi$ at $T_{3/2}$, stabilized to obtain $\Omega_\chi h^2 \simeq 0.1$ at a temperature $T_{3/2}$ respecting $H(T_{3/2}) = \Gamma_{\tilde{G}\to\chi\chi}$. Several comments are in order. First, it should be noted that the entropy produced by the decay of the gravitino is negligible since the energy density of the gravitino at the decay time is subdominant. Second, it should be also noted that the annihilation of the wino after the non-thermal production is negligible, since the yield of the non-thermally produced wino is small enough. Indeed,

$$Y_\chi^{decay} \ll Y_\chi^{annihilation} \simeq \frac{H(T_{3/2})}{s\langle\sigma v\rangle} \simeq 10^{-9} \times \left(\frac{10^{-24}\ \text{cm}^3\text{s}^{-1}}{\langle\sigma v\rangle}\right)\left(\frac{1\ \text{MeV}}{T_{3/2}}\right).$$

The dark matter is thus largely diluted and not able to annihilate anymore (no return to the thermal bath). A summary is shown in Fig. 3.24.

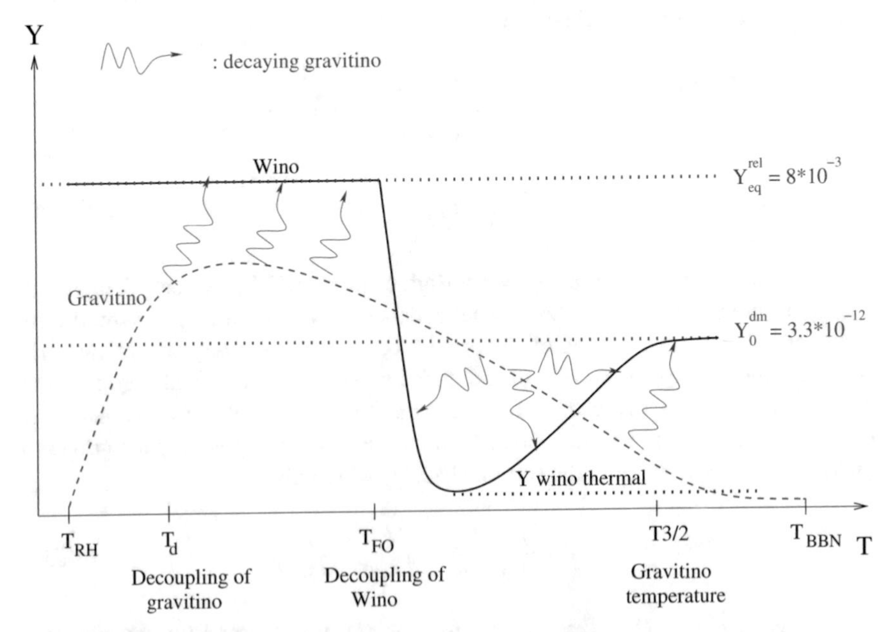

Fig. 3.24 Evolution of the yields Y of a wino dark matter and through decay of the gravitino

3.7 Extracting Information from the CMB Spectrum

3.7.1 Generalities

The study of the anisotropies in the CMB spectrum is far beyond the scope of this textbook. The literature is full of very complete and precise accounts on the subject. The reader who is especially interested can, for instance, look at [16]. In this section, we just want to give some intuitive insights about how physicists extract information from the CMB measurement. Therefore, the reader should not expect here a rigorous study of two-point correlation functions, but rather qualitative estimates. We suggest to start by the Sect. 5.12.3 in order to approach our discussion with confidence.

The saga of the CMB prediction and measurement is a long story that is admirably told in [17]. We already have shown the Gamow's way to find the present microwave temperature of $T_0 \simeq 2.7$ K from the BBN mechanism, see Eq. (3.119). But what about the anisotropy of the CMB? This issue was already raised in the late 60's. Indeed, the first anisotropy that was predicted (and observed) is the one generated by the motion of the Earth *inside* the homogenous radiation, mainly due to the motion of the Earth in the solar system, the Sun in the Milky Way, the Milky Way in the local group… We can estimate the velocity of the Earth with respect to the CMB of $v \sim 300$ km/s $\sim 10^{-3}\ c$. This corresponds to a dipole anisotropy (also called "aether drift") between the temperature pointing to the direction of motion, and the temperature pointing backward, of

$$\Delta T \sim 2\, T_0 \frac{v}{c} \quad \Rightarrow \quad \frac{\Delta T}{T_0} \simeq 10^{-3}. \tag{3.253}$$

Exercise Show that the received frequency ν of an emitted photon of original frequency ν', from an angle θ with respect to the direction of motion of the observer with velocity v is

$$\nu = \frac{1}{\Delta t} = \sqrt{1 - \frac{v^2}{c^2}} \times \frac{\nu'}{1 - \frac{v}{c}\cos\theta}. \tag{3.254}$$

Hint: have a look at Sect. 5.6.2. Deduce then that

$$T = T' \frac{\sqrt{1 - \frac{v^2}{c^2}}}{1 - \frac{v}{c}\cos\theta} \simeq T\left(1 + \frac{v}{c}\cos\theta\right). \tag{3.255}$$

Once the spectrum has been renormalized by the aether drift, one is left with two other possible sources of anisotropy: the one generated by the Sachs-Wolfe effect described in Sect. 5.12.3 and more specifically by Eq. (5.208)

$$\frac{\delta T_{SW}}{T_0} \simeq 2 \times 10^{-5} \quad \Rightarrow \quad \delta T_{SW} \simeq 50\ \mu\text{K}, \tag{3.256}$$

at large scale ($\gtrsim$2 degrees in the sky, reminding that the moon covers half a degree) corresponding to the difference of temperature needed to form the actual patchwork of large scale structures observed nowadays. This level of anisotropy is clearly visible on the left part of the spectrum in Fig. 1.6.

The second effect is related to the horizon problem we discussed in Sect. 2.2.1. Two photons which were separated by a distance smaller than the horizon size at decoupling time could have been in contact and could have shared information, by mutual scattering. We then expect a larger correlation between points below the horizon size. We know the radius of our Universe is roughly 50 Glyrs, Eq. (2.29). The CMB being emitted at a redshift $z_{CMB} \sim 1000$, we deduce that the radius of the Universe at the decoupling time is roughly $R_{CMB} = 50$ Mlyrs. The horizon diameter being $d_H = 2 \times ct_{CMB} \simeq 760000$ lyrs, we deduce that the angle under which is seen the horizon is

$$\theta_{CMB} = \frac{d_{CMB}}{R_{CMB}} \simeq \frac{7.6 \times 10^5}{50 \times 10^6} \simeq 1.5 \times 10^{-2} \text{ radian} \simeq 1 \text{ degree.} \tag{3.257}$$

We clearly see in Fig. 1.6 that the behavior *above* 1 degree is radically different that the behavior *below* 1 degree. At small angle we are in presence of photons which were part of the primordial plasma, in equilibrium with baryons, and probably dark matter, before the decoupling time. It means that the spectrum we observe is the one expected from oscillations present in any plasma, with wavelengths depending on the constitution of the plasma.

3.7.2 To Find the Components of the Universe

The astrophysicists play with a set of parameters to extract the information about the composition of the Universe. The best fit gives the value of H_0, the present Hubble rate, and the Ω_i's we gave in appendix. Let us detail the influence of each of them one by one.

3.7.2.1 Influence of the Matter, Ω_m

Suppose first a flat Universe with a null cosmological constant. It is easy to understand that the distance traveled by the light, before the decoupling time, depends on the rate of expansion H_0 and on the matter content. We already saw that the matter already dominated the Universe at decoupling time (Eq. 3.57), we then can write following Eq. (2.25)

$$d_{CMB} = a_{CMB} \times R_0 \int_0^{\chi_{CMB}} d\chi = a_{CMB} \int_0^{a_{CMB}} \frac{cda}{a^2 H(a)} = \frac{2\,c\,a_{CMB}^{3/2}}{H_0\,\sqrt{\Omega_m^0}}, \tag{3.258}$$

R_0, H_0, and Ω_m^0 being the present radius of the Universe, the present Hubble parameter, and the present matter relic density, whereas a_{CMB} is the scale factor at decoupling time. In the meantime, R_{CMB} keeping the same value, $R_{CMB} = a_{CMB} \times R_0$, we obtain

$$\theta_{CMB} = \frac{d_{CMB}}{R_{CMB}} = \frac{2\,c\sqrt{a_{CMB}}}{R_0 H_0 \sqrt{\Omega_m^0}} = \frac{2\,c}{R_0 H_0 \sqrt{\Omega_m^0}\sqrt{1+z_{CMB}}}. \tag{3.259}$$

We understand then that a *decrease* in the value of Ω_m^0 would shift all the spectrum of Fig. 1.6 to the *left* (larger angle), whereas an *increase* would shift it to the *right* (smaller angle). This can be physically understood by the fact that a Universe with more matter has a tendency to grow "faster" due to a larger Hubble rate and then needs a shorter time to reach the same size, meaning a shorter horizon size (and vice versa). We illustrate our interpretation in Fig. 3.25.

3.7.2.2 Influence of the Curvature, Ω_k

The influence of the curvature of space-time on the spectrum of CMB anisotropies is very similar to the case just discussed. Indeed, a modification of the curvature (positive or negative) will have as a direct consequence a modification of the observer's viewing angle. Indeed, this strongly resembles a deformation effect that we find in general relativity, reproducing magnifying effects on the trajectory of

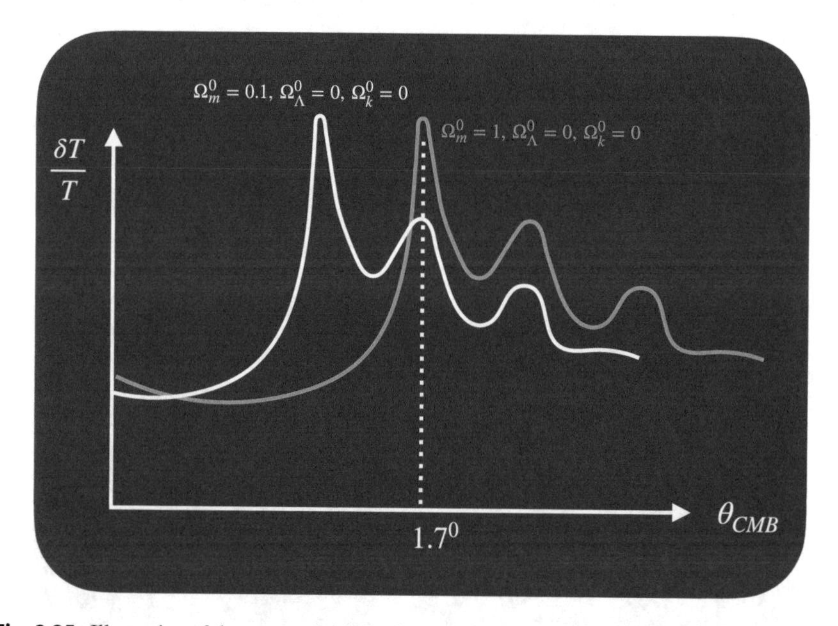

Fig. 3.25 Illustration of the expected shift in the spectrum in a flat Universe, without cosmological constant, for different values of Ω_m^0: 0.1 and 1

light rays. For a curvature $k = -1; 0; 1$, the angle becomes

$$\theta^k_{CMB} = \frac{d_{CMB}}{S_k R_{CMB}}, \tag{3.260}$$

with S_k defined in Eq. (2.21), in other words

$$\begin{aligned}
\theta^{k=-1}_{CMB} & \frac{2\,c}{R_0 H_0 \sqrt{\Omega^0_m}\,\sinh(1/(1+z_{CMB}))} \frac{1}{\sqrt{1+z_{CMB}}} \\
\theta^{k=0}_{CMB} &= \frac{2\,c}{R_0 H_0 \sqrt{\Omega^0_m}} \frac{1}{\sqrt{1+z_{CMB}}} \\
\theta^{k=1}_{CMB} &= \frac{2\,c}{R_0 H_0 \sqrt{\Omega^0_m}\,\sin(1/(1+z_{CMB}))} \frac{1}{\sqrt{1+z_{CMB}}}.
\end{aligned} \tag{3.261}$$

Exercise Recover the previous set of equations.

We illustrate our result in Fig. 3.26 where we plotted the expected spectrum for an open ($k = -1$), flat ($k = 0$) and close ($k = +1$) Universe. In the case of a close space ($k = +1$), we expect to see the illusion of a larger horizon, coming from the convex curvature of the light ray, and vice versa for an open space.

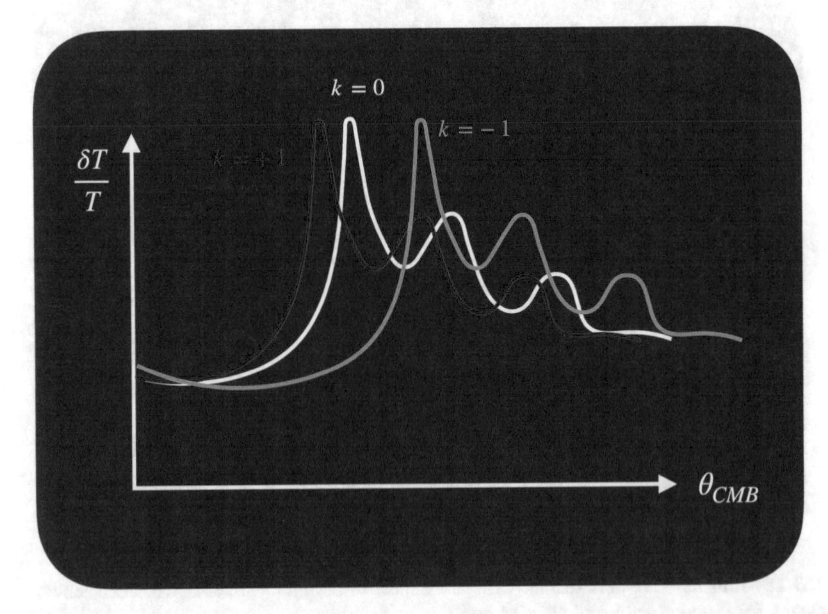

Fig. 3.26 Illustration of the expected shift in the spectrum in a for different curvature of the Universe: open ($k = -1$), flat ($k = 0$) and closed ($k = +1$)

3.7.2.3 Influence of the Cosmological Constant, Ω_Λ

If one adds to the game the presence of a cosmological constant, the distance traveled by the light to reach its horizon at t_{CMB} becomes a little bit more complex:

$$d_{CMB} = \frac{a_{CMB} \times c}{H_0\sqrt{\Omega_m^0}} \int_0^{a_{CMB}} \frac{da}{\sqrt{a + \frac{\Omega_\Lambda^0}{\Omega_m^0} a^4}}. \tag{3.262}$$

On the other hand, the radius of the Universe nowadays should also be modified, compared to the one computed with the approximation of a matter dominated Universe. We can write for the radius at CMB:

$$R_{CMB} = \frac{a_{CMB} \times c}{H_0\sqrt{\Omega_m^0}} \int_0^{1} \frac{da}{\sqrt{a + \frac{\Omega_\Lambda^0}{\Omega_m^0} a^4}} \tag{3.263}$$

where we used Eq. (2.64). Between 0 and a_{CMB}, the Universe is never dominated by the cosmological constant, one can then safely neglect the term $\frac{\Omega_\Lambda^0}{\Omega_m^0}$, which is not the case when computing the radius R_{CMB}. We then obtain

$$\theta_{CMB} = \frac{d_{CMB}}{R_{CMB}} = \frac{2\, a_{CMB}^{3/2}}{\int_0^1 \frac{da}{\sqrt{a + \frac{\Omega_\Lambda^0}{\Omega_m^0} a^4}}}. \tag{3.264}$$

We illustrate in Fig. 3.27 the influence of $\frac{\Omega_\Lambda^0}{\Omega_m^0}$ on the position of the spectrum. We clearly see that increasing the cosmological constant component decrease the radius of the Universe and increase the angle under which the horizon at decoupling time is seen.

3.7.2.4 Influence of the Baryons, Ω_b

Finally, the presence of the baryons also affects the spectrum. Combining it with the determination of Ω_m^0, we can deduce the cold dark matter density, Ω_{cdm}^0. The effect of baryons is slightly more subtle than those described above. The presence of baryon affects first the horizon size, the same way the curvature does. Indeed, until now, we considered that the photons were traveling in vacuum. Whereas that is true from the decoupling time to now, that is *not* the case before, because they are in a medium governed by mutual interactions, and one should consider its *sound speed* and not its *vacuum celerity* to compute the horizon size. The information travels at this epoch at the sound speed c_s, not at c. We detail the computation of the sound speed around Eq. (5.190). If we suppose that the Universe is only composed

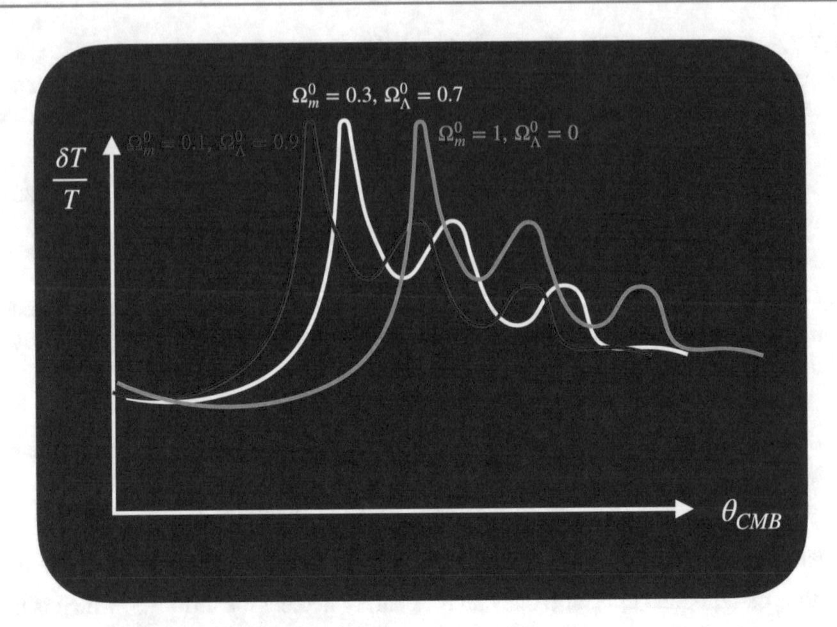

Fig. 3.27 Influence of the presence of the cosmological constant on the CMB spectrum for different values of Ω_Λ

of baryons and photons, we should then use for d_{CMB}

$$d_{CMB} = a_{CMB} \int_0^{a_{CMB}} \frac{cdt}{a\sqrt{1 + \frac{3}{4}\frac{\rho_b}{\rho_R}}}, \tag{3.265}$$

where ρ_b and ρ_R are, respectively, the baryon and photon energy density. Indeed, we suppose that the neutrino has decoupled from the thermal bath at decoupling time, thus not affecting the sound speed of the photons.

However, this effect is mild, as the proportion of baryons to photons is not gigantic at this epoch. The more interesting effect induced by the presence of baryons is on the *amplitude* of the first peak. Indeed, in a gravitational potential generated (for instance) by the presence of dark matter (but could also be only baryonic) Φ, the conservation of energy imposes the pressure to adapt following

$$\delta P + (\rho + P)\Phi = 0, \tag{3.266}$$

with $\rho = \rho_b + \rho_R$, $P = P_b + P_R = P_R$ and where we neglected the expansion of the Universe and supposed adiabatic evolution. With $\delta P = \delta P_R = \frac{1}{3}\delta\rho_R$, we can rewrite Eq. (3.266)

$$\frac{\delta\rho_R}{\rho_R(1 + \frac{3}{4}\frac{\rho_b}{\rho_R})} = -4\Phi \tag{3.267}$$

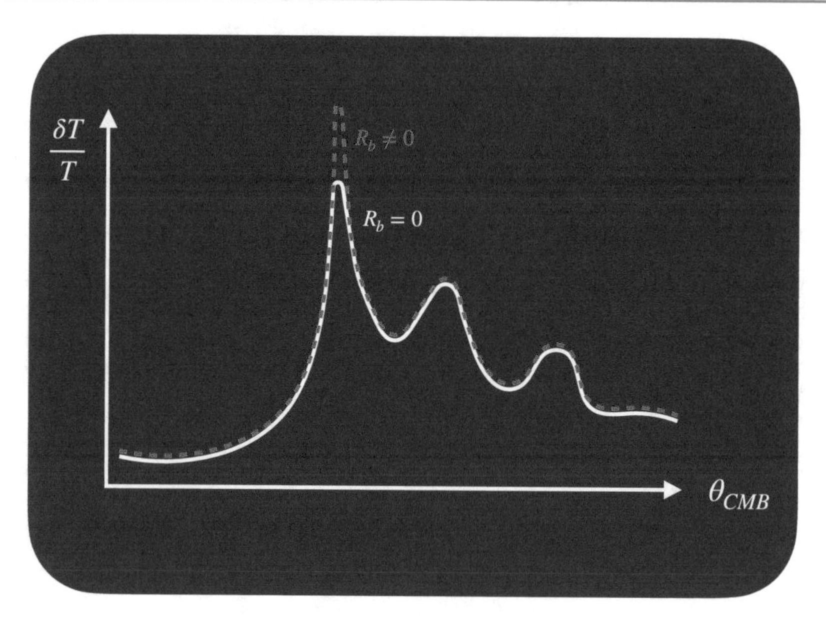

Fig. 3.28 Influence of the baryonic component of the primordial plasma on the CMB spectrum

or

$$\frac{\delta T}{T} = \frac{1}{4}\frac{\delta\rho_R}{\rho_R} = \Phi\left(1 + \frac{3}{4}\frac{\rho_b}{\rho_R}\right). \tag{3.268}$$

We then see that the relative abundance between baryons and photons will modify the *height* of the peaks, as one can see in Fig. 3.28.

References

1. J. Martin, C. R. Phys. **13**, 566–665 (2012). https://doi.org/10.1016/j.crhy.2012.04.008 [arXiv:1205.3365 [astro-ph.CO]]
2. M. Pospelov, J. Pradler, Ann. Rev. Nucl. Part. Sci. **60**, 539–568 (2010). https://doi.org/10.1146/annurev.nucl.012809.104521 [arXiv:1011.1054 [hep-ph]]
3. M. Tanabashi et al. [Particle Data Group], Phys. Rev. D **98**(3), 030001 (2018). https://doi.org/10.1103/PhysRevD.98.030001
4. R.A. Alpher, H. Bethe, G. Gamow, Phys. Rev. **73**, 803–804 (1948). https://doi.org/10.1103/PhysRev.73.803
5. P.J.E. Peebles, Eur. Phys. J. H **39**, 205–223 (2014). https://doi.org/10.1140/epjh/e2014-50002-y [arXiv:1310.2146 [physics.hist-ph]]
6. P.J.E. Peebles, *Cosmology's Century: An Inside History of Our Modern Understanding of the Universe* (Princeton University Press, Princeton, 2020)
7. M.E. Burbidge, G.R. Burbidge, W.A. Fowler, F. Hoyle, Rev. Mod. Phys. **29**, 547–650 (1957). https://doi.org/10.1103/RevModPhys.29.547

8. B.W. Lee, S. Weinberg, Phys. Rev. Lett. **39**, 165–168 (1977). https://doi.org/10.1103/PhysRevLett.39.165
9. P. Hut, Phys. Lett. B **69**, 85 (1977). https://doi.org/10.1016/0370-2693(77)90139-3
10. K. Griest, D. Seckel, Phys. Rev. D **43**, 3191–3203 (1991). https://doi.org/10.1103/PhysRevD.43.3191
11. L.F. Abbott, P. Sikivie, Phys. Lett. B **120**, 133–136 (1983). https://doi.org/10.1016/0370-2693(83)90638-X
12. J. Wess, J. Bagger, *Supersymmetry and Supergravity* (Princeton University Press, Princeton, 1992)
13. P. Binetruy, *Supersymmetry: Theory, Experiment, and Cosmology* (Oxford University Press, Oxford, 2012)
14. D.V. Volkov, V.P. Akulov, JETP Lett. **16**, 438–440 (1972)
15. W. Buchmuller, V. Domcke and K. Schmitz, Phys. Lett. B **713**, 63–67 (2012). https://doi.org/10.1016/j.physletb.2012.05.042 [arXiv:1203.0285 [hep-ph]]
16. R. Durrer, *The Cosmic Microwave Background* (Cambridge University Press, Cambridge, 2020). ISBN 978-1-316-47152-4. https://doi.org/10.1017/9781316471524
17. P. James, E. Peebles, L.A. Page, R.B. Partridge, *Finding the Big Bang* (Cambridge University Press, Cambridge, 2009)

Part II

Modern Times $[T_{CMB} \to T_0]$

Direct Detection [T_0]

4

Abstract

Once we understood how dark matter can be produced in the early Universe, from the inflationary phase to the thermal one, passing through the reheating process, it is time to question the possibility to detect this dark component largely present in our galaxy. The most natural way to detect it is through direct interaction with a nucleus on Earth. Direct Detection (DD) was first proposed by Goodman and Witten (Phys Rev D 31: 3059, 1985). The idea was to study the scattering of dark matter particles from the Galactic halo with targets on earth and extract informations about their masses and couplings to the Standard Model. Because dark matter is weakly interacting with standard matter, such experiments require large detectors (from 10 kilos for the first version several tons for the future projects) and deeply buried underneath the surface of the Earth to avoid contamination from cosmic rays. This is similar to neutrino detection experiments in the principle, except that dark matter is supposed to be quite massive. While neutrino scatterings generate a Cherenkov shower, a WIMP bumps into a heavy nuclei and experimentalists try to measure the energy of recoil produced by the interaction. A nice and pedagogical review can be found in Arcadi et al. (Eur Phys J C 78(3): 203, 2018).

4.1 Generality

The general strategy of direct detection is to measure the recoil energy of a target nucleus produced by its elastic collision with a dark matter candidate, to distinguish it from the background, before comparing it with theoretical predictions (parameterized by the dark matter mass and its coupling to matter). A nice and pedagogical review can be found in [1]. Interactions with a nucleus act on three

Y. Mambrini, *Particles in the Dark Universe*,
https://doi.org/10.1007/978-3-030-78139-2_4

steps: first the interaction with the partons (quarks and gluons) is given by the Lagrangian and the calculation of Feynman amplitudes. The second step is to deduce from the parton interaction and the nucleon interaction that depends strongly on the structure of the nucleons. Results from lattice QCD or chiral perturbation theories $\chi PT$are the fundamental tools to extract the component of the nucleons (and often both approaches disagree). The third and last step is to pass from the nucleon level to the nuclear level. At this stage, the distribution function of the nucleon inside the nucleus is crucial. Different form factors (Woods–Saxon, for instance) enter in the game. We will first concentrate on each step one by one. The physical quantity connecting experiments to theoreticians' community is the elastic nuclear recoil spectrum dR/dE_R, where R is the recoil event rate (the number of events per unit of time and per unit of mass of detector) and E_R is the energy of the recoiling nucleus. The majority of experiments are based on ionization, scintillation, phonon techniques, or some combinations of these. They have in common the same basic theoretical interpretation: the differential energy spectrum of such nuclear recoils is expected to be featureless and smoothly decreasing, with (for the simplest case of a detector stationary in the galactic frame) the typical form:

$$\frac{dR}{dE_R} = \frac{R_0}{E_0 k} e^{-E_R/E_0 k}, \tag{4.1}$$

where E_R is the recoil energy, E_0 is the most probable incident kinetic energy of a dark matter particle of mass m_χ, k is a kinematic factor $4m_\chi M_T/(m_\chi + M_T)^2$ for a target nucleus of mass M_T, R is the event rate, and R_0 is the total event rate.[1] We will reconstruct this expression step by step in the following sections (see 4.31 for the complete form). Since typical galactic dark matter velocities are of order $v \simeq 10^{-3}c$, values of m_χ in the 10–1000 GeVc^{-2} range would give typical recoil energies in the range 5–500 keV ($\sim \frac{1}{2} m_\chi v^2$). All the experimental efforts lie on discriminating in Eq. (4.1) the signal from the background composed for instance of neutrons produced in muons cosmic rays, or to develop methods to distinguish nuclear recoils from electron recoils that allow to reject gamma and beta decay backgrounds. More generically, one can deduce the differential rate R per energy of recoil E_R, dR/dE_R for the interaction with an incoming particle of mass m_χ and velocity v_χ with a distribution $f(v_\chi)$ on a nucleus of mass $M_N \simeq Am_p$ per kilogram of target (m_p being the proton mass):

$$dR = \frac{N_0}{A} \frac{\rho_0}{m_\chi} \left(v_\chi \frac{d\sigma}{d|\mathbf{q}|^2} \right) d|\mathbf{q}|^2 \, f(\mathbf{v}_\chi) d^3\mathbf{v}_\chi, \tag{4.2}$$

[1] See Eq. (4.15) with $\theta^* = \pi$ to understand the appearance of the factor k in (4.1).

$\mathcal{N}_0$ being the Avogadro number[2] (6.02×10^{26} kg^{-1}), ρ_0 the local mass density of the dark matter, and $\mathbf{q}$ the momentum transferred from the WIMP to the nucleus. We will compute each of the terms of this equation in the following sections.

In this section we will reconstruct step by step Eq. (4.1) and show its validity limits. We will also correct its, adding the dependence on the Earth velocity, escape velocity or form factor of the nucleus. We will then go into the details of the experimental strategies settled nowadays to detect such events. The unit for differential events or background rates is conventionally expressed as 1 event keV^{-1} kg^{-1} day^{-1} called '*differential rate unit*' (dru). When integrated over the recoil energy, one can talk about "*total rate unit*" (tru) expressed in events kg^{-1} day^{-1}. When integrated between two energies (threshold energy and escape velocity for instance), we usually refer to "*integrated rate unit*" (iru) also expressed in events kg^{-1} day^{-1}.

4.2 Velocity Distribution of Dark Matter: $f(v)$

The differential particle density, the number of particles with velocities on Earth (hitting the detector) comprised between $\mathbf{v}$ and $\mathbf{v} \pm d\mathbf{v}$, can be written as

$$dn = \frac{n_0}{C} f(\mathbf{v}, \mathbf{v}_E) d^3\mathbf{v}, \tag{4.3}$$

with $\mathbf{v}_E = \mathbf{v}_{Earth/Halo}$ the Earth velocity relative to the halo (see Sect. 5.1.3 for details) and where C is a normalization constant such that $\int_0^{v_{esc}} dn = n_0$, v_{esc} being the *local* escape velocity; in other words,

$$C = \int_0^{2\pi} d\phi \int_{-1}^{+1} d\cos\theta \int_0^{v_{esc}} f(\mathbf{v}, \mathbf{v}_E) v^2 dv. \tag{4.4}$$

n_0 is the mean dark matter particle number density (ρ_0/m_χ), whereas $\mathbf{v} = \mathbf{v}_{\chi/Earth}$ is the velocity of the dark matter in the reference frame of the experiment, the velocity that is *effectively* measured by the collaboration (through the recoil of the nucleus for instance). As a good approximation, one can represent the annual modulation of the orbit of the Earth around the Sun by

$$v_E = 232 + 15 \cos\left(2\pi \frac{t - 152.5 \text{ days}}{365.25 \text{ days}}\right) \text{ km s}^{-1}. \tag{4.5}$$

[2] As a reminder, the Avogadro number 6.02×10^{26} kg^{-1} corresponds by definition to the number of atoms of Carbon 12 in 12 grams of Carbon, or in other words, the number of mass (nucleons) per gram of material. $\frac{\mathcal{N}_0}{A}$ corresponds then to the number of nucleus of atomic mass A in 1 gram of substance.

Considering that the dark matter statistical distribution of its velocity in the halo follows a Maxwellian distribution with a most probable velocity[3] v_0, and observing that $\mathbf{v}_{\chi/Halo} = \mathbf{v}_{\chi/Earth} + \mathbf{v}_{Earth/Halo} = \mathbf{v} + \mathbf{v}_E$, one obtains

$$f(\mathbf{v}, \mathbf{v}_E) = e^{-\frac{1}{2}m_\chi v^2_{\chi/Halo}/kT} \equiv e^{-v^2_{\chi/Halo}/v_0^2} = e^{-(\mathbf{v}+\mathbf{v}_E)^2/v_0^2}. \tag{4.6}$$

For $v_{esc} = \infty$ and using Eq. (A.122) we can compute

$$\frac{1}{C}\int_0^{v_{esc}=\infty} d\cos\theta d\phi e^{-\frac{v^2_{\chi/Halo}}{v_0^2}} v^2_{\chi/Halo} dv_{\chi/Halo} = 1 \;\Rightarrow\; C = (\pi v_0^2)^{3/2},$$

which gives[4]

$$dn = \frac{\rho_0}{m_\chi}\frac{e^{-(\mathbf{v}+\mathbf{v}_E)^2/v_0^2}}{(\pi v_0^2)^{3/2}} d\phi d\cos\theta v^2 dv. \tag{4.7}$$

As we could have expected, the number density of dark matter ($n_0 = \int dn$) is independent of the Earth velocity. The (normalized) distribution

$$v^2 f(\mathbf{v}, \mathbf{v}_E) = \frac{v^2}{(\pi v_0^2)^{3/2}} e^{-(\mathbf{v}+\mathbf{v}_E)^2/v_0^2} \tag{4.8}$$

is maximum for $v = v_0$. Considering that the Earth is moving with a velocity $v_E \simeq v_0$ around the galactic center, it corresponds to an absolute dark matter velocity $\mathbf{v}_{\chi/Halo} = \mathbf{v} + \mathbf{v}_E \simeq 0$: a static halo. We show in Fig. 4.1 some examples of distributions for different values of v_0.

If one would have taken a particular value for v_{esc} (and not ∞), the result one would have obtained is

$$C_{v_{esc}\neq\infty} = (\pi v_0^2)^{3/2}\left[\mathrm{erf}\left(\frac{v_{esc}}{v_0}\right) - \frac{2}{\pi^{1/2}}\frac{v_{esc}}{v_0} e^{-v_{esc}^2/v_0^2}\right], \tag{4.9}$$

Exercise Calculate this value of $C_{v_{esc}\neq\infty}$.

with erf the Gauss error function (see A.128 for details). If we take $v_0 = 230$ km s^{-1} and $v_{esc} = 600$ km s^{-1}, we obtain $C_{v_{esc}=\infty}/C_{v_{esc}=600} \simeq 0.9965$, which can justify in the majority of cases the approximation $v_{esc} = \infty$. For our calculation, we made

[3]Notice that v_0 is the most probable velocity, whereas the mean velocity is $\langle v^2\rangle = \frac{3}{2}v_0^2$.

[4]Remark that $(\mathbf{v} + \mathbf{v}_E)^2 = v^2 + v_E^2 + 2vv_E\cos\theta$.

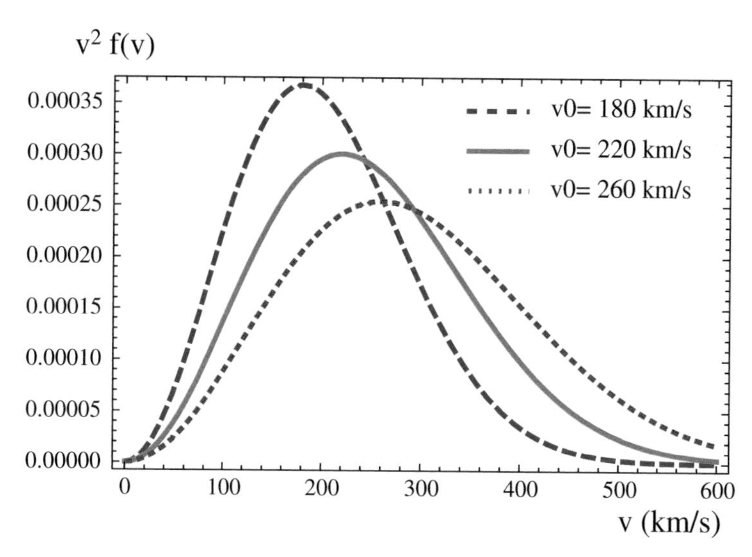

Fig. 4.1 Examples of distribution function $\frac{v^2}{(\pi v_0^2)^{3/2}} e^{-v^2/v_0^2}$ for different values of v_0

the assumption that the dark matter in the halo has a "sufficiently" Maxwellian velocity distribution. The Maxwell–Boltzmann distribution $\propto e^{-m_\chi v^2/2kT}$ describes the velocities of particles that move freely up to few collisions. We assumed that the WIMPs are isothermal and isotropically distributed in phase space. It is important to notice that this assumption is not exact and there can exist deviations from this Maxwellian distribution. Noticing that

$$\frac{1}{2} m_\chi v_0^2 = kT \quad \Rightarrow \quad v_0 \simeq 220\,\text{km/s} \simeq 0.75 \times 10^{-3} c, \tag{4.10}$$

which shows that dark matter is highly non-relativistic in the Milky Way (and justifies using the Maxwell–Boltzmann distribution instead of Fermi–Dirac or Bose–Einstein one).

Usually, estimation of the local density ρ_0 is in the range 0.2 GeV cm^{-3} $< \rho_0 <$ 0.4 GeV cm^{-3} leading to the adoption of $\rho_0 = 0.3$ GeV cm^{-3} as the central value. However, the main result of a more recent analysis has given a novel determination of the local dark matter halo density, which assumes spherical symmetry and either an Einasto or NFW density profile is found to be around 0.39 GeV cm^{-3} with a 1σ error bar of about 7%; more precisely, it was found $\rho_0 = 0.385 \pm 0.027$ GeV cm^{-3} for the Einasto profile and $\rho_0 = 0.389 \pm 0.025$ GeV cm^{-3} for the NFW. This is in contrast to the standard assumption that ρ_0 is about 0.3 GeV cm^{-3} with an uncertainty of factors of 2 to 3. In any case, the difference in the event rates corresponds to a simple rescaling of ρ_0.

4.3 Measuring a Differential Rate: $\frac{d\sigma}{d|\mathbf{q}|^2}$

4.3.1 Kinematics

One should compute at the first place the energy transferred to a nucleon of mass M_N, at rest, from a collision with a dark matter particle of mass m_χ hitting the nucleus with a velocity $\mathbf{v}_\chi$. We first compute in the center of mass frame with velocity (see Fig. 4.2 for illustration):

$$\mathbf{v}_{CM} = \frac{m_\chi\,\mathbf{v}_\chi + \mathbf{0}}{m_\chi + M_N} = \frac{\mu}{M_N}\mathbf{v}_\chi \tag{4.11}$$

$$\Rightarrow\ \mathbf{V}_N = \mathbf{v}_N - \mathbf{v}_{CM} = \mathbf{0} - \mathbf{v}_{CM} = -\frac{\mu}{M_N}\mathbf{v}_\chi$$

$$\mathbf{V}_\chi = \mathbf{v}_\chi - \mathbf{v}_{CM} = \frac{\mu}{m_\chi}\mathbf{v}_\chi$$

with $\mu = \frac{m_\chi M_N}{m_\chi + M_N}$ is the reduced mass and $\mathbf{V}_i$ ($\mathbf{V}'_i$) the velocity of the nucleus before (after) the collision in the center of mass frame. In this frame, $\mathbf{P}_\chi + \mathbf{P}_N = \mathbf{0} = \mathbf{P}'_\chi + \mathbf{P}'_N$, the last equality being the conservation of momentum implying $\mathbf{V}'_\chi = -\frac{M_N}{m_\chi}\mathbf{V}'_N$. The kinetic energy conservation gives

$$\frac{1}{2}m_\chi \mathbf{V}_\chi^2 + \frac{1}{2}M_N\mathbf{V}_N^2 = \frac{\mu}{2}v_\chi^2 = \frac{1}{2}m_\chi {\mathbf{V}'}_\chi^2 + \frac{1}{2}M_N{\mathbf{V}'}_N^2 = \frac{M_N}{2m_\chi}[m_\chi + M_N]{\mathbf{V}'}_N^2$$

$$\Rightarrow\ (V'_N)^2 = \frac{\mu^2}{M_N^2}v_\chi^2 \quad \Rightarrow (v'_N)^2 = |\mathbf{V}'_N - \mathbf{v}_{CM}|^2 = 2v_\chi^2\frac{\mu^2}{M_N^2}(1-\cos\theta^*) \tag{4.12}$$

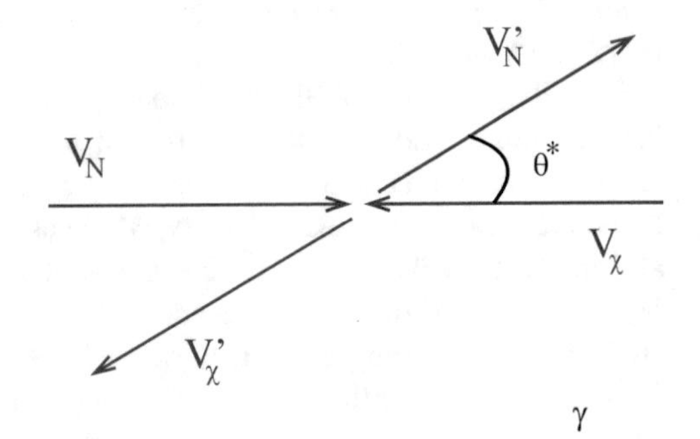

Fig. 4.2 Collision between the nucleus N and the WIMP χ in the center of mass frame

with θ^* the angle of scattering in the rest frame. From Eq. (4.12), one can deduce the nucleus recoil energy in the laboratory frame (the *measured* one),

$$E_R = \frac{|\mathbf{p}'_{\mathbf{N}}|^2}{2M_N} = \frac{|M_N\, v'_N|^2}{2M_N} = \frac{\mu^2 v_\chi^2}{M_N}(1-\cos\theta^*). \tag{4.13}$$

The maximal recoil energy that can be generated by the elastic collision is when the nucleus moves backward after the hit ($\cos\theta^* = -1$) and is $E_R^{max} = \frac{2\mu^2 v_\chi^2}{M_N}$. Another way of understanding the phenomenon is to observe that for a fixed (measured) energy of recoil E_R the minimum velocity able to deposit E_R corresponds to $\cos\theta^* = -1$, which gives

$$v_\chi^{\min} = \sqrt{\frac{E_R M_N}{2\mu^2}}. \tag{4.14}$$

Usually, each experiment possesses a threshold energy E_{th}, i.e., a minimum energy that it can measure. The minimum velocity required to measure an event is thus $v_\chi^{min} = \sqrt{\frac{E_{th} M_N}{2\mu^2}}$.

As a summary, we obtained

$$\begin{aligned} \mathbf{P'}_N &= -\mu \mathbf{v}_\chi\ ;\ \ E'_N = E_R = \frac{\mu^2 v_\chi^2}{M_N}(1-\cos\theta^*)\ ; \\ E_{tot} &= E'_{tot} = \frac{\mu}{2} v_\chi^2\ ;\ \ |\mathbf{q}|^2 = 2\mu^2 v_\chi^2 (1-\cos\theta^*), \end{aligned} \tag{4.15}$$

$|\mathbf{q}| = \sqrt{2 M_N E_R}$ being the momentum transferred. One can write Eq. (4.13) as

$$E_R = 2\frac{m_\chi M_N}{(m_\chi + M_N)^2} E_\chi (1-\cos\theta^*) \tag{4.16}$$

with $E_\chi = \frac{1}{2} m_\chi v_\chi^2$ the initial energy of the collision. The function $\frac{m_\chi M_N}{(m_\chi + M_N)^2}$ is maximum (for a given m_χ) for $M_N = m_\chi$ (classical ping-pong ball versus bowling ball situation). It means that for a 100 GeV WIMP, the energy of recoil will be maximized if the nucleus is also of the order of $\simeq$ 100 GeV. This explains the strategy of the present experiments for the detection of WIMPs, to use relatively heavy nuclear targets.

4.3.2 Differential Rate

For a given recoil energy, one needs to compute $\frac{d\sigma}{dE_R}$. Notice that we are in non-relativistic conditions, which means that we can apply the Fermi's golden rule:[5]

$$d\Gamma_{fi} = 2\pi\ \rho(E_f)\ |\mathcal{M}_{fi}|^2 \tag{4.17}$$

with

$$\rho(E_f) = \frac{dn}{dP'_N}\frac{dP'_N}{dE_f} \quad \text{and} \quad dn = \frac{d^3\mathbf{P'_N}}{(2\pi)^3} = \frac{(P'_N)^2 dP'_N d\Omega}{(2\pi)^3},$$

where Γ_{fi} is the number of transitions[6] per unit of time from initial state $|i\rangle$ to final state $|f\rangle$, $\mathcal{M}_{fi}$ the transition matrix element, and $\rho(E_f)$ the density of final state, $E_f = E'_N + E'_\chi$. From Eq. (4.15) one deduces $d\Omega = 2\pi d(\cos\theta^*) = \frac{\pi\ d|\mathbf{q}|^2}{\mu^2\ v_\chi^2}$, $P'_N = M_N V'_N = \mu v_\chi$, $E_f = \frac{\mu}{2} v_\chi^2 = \frac{(P'_N)^2}{2\mu}$, and so $dE_f = \mu^{-1} P'_N dP'_N$. Combining all these relations with Eq. (4.17) we obtain

$$d\Gamma_{if} = \frac{|\mathcal{M}_{fi}|^2 d|\mathbf{q}|^2}{4\pi\ v_\chi}. \tag{4.18}$$

$d\Gamma_{if}$ being the rate of the interaction equals the cross section times the velocity:

$$d\Gamma_{if} = v_\chi \times d\sigma \ \Rightarrow \ \frac{d\sigma}{d|\mathbf{q}|^2} = \frac{|\mathcal{M}_{fi}|^2}{4\pi\ v_\chi^2}, \tag{4.19}$$

which is a less conventional (but more practical in our case) form of the Fermi's golden rule. A quantum field theory treatment can be found in Eq. (B.154).

4.4 Structure Function of the Nucleus: $F(q)$

One has to keep in mind, when using Eq. (4.19), that the transition matrix element should be taken between two *nuclei* final states. However, a WIMP hits protons and neutrons of the nucleus (in fact quarks at a more microscopical level as we will explain later on). For a nucleus of nuclear mass number A of density $\rho(\mathbf{r})$, with a coupling WIMP-nucleons f and nucleon wave functions $\psi_{i,f}(r) = e^{-i\mathbf{p_{i,f}}.\mathbf{r}}$, one

[5] See Sect. B.4.1 and Eq. (B.75) for details.

[6] Due to the very low velocities in consideration for direct detection processes ($v_\chi \simeq 200$ km s^{-1}), we will work in classical quantum mechanics framework. We could work in relativistic limits with the changes $|\mathcal{M}|^2 \to |\mathcal{M}|^2/2E_\chi 2E_N$ and $2\pi d^3P/(2\pi)^3 \to (2\pi)^4 d^3P'_\chi d^3P'_N/(2\pi)^6 2E'_\chi 2E'_N$.

should then write

$$\mathcal{M}_{fi} = \int d^3\mathbf{r}\psi_f^*(r)V(\mathbf{r})\psi_i(r) \text{ with } V(\mathbf{r}) = A\rho(\mathbf{r})f$$

$$\Rightarrow \mathcal{M}_{fi} = Af\int d^3\mathbf{r}\rho(\mathbf{r})e^{-i\mathbf{q}.\mathbf{r}} = AfF(q) \tag{4.20}$$

with $\mathbf{q} = \mathbf{p}_f - \mathbf{p}_i$ and $F(q)$ the form factor of the nucleus, defined as the Fourier transform of wave function of the target nucleus, normalized to unity when no momentum is transferred,

$$F(q) = \int \rho(\mathbf{r})e^{i\mathbf{q}.\mathbf{r}}d^3\mathbf{r} = \frac{4\pi}{q}\int_0^\infty r\sin(qr)\rho(r)dr, \tag{4.21}$$

where we supposed a spherical distribution for ρ.

Exercise Recover the expression (4.21).

If protons and neutrons have different coupling coefficients, the transition matrix element has two parts and can be written as

$$\mathcal{M}_{fi} = \left[Zf_p + (A-Z)f_n\right]F(q). \tag{4.22}$$

The form factor represents the effect of the coherence of the interaction with the nucleon. When the momentum transfer $|\mathbf{q}| = \sqrt{2M_N E_R}$ becomes large enough such that the wavelength becomes similar to the size of the nucleus $r_N = (a_N A^{1/3} + b_N)$ (a_N and b_N being parameters experimentally determined), the structure of the nucleus becomes important. In other words when

$$\begin{aligned} q\, r_N \text{ dimensionless } &\simeq (2AE_R)^{1/2}(a_N A^{1/3} + b_N) \\ &\simeq 6.92 \times 10^{-3} A^{1/2} E_R^{1/2}(a_N A^{1/3} + b_N) \end{aligned} \tag{4.23}$$

with E_R in keV, and a_N and b_N expressed in fermi. To find this form factor, one needs to know the distribution function $\rho(\mathbf{r})$ of the nucleons in the nucleus. There exist different options:

- The simplest option is to consider a sum of delta Dirac functions, assuming point-like nucleons $\rho(\mathbf{r}) \propto \sum_i \delta^3(\mathbf{r} - \mathbf{r_i})$.
- A more (intermediate) realistic distribution is a thin shell of effective radius r_N: $\rho(\mathbf{r}) = \frac{\delta(\mathbf{r}-\mathbf{r_N})}{4\pi r^2}$. In this case, it is equivalent to imagine that the nucleons are homogeneously dispersed around the nucleus. Implementing the density in

Eq. (4.21) we then obtain

$$F(qr_N) = \frac{\sin(qr_N)}{qr_N}. \tag{4.24}$$

- Another commonly used approximation is

$$F(qr_N) = e^{-(qr_n)^2/10}, \tag{4.25}$$

which is the exact form factor for a Gaussian density distribution $\rho(\mathbf{r})$ with mean square radius $\sqrt{(3/5)}\, r_N$, $r_N = 1.2A^{1/3}$ fm.
- Helm found a more realistic form of the nucleons in the nucleus after studying the scattering of electrons from different nuclear targets [2]. Helm has approximated the nucleus as a solid core with a nearly constant density and a surface of thickness s:

$$\rho(\mathbf{r}) = \int d^3\mathbf{r}'\, \rho_0(\mathbf{r}')\, \rho_1(\mathbf{r} - \mathbf{r}'). \tag{4.26}$$

ρ_0 defines the radius of the target nuclei and is given by a constant inside a sphere of radius $r_N \simeq 1.2A^{1/3}$ fm and is zero outside the sphere. ρ_1 defines the surface thickness of the order of 1 fm by $\rho_1(\mathbf{r}) = e^{-|\mathbf{r}|^2/2s^2}$. ρ_0 and ρ_1 are normalized such that $\int d^3\mathbf{r}' \rho_{0,1}(\mathbf{r}') = 1$. After implementing the density function in Eq. (4.21) one obtains the Woods–Saxon form

$$F(q) = \frac{3 j_1(qr_N)}{qr_N} e^{-(qs)^2/2}, \tag{4.27}$$

where j_1 is a spherical Bessel function of the first kind. A least squares fit from muon scattering data gives

$$r_N^2 = c^2 + \frac{7}{3}\pi^2 a^2 - 5s^2, \quad c \simeq 1.23A^{1/3} - 0.60 \text{ fm}, \quad a \simeq 0.52 \text{ fm}, \; s \simeq 0.9 \text{ fm}.$$

This result is quite similar to the Fermi distribution:

$$\rho_{Fermi}(r) = \rho_0 \left[1 + e^{(r-c)/a}\right]^{-1}. \tag{4.28}$$

We show in Fig. 4.3 the evolution of the coherence factor F as a function of the recoil energy $E_R = q^2/2M_N$ for different nuclei used in several direct detection experiments (Sodium for DAMA, Germanium for CDMS, and Xenon for XENON1T, LZ, or PANDAX). We notice that the effect of the form factor becomes stronger for higher energies of recoil, where the condition $qr_N \lesssim 1$ becomes not valid anymore. The exchanged wavelength begins to reach the size of the nucleus.

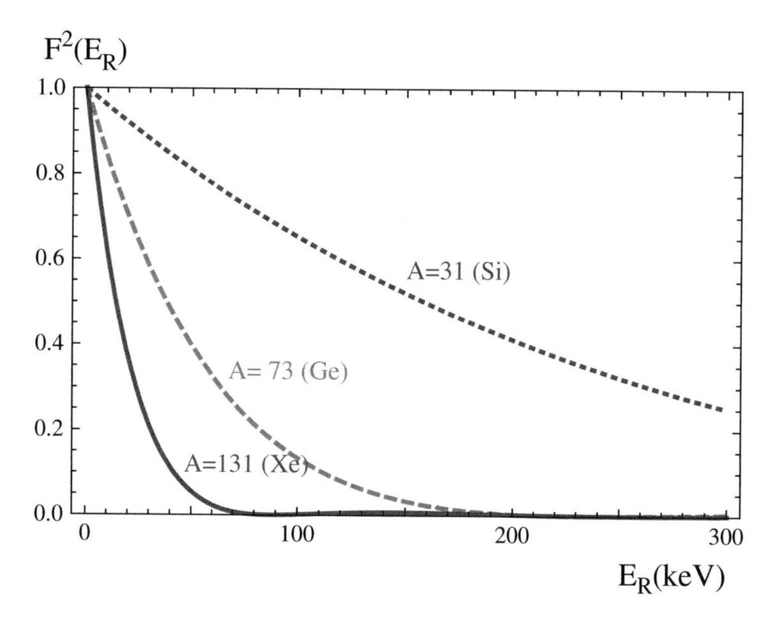

Fig. 4.3 Woods–Saxon form factor as a function of the energy of recoil of the nucleus for several types of nuclei used in different experiments: Silicium (dotted blue), Germanium (dashed green), and Xenon (-full red)

We also clearly see the influence of the number of nucleons in the process: the larger is the mass number A, the larger is the radius of the nucleus and the stronger becomes the decoherence effect. For a Xenon nucleus, the form factor already vanishes for an energy of recoil of 50–100 keV.

4.5 Computing a Rate

After defining the differential cross section at zero momentum transfer σ_0 and remembering that F(q=0) = 1, we can rewrite Eq. (4.19) as a function of σ_0:

$$\sigma_0 = \int_0^{q^2_{max}=4\mu^2 v_\chi^2} \frac{d\sigma}{d|\mathbf{q}|^2} d|\mathbf{q}|^2 = \frac{(\mu A f)^2}{\pi} \quad \Rightarrow \quad \frac{d\sigma}{d|\mathbf{q}|^2} = \sigma_0 \frac{F^2(E_R)}{4\mu^2 v_\chi^2}, \tag{4.29}$$

and when we combine Eqs. (4.2), (4.19), and (4.20), we can write

$$\frac{dR}{dE_R} = \frac{N_0 M_N}{A} \frac{\rho_0}{m_\chi} \frac{\sigma_0 \, F^2(E_R)}{2\mu^2 v_\chi} f(\mathbf{v}_\chi, \mathbf{v_E}) d^3\mathbf{v}_\chi. \tag{4.30}$$

To have an idea of the evolution of the rate as a function of the recoil energy, one can build a toy model with $v_E = 0$, $v_{esc} = \infty$. In this case, assuming a Maxwellian velocity distribution and integrating on the WIMP velocity from $v_{min} = \sqrt{\frac{E_R M_N}{2\mu^2}}$ to ∞ using Eq. (A.122), one obtains

$$f(v) = \frac{1}{(v_0^2\pi)^{3/2}} e^{-v^2/v_0^2} \tag{4.31}$$

$$\Rightarrow \frac{dR}{dE_R} = \frac{N_0 M_N}{A} \frac{\rho_0 \sigma_0}{m_\chi \mu^2} \frac{F^2(E_R)}{v_0 \pi^{1/2}} e^{-E_R M_N/2\mu^2 v_0^2} \equiv \frac{R_0}{E_0 k} e^{-E_R/E_0 k},$$

which is the form of Eq. (4.1) that we have reconstructed. If moreover we neglect the influence of the form factor ($F(E_R) \simeq 1$), we can compute an analytical expression of the total rate after integrating on the recoil energy E_R from the threshold energy to an energy of recoil E_R and we have

$$R(E_R) = \frac{2}{\pi^{1/2}} \frac{N_0}{A m_\chi} \rho_0 \sigma_0 v_0 \left[e^{-E_{th} M_N/2\mu^2 v_0^2} - e^{-E_R M_N/2\mu^2 v_0^2} \right]$$

$$= R_0 \left[e^{-E_{th} M_N/2\mu^2 v_0^2} - e^{-E_R M_N/2\mu^2 v_0^2} \right] \quad \text{with} \tag{4.32}$$

$$R_0 \simeq \frac{5 \times 10^{-6}}{A} \left(\frac{100\ \text{GeV}}{m_\chi} \right) \left(\frac{\rho_0}{0.4\ \text{GeVcm}^{-3}} \right) \left(\frac{\sigma_0}{10^{-42}\ \text{cm}^2} \right) \left(\frac{v_0}{220\ \text{kms}^{-1}} \right) \times \frac{\text{events}}{\text{day kg}}$$

R_0 corresponding to the total number of events expected (integration on E_R from 0 to ∞). In fact, the majority of the events appear at low energy of recoil due to the Boltzmann suppression. Lowering threshold energy is thus a fundamental keypoint in the experimental process to increase the number of events, especially for light dark matter candidates. A first glance at this number shows that the rate is *very* weak and the detection is thus a difficult task from the experimental front. If we take as an example a 100 GeV WIMP hitting a 100 kg experience of Xenon ($A = 131$) with a cross section of 1 pb (10^{-36} cm^2), we expect around 5 events per day. An important point is that σ_0 is the WIMP-*nucleus* cross section. However, to be able to compare different experiments, the interesting cross section is the WIMP-*nucleon* one (proton or neutron) that we will note σ_0^n. By rescaling, we can easily show that σ_0 can be written as

$$\sigma_0 = \frac{\mu^2}{\mu_n^2} A^2 \sigma_0^n, \tag{4.33}$$

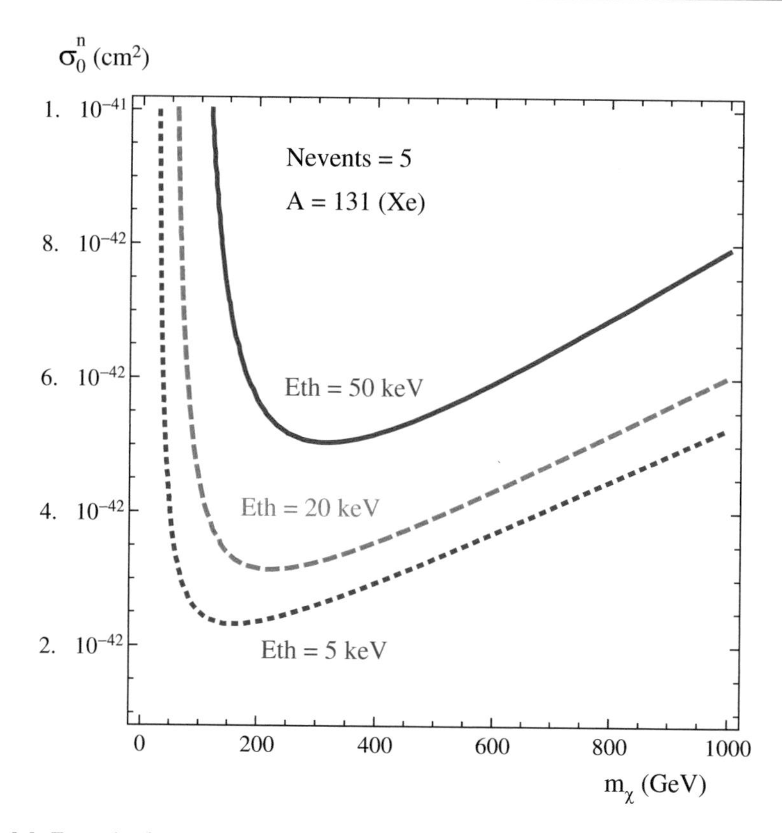

Fig. 4.4 Example of limits obtained from Eq. (4.32) if we suppose 5 events observation on 100 kg of Xenon for different threshold energies: 5 keV (dotted blue), 20 keV (dashed green), and 50 keV (full red)

with $\mu_n = M_n m_\chi/(M_n + m_\chi)$, "$n$" standing for $n$ucleon. A 1 pb cross section with a nucleus of Xenon corresponds to a 1.7×10^{-6} pb cross section with a nucleon. The exclusion limits given by the different experiences are based on a number of events expected but finally not observed. We can plot the (total) iso-events expected from a collision with a nucleus of mass A, above a threshold energy E_{th}—taking $E_R = \infty$ in Eq. (4.32)—in a plane (μ_χ, σ_0^n) values of A and E_{th}. We show the result in Fig. 4.4. All the points lying on the line predict 5 events for a 100 kg Xenon-like experiment. Even if the model we were working on is a "toy" model (without taking into account the velocity of the Earth or the form factor of the nucleus), it is interesting to see that the behavior and the order of magnitude we obtained are quite reasonable. The behavior for large mass of dark matter is mainly governed by R_0 in Eq. (4.32) as the exponential factor is suppressed. The rate thus evolves proportionally to $1/m_\chi$ as we can see in Fig. 4.4. For light dark matter mass however, the Boltzmann suppression becomes dominant. It is indeed more difficult to find high velocities able to generate sufficient kinetic energy and compensate the low mass of the dark matter. In this region, the queue of the Boltzmann distribution and the different effects of the Earth velocities become important.

4.6 Being More Realistic

4.6.1 Taking into Account the Earth Velocity

We now have developed all the necessary tools to compute the differential rate taking into account the form factor of the nucleus and the kinematics of the Earth in the galactic plane. If we do not neglect the velocity of the Earth, we can have a more precise analytical expression of the differential rate dR/dE_R noticing that $|\mathbf{v}_\chi + \mathbf{v}_E|^2 = v_\chi^2 + v_E^2 + 2v_\chi v_E \cos\theta$

$$\frac{dR}{dE_R} = \frac{N_0 M_N}{A} \frac{\rho_0 \sigma_0 F^2(E_R)}{2m_\chi \mu^2 v_0 v_E \sqrt{\pi}} \int_{v_\chi^{min}}^{\infty} \left(e^{(v-v_E)^2/v_0^2} - e^{-(v+v_E)^2/v_0^2} \right) dv$$
$$= \frac{N_0 M_N}{A} \frac{\rho_0 \sigma_0 F^2(E_R)}{4m_\chi \mu^2 v_E} \left[\text{erf}\left(\frac{v_\chi^{min} + v_E}{v_0} \right) - \text{erf}\left(\frac{v_\chi^{min} - v_E}{v_0} \right) \right], \quad (4.34)$$

where v_χ^{min}, given by Eq. (4.14), is the minimum WIMP velocity able to induce a nuclear recoil energy above the threshold. We show in Fig. 4.5 the differential rate as a function of the energy of recoil for different dark matter masses (10 and 100 GeV). We notice that a 10 GeV dark matter gives a measurable spectrum at very low energy of recoil ($\lesssim$10 keV), whereas the spectrum is harder for heavier WIMPs. For light WIMP, the shape of the spectrum is mainly independent of the form factor $F(E_R)$ because at the energy considered ($\lesssim$10 keV) $F(E_R) \simeq 1$. The differential rate depends mainly on the factor $\left[\text{erf}\left(\frac{v_\chi^{min}+v_E}{v_0} \right) - \text{erf}\left(\frac{v_\chi^{min}-v_E}{v_0} \right) \right]$ of Eq. (4.34), which converges rapidly to 0 as $v_{min} \gg v_E$ for $E_R \gtrsim 10$ keV. Physically speaking, it means that the number of WIMPs able to give a 10 keV energy of recoil to the nucleus is very low because of the low mass of the dark matter: the population of WIMP with high velocity is Boltzmann suppressed, as we have seen in Eq. (3.28). We do not see such a strong phenomenon in the case of a 100 GeV WIMP on the right panel of Fig. 4.5. We also plotted the differential rate *without* taking into account the form factor. As expected, the effect of $F(E_R)$ is much stronger for heavier nuclear targets (due to the decoherence effect $q\, r_N \lesssim 1$ on "bigger" targets) and is even the dominant process in the case of the Xenon nucleus above $\simeq$20 keV, whereas the effect can be neglected for Silicium targets.

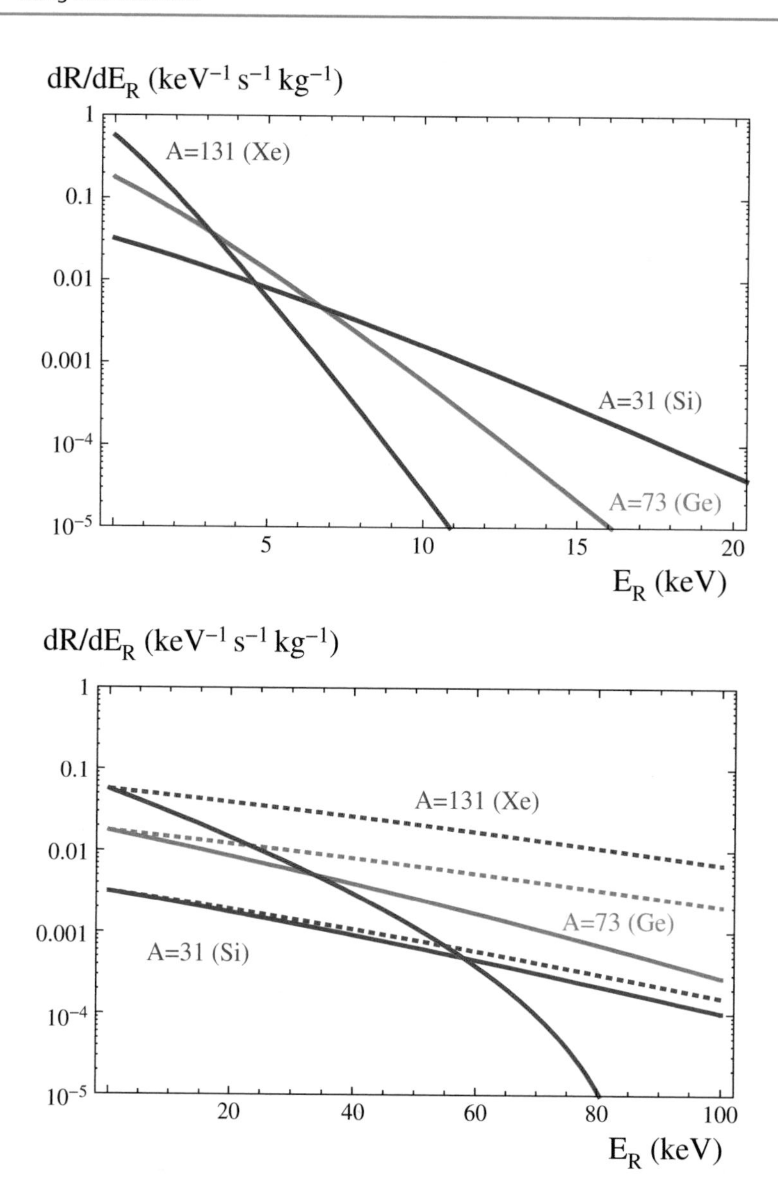

Fig. 4.5 Examples of differential fluxes obtained for a 10 GeV WIMP (top) and a 100 GeV WIMP (bottom) hitting different nuclei: Silicium (blue), Germanium (green), and Xenon (red). The effect of the form factor is shown in the bottom panel where the differential rate is shown without taking into account the form factor in dotted lines

4.6.2 Annual Modulation of the Signal

Up to now we did not take into account that the Earth is rotating annually around the Sun. As we can see in the left panel of Fig. 4.6 and using Eq. (4.5), the absolute velocity of the Earth with respect to the Local Standard of Rest—see Sect. 5.1.3—varies from 217 km/s to 247 km/s being maximum on the 2nd of June. This variation induces a modulation of $\simeq 3\%$ in the signal measurable by experiments like DAMA. We represented the case of a Germanium target in the right panel of Fig. 4.6.

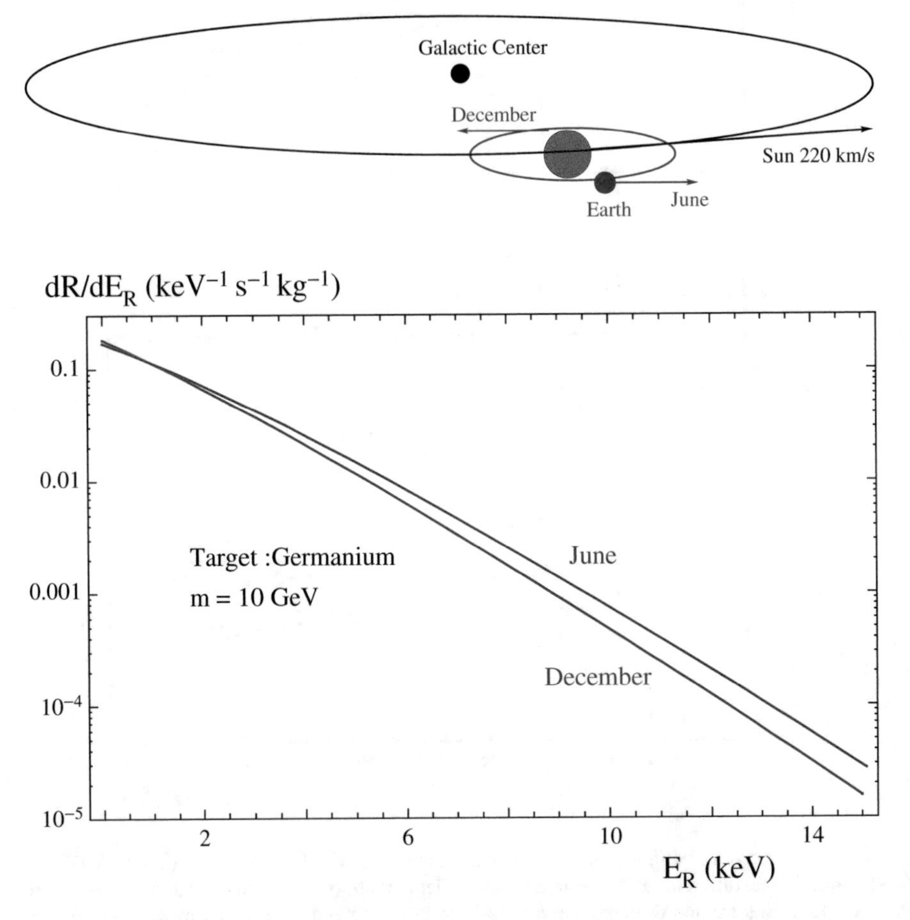

Fig. 4.6 Top: Rotation of the Earth around the galactic center. Bottom: Spectrum as a function of the energy of recoil on a Germanium target for a 10 GeV WIMP in June and December

4.7 Influence of the Structure of the Nucleons**

Since several years, it is known that the uncertainties generated by the quark contents of the nucleons can be as important (if not more) than astrophysical uncertainties. Some authors pointed out this issue and analyzed it to supersymmetric models, in effective operator approach or even in the scalar extension of the SM, but rarely taking into account the latest lattice results. Indeed, due to its large Yukawa coupling, the strange quark and its content in the nucleon are of particular interest in the elastic scattering of the dark matter on the proton. We learnt in the previous section that the spin-independent part of the cross section at zero momentum (4.29) can be written as

$$\sigma_0 = \frac{\mu^2}{\pi}\left[Z f_p + (A - Z) f_n\right]^2. \tag{4.35}$$

The couplings appearing in Eq. (4.35) are the effective couplings of dark matter to *nucleons*, whereas we know at the microscopic level the dark matter couplings to the *partons* (quarks or gluons). Due to the low energy scale of the interaction in the elastic scattering (the velocity of the dark matter χ is around 300 km/s $\simeq 10^{-3}c$), we can define the effective dark matter–quark couplings as $\mathcal{L}_{qq\chi\chi} = f_q \bar{q} q \bar{\chi}\chi$, f_q being a dimensionful coefficient proportional to $1/m_h^2$ in the case of t-channel Higgs exchange for instance.[7] The mean interaction of the dark matter particle with a nucleon N of mass m_N can then be written as $f_q\langle N|\bar{q}q\bar{\chi}\chi|N\rangle = f_q\langle N|\bar{q}q|N\rangle\bar{\chi}\chi$. If one defines $\langle N|\bar{q}q|N\rangle = \frac{m_N}{m_q} f_q^N$, the effective quark–nucleon coupling becomes

$$f_{\chi N}^{eff} = \frac{m_N}{m_q} f_q f_q^N. \tag{4.36}$$

This effective coupling can be understood as the probability for a particle χ of hitting a quark q (f_q) multiplied by the probability to find a quark q in the nucleon N (f_q^N). Summing on all the quarks constituting the nucleon, we can write[8] the $f_{N=p,n}^\chi$ appearing in Eq. (4.35)

$$f_N^\chi = m_N \left(\sum_{q=u,d,s} f_q^N \frac{\mathcal{A}_q^\chi}{m_q} + \frac{2}{27} f_H^N \sum_{q=c,b,t} \frac{\mathcal{A}_q^\chi}{m_q} \right) \tag{4.37}$$

[7]This corresponds to the Higgs propagator for a $t = 0$ exchanged momentum, which corresponds to the definition of σ_0.

[8]The partons content of a nucleon can be derived from computations of anomalies of the trace of energy–momentum tensor in QCD: $M_N\langle N|N\rangle = \langle N|\sum_{q=u,d,s} m_q\bar{q}q - \frac{9}{8\pi}\alpha_s G_{\mu\nu}^a G_a^{\mu\nu}|N\rangle$, $G_{\mu\nu}^a$ being the gluon tensor [3]. That is from where is extracted the factor $\frac{2}{27}$ in Eq. (4.37).

with $\mathcal{A}_q^{\chi}$ the scattering amplitude on a single quark q and $f_q^N = (m_q/m_N)\langle N|\bar{q}q|N\rangle$ is the reduced (dimensionless) sigma terms of the nucleon N, and $f_{\mathbf{H}}^N = 1-\sum_{q=u,d,s} f_q^N$ is the **H**eavy quark content of the nucleon (through loop interaction with the gluon content of the nucleon), see [3] and Sect. 4.9.3 for details. The superscript χ indicates that the coupling of the dark matter to quarks q can be dependent on the nature of q.

There are different ways of extracting the reduced dimensionless nucleon sigma terms $f_q^N \equiv (m_q/m_N)\langle N|\bar{q}q|N\rangle$. They can be derived by phenomenological estimates of the $\pi - N$ scattering $\Sigma_{\pi N}$ (see [4,5] and references therein for a review):

$$\Sigma_{\pi N} \equiv m_N f_l = m_l \langle N|\bar{u}u + \bar{d}d|N\rangle \tag{4.38}$$

with $m_l = (m_u + m_d)/2$. While an early experimental extraction gave $\Sigma_{\pi N} = 45 \pm 8$ MeV, a more recent determination obtained $\Sigma_{\pi N} = 64 \pm 7$ MeV. A complete summary of the different values of $\Sigma_{\pi N}$ can be found in Table 1 of [5], where they also give a weighted mean of 21 determinations:

$$\Sigma_{\pi N} = 46.1 \pm 2.2 \text{ MeV}. \tag{4.39}$$

On the other hand, the study of the breaking of $SU(3)$ within the baryon octet and the observation of the spectrum lead to derive a constraint on the non-singlet combination $\tilde{\sigma} = m_l \langle N|\bar{u}u + \bar{d}d - 2\bar{s}s|N\rangle$. $\tilde{\sigma}$ can be extracted phenomenologically from the octet baryon mass splittings, normalized to corrections that can be calculated in chiral effective field theory. Several studies lead to a value $\tilde{\sigma} = 36 \pm 7$ MeV, whereas some others obtained $\tilde{\sigma} = 50$ MeV, still in the limit allowed by chiral perturbation theories [5]. By introducing

$$z = \frac{\langle N|\bar{u}u - \bar{s}s|N\rangle}{\langle N|\bar{d}d - \bar{s}s|N\rangle} = 1.49, \tag{4.40}$$

one deduces

$$\begin{aligned} f_d^N &= \frac{m_d}{m_N} \frac{\Sigma_{\pi N}}{m_u + m_d} \frac{y(z-1)+2}{1+z} \\ f_u^N &= \frac{m_u}{m_N} \frac{\Sigma_{\pi N}}{m_u + m_d} \frac{y(1-z)+2z}{1+z} \\ f_s^N &= \frac{m_s}{m_N} \frac{\Sigma_{\pi N}}{m_u + m_d}\, y, \end{aligned} \tag{4.41}$$

where

$$y = 2\frac{\langle N|\bar{s}s|N\rangle}{\langle N|\bar{u}u + \bar{d}d|N\rangle} = 1 - \frac{\tilde{\sigma}}{\Sigma_{\pi N}} \tag{4.42}$$

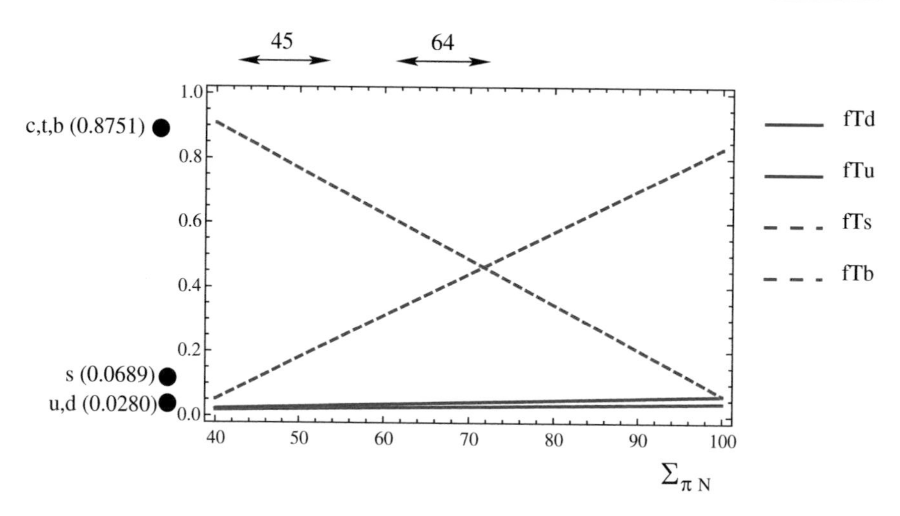

Fig. 4.7 Sigma commutator of the proton ($f_q^P = \frac{m_q}{m_p}\langle P|\bar{q}q|P\rangle$) with two different phenomenological measurements of $\Sigma_{\pi N} = 45\pm8$ MeV and 64 ± 7 MeV. We also showed the mean evaluation from lattice results (points in the left)

represents the strange fraction in the nucleon. We show in Fig. 4.7 the dependence of f_q^N as a function of $\Sigma_{\pi N}$. The two extreme values are obtained with the lower bound of $\Sigma_{\pi N}$ at 1σ (37 MeV) and the higher bounds (71 MeV), which give for $(m_u, m_d, m_s, m_c, m_b, m_t, m_p) = (2.76, 5., 94.5, 1250, 4200, 171400, 938.3)$ [MeV]:

$$
\begin{aligned}
f_u^{min} &= 0.016 \quad & f_u^{max} &= 0.030 \\
f_d^{min} &= 0.020 \quad & f_d^{max} &= 0.044 \\
f_s^{min} &= 0.013 \quad & f_s^{max} &= 0.454.
\end{aligned}
\tag{4.43}
$$

A quick look at the figure shows us how important is to know the strange content of the nucleon because it varies a lot in the phenomenologically viable region of $\Sigma_{\pi N}$ that can affect drastically the direct detection prediction. For the heaviest quarks, the factor $\frac{m_N}{m_{q=c,b,t}}$ reduces considerably their influence on the coupling f_N as we can see in Eq. (4.37).

These limitations on the phenomenological estimation of the strange structure of the nucleon clearly open the way for lattice QCD to offer significant improvements. Using the Feynman–Hellman relation $f_q^N = (m_q/m_N)\partial m_N/\partial m_q$, different authors have extracted the light quark and strangeness sigma terms (see [4] for a clear review). The last results obtained by the authors (labeled “Young” from now on) provide stringent limits on the strange quark sigma terms. The lattice results for f_s^N agree that the size is substantially smaller than that has been previously thought:

$$
f_s^{\text{Young}} = 0.033 \pm 0.022 \tag{4.44}
$$

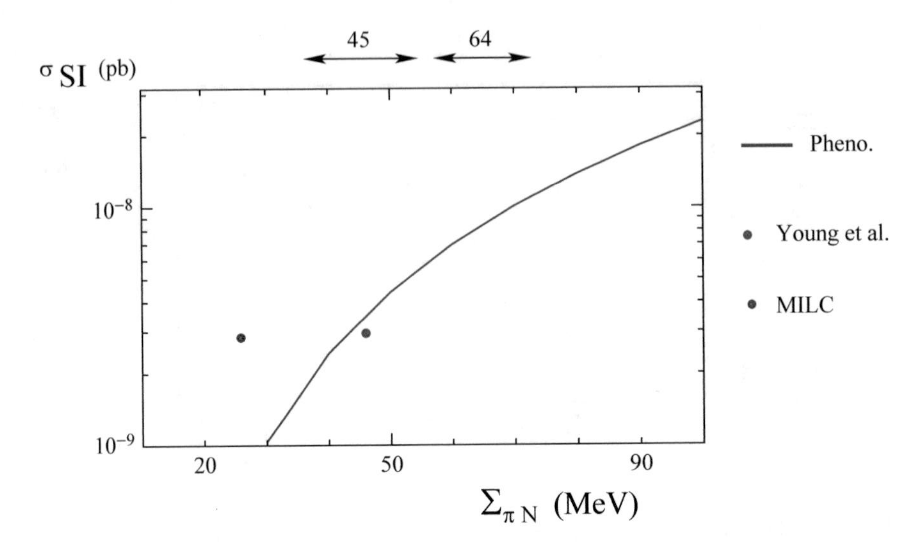

Fig. 4.8 Example of spin-independent elastic scattering cross section as a function of the pion–nucleon sigma term $\Sigma_{\pi N}$ for a scalar dark matter $m_\chi = 90$ GeV. We also represented the central values of the cross section for the lattice simulations [6] labeled "MILC" and "Young"

This result tends to favor the smaller phenomenological evaluation of $\Sigma_{\pi N}$. In the following, we will consider the central values of f_q^N extracted from the Young et al. analysis and referred it to the "lattice" one: $f_u^N = f_d^N = f_l^N = 0.050$, and $f_s^N = 0.033$, and the maximum and minimum values for f_q^N given by phenomenological references (Eq. 4.43).

As we can see in Fig. 4.8, these uncertainties have a strong impact on the direct detection cross section on the nucleon, σ_{SI} (*SI* for *S*pin *I*ndependent) up to one order of magnitude. We also plotted the values of σ_{SI} obtained by the two lattice groups corresponding to the central values $(\Sigma_{\pi N}, \sigma_{SI}) = (26$ MeV, 2.84×10^{-9} pb) and (47 MeV; 2.95×10^{-9} pb) [6]. We clearly see that the lattice results are in much more accordance with the lower bound on $\Sigma_{\pi N}$: $\sigma_{SI}^{min}(\Sigma_{\pi N} = 37 \text{ MeV}) = 1.93 \times 10^{-9}$ pb, whereas $\sigma_{SI}^{max}(\Sigma_{\pi N} = 71 \text{ MeV}) = 1.05 \times 10^{-8}$ pb. We compiled all the necessary values of f_i in the following table (see Sect. 4.9.3 and Eq. (4.86) for a detailed calculation of f_H^N and $f_N = \sum f_l^N + 3 \times f_H^N = \frac{7}{9} \sum f_l^N + \frac{2}{9}$).

f_i^N	Lattice	Min	Max
f_u^N	0.050	0.016	0.030
f_d^N	0.050	0.020	0.044
f_s^N	0.033	0.012	0.454
$f_N = \sum f_l^N + 3 \times f_H^N$	0.326	0.260	0.629

In the rest of the section, we will always present our results with the evaluation of f_s^N given by the maximum and minimum allowed values for $\Sigma_{\pi N}$ and the lattice extraction of Young *et al.* More recently, a general result, taking a mean over more than 20 analyses, gave these values for f_i^N [5]:

Nucleon	f_u^N	f_d^N	f_s^N	f_G^N	f_c^N	f_b^N	f_t^N
Proton	0.018(5)	0.027(7)	0.037(17)	0.917(19)	0.078(2)	0.072(2)	0.069(1)
Neutron	0.013(3)	0.040(10)	0.037(17)	0.910(20)	0.078(2)	0.071(2)	0.068(2)

which is quite in accordance with the values we obtained in Eq. (4.43).

4.8 Spinorial Effect

When the dark matter candidate has a spin, there is possibility to measure spin–spin interaction between the dark matter and the nucleus.[9] Indeed, experimentalists cannot control the energy spectrum of the incident WIMPs but can control the target material. The choice of the target nuclei has a major impact on the type of WIMP scatterings allowed: the interaction can be spin-dependent or spin-independent. The spin-independent interaction occurs when a WIMP scatters off *all* the nucleons in a nucleus, for a sufficiently long exchanged wavelength. On the contrary, for a spin-dependent interaction, the incident WIMP *must* have a spin, and the target nucleus must *also* have net spin from an unpaired nucleon: the unpaired nucleon and the WIMP interact. As we already discussed, the spin-independent interaction will be obviously enhanced by a coherence factor of A^2 if $qR \lesssim 1$ (Heisenberg uncertainty principle, in other words when the wavelength of the exchanged energy is larger than the size of the nucleus—not nucleons of course), which is not the case in spin-dependent interactions if we suppose that the dark matter interacts only with the unpaired nucleons (the *independent single-particle shell model*). In this section, we will extract the spin-dependent interaction of the dark matter on the nucleus step by step: from the partons level to the nucleons one and then the nucleus level. At the partons level, the operator responsible for spin-dependent dark matter–nucleus $(\chi - N)$ interactions is generated by the Lagrangian

$$\mathcal{L}_{SD} = \sum_{q=u,d,s} \alpha_q \, (\bar{\chi}\gamma^\mu\gamma^5\chi)(\bar{q}\gamma_\mu\gamma^5 q), \tag{4.45}$$

[9] One of the clearest presentations of the phenomenon is present in the first article proposing the direct detection principle [7] that is also a very pedagogical work.

where α_q is the coupling determined by the model under study. We can then deduce from Sect. B.3.3, especially Eq. (B.22)

$$\langle \mathbf{n}|\bar{q}\gamma_\mu\gamma^5 q|\mathbf{n}\rangle = 2s_\mu^{(\mathbf{n})}\Delta q^{(\mathbf{n})} \tag{4.46}$$

n standing for **n**ucleon (p or n), and where $s_\mu^{(\mathbf{n})}$ is the spin of the nucleon, whereas the $\Delta q^{(\mathbf{n})}$'s are extracted from polarized deep inelastic scatterings measurements. They represent the fractional spin of the proton carried by quarks of type q. Assuming flavor SU(3) symmetry, one can extract two independent combinations out from three Δq, denoted by F and D, from the observed semileptonic decay rates of baryon in the lowest octet. Fits to all the decays fix F and D to

$$\begin{aligned} F &= \frac{1}{2}(\Delta u - \Delta s) = 0.47 \pm 0.04, \\ D &= \frac{1}{2}(\Delta u - 2\Delta d + \Delta s) = 0.81 \pm 0.03. \end{aligned} \tag{4.47}$$

A third combination, $\frac{1}{9}(4\Delta u + \Delta d + \Delta s) = 0.175 \pm 0.018$, has been measured in polarized electron and muon scattering experiments, but this result should be regarded with caution due to the lack of precision of the experiment. Combining the three measurements gives us

$$\begin{aligned} \Delta u^{(p)} &= \Delta d^{(n)} = 0.78 \pm 0.08, \\ \Delta d^{(p)} &= \Delta u^{(n)} = -0.50 \pm 0.08, \\ \Delta s^{(p)} &= \Delta s^{(n)} = -0.16 \pm 0.08. \end{aligned} \tag{4.48}$$

The conventional non-relativistic QCD partons model, however, predicts instead $\Delta u = 0.97$, $\Delta d = -0.28$, $\Delta s = 0$, though with reasonable modification the model can be made consistent with the scattering data. Global QCD analysis for the g_1 structure function including $O(\alpha_s^3)$ corrections corresponds to the following values of spin nucleon parameters:

$$\begin{aligned} \Delta u^{(p)} &= \Delta d^{(n)} = 0.78 \pm 0.02, \\ \Delta d^{(p)} &= \Delta u^{(n)} = -0.48 \pm 0.02, \\ \Delta s^{(p)} &= \Delta s^{(n)} = -0.15 \pm 0.02. \end{aligned} \tag{4.49}$$

The coefficient of the effective dark matter–nucleon interaction can thus be written as

$$\mathcal{L}_{SD} = 2\sqrt{2}G_F\, a_{\mathbf{n}}(\bar{\chi}\gamma^\mu\gamma^5\chi)(\bar{\mathbf{n}}s_\mu^{(\mathbf{n})}\mathbf{n}) \;\; \text{with} \;\; a_{\mathbf{n}} = \sum_{q=u,d,s}\frac{\alpha_q}{\sqrt{2}G_F}\,\Delta q^{(\mathbf{n})}. \tag{4.50}$$

After summing on all the nucleons **n** composing the nucleus N and noticing that, in the non-relativistic limit, $\bar{\chi}\gamma^{\mu}\gamma^{5}\chi = 2s_{\chi}^{\mu}$, $\mathbf{s}_{\chi}$ being the spin of the dark matter, we obtain

$$\mathcal{M} = 4\sqrt{2}G_F\ \mathbf{s}_{\chi}.\langle N||a_p\mathbf{S}^p + a_n\mathbf{S}^n||N\rangle \tag{4.51}$$

with $\mathbf{S}^{p,n} = \sum_i \mathbf{s}^{p_i,n_i}$. After applying the Wigner–Eckart theorem to the reduced matrix element of the spin $\langle N||\mathbf{S}||N\rangle$

$$\langle N;\ J||\mathbf{S}||N;\ J\rangle = \frac{\sqrt{J(2J+1)(J+1)}}{M_J}\langle N;\ J,\ M_J|\mathbf{S}|N;\ J,\ M_J\rangle \tag{4.52}$$

and in accordance with the convention where the z component of the angular momentum J and spin operators S are evaluated in the maximal M_J state, e.g., $\langle\mathbf{S}\rangle = \langle N|\mathbf{S}|N\rangle = \langle J, M_J = J|S_z|J, M_J = J\rangle$, we finally can write

$$\mathcal{M} = 4\sqrt{2}G_F\ \mathbf{s}_{\chi}.\sqrt{\frac{(2J+1)(J+1)}{J}}\left(a_p\langle N|\mathbf{S}^p|N\rangle + a_n\langle N|\mathbf{S}^n|N\rangle\right) \tag{4.53}$$

from the Fermi's golden rule and after averaging over the initial spin/momentum

$$\begin{aligned}\frac{d\sigma_{SD}}{dE_R} &= 2M_N\frac{d\sigma}{d|q|^2} = \frac{2M_N|\mathcal{M}|^2}{\pi v^2(2J+1)}\\ &= \frac{16M_N G_F^2 J(J+1)}{\pi\ v^2}\left[a_p\frac{\langle N|\mathbf{S}^p|N\rangle}{J} + a_n\frac{\langle N|\mathbf{S}^n|N\rangle}{J}\right]^2\\ &= \frac{16M_N G_F^2 J(J+1)}{\pi\ v^2}\Lambda^2 = 2M_N\frac{\sigma_{SD}^0}{4\mu^2v^2}\end{aligned} \tag{4.54}$$

with

$$\Lambda = a_p\frac{\langle N|\mathbf{S}^p|N\rangle}{J} + a_n\frac{\langle N|\mathbf{S}^n|N\rangle}{J}\quad \text{and}\quad \sigma_{SD}^0 = \frac{32G_F^2}{\pi}\mu^2 J(J+1)\Lambda^2. \tag{4.55}$$

A typical cross section for weakly interacting particles is 10^{-4} pb. From the experimental point of view, the PICASSO (Project in Canada to Search for Supersymmetric Objects) experiment is a direct dark matter search experiment that is located at SNOLAB in Canada. It uses bubble detectors with Freon as the active mass. PICASSO is predominantly sensitive to spin-dependent interactions of WIMPs with the fluorine atoms in the Freon.

4.9 More About the Effective Approach

4.9.1 Validity of the Approach

To compute a differential or absolute rate in Sect. 4.7 we took an effective approach where we integrated out the massive fields exchanged in the interaction process. This is only possible if the energy Q transferred from the dark matter to the nucleus in the interaction is negligible compared the mass scale of the intermediate particle. We computed in Eq. (4.13) the maximum recoiled energy transferred, $E_R^{max} = \frac{2\mu^2 v_\chi^2}{M_N}$, which gives for the maximum momentum of the intermediate state

$$Q^{max} = \sqrt{2M_N \times E_R^{max}} = 2\mu v_\chi = 2\frac{M_N m_\chi}{M_N + m_\chi} v_\chi \simeq m_\chi v_\chi \simeq 100, \text{ MeV} \tag{4.56}$$

where we took for the *nucleus* mass $M_N = m_\chi = 100$ GeV and $v_\chi = 300$ km/s for a numerical approximation. We thus see that any mediator M at the electroweak scale ($M_M \simeq 100$ GeV like a Z or Higgs boson for instance) is sufficiently heavy to justify the approximation in the propagator $(Q^2 - M_M^2)^2 \simeq M_M^4$: the effective approximation is valid. We can also notice that the energy of recoil of a 100 GeV nucleus is then of the order of

$$E_R^{max} \simeq \frac{1}{2} m_\chi v_\chi^2 \simeq \frac{1}{2}\left(\frac{m_\chi}{1 \text{ GeV}}\right) \text{keV}, \tag{4.57}$$

which means about 50 keV for a 100 GeV dark matter particle (still considering $v_\chi \simeq 300$ km/s), justifying the low threshold needed in direct detection experiments (around the keV scale).

4.9.2 Effective Operators

In this section, we propose to calculate scattering amplitudes for couplings of different nature, at the effective level. It is then easy to adapt our results in the framework of microscopic models and then to use the results of Appendix B.4.5.

4.9.2.1 Generalities

It is possible to write the more general form of a Lagrangian for a fermion field in the effective approach

$$\mathcal{L}_F = S_F\, \bar\chi\chi\bar f f + V_F\, \bar\chi\gamma^\mu\chi\bar f\gamma_\mu f + A_F\, \bar\chi\gamma^\mu\gamma^5\chi\bar f\gamma_\mu\gamma^5 f - C_F\, \bar\chi\sigma^{\mu\nu}\chi\bar f\sigma_{\mu\nu} f, \tag{4.58}$$

f being the Standard Model fermions (quarks and leptons), χ the dark matter field, and $\sigma^{\mu\nu} = \frac{1}{2}(\gamma^\mu\gamma^\nu - \gamma^\nu\gamma^\mu)$ is the antisymmetric current tensor. A lot of operators are absent because they disappear in the limit of null momentum transfer. Indeed, $\bar{u}(p_2)\gamma^5 u(p_1)$ is proportional to $p_2 - p_1$ and so is negligible in direct detection prospects. Let us demonstrate it with the use of Eq. (B.24):

$$\bar{u}(p_2)\gamma^5 u(p_1) = \bar{u}(p_2)\frac{\not{p}_2}{2m_\chi}\gamma^5 u(p_1) + \bar{u}(p_2)\gamma^5 \frac{\not{p}_1}{2m_\chi} u(p_1)$$

$$= \frac{(p_2 - p_1)_\nu}{2m_\chi}\bar{u}(p_2)\gamma^\nu\gamma^5 u(p_1) \propto (p2 - p1)_\nu \gamma^\mu \simeq 0. \quad (4.59)$$

With the same argument we can show that $\bar{u}(p_2)\gamma^\mu u(p_1)$ is not null only for the time component ($\mu = 0$), whereas $\bar{u}(p_2)\gamma^\mu\gamma^5 u(p_1)$ is null for the time component. As a consequence, mixing terms of the form $\bar{u}(p_2)\gamma^\mu\gamma^5 u(p_1) \times \bar{u}(p_2)\gamma_\mu u(p_1)$ are absent. Let us demonstrate it:

$$\bar{u}(p_2)\gamma^\mu u(p_1) = \bar{u}(p_2)\frac{\not{p}_2}{2m_\chi}\gamma^\mu u(p_1) + \bar{u}(p_2)\gamma^\mu \frac{\not{p}_1}{2m_\chi} u(p_1)$$

$$= \frac{(p_2)_\nu}{2m_\chi}\bar{u}(p_2)\gamma^\nu\gamma^\mu u(p_1) + \frac{(p_1)_\nu}{2m_\chi}\bar{u}(p_2)\gamma^\mu\gamma^\nu u(p_1)$$

$$= \frac{(p_2)_\nu}{2m_\chi}\bar{u}(p_2)(-\gamma^\mu\gamma^\nu + 2\eta^{\mu\nu})u(p_1) + \frac{(p_1)_\nu}{2m_\chi}\bar{u}(p_2)\gamma^\mu\gamma^\nu u(p_1)$$

$$= \frac{(p_1 - p_2)_\nu}{2m_\chi}\bar{u}(p_2)\gamma^\mu\gamma^\nu u(p_1) + \frac{p_2^\mu}{m_\chi}\bar{u}(p_2)u(p_1) \simeq \frac{p_2^\mu}{m_\chi}\bar{u}(p_2)u(p_1). \quad (4.60)$$

Only the time component is non-zero in the non-relativistic case ($p_i \ll E_i$). In the case of the operator $\bar{u}(p_2)\gamma^\mu\gamma^5 u(p_1)$, one can write

$$\bar{u}(p_2)\gamma^\mu\gamma^5 u(p_1) = [..] = \frac{(p_1 + p_2)_\nu}{2m_\chi}\bar{u}(p_2)\gamma^\mu\gamma^5\gamma^\nu u(p_1) + \frac{(p_1 + p_2)^\mu}{2m_\chi}\bar{u}(p_2)\gamma^5 u(p_1)$$

$$\simeq \frac{E_1 + E_2}{2m_\chi}\bar{u}(p_2)\gamma^\mu\gamma^5\gamma^0 u(p_1) \simeq -\bar{u}(p_2)\gamma^\mu\gamma^0\gamma^5 u(p_1) \quad (4.61)$$

which vanishes in the zero momentum transfer limit for the time component $\mu = 0$. This result shows that terms of the form $\bar{u}(p_2)\gamma^\mu\gamma^5 u(p_1) \times \bar{u}(p_2)\gamma^\mu u(p_1)$ can be neglected in the computation of direct detection processes.

4.9.2.2 Scalar Coefficient: Generalities

As we discussed in the previous section, the scalar coefficient of Eq. (4.58), $S_F \bar{\chi}\chi \bar{f} f$, is in fact a sum over partonic interactions

$$\mathcal{L}_S = S_F \ \bar{\chi}\chi \bar{f} f = \sum_q \lambda_q \bar{\chi}\chi \bar{q} q \tag{4.62}$$

$$\Rightarrow \langle N|\mathcal{L}_S|N\rangle = \sum_q \lambda_q \langle N|\bar{q}q|N\rangle \bar{\chi}\chi = \sum_q \lambda_q \frac{m_N}{m_q} f_q \ \bar{N} N \bar{\chi}\chi$$

with $f_q = \frac{m_q}{m_N}\langle N|\bar{q}q|N\rangle \equiv \frac{\sigma_q}{m_N}$ that we already introduced in Eq. (4.37) and which computation[10] is reserved for Sect. 4.9.3. We can then compare Eq. (4.62) to Eq. (4.58) to express the effective scalar coupling to the nucleon $S_F = \sum_q \lambda_q \frac{m_N}{m_q} f_q$. From this effective Lagrangian, we can compute the amplitude square $|\mathcal{M}_N|^2$ and the cross scattering cross section of χ on the nucleon N:

$$|\mathcal{M}_N|^2 = Tr\Big[(\not{p}_N + m_N)(\not{p}'_N + m_N)\Big] Tr\Big[(\not{p}_\chi + m_\chi)(\not{p}'_\chi + m_\chi)\Big]$$
$$\times \left(\sum_q \lambda_q f_q \frac{m_N}{m_q}\right)^2 \simeq 64 \left(\sum_q \lambda_q f_q \frac{m_N}{m_q}\right)^2 m_N^2 m_\chi^2. \tag{4.63}$$

The cross section is then (using Eq. B.154)

$$\sigma_{\chi N}^{SI} = \frac{|\mathcal{M}|^2}{4} \frac{1}{16\pi (m_N + m_\chi)^2} = \left(\sum_q \lambda_q f_q \frac{m_N}{m_q}\right)^2 \frac{m_N^2 m_\chi^2}{\pi (m_N + m_\chi)^2}. \tag{4.64}$$

If the particle is a Majorana fermion, one needs to multiply the amplitude by 2 (and thus the cross section by 4) because one needs to add the two symmetric diagrams by the exchange of ingoing/outgoing dark matter particle, and the factor 1/4 comes from the average over the incoming spin particles. To obtain the scattering amplitude on a nucleus made of Z protons p and $(A - Z)$ neutron n,

[10] The physical interpretation of $\langle N|\bar{q}q|N\rangle$ is the creation of a quark–antiquark pair from the nucleons N. This is a measurement of the "number" or probability of presence of the sea quarks in the nucleons and does not take into account the valence quarks (contrarily to the vectorial interaction as we will see later).

one obtains (considering $m_p \simeq m_n = m_N$)

$$|\mathcal{M}|^2 = 64 m_\chi^2 m_N^2 \left(\lambda_p Z + \lambda_n (A - Z)\right)^2 \tag{4.65}$$

with $\lambda_N = \sum_q \lambda_q f_q^N \frac{m_N}{m_q}$ and $f_q^N = \frac{m_q}{m_N} \langle N|\bar{q}q|N\rangle$.

4.9.2.3 Scalar Coefficient: Application

Let us apply the previous result to the computation of the cross section in models with Higgs (scalar) portal in the case of a spin 0 or spin 1/2 dark matter. If one includes the Higgs propagator in an effective Lagrangian at the quark level, we can write after the electroweak symmetry breaking[11]

$$\begin{aligned}
\mathcal{L}_S &= y_\chi H \bar{\chi}\chi + \sum_q y_q H \bar{q} q \\
\rightarrow \quad \mathcal{L}_S^{\text{eff}} &\sim \sum_q \frac{y_q y_\chi}{(p^2 - m_H^2)} \bar{\chi}\chi\bar{q}q \quad \simeq \quad -\sum_q \frac{y_q y_\chi}{m_H^2} \bar{\chi}\chi\bar{q}q \\
\Rightarrow \quad \mathcal{L}_S^N &= -\sum_q \frac{y_q y_\chi}{m_H^2} \frac{m_N}{m_q} f_q \, \bar{\chi}\chi\bar{N}N,
\end{aligned} \tag{4.66}$$

where y_i are the Yukawa-like couplings and $p = p_q' - p_q$ the transferred energy in the process, negligible compared to the Higgs mass. We can consider the Higgs as a general extra-scalar particle. In the case of a Standard Model Higgs, we know that $y_q = \frac{m_q}{v} = \frac{g m_q}{2 M_W}$ ($\langle H\rangle = v/\sqrt{2}$). We will compute the elastic scattering cross section for a scalar and a fermionic dark matter by first calculating the amplitude squared in the case of a scalar dark matter,

$$\begin{aligned}
|\mathcal{M}_S|^2 &= (4) \times \frac{y_\chi^2 m_N^2}{m_H^4} Tr\left[(\not{p}_N + m_N)(\not{p}_N' + m_N)\right] \left(\sum_q \frac{y_q}{m_q} f_q\right)^2 = \\
&(4) \times \frac{8 m_N^4 y_\chi^2}{m_H^4} \left(\sum_q \frac{y_q}{m_q} f_q\right)^2
\end{aligned} \tag{4.67}$$

in the limit $\not{p}_N - \not{p}_N' \ll M_N$. The first factor (4) is a symmetry factor that should be present if $\bar{\chi} = \chi$ (a factor 2 from the amplitude if χ is real). From Eq. (B.154)

[11] Be careful, in our notation H is the *physical* Higgs field after the symmetry breaking, which means $H \to (H + v)/\sqrt{2}$, compared with the definition of the Yukawa coupling. The y_q is the Standard Model Yukawa coupling (defined by $\mathcal{L} = y_q^{SM} (\frac{H+v}{\sqrt{2}})$ divided by $\sqrt{2}$). In other words, the $\sqrt{2}$ has been absorbed in a new definition of y_q: $y_q = \frac{y_q^{SM}}{\sqrt{2}}$.

we can compute the spin-independent cross section on the nucleon N

$$\sigma_{\chi N}^{\rm SI} = \frac{1}{2}\frac{|\mathcal{M}_S|^2}{16\pi(m_N+m_\chi)^2} = \frac{(4)\times m_N^4 y_\chi^2}{4\pi(m_N+m_\chi)^2 m_H^4}\left(\sum_q \frac{y_q}{m_q} f_q\right)^2 \quad (4.68)$$

the first $1/2$ factor comes from the mean on the nucleon spin.

In a microscopic (gauge invariant) model, the effective coupling $y_\chi H\bar{\chi}\chi$ is generated through the invariant term (before the symmetry breaking) $-\frac{1}{4}\lambda_{hSS}H^\dagger H\bar{\chi}\chi$, which gives after the symmetry breaking $H\to(v+H)/\sqrt{2}\to\ -\frac{1}{4}v\lambda_{hSS}H\bar{\chi}\chi = -\frac{M_W}{2g}\lambda_{hSS}H\bar{\chi}\chi$ after the $SU(2)\times U(1)$ breaking. We then obtain using $m_q = y_q v$

$$\sigma_{\chi N}^{\rm SI} = \frac{(4)\times m_N^4\lambda_{hSS}^2}{64\pi(m_N+m_\chi)^2 m_H^4}\left(\sum_q f_q\right)^2 \quad (4.69)$$

$$\Rightarrow\ \sigma_{\chi A}^{\rm SI} = \frac{(4)\times m_N^4\lambda_{hSS}^2}{64\pi(m_N+m_\chi)^2 m_H^4}(f_p Z+(A-Z)f_n)^2,$$

where $\sigma_{\chi A}^{\rm SI}$ represents the cross section on the nucleus and $f_{N=p,n} = \sum_q f_q^N = \sum_q \frac{m_q}{m_N}\langle N|\bar{q}q|N\rangle$. You can see another computation in Sect. B.4.5. For a 100 GeV dark matter and a typical coupling $\lambda_{hSS} = 10^{-2}$, we obtain a nucleon cross section of the order of $\sigma_{\chi N}^{\rm SI}\simeq 2\times10^{-19}\ \text{GeV}^{-2}\simeq 10^{-46}\text{cm}^2\simeq 10^{-22}$ barns = 10^{-10} pb, which is in the limits of actual experiments. In the case of a fermionic dark matter we have

$$|\mathcal{M}_F|^2 = (4)\times\frac{y_\chi^2 M_N^2}{m_H^4}Tr\Big[(\not p_\chi+m_\chi)(\not p'_\chi+m_\chi)\Big]$$
$$\times Tr\Big[(\not p_N+m_N)(\not p'_N+m_N)\Big]$$
$$\times\left(\sum_q\frac{y_q}{m_q}f_q\right)^2 = (4)\times 64\frac{y_\chi^2 M_N^4 m_\chi^2}{m_H^4}\left(\sum_q\frac{y_q}{m_q}f_q\right)^2 \quad (4.70)$$

$$\Rightarrow\ \sigma_{N\chi}^{\rm SI} = \frac{(4)}{4}\frac{|\mathcal{M}_F|^2}{16\pi(M_N+m_\chi)^2} = (4)$$
$$\times\frac{y_\chi^2 M_N^4 m_\chi^2}{\pi m_H^4(M_N+m_\chi)^2}\left(\sum_q\frac{y_q}{m_q}f_q\right)^2$$

the factor $(4)\times$ being present if we have a Majorana dark matter. In the case of a microscopic model, using $y_q = \frac{g m_q}{2M_W}$ and $y_\chi = -\frac{1}{4}\frac{\lambda_{hff}}{\Lambda}\frac{2M_W}{g}$ generated by the gauge invariant Lagrangian $-\frac{\lambda_{hff}}{4\Lambda}H^\dagger H\bar{\chi}\chi$, with Λ an effective BSM scale, we obtain

$$\sigma^{\rm SI}_{\chi N} = (4)\times\frac{\lambda^2_{hff}}{16\pi\Lambda^2}\frac{M_N^4 m_\chi^2}{m_H^4(M_N+m_\chi)^2}\left(\sum_q f_q\right)^2. \tag{4.71}$$

The sum $\sum f_q$ is performed on all the heavy and light quarks and is explicitly computed in Sect. 4.9.3, Eq. (4.86).

4.9.2.4 Vector Coefficient: Generalities

The operator $\bar{\chi}\gamma^\mu\chi$ is odd under the charge conjugation and is not present if the dark matter is a Majorana particle (see Sect. 4.9.2.6). However, if the dark matter is a Dirac one, the operator $\bar{\chi}\gamma^\mu\chi$ does not vanish and contributes to the direct detection process. It is typically the case for the exchange of a vectorial particle (Z or Z'). Indeed, if the particle is not a Majorana particle, its charge conjugate is not itself and the detector detects both signals without distinguishing them. Using the Dirac equation (B.24) and $\gamma^0(\gamma^\nu)^\dagger\gamma^0 = \gamma^\nu$, we have seen that that the operators $\bar{\chi}\gamma^\mu\chi$ or $\bar{q}\gamma^\mu q$ are in fact proportional to the momentum, see Eq. (4.60). If the interacting particles are non-relativistic, only the time component is non-vanishing. We can then write $\bar{q}\gamma^0 q = q^\dagger q$, and after the decomposition on creation and annihilation operators (see the Peskin and Schroeder [8] pages 60–62 for details), we obtain

$$q(x) = \int\frac{d^3p}{(2\pi)^{3/2}}\frac{1}{\sqrt{2E_p}}\sum_s(a_p^s u^s(p)e^{-ip.x} + (b_p^s)^\dagger v^s(p)e^{ip.x}) \tag{4.72}$$

$$\Rightarrow \int d^3x q^\dagger(x)q(x) = \int\frac{d^3p}{(2\pi)^3}\sum_s((a_p^s)^\dagger a_p^s + b_{-p}^s(b_{-p}^s)^\dagger)$$

$$= \int\frac{d^3p}{(2\pi)^3}\sum_s((a_p^s)^\dagger a_p^s - (b_p^s)^\dagger b_p^s),$$

and the minus sign in the last equation comes from the anticommutation relation between two fermionic states. a_p^s $[(a_p^s)^\dagger]$ being the annihilation (creation) operator of a particle of momentum p and spin s, and b_p^s $[(b_p^s)^\dagger]$ the annihilation (creation) operator of an *anti*particle of momentum p and spin s, see (B.77) for details. For simplification in the following, we will forget the sum on the spin and note a_p^q (b_p^q) the annihilation operator of (anti)quark q $(\bar{q})$ of momentum p. We can then define a nucleon state as a normalized sum of states composed by valence quarks, valence quarks plus a pair of quark–antiquark of the sea, valence quarks plus 2 pairs of quark–antiquark... We will take the case of a proton and up-type quarks in the sea. The generalization for a neutron and a complete sea of quarks is straightforward. If we write[12]

$$|P\rangle = |u_{p_1} u_{p_2} d_{p_3}\rangle \left(1 + |\bar{u}_{p_4} u_{p_5}\rangle + |\bar{u}_{p_4} u_{p_5} \bar{u}_{p_6} u_{p_7}\rangle + \ldots\right) \quad \text{with} \quad \langle P|P\rangle = 1, \tag{4.73}$$

where the first state represents the valence-quark states and the states in parenthesis are the sea quarks,

$$|u_{p_1} u_{p_2} d_{p_3}\rangle = (a^u_{p_1})^\dagger (a^u_{p_2})^\dagger (a^d_{p_3})^\dagger |0\rangle \quad \text{and} \quad |\bar{u}_{p_4} u_{p_5}\rangle = (b^u_{p_4})^\dagger (a^u_{p_5})^\dagger |0\rangle. \tag{4.74}$$

Using the anticommutation relation $a^u_p (a^u_{p_i})^\dagger = -(a^u_{p_i})^\dagger a^u_p + (2\pi)^3 \delta^3(p - p_i)$, we can easily show (do not forget there is an integration on the momentum on Eq. (4.72) eliminating the δ function and using anticommutation relations to reorder p_1, p_2, p_3

$$\begin{aligned}
&\int \frac{d^3p}{(2\pi)^3} (a^u_p)^\dagger a^u_p (a^u_{p_1})^\dagger (a^u_{p_2})^\dagger (a^d_{p_3})^\dagger |0\rangle \\
&= (a^u_{p_1})^\dagger (a^u_{p_2})^\dagger (a^d_{p_3})^\dagger |0\rangle - \int \frac{d^3p}{(2\pi)^3} (a^u_p)^\dagger (a^u_{p_1})^\dagger (a^u_p)^\dagger (a^u_{p_2})^\dagger (a^d_{p_3})^\dagger |0\rangle \\
&= |u_{p_1} u_{p_2} d_{p_3}\rangle - (a^u_{p2})^\dagger (a^u_{p_1})^\dagger (a^d_{p_3})^\dagger |0\rangle - 0 = 2\, |u_{p_1} u_{p_2} d_{p_3}\rangle \\
&\Rightarrow \int \frac{d^3p}{(2\pi)^3} \left[(a^u_p)^\dagger a^u_p - (b^u_p)^\dagger b^u_p \right] |u_{p_1} u_{p_2} d_{p_3}\rangle = 2 |u_{p_1} u_{p_2} d_{p_3}\rangle.
\end{aligned} \tag{4.75}$$

With the same method, we can easily show that

$$\begin{aligned}
&\int \frac{d^3p}{(2\pi)^3} (a^u_p)^\dagger a^u_p |u_{p_1} u_{p_2} d_{p_3}\rangle |\bar{u}_{p_4} u_{p_5}\rangle \\
&= 2 |u_{p_1} u_{p_2} d_{p_3}\rangle |\bar{u}_{p_4} u_{p_5}\rangle + |u_{p_1} u_{p_2} d_{p_3}\rangle |\bar{u}_{p_4} u_{p_5}\rangle \\
&\int \frac{d^3p}{(2\pi)^3} (b^u_p)^\dagger b^u_p |u_{p_1} u_{p_2} d_{p_3}\rangle |\bar{u}_{p_4} u_{p_5}\rangle \\
&= |u_{p_1} u_{p_2} d_{p_3}\rangle |\bar{u}_{p_4} u_{p_5}\rangle \\
&\Rightarrow \int \frac{d^3p}{(2\pi)^3} \left[(a^u_p)^\dagger a^u_p - (b^u_p)^\dagger b^u_p \right] |u_{p_1} u_{p_2} d_{p_3}\rangle |\bar{u}_{p_4} u_{p_5}\rangle = 2 |u_{p_1} u_{p_2} d_{p_3}\rangle |\bar{u}_{p_4} u_{p_5}\rangle.
\end{aligned} \tag{4.76}$$

---

[12]We will not include the gluonic contribution in the expression of the proton state as the creation/annihilation operators in $q^\dagger q$ do not create or annihilate gluons.

Combining the previous results, we can deduce

$$\int \frac{d^3p}{(2\pi)^3}\left[(a^u_p)^\dagger a^u_p - (b^u_p)^\dagger b^u_p\right]|P\rangle = 2|P\rangle \quad \Rightarrow \quad \int d^3x \langle P|u^\dagger u|P\rangle = 2.$$

We can then obviously prove, following the same procedure that

$$\int d^3x \langle P|d^\dagger d|P\rangle = 1, \quad \int d^3x \langle Ne|u^\dagger u|Ne\rangle = 1, \quad \int d^3x \langle Ne|d^\dagger d|Ne\rangle = 2,$$

$|Ne\rangle$ being the neutron state. As a conclusion, we have proved that the operator $\langle N|q^\dagger q|N\rangle$ measures the contents of the *valence* quark q in the nucleon N. We can then deduce the effective dark matter–nucleon coupling from the dark matter–quark coupling

$$\begin{aligned}\mathcal{L}^\chi_V &= \sum_q f_q\, \bar{\chi}\gamma^\mu\chi\bar{q}\gamma_\mu q \quad \Rightarrow \quad \langle N|\mathcal{L}_V|N\rangle = \sum_q f_q \langle N|\bar{q}\gamma^\mu q|N\rangle\bar{\chi}\gamma_\mu\chi \\ &= \sum_q f_q f^N_q\, \bar{\chi}\gamma^\mu\chi\bar{N}\gamma_\mu N, \quad \text{with } f^p_u = f^n_d = 2 \text{ and } f^p_d = f^n_u = 1.\end{aligned}$$

The particle χ being a Dirac particle, its anti-matter $\tilde{\chi}$, has also interaction with the nucleus. However, contrarily to the scalar interaction, we can easily show using Eq. (B.48) that the operator $\bar{\tilde{\chi}}\gamma^\mu\tilde{\chi} = -\bar{\chi}\gamma^\mu\chi$

$$\mathcal{L}^{\tilde{\chi}}_V = \sum_q f_q f^N_q\, \bar{\chi}\gamma^\mu\chi\bar{N}\gamma_\mu N. \tag{4.77}$$

We notice that the interaction Lagrangian for the anti-dark matter particle is the opposite to the one with the dark matter interaction. However, these two processes are independent without interference between them. When one computes an interaction cross section, one should add both processes, proportional to the local density of (anti)dark matter, respectively. This happens typically for asymmetric dark matter where the density of the anti-dark matter is much less important that the dark matter one.

4.9.2.5 Vector Coefficient: Application

We can also apply the previous calculation to compute the effective Lagrangian in the case of the exchange of a massive vectorial particle V_μ

$$\mathcal{L} = y_\chi V_\mu\bar{\chi}\gamma^\mu\chi + \sum_q y_q V_\nu\bar{q}\gamma^\nu q \quad \Rightarrow \quad \mathcal{L}^{\text{eff}} = \frac{y_\chi y_q}{M_V^2}\bar{\chi}\gamma^\mu\chi\, \bar{q}\gamma_\mu q. \tag{4.78}$$

We will concentrate on fermionic dark matter candidates; we let the reader make the computation for a scalar dark matter. The amplitude squared is then (we let the reader check it with the trace formulae in the appendix)

$$|\mathcal{M}|^2 = |\bar{u}(p_\chi)\gamma^\mu u(p_\chi)\,\bar{u}(p_N)\gamma_\mu u(p_N)|^2 = 64\, m_\chi^2 M_N^2, \tag{4.79}$$

which is the same amplitude squared than in the case of the exchange of a massive scalar we computed in Eq. (4.70). We can then use the results already obtained.

Exercise Redo the same calculation for axial-vector-type coefficients ($\gamma^\mu\gamma^5$) and magnetic ($\sigma^{\mu\nu} = \frac{1}{2}[\gamma^\mu, \gamma^\nu]$) interactions.

4.9.2.6 Majorana Case

In the case of a Majorana dark matter particles, the anti-matter field $\tilde{\chi} = \chi$. The expressions are even simpler as some operators vanish. The operator $\bar{\chi}\gamma^\mu\chi$ for instance can be written as

$$\bar{\chi}\gamma^\mu\chi = \frac{1}{2}\left(\bar{\chi}\gamma^\mu\chi + \bar{\tilde{\chi}}\gamma^\mu\tilde{\chi}\right) = \frac{1}{2}\left(\bar{\chi}\gamma^\mu\chi - \bar{\chi}\gamma^\mu\chi\right) = 0,$$

where we used the relation Eq. (B.48). We also obtain a null contribution for the $\sigma_{\mu\nu}$ term. The only operators that we need to take into consideration are then

$$S_F\bar{\chi}\chi\bar{f}f + A_F\bar{\chi}\gamma^\mu\gamma^5\chi\bar{f}\gamma_\mu\gamma^5 f, \tag{4.80}$$

which gives, respectively, the spin-independent and spin-dependent amplitudes.

4.9.3 Gluons and Heavy Quarks Contributions*

We can go into a more detailed study of the quark content of the nucleon. Indeed, in equation (4.58) the fermion f entering in the interaction should be the nucleon. However, as we discussed, the effective interaction at a microscopic level is on the partons (quarks and gluons) and not directly on the nucleon. Concentrating on the scalar case first, one needs to compute

$$\begin{aligned}\mathcal{L}_{eff} &= \sum_f \lambda_f \langle N;\chi|\bar{\chi}\chi\bar{f}f|\chi;N\rangle = \sum_f \lambda_f \langle\chi|\bar{\chi}\chi|\chi\rangle\langle N|\bar{f}f|N\rangle \\ &= \sum_{q=u,d,s} \lambda_q\bar{\chi}\chi\langle N|\bar{q}q|N\rangle + \sum_{Q=c,b,t} \lambda_Q\bar{\chi}\chi\langle N|\bar{Q}Q|N\rangle,\end{aligned} \tag{4.81}$$

where we have separated light ($q = u, d, s$) and heavy ($Q = c, b, t$) quark states. The values of $\langle N|\bar{q}q|N\rangle$ can be deduced directly by lattice QCD or perturbative QCD approaches. By these methods, one can usually extract the quantity $\sigma_q = m_q\langle N|\bar{q}q|N\rangle$ (in MeV) or equivalently $f_q = \frac{m_q}{M_N}\langle N|\bar{q}q|N\rangle$ (dimensionless). However, there is no equivalent method to compute $\langle N|\bar{Q}Q|N\rangle$. In order to obtain such a quantity, one needs first to compute the trace of the energy–momentum tensor $\theta_{\mu\mu}$ in the nucleon state [3]:

$$
\begin{aligned}
& M_N\langle N|N\rangle \\
& = \langle N|\theta_{\mu\mu}|N\rangle = \langle N|\sum_{q=u,d,s} m_q\bar{q}q + \sum_{Q=c,b,t} m_Q\bar{Q}Q + \frac{\beta(\alpha_s)}{4\alpha_s}G^a_{\mu\nu}G^{\mu\nu}_a|N\rangle \\
& = \langle N|\sum_{q=u,d,s} m_q\bar{q}q + \sum_{Q=c,b,t} m_Q\bar{Q}Q - \frac{\alpha_s}{8\pi}\left[11-\frac{2}{3}n_{f=q+Q}\right]G^a_{\mu\nu}G^{\mu\nu}_a|N\rangle
\end{aligned}
\tag{4.82}
$$

for n_f quark family, where $G^a_{\mu\nu}$ is the gluonic field tensor and $\beta(\alpha_s) = \mu\frac{\partial\alpha_s}{\partial\mu} = -\frac{\alpha_s^2}{2\pi}\left[11-\frac{2}{3}n_f\right]$ is the β-function of the strong $SU(3)_c$ gauge coupling, μ being the energy scale. The contribution proportional to $G^a_{\mu\nu}G^{\mu\nu}_a$ in Eq. (4.82) comes from the triangle anomalies in $\theta_{\mu\mu}$ [3]. We will separate the energy momentum into heavy quark states and light quark states. Indeed, one can make one step further and get rid of the heavy quarks in Eq. (4.82) because there are no valence heavy quarks in the nucleon and they can enter only via a virtual state, at short distances of order $\frac{1}{2m_Q}$. It is equivalent to develop a quark propagator $\frac{1}{\not{P}-m_Q}$ in the regime where $m_Q \gg P$. The heavy quark term at the heavy mass limit gives (at the first order) [9]

$$
\sum_{Q=c,b,t} m_Q\bar{Q}Q \to -\frac{2}{3}\frac{\alpha_s}{8\pi}n_Q G^a_{\mu\nu}G^{\mu\nu}_a + O(\alpha_s^2),
\tag{4.83}
$$

n_Q being the number of heavy quarks. As we can see, the limit obtained in Eq. (4.83) cancels exactly the anomaly contribution of the heavy quarks in the energy–momentum tensor in Eq. (4.82) and is independent of the mass of the quark (at first order). We are then left with only the light quark contribution; in other words

$$
M_N\langle N|N\rangle = M_N = \left\langle N|\sum_{q=u,d,s} m_q\bar{q}q - \frac{9\alpha_s}{8\pi}G^a_{\mu\nu}G^{\mu\nu}_a|N\right\rangle.
\tag{4.84}
$$

Combining Eqs. (4.83) and (4.84), we deduce

$$
\begin{aligned}
& \langle N|\sum_{Q=c,b,t} m_Q\bar{Q}Q|N\rangle = \frac{2n_Q M_N}{27} - \frac{2n_Q}{27}\langle N|\sum_{q=u,d,s} m_q\bar{q}q|N\rangle \\
& \Rightarrow\ f_Q = \frac{m_Q}{M_N}\langle N|\bar{Q}Q|N\rangle = \frac{2}{27}\left[1-\sum_{q=u,d,s} f_q\right],
\end{aligned}
\tag{4.85}
$$

which is the same contribution for all heavy quarks as their mass never appears in the limit $m_Q \to \infty$. We can also understand it as their contributions are generated by triangle anomalies that are mass-independent (topological) quantities. The useful quantity in the computation of the direct detection cross section is $\sum f_q$ (see Eq. 4.70 for instance). We then have

$$f_N = \sum_{q=u;d;s} f_q + \sum_{Q=c,b,t} f_Q = \sum_q f_q + 3 \times \frac{2}{27}\left[1 - \sum_q f_q\right] = \frac{7}{9}\sum_q f_q + \frac{2}{9}. \tag{4.86}$$

References

1. G. Arcadi, M. Dutra, P. Ghosh, M. Lindner, Y. Mambrini, M. Pierre, S. Profumo, F.S. Queiroz, Eur. Phys. J. C **78**(3), 203 (2018). https://doi.org/10.1140/epjc/s10052-018-5662-y [arXiv:1703.07364 [hep-ph]]
2. R.H. Helm, Phys. Rev. **104**, 1466–1475 (1956). https://doi.org/10.1103/PhysRev.104.1466
3. M.A. Shifman, A. Vainshtein, V.I. Zakharov, Phys. Lett. B **78**, 443–446 (1978). https://doi.org/10.1016/0370-2693(78)90481-1
4. R.D. Young, A.W. Thomas, Nucl. Phys. A **844**, 266C–271C (2010)
5. J. Ellis, N. Nagata, K.A. Olive, Eur. Phys. J. C **78**(7), 569 (2018). https://doi.org/10.1140/epjc/s10052-018-6047-y [arXiv:1805.09795 [hep-ph]]
6. R.D. Young, A.W. Thomas, Phys. Rev. D **81**, 014503 (2010). https://doi.org/10.1103/PhysRevD.81.014503 [arXiv:0901.3310 [hep-lat]]
7. M.W. Goodman, E. Witten, Phys. Rev. D **31**, 3059 (1985). https://doi.org/10.1103/PhysRevD.31.3059
8. M.E. Peskin, D.V. Schroeder, *An Introduction to Quantum Field Theory* (Westview Press, Boulder, 1995)
9. A.I. Vainshtein, V.I. Zakharov, M.A. Shifman, JETP Lett. **22**, 55–56 (1975)

In the Galaxies [T_0]

5

Abstract

As we have seen, the direct detection experiments, which are based in large majority on the neutrino detectors principles, have not yet given any hints of dark matter presence in the neighborhood of the Earth. Another possibility is to look for indirect effects of dark matter particles in our galaxy, or dwarf galaxies surrounding the Milky Way. The main idea underlying this kind of research is the possibility for dark matter to annihilate (or decay) in the interstellar medium, or generally near sources of gravity like the dynamical galactic centers, where the population of dark matter is the largest due to the gravitational well. The products of annihilation can be charged (like an electron-positron pair) or neutral (neutrino or photon final states). In this case, satellites or telescopes look directly for the product of annihilation. Another possibility is to observe the radiation emitted by these products, like synchrotron or bremsstrahlung effects for instance.

The indirect effects of the presence of dark matter in the sky range from keV (or X-ray) spectrum or lines when it concerns radiative processes, to GeV scale for WIMP annihilation. It can even reach PeV scale (10^6 GeV) or even EeV scale (10^9 GeV) for superheavy dark matter candidates that can be produced in the very early stage of the Universe as we saw in first part of the book. These kinds of signals are observable by the Icecube collaboration or the ANITA balloon in the South Pole, which track ultra-high energy neutrinos. But the final products at such energies are limited by their interaction with the CMB or their loss of energies during their propagation in the interstellar medium. For instance, we will see that a $\sim$ GeV positron loses 90% of its energy after a little bit more than a 3 light-years journey in the Milky Way, whereas a 100 TeV photon is stopped by the CMB, through the production onshell of e^+e^- pairs.

Y. Mambrini, *Particles in the Dark Universe*,
https://doi.org/10.1007/978-3-030-78139-2_5

Exercise Recover the Greisen–Zatsepin–Kuzmin limit (GZK cut) of 5×10^{10} GeV corresponding to a proton hitting a photon from the CMB producing a Δ^+ onshell. With the same kind of reasoning, compute the mean free path of a photon in the CMB medium (production of e^+e^- pair onshell) and of a neutrino in the CMB (production of a e^- W^+ pair onshell) as a function of their energies.

In order to observe indirectly particles in the Universe, one needs to understand how such particle populates the Universe, how they are produced, and how they propagate. The density of dark matter in the Milky Way, especially its spatial distribution function, is one of the keypoints of its understanding. We will describe in the first sections of this chapter the different elements one should take into consideration when studying indirect detection of dark matter, especially the astrophysical knowledge (and uncertainties) needed to lead such analysis. We will then enter in more detail into the possible detection modes and strategy set up by experimentalists and theoreticians to actually observe such particles indirectly.

5.1 The Anatomy of the Milky Way

5.1.1 Internal Characteristics

The unit of distance commonly used in astronomy is the parsec. The parsec (parallax of one arcsecond, symbol: pc) is a unit of length, equal to just under 31 trillion kilometers (30.857×10^{15} meters), or about 3.26 light-years.[1] It is defined as the length of the adjacent side of an imaginary right triangle in space. The two dimensions that specify this triangle are the parallax angle (defined as 1 arcsecond) and the opposite side (which is defined as 1 astronomical unit (AU), the distance from the Earth to the Sun). Given these two measurements, along with the rules of trigonometry, the length of the adjacent side (the parsec) can be found.

Exercise Recover the value of 1 parsec with the geometrical definition given above.

The stellar disk of the Milky Way galaxy is approximately 100,000 light-years (30 kpc, 9.5×10^{17} km) in diameter and is considered to be, on average, about 1000 light-years (0.3 kpc, 9.5×10^{15} km) thick, see Fig. 5.1. It is estimated to contain at least 200 billion stars and possibly up to 400 billion stars, the exact figure depending on the number of very low-mass stars, which is highly uncertain.

[1] A light-year or light year (symbol: ly) is a unit of length, equal to just under 10 trillion (i.e. 10^{13}) kilometers. As defined by the International Astronomical Union (IAU), a light-year is the distance that light travels in a vacuum in one Julian year.

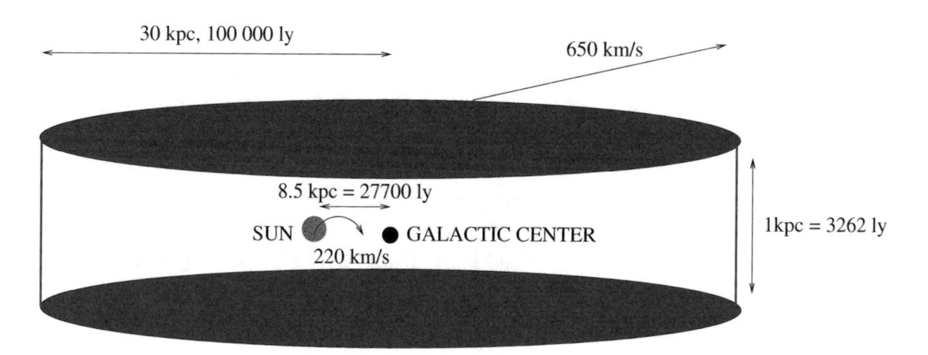

Fig. 5.1 Anatomy of the Milky Way

The center of the Milky Way is formed by a population of old star, in a spherical distribution, called *the bulge*. The Sun (and therefore the Earth and the solar system) lies close to the inner rim of the Milky Way galaxy's Orion Arm at a distance of 7.5–8.5 kpc (25,000–28,000 light-years) from the galactic center. The Earth is turning around the Sun at about 30 km/s, whereas the orbital speed of the solar system around the center of the galaxy (hosting its supermassive black hole Sagittarius A^*) is approximately 251 km/s. In the meantime, the Milky Way converges toward its twin galaxy Andromeda, situated at about 2.5 million light-years from us, at a speed of 650 km/s. A striking point when dealing with these numbers is the relative homogeneity of the astrophysical velocities. Indeed, a 300 km/s proton possesses a keV kinetic energy ($\sim \frac{1}{2} m_p \times \frac{v^2}{c^2}$GeV). This explains why typical energies due to interactions in the interstellar medium turn around this scale. As we also learnt when we treated the case of the direct detection experiments, it is also the typical energy of recoil we expect, due to the motion of our solar system in the Galactic halo.

5.1.2 The Color of the Sky: The Diffuse Gamma Ray Background

When we look up on a clear night, we see the beautiful twinkling of stars and planets amid a black sky. If we could see the same sight with X-ray or gamma-ray eyes, we would still see bright points, but the sky would no longer be dark. Instead, it would glow faintly. This is the diffuse high energy background: X-ray and gamma-ray light from all over the sky. By looking at different wavelengths of X-rays and gamma rays to different energies, we can find out what causes this background glow.

5.1.2.1 X-Ray Diffuse Background

At low X-ray energies ($\lesssim \frac{1}{4}$ keV), the sky glows with radiation from hot gas filling some of the space between the stars. This gas has a temperature of about 1 million degrees and is heated in two ways: by supernovae, which leave shining remnants of hot gas behind, and by the hot winds of massive young stars, which heat surrounding gas forms stellar wind bubbles. At higher X-ray energies ($\gtrsim \frac{1}{2}$ keV), the source of the diffuse background changes considerably. While the emission from supernova remnants and stellar wind bubbles is still visible, it is less dominant than at lower energies. Much of the background radiation becomes isotropic (i.e. it looks the same in all directions). Scientists believe the radiation comes from outside our Milky Way galaxy, since radiation from within the galaxy would be brighter in some places and dimmer in others, due to our galaxy's shape.

Above 1 keV, most of the "diffuse" background is not truly diffuse in origin at all but comes from many distant extragalactic objects. We know this from "deep" observations of the diffuse X-ray background. In astronomy, a "deep" observation means that a detector points to a given point in space for a very long time. Using the deepest ROSAT observations, over 60% of the 1–2 keV diffuse background has been resolved into very distant, separate sources, typically quasars.

5.1.2.2 Gamma-Ray Diffuse Background

While there are many individual point sources of gamma rays, there is also a significant amount of gamma-ray light from gas in the Milky Way. Gamma-ray light from this gas stretches out in a band across the sky, comprising much of the diffuse gamma-ray background. The remaining gamma-ray background light we see comes from far outside our galaxy: it is the faint glow of the rest of the Universe, covering all the sky beyond the Milky Way.

Yet another source of diffuse gamma-ray radiation in our galaxy is the violent interaction of matter and anti-matter. Certain galactic phenomena produce positrons. When positrons and electrons come into contact, they annihilate and release a burst of gamma-ray energy of exactly 511 keV, corresponding to the rest mass of the electron (and positron). These gamma rays are part of the diffuse gamma-ray background in our galaxy.

The diffuse gamma-ray emission from outside our galaxy may be due to the combined light of far-away individual objects. The objects are most likely active galactic nuclei (AGN) like blazars, starburst galaxies, and millisecond pulsars. These are very similar to quasars, the main source of the X-ray background outside our galaxy. The diffuse gamma-ray background was forecast to be one of the more robust constraints of annihilating WIMP dark matter, and determining its spectrum/amplitude is a challenge to restrict the dark matter parameter space.

5.1.3 Galactic Coordinates, Velocity of the Sun and of the Earth

To observe a position in the sky, or when satellites like COBE, WMAP, FERMI, or PLANCK give a map of the sky, they usually work in galactic coordinates: galactic longitude (l) and galactic latitude (b). The longitude measures the angular distance of an object from the Sun eastward along the galactic equator from the galactic center, whereas the latitude measures the angular distance of an object perpendicular

to the galactic equator, positive to the north, negative to the south. This is well illustrated in Fig. 5.2.
The Milky Way is at his densest near Sagittarius as it is where the galactic center lies. The center of Sagittarius, a black hole called Sagittarius A^*, is often considered as the real galactic center. Its coordinates are in fact:[2] ($l = 359^0\ 56'\ 39.5''$, $b = -0^0\ 2'\ 46.3''$), which is in fact a little bit on the south-east of the galactic center. We list in the following Table 5.1 some coordinates of most common objects, especially dwarf galaxies surrounding the Milky Way and important source of observation for dark matter candidates due to their few amount of baryonic matter.

Exercise Find in this list, the nearest dwarf galaxy from the galactic center, and the farthest one.

The velocity of the Sun is defined from the Local Standard of Rest (LSR) and is usually designated by U, V, and W, given in km/s, with U positive in the direction of the galactic center, V positive in the direction of galactic rotation, and W positive in the direction of the North Galactic Pole. If one defines a point in space that is moving on a perfectly circular orbit around the center of the galaxy at the Sun's galactocentric distance, all velocities of stars are described relative to this point, which is known as the Local Standard of Rest. The velocity of the Local Standard of Rest is then given by ($U_{LSR} = 0$; $V_{LSR} = 0$; $W_{LSR} = 0$). The peculiar motion of the Sun with respect to the LSR is then measured and gives $(U, V, W) = (10.00 \pm 0.36, 5.23 \pm 0.62, 7.17 \pm 0.38)$ km/s. We show an illustration of this system of coordinates in Fig. 5.3. The Sun is thus deriving slowly toward the galactic center, moving a little bit faster than the entire disk, with some components toward the north pole of the Milky Way.

We can then deduce the absolute velocity of the Earth in the Milky Way (and as a consequence, the velocity of the Earth relative to the dark halo):

$$\mathbf{v}_E = \mathbf{v}_r + \mathbf{v}_S + \mathbf{v}_e \tag{5.1}$$

with, in galactic coordinates:

- the velocity of the rotation of the galaxy at the solar neighborhood $\mathbf{v}_r = (0, 230, 0)$ km s^{-1};
- the Sun's proper motion (motion relative to nearby stars) $\mathbf{v}_S = (10, 5, 7)$ km s^{-1};
- the Earth's orbital velocity relative to the Sun. $\mathbf{v}_e = 15 \cos\left(2\pi \frac{t - 152.5 \text{ days}}{365.25 \text{ days}}\right)$ km s^{-1}.

[2] $1^0 = 60'$ (arcminutes) $= 3600''$ (arcseconds).

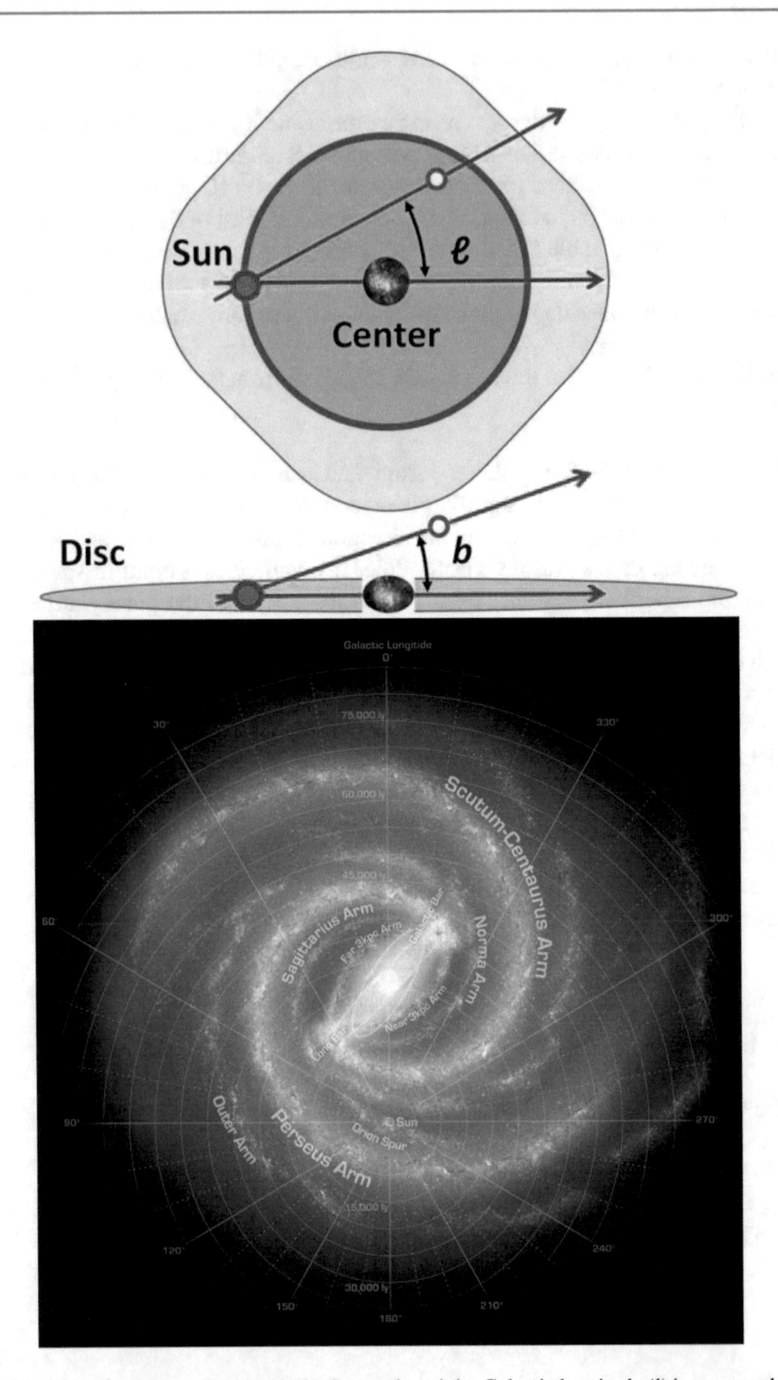

Fig. 5.2 The galactic coordinates use the Sun as the origin. Galactic longitude (l) is measured with primary direction from the Sun to the center of the galaxy in the galactic plane, while the galactic latitude (b) measures the angle of the object above the galactic plane (Brews ohare licensed under the Creative Commons Attribution-Share Alike 3.0 Unported license; NASA/JPL-Caltech/ESO/R. Hurt)

Table 5.1 Galactic coordinates of some dwarf spheroidal galaxies used to constrain dark matter interaction

Object	Longitude (l)	Latitude (b)	Distance (from the Sun/kpc)
Carina	260.1	−22.2	103 ± 4
Draco	86.4	34.7	84 ± 8
Fornax	237.1	−65.7	138 ± 9
LeoI	226.0	49.1	247 ± 19
LeoII	220.2	67.2	216 ± 9
Sculptor	287.5	−83.2	87 ± 5
Sextans	243.5	42.3	88 ± 4
Ursa Minor	105.0	44.8	74 ± 12

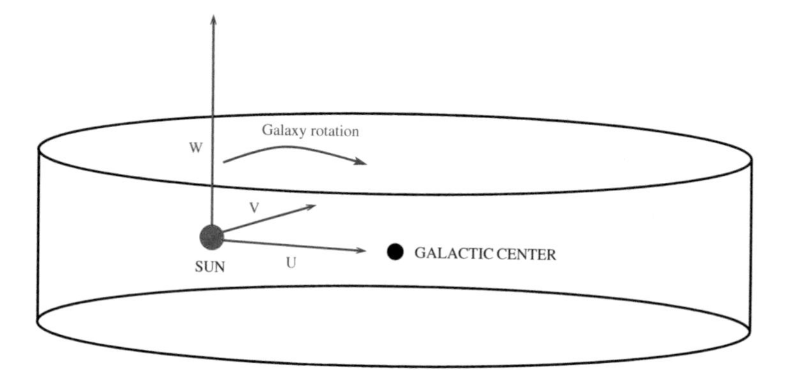

Fig. 5.3 (U, V, W) system of coordinates to define the peculiar motion of the Sun and stars relatively to the Local Standard of Rest (LSR). Sometimes U is defined pointing out from the galactic center (given negative values for the Sun's velocity U-component)

5.2 Computation of a Flux

To describe the distribution of dark matter in the Milky Way, observations of the sky are clearly not sufficient. Not only because we still did not observe any dark matter signal, but also because N-body simulations are much more efficient for that. In two words, the principle of a simulation is to put all the known ingredients (and there are a lot: physical laws, proportion between dark and baryonic matter, initial velocities or orbital momentum, self-interaction of the dark matter or not, self-interaction of baryonic matter, feedback reactions of the gravitational well formed by the dark matter halo, entropy injection due to dark matter motions, etc.); let run a code during days, months, years..., look at the results, extract any kind of simulated galaxies that are similar to the Milky Way in terms of mass, components, size, orbital momentum, and look at the profile of dark matter obtained in these selected galaxies. Then, the physicists make a mean of all these distributions and try to fit them with a usable universal function.

Table 5.2 NFW and Moore et al. density profiles without and with adiabatic compression (NFW$_c$ and Moore$_c$, respectively) with the corresponding parameters, and values of $\bar{J}(\Delta\Omega)$ (mean of J over the solid angle, see Sect. 5.8 for details)

	a (kpc)	α	β	γ	$\bar{J}(10^{-3}\text{sr})$	$\bar{J}(10^{-5}\text{sr})$
NFW	20	1	3	1	1.214×10^3	1.264×10^4
NFW$_c$	20	0.8	2.7	1.45	1.755×10^5	1.205×10^7
Moore et al.	28	1.5	3	1.5	1.603×10^5	5.531×10^6
Moore$_c$	28	0.8	2.7	1.65	1.242×10^7	5.262×10^8

Considering all the ingredients that appear in the recipe, one can easily understand why different simulation teams will give different distribution functions. All depend on which kinds of ingredients were thrown in the "soup." As a simple example, in the earliest times of simulations, the analysis showed a tendency for the dark matter to concentrate a lot near the galactic center, which was in tension with certain observations of Sun motions around Sagittarius. Nowadays, when taking into account the injection of entropy by the baryons populating the galactic center, the profiles are much less steeper, and the dark matter concentration much more in accordance with the observations. We will see another example in Sect. 5.8.3 concerning the adiabatic compression effect.

Various profiles have been proposed in the literature and can be parameterized as

$$\rho(r) = \frac{\rho_0[1 + (R_0/a)^\alpha]^{(\beta-\gamma)/\alpha}}{(r/R_0)^\gamma[1 + (r/a)^\alpha]^{(\beta-\gamma)/\alpha}}, \tag{5.2}$$

where ρ_0 is the local (solar neighborhood) measured halo density ($\simeq 0.3$ GeV/cm^3), R_0 the Sun to galactic center distance ($\simeq 8.5$ kpc), and a is a characteristic length given by the resulting simulation, as the powers α, β, and γ. Some results of simulations are given in Table 5.2. The name of the profile corresponds usually to the name of the author of the simulations (or their initials, NFW for Navarro, Frenk, and White).

The calculation of the flux produced by the annihilation of dark matter at a distance r from the galactic center along the line of sight forming an angle ψ with the (Sun)-(galactic-center) axe (see Fig. 5.4) necessitates to know the integral[3]

$$J(\psi) = \frac{1}{8.5\text{kpc}}\left(\frac{1}{0.3\text{ GeV cm}^{-3}}\right)^2 \int_{\text{line of sight}} \rho^2(r(l,\psi))dl \tag{5.3}$$

in a solid angle Ω.

[3]In the case of a dark matter decay, one needs to replace ρ^2 by ρ in the integral.

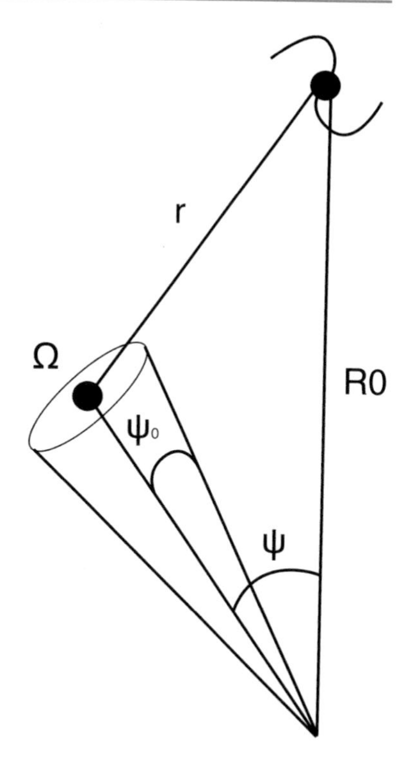

Fig. 5.4 Topology for the measurement of a flux from the galactic halo. In this example, the experience is pointed at an angle Ψ from the galactic center, inside a cone defined by the solid angle $\Delta\Omega = \Psi_0$ along with is the line of sight l

One can write $r = \sqrt{l^2 + R_0^2 - 2R_0 l \cos\psi}$ re-expressing Eq. (5.2) as $\rho(l, \cos\psi)$, and for a solid angle Ω one has $d\Omega = d\phi \sin\psi d\psi \Rightarrow \cos\psi_0 = 1 - \Omega/2\pi$, which gives

$$\int_{\text{line of sight}} \rho^2(r) dl = \int_{\cos(\psi+\psi_0)}^{\cos\psi} d\cos\psi' \int_0^{R_0 - 10^{-5}\text{kpc}} dl \rho^2(l, \cos\psi') + \int_{\cos(\psi+\psi_0)}^{\cos\psi} d\cos\psi' \int_{R_0+10^{-5}\text{kpc}}^{100\ \text{kpc}} dl \rho^2(l, \cos\psi'), \quad (5.4)$$

10^{-5} kpc being the limit radius of the integral, where one cannot say anything about the pure galactic center where lies a possible black hole (for $\psi = 0$), and the upper bound of the integral, 100 kpc, corresponds to the end of the halo core. Here you can find an example of mathematica code in the case of the galactic center in the NFW model within a solid angle Ω of 10^{-3}.

```
rho0 = 0.3;
alpha = 1;
beta = 3;
gamma = 1;
a = 20.;
```

```
R0 = 8.;
r = Sqrt[l^2 + R0^2 - 2  R0 l cpsi];
fr = rho0 *(1 + (R0/a)^alpha)^((beta - gamma)/alpha)/((r/R0)^
        gamma  (1 + (r/a)^alpha)^((beta - gamma)/alpha));
fr /. r -> Sqrt[l^2 + R0^2 - 2  R0 l cpsi];
fl = %
rhol[l_, cpsi_] = fl
dOmega = 0.001;
cpsi0 = 1 - dOmega/(2 Pi);
ftointegrate = fl^2/(1 - cpsi0)/8.5*(1/0.3)^2;
Jave = NIntegrate[
   ftointegrate, {cpsi, cpsi0, 1.}, {l, 0, R0 - 0.00001}] +
  NIntegrate[
   ftointegrate, {cpsi, cpsi0, 1.}, {l, R0 + 0.00001, 100}] +
  0.00002*rhol[0.00001, 0]

 Out =  1223.31
```

The third contribution in the integral is just the center part supposing that the density is constant in the galactic center region. This hypothesis is not really important as the 2×10^{-5} kpc region is very small. We give some values of the flux parameter J for different profiles and solid angle in Table 5.2.

5.3 Example of the Isothermal Profile

It is legitimate to ask how such a form of equation as Eq. (5.2) can appear from simulations or arise from N-body simulations based on primeval hypothesis. To illustrate it, we will take a concrete example, which was one of the more popular profiles in the 80s–90s before more complex simulations leading to NFW types emerged. This profile is named *isothermal profile*,

$$\rho_{iso}(r) = \frac{A}{(r^2 + r_c^2)}, \tag{5.5}$$

where r_c is a core radius and A a constant, both determined by the observations and/or simulations. Considering a cold dark matter profile as a set of collisionless particles, we cannot treat them as we did when we studied the thermal bath, but we can consider that statistically they follow a Boltzmann velocity distribution. We can then write for a set of particles in a potential $\Phi(r)$

$$\rho(r) = \int d^3v f(v, r) \tag{5.6}$$

with $f(v, r) = Ce^{-\frac{E}{kT}}$, C a constant, and

$$E = \frac{1}{2}mv^2 + m\Phi(r). \tag{5.7}$$

Supposing that the potential does not depend on v, we deduce

$$\rho = \rho_0 e^{-\frac{m\Phi(r)}{kT}} \quad \Rightarrow \quad \Phi(r) = -\frac{kT}{m} \ln\left(\frac{\rho}{\rho_0}\right). \tag{5.8}$$

From the Poisson's equation (see the box below),

$$\Delta\Phi = \frac{1}{r^2}\frac{\partial}{\partial r}r^2\frac{\partial}{\partial r}\Phi = 4\pi G\rho, \tag{5.9}$$

we deduce, using Eq. (A.137),

$$\frac{\partial}{\partial r}r^2\frac{\partial}{\partial r}\ln\rho = -\frac{m}{kT}4\pi G r^2 \rho \tag{5.10}$$

or

$$\rho(r) = \rho_{iso}(r) = \frac{kT}{2\pi G m r^2}. \tag{5.11}$$

Exercise Supposing $\rho = \frac{A}{r^2}$, recover the result above showing that $A = \frac{kT}{2\pi G m}$.

The density ρ_{iso} being singular at $r = 0$, we cure the pathology by introducing a cutoff at a radius r_c, which value is chosen to fit with the observations:

$$\rho_{iso}(r) = \frac{kT}{2\pi G m (r^2 + r_c^2)}. \tag{5.12}$$

This profile is thus constant for $r \ll r_c$ and follows $\rho \propto 1/r^2$ for $r \gg r_c$, recovering the property of a constant velocity behavior far away from the core of the structure that we discussed after Eq. (1.3). Notice that the profile is of the form (5.2) with $(\alpha, \beta, \gamma) = (2, 2, 0)$.

The name *isothermal* came back from a series of lectures Lord Kelvin gave at Baltimore in 1884[4] where he observed that if you consider an ideal gas in a sphere of density ρ, the force exerted on a surface dS of the external surface can be written as

$$dF = \frac{GM \times \rho \times dS \times dr}{r^2} = dP \times dS, \tag{5.13}$$

[4] Also published in 1907 by Robert Emden (brother-in-law of Karl Schwarzschild) in a book called «Gaskugeln».

or

$$\frac{dP}{dr} = \frac{GM \times \rho}{r^2} \tag{5.14}$$

P being the corresponding pressure. Considering a gas at a constant temperature T (*isothermal*) we also know that

$$PV = nRT = NkT \quad \Rightarrow \quad P = \frac{\rho}{m} kT, \tag{5.15}$$

where N is the total number of particles in the sphere. Combining Eqs. (5.14) and (5.15), with $M = \frac{4}{3}\pi r^3 \rho$, we obtain

$$\frac{d\rho}{\rho^2} = \frac{4\pi}{3} \frac{mG}{kT} r dr, \tag{5.16}$$

which gives

$$\rho(r) = \frac{3kT}{2\pi G m r^2} \propto \frac{1}{r^2}, \tag{5.17}$$

hence the name given to such profiles where $\rho \propto 1/r^2$.

The Poisson's Equation

The Poisson equation is a consequence of the Gauss's law for gravity that stated that *the gravitational flux through any closed surface is proportional to the enclosed mass*. The proof is straightforward, once we consider the Newton law of gravity for an acceleration **g**:

$$\mathbf{g}(\mathbf{r}) = \frac{-GM}{r^2} \frac{\mathbf{r}}{r}, \tag{5.18}$$

which gives in a continuous distribution of mass

$$\mathbf{g}(\mathbf{r}) = -G \int \rho(\mathbf{r}_i) \frac{\mathbf{r} - \mathbf{r}_i}{|\mathbf{r} - \mathbf{r}_i|^3} d^3\mathbf{r}_i. \tag{5.19}$$

Using

$$\nabla . \left(\frac{\mathbf{r}}{|r|^3} \right) = 4\pi \delta(\mathbf{r}), \tag{5.20}$$

(continued)

we obtain

$$\nabla.\mathbf{g} = -4\pi G\rho(\mathbf{r}), \tag{5.21}$$

which is the differential formulation of the Gauss's law for gravity. The gravitational field being conservative can be expressed as a function of a scalar potential Φ:

$$\mathbf{g} = -\nabla\Phi, \tag{5.22}$$

which gives, combined with the Gauss's law,

$$\Delta\Phi = 4\pi G\rho, \tag{5.23}$$

which is the Poisson's equation.

5.4 Radiative Processes in Astrophysics Part I: The Non-Relativistic Case

To compute dark matter effects in the galaxies, and more precisely the effects generated by their products of annihilation or decay, one first needs to understand the properties of charged particles (product of the annihilation) propagating in a charged medium (the interstellar gas for instance). We present in this section the main processes involved in the loss of energy undergone by a charged particle or radiation from its interactions with a medium. The typical example is of course the interaction of a cosmic ray—charged particles—on the CMB or the radiation emitted by the crossing of such particle in the galactic magnetic field. We will first describe such phenomena in the non-relativistic approximation ($v \ll c$) and then apply the formulae in the relativistic limit. We refer the reader to the very complete Jackson [1] or Longair [2] books for more details. For nucleon or electron propagation in the neutral matter of the interstellar medium (90% H and 10% He) the relevant energy losses are due to electromagnetic and nuclear effects depending on the specific type of interaction. The most relevant processes of the first group are ionization and Coulomb scattering, while for the second the relevant ones are spallation, fragmentation, and radioactive decay. For electromagnetic processes that involve electrons, even bremsstrahlung in the neutral and ionized medium, as well as Compton and synchrotron losses, becomes important. Although all these processes are well known and are often explained during academic courses, the formulae for the different cases are rather scattered throughout the literature. We try to present in this section the main results concerning the energy losses in the interstellar medium and give formulae that can be used in any study concerning the subject.

5.4.1 Maxwell Equations

The Maxwell equations are the first ingredients we need to understand the principle underlying the propagation and interaction of charged particles. They were discovered by Maxwell between 1861 and 1862 and consist of a set of 4 linear equations describing the space-time evolution of the electromagnetic fields **E** and **B**:

$$\nabla.\mathbf{E} = \frac{\rho}{\epsilon_0} \quad [\text{Maxwell} - \text{Gauss}]\,; \quad \nabla.\mathbf{B} = 0 \quad [\text{Maxwell} - \text{Thomson}] \tag{5.24}$$

$$\nabla \times \mathbf{E} = -\frac{\partial \mathbf{B}}{\partial t} \quad [\text{Maxwell} - \text{Faraday}];$$

$$\nabla \times \mathbf{B} = \mu_0\,\mathbf{j} + \mu_0\epsilon_0 \frac{\partial \mathbf{E}}{\partial t} \quad [\text{Maxwell} - \text{Ampere}]$$

with $c^2 = 1/\mu_0\epsilon_0$, ρ the density of charges, and $\mathbf{j} = \sum e_i \mathbf{v}_i$, the density of current. Combining cleverly the equations, one can write

$$\mathbf{j}.\mathbf{E} + \frac{\partial}{\partial t}\left(\frac{\epsilon_0}{2}|\mathbf{E}|^2 + \frac{1}{2\mu_0}|\mathbf{B}|^2\right) = -\nabla.\left(\frac{\mathbf{E}\wedge\mathbf{B}}{\mu_0}\right) = -\nabla.\mathbf{\Pi} \tag{5.25}$$

with $\mathbf{\Pi} = (\mathbf{E}\times\mathbf{B})/\mu_0$ is the Poynting vector whose module $|\mathbf{\Pi}|$ gives the power flux (energy emitted per second per m^2). $\mathbf{j}.\mathbf{E} = \sum e_i \mathbf{v}_i.\mathbf{E}$ is the density of mechanical energy (energy emitted by a moving charged particle in an electric field **E**), whereas $\frac{\epsilon_0}{2}|\mathbf{E}|^2 + \frac{1}{2\mu_0}|\mathbf{B}|^2$ is the density of energy of the radiated electric and magnetic field.

These equations are written in the SI system of units (International System). Sometimes one can find expressions of the fields in another popular system of units, the Gaussian units, that is part of the CGS (Centimeter–Gram–Second) system. When using CGS units, it is conventional to use a slightly different definition of electric field $\mathbf{E}_{CGS} = c^{-1}\,\mathbf{E}_{SI}$. This implies that the modified electric and magnetic fields have the same units (in the SI convention this is not the case). Then it uses a unit of charge defined in such a way that the permittivity of the vacuum $\epsilon_0 = 1/(4\pi c)$, hence $\mu_0 = 4\pi/c$. Using these different conventions, the Maxwell equations become

$$\nabla.\mathbf{E} = 4\pi\rho\,; \quad \nabla.\mathbf{B} = 0 \tag{5.26}$$

$$\nabla \times \mathbf{E} = -\frac{1}{c}\frac{\partial \mathbf{B}}{\partial t}\,; \quad \nabla \times \mathbf{B} = \frac{1}{c}\left(4\pi\mathbf{j} + \frac{\partial \mathbf{E}}{\partial t}\right).$$

5.4.2 Loss of Energy of a Moving Charged Particle

When a charged particle travels in a medium, it can lose energy by transferring part of its kinetic energy to a nearby atom, causing ionization or excitation (Coulomb and ionization losses). Another possibility for the particle is to be excited by an external radiation (Thomson scattering). The particle can also radiate energy in the presence of a magnetic field (cyclotron radiation) or electric field (bremsstrahlung radiation). We study each of these cases in the following sections, in a non-relativistic treatment first. We let the relativistic treatment to Sect. 5.6. We summarized the processes in Fig. 5.5.

5.4.2.1 Larmor's Formula

We will not dig too much into the detail study of radiative fields (see the Jackson book [1] for more details). A particle of charge e and velocity $\mathbf{v}$ generates an electric field $\mathbf{E}$ and a magnetic field $\mathbf{B}$

$$\mathbf{E}(\mathbf{r},t) = \frac{e}{4\pi\epsilon_0}\left[\frac{(\mathbf{n}-\boldsymbol{\beta})(1-\beta^2)}{\kappa^3 R^2}\right] + \frac{e}{4\pi\epsilon_0 c}\left[\frac{\mathbf{n}}{\kappa^3 R}\times[(\mathbf{n}-\boldsymbol{\beta})\times\dot{\boldsymbol{\beta}}]\right] = \mathbf{E}_r + \mathbf{E}_\theta$$
$$\mathbf{B}(\mathbf{r},t) = \frac{1}{c}\left[\mathbf{n}\times\mathbf{E}\right], \tag{5.27}$$

with $\mathbf{n} = \mathbf{R}/R$, $\boldsymbol{\beta} \equiv \mathbf{v}/c$ and $\kappa \equiv 1 - \mathbf{n}.\boldsymbol{\beta}$ (see Fig. 5.7). Notice in this expression the presence of the *retarded* effect illustrated by $\mathbf{n} - \boldsymbol{\beta}$. Indeed, the direction of the radiated field at a time t is not exactly radial when a charge is moving: the field was produced at a time $t_{rad} < t$ when the charge was at a position $c\boldsymbol{\beta}(t - t_{rad})$ compared to its position $R\mathbf{n}$ at t.

Exercise Recover the very first term of Eq. (5.27) and the retarded effect.

The electric field can be divided into the first term, the *velocity field*, or radial field $\mathbf{E}_r$, which falls off as $1/R^2$ and is just the generalization of the Coulomb law to moving particles. The second term the *acceleration field* $\mathbf{E}_\theta$, which falls off as $1/R$, is proportional to the particle's acceleration and is perpendicular to $\mathbf{n}$. This electric field, along with the corresponding magnetic field, constitutes the *radiation field*. For large values of R, the dominant term is $\mathbf{E}_\theta$, the acceleration field. If we look at the radiated field in a region during a time Δt at a distance $R \gg v \times \Delta t$, we can neglect the effect of the retarded emission and consider that the retarded effects are negligible: $R \simeq$ constant during the process. In this limit, and considering the non-relativistic case ($\beta \ll 1$), we can simplify Eq. (5.27) to

$$\mathbf{E}_{rad} = \frac{e}{4\pi R\epsilon_0 c^2}\,\mathbf{n}\times(\mathbf{n}\times\dot{\mathbf{v}})$$
$$\mathbf{B}_{rad} = \frac{1}{c}\,\mathbf{n}\times\mathbf{E}_{rad}. \tag{5.28}$$

We illustrate the dependence of the radiated field strength as a function of the direction of the acceleration in Fig. 5.6. Notice that the value of the radiated field cancels for $\theta = 0$ and is maximal for $\theta = \frac{\pi}{2}$, as a direct consequence of $\mathbf{n} \wedge \dot{\mathbf{v}}$

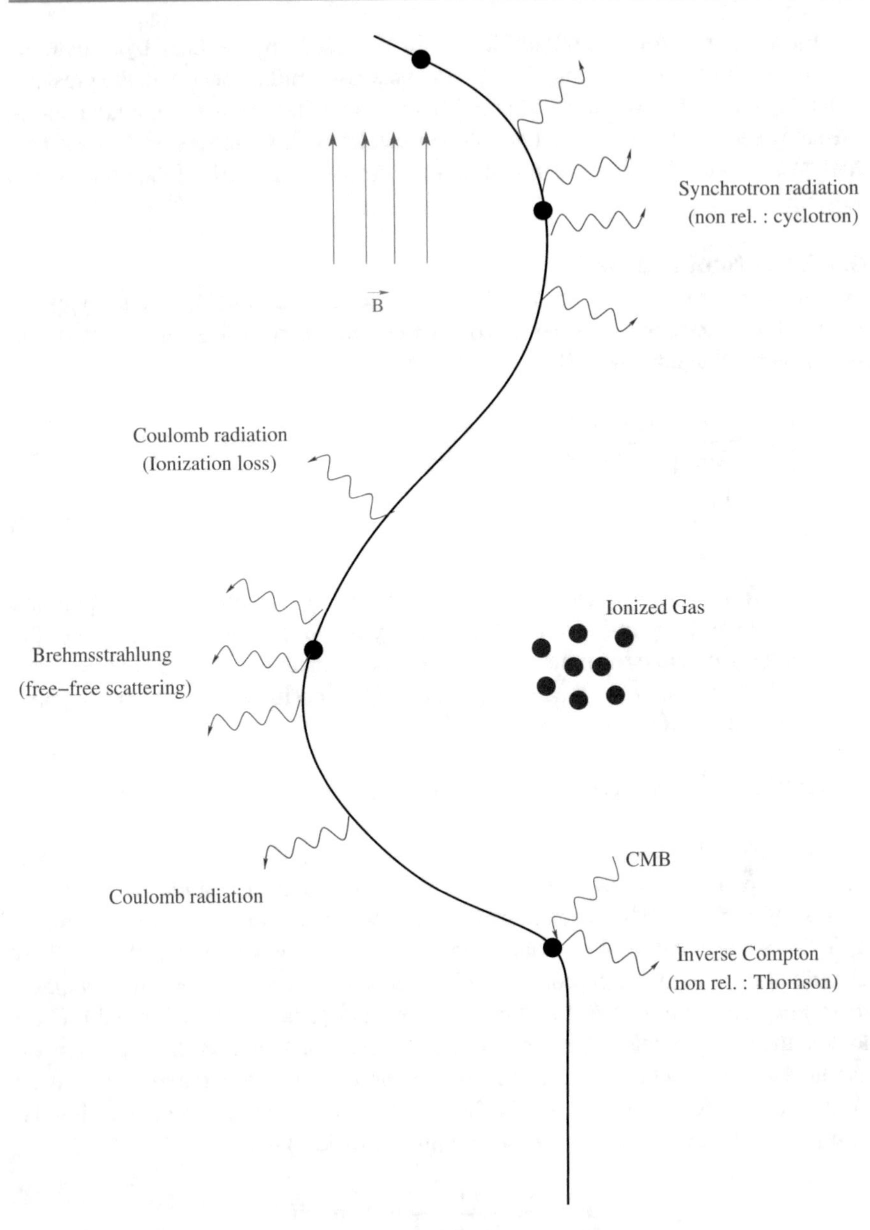

Fig. 5.5 Summary of energy loss of a charged particle in an astrophysical medium

in Eq. (5.28). It is also in this direction ($\theta = \frac{\pi}{2}$) that the radiated energy will be maximal.

The energy W radiated during a time dt is then the density of energy radiated by the charge multiplied by the volume dV of the radiation covered during the time

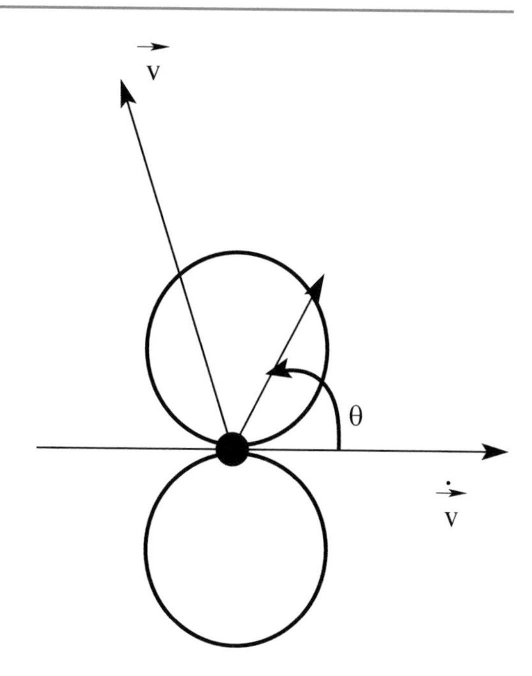

Fig. 5.6 Radiated field strength of an accelerated particle

dt, $dV = R^2 d\Omega \times c dt$, $d\Omega$ being the solid angle from where is seen the radiating particle at the retarded time (see Fig. 5.7) is:

$$d^2W = \left(\frac{\epsilon_0}{2}|\mathbf{E}_{rad}|^2 + \frac{1}{2\mu_0}|\mathbf{B}_{rad}|^2\right) \times R^2 d\Omega\, c dt, \tag{5.29}$$

which gives using Eq. (5.28)

$$\frac{d^2W}{dt d\Omega} = \frac{e^2 \sin^2\theta |\dot{\mathbf{v}}|^2}{16\pi^2 \epsilon_0 c^3}, \tag{5.30}$$

θ being the angle between $\dot{\mathbf{v}}$ and $\mathbf{n}$. After integration over $d\Omega = 2\pi d\cos\theta$, one obtains the power P emitted by the radiating charge:

$$P = \frac{dW}{dt} = \frac{e^2|\dot{\mathbf{v}}|^2}{6\pi\epsilon_0 c^3}, \tag{5.31}$$

which is known under the name of Larmor's formula.

We can make some remarks about the morphology of radiated field. Notice first that the acceleration is the proper acceleration of the particle and the loss rate is measured in its instantaneous rest frame. The electric field strength varies as $\sin\theta$, and the power radiated per solid angle is proportional to $\sin^2\theta$, where θ is the angle with respect to the acceleration vector of the particle. As a consequence, there is no radiation along the acceleration vector and the field strength is greater at right angle

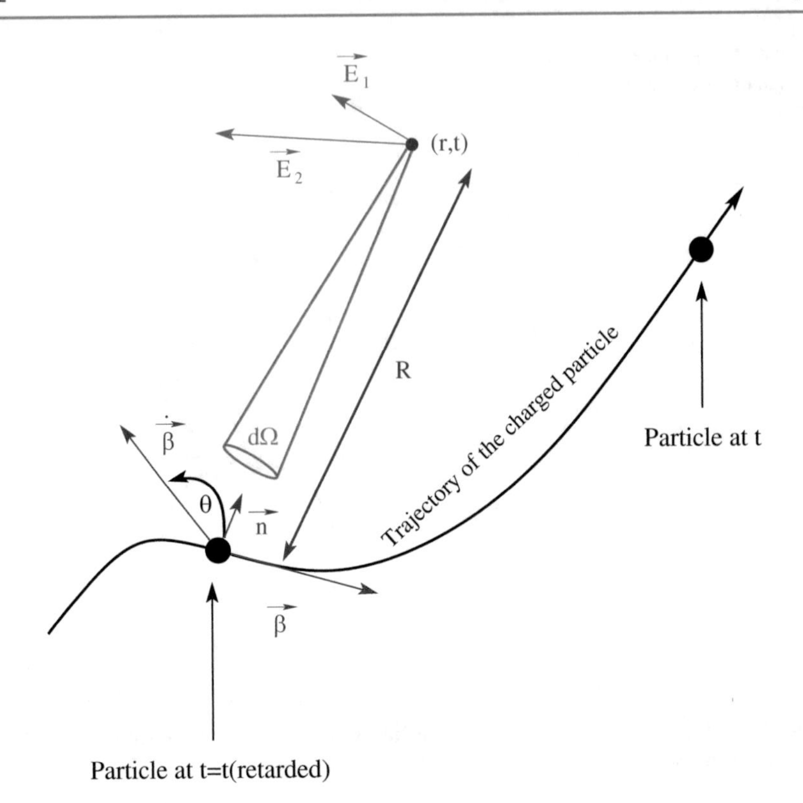

Fig. 5.7 Example of the electric field $\mathbf{E} = \mathbf{E}_1 + \mathbf{E}_2$ radiated by a charged particle at a distance R from the position of the particle at the retarded time

to the acceleration vector. The radiation is polarized with the electric field vector lying in the direction of the acceleration vector of the particle, as projected onto a sphere at a distance r from the charged particle, see Fig. 5.6. In the CGS units system we would have obtained

$$P_{CGS} = \frac{2e^2|\dot{\mathbf{v}}|^2}{3c^3}. \tag{5.32}$$

5.4.2.2 Case of a Rotating Particle

We can then compute the frequency of the radiated field E_{rad} in a specific example of a charged particle rotating around an axe. With the convention of Fig. 5.8 and using Eq. (5.28) one can express

$$\mathbf{r} = a(\cos\omega t\ \mathbf{i} + \sin\omega t\ \mathbf{j}) \ \Rightarrow\ \dot{\mathbf{v}} = -v\ \omega(\cos\omega t\ \mathbf{i} + \sin\omega t\ \mathbf{j})$$
$$\Rightarrow\ \mathbf{E}_{rad} = -\frac{ev\omega}{4\pi R\epsilon_0 c^2}(\cos\omega t\ \mathbf{a_1} + \cos\theta\ \sin\omega t\ \mathbf{a_2}). \tag{5.33}$$

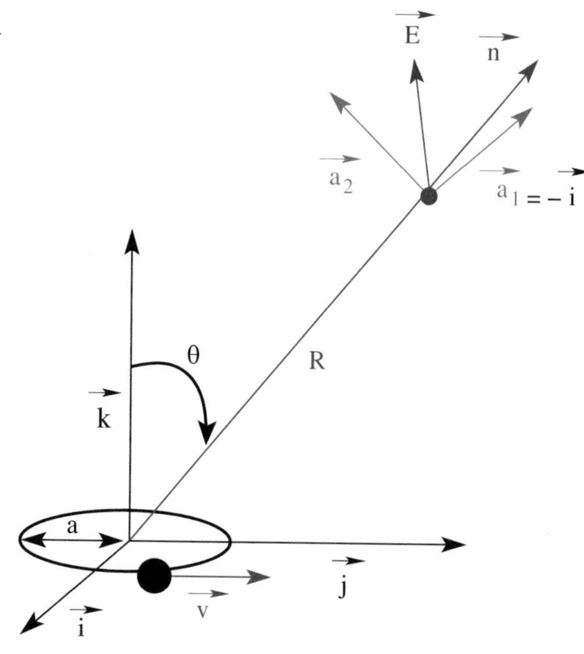

Fig. 5.8 Geometry for polarization decomposition of radiation emitted by a circulating charge. The unitary vectors $\mathbf{a}_1$ and $\mathbf{a}_2$ are in the plane perpendicular to $\mathbf{n}$ and form an orthonormal basis with $\mathbf{n}$ ($\mathbf{a}_1 = -\mathbf{i}$, $\mathbf{a}_2 = -\cos\theta\mathbf{j} + \sin\theta\,\mathbf{k}$, $\mathbf{n} = \sin\theta\,\mathbf{j} + \cos\theta\,\mathbf{k}$)

We then can conclude (which is an important and no so-intuitive result) that a circulating particle will radiate a *monochromatic* electromagnetic field of the same frequency as that of the one of the circulating charged particles. This will be a useful result when studying cyclotron and synchrotron radiation. This result is not exact anymore when radiation of higher order than dipole is included.

5.4.3 Coulomb and Ionization Losses

When a high energy charged particle passes through a medium or near other charged particles like electrons bounded in atoms, it transfers part of its kinetic energy to the electrons, causing ionization or excitation of the atoms. Indeed, the electromagnetic force of the traveling particle can give to the electron orbiting on atoms a sufficiently large momentum to extract it from its orbit (or to make it pass to an upper level) and ionize the gas (or excite it). From an historical perspective, in 1895, the passing of electrons (called "cathode rays" at this time) through a tube of Crookes was the source of the first X-ray observed by Rontgen through the ionization created on the target. In this section we will concentrate on the influence on the propagation of a charged particle in mediums made of electrons, and its energy lost along its way. The result we obtain is quite general and can be applied to several types of radiative processes. Technically speaking, we call Coulomb scattering when the interaction is on free electron, and ionization losses when the interaction takes place on bound electrons, the binding energy corresponding to the ionization potential.

Consider first the collision of a high energy nucleus of mass M, charge Ze, and velocity v with a stationary electron of mass m_e. Only a small fraction of the nucleus kinetic energy is transferred to the electron. We can understand it easily, remembering that the maximal velocity of an electron after being hit by the nucleus, v_e^{max}, is when all the momentum is transferred to the electron, which escapes at an angle of 0 degree.

Exercise Using the conservation of energy and momentum, $\frac{P_1^2}{2M} = \frac{P_1'^2}{2M} + \frac{P_2'^2}{2m_e}$ and $\mathbf{P}_1' = \mathbf{P}_1 - \mathbf{P}_2'$, show that $V_2' = v_e = \frac{2M}{m_e+M} v \cos\theta$, with θ the angle between $\mathbf{P}_1$ and $\mathbf{P}_2'$ and $v = V_1$ the initial velocity of the nucleus of mass M (hints: compute first $\mathbf{P}_1' . \mathbf{P}_1$).

If one considers solid spheres collision, $v_e^{max} = \frac{2M}{m_e+M} v$, which tends to $2v$ for $m_e \ll M$. The fractional kinetic energy loss is then less than $\frac{1}{2} m_e (2v)^2 / \frac{1}{2} M v^2 = 4m_e/M$. We can then consider that the trajectory of the hitting nucleus is not influenced by the presence of the electron.

To compute the energy transferred to an electron bounded in an atom, we compute first the momentum received by the electron from the electromagnetic field created by the traveling particle. The force can be repulsive or attractive depending on the charge of the moving particle; it does not affect the result as in both cases, the electron is expelled from the atom. From Fig. 5.9 and remarking that by symmetry only the component of the force perpendicular to the motion contributes to the electronic momentum,

$$\Delta p = \int_{-\infty}^{+\infty} |F| dt = \int_0^{-\pi} \frac{Ze^2}{4\pi\epsilon_0 (b \sin\theta)^2} \sin\theta \, \frac{1}{v} d\left(\frac{b}{\tan\theta}\right) = \frac{Ze^2}{2\pi\epsilon_0 b v}, \tag{5.34}$$

which gives for the kinetic energy transferred to the electron

$$\Delta E_c = \frac{\Delta p^2}{2m_e} = \frac{Z^2 e^4}{8\pi^2 \epsilon_0^2 b^2 v^2 m_e}. \tag{5.35}$$

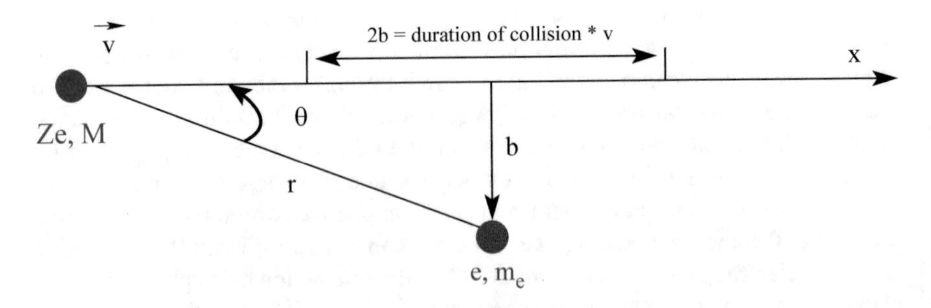

Fig. 5.9 Interaction of a high energy particle of charge Ze with an electron at rest

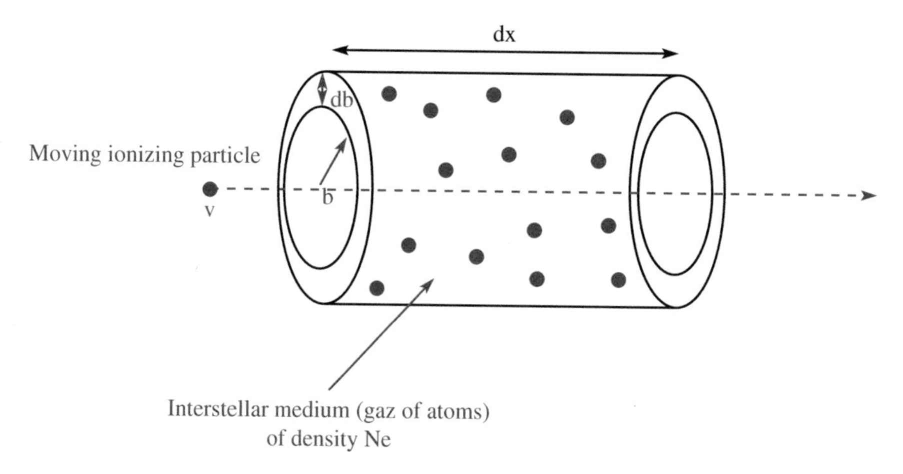

Fig. 5.10 Moving particle in an interstellar medium of density N_e

The moving particle loses obviously the equivalent energy. When traveling in a medium like the interstellar medium, one has to sum over all the atoms in the medium. The number of electrons in a gas of density N_e at a distance between b and $b + db$ from the axe of the moving particle (called impact parameter) is along a length dx (see Fig. 5.10) $2\pi b db dx \times N_e$. We can then compute the total loss of energy on the way

$$\frac{dE_{Ze}^{\text{loss}}}{dx} = -\frac{Z^2 e^4 N_e}{4\pi\epsilon_0^2 v^2 m_e} \int_{b_{\min}}^{b_{\max}} \frac{db}{b} = -\frac{Z^2 e^4 N_e}{4\pi\epsilon_0^2 v^2 m_e} \ln\left(\frac{b_{\max}}{b_{\min}}\right), \tag{5.36}$$

the minus sign representing the fact that the energy is lost by the traveling particle. The computing of $b_{\min}$ and $b_{\max}$ requires a special treatment. $b_{\max}$ represents the distance at which the influence of the traveling particle on the electron is negligible. It corresponds roughly to the time when the orbital period is lower than the typical interaction time. In other words, if the electron takes more time to move around the nucleus than to interact with the moving particle, the electromagnetic influence of the later becomes weak. If one write τ the interacting time and ν_0 the frequency of the rotating electron in the atom ($\nu_0 = \omega_0/2\pi$), it corresponds to

$$\tau \simeq \frac{2b}{v} < \frac{1}{\nu_0} \Rightarrow b < \frac{v}{2\nu_0} = b_{\max}. \tag{5.37}$$

The lower limit $b_{\min}$ can be obtained if we suppose, by a quantum treatment and the application of the uncertainty principle, that the maximum energy transfer is $\Delta p_{\max} = 2m_e v$ (because as we discussed earlier, the maximum velocity transferred to the electron is $2v$) from $\Delta p \Delta x \gtrsim \hbar$ (Heisenberg principle) we have $\Delta x \gtrsim \hbar/2m_e v$. We can then write

$$b_{\min} = \frac{\hbar}{2m_e v}, \tag{5.38}$$

which implies

$$\frac{dE}{dx} = -\frac{Z^2 e^4 N_e}{4\pi\epsilon_0^2 v^2 m_e} \ln\left(\frac{2\pi m_e v^2}{\hbar\omega_0}\right) = -\frac{Z^2 e^4 N_e}{4\pi\epsilon_0^2 v^2 m_e} \ln\left(\frac{\pi m_e v^2}{I}\right), \tag{5.39}$$

where I is the ground state binding energy of the electron in the atom ($I = \frac{1}{2}\hbar\omega_0$). But the system being more complex, it usually contains different atoms with different kinds of excited levels, so one should take into account a mean ionization potential $\bar{I}$ instead of I. Integrating the π factor and a factor 2 for physical interpretation (see below) in the definition of I we finally obtain

$$\frac{dE}{dx} = -\frac{Z^2 e^4 N_e}{4\pi\epsilon_0^2 v^2 m_e} \ln\left(\frac{2m_e v^2}{\bar{I}}\right) \Rightarrow \langle P \rangle = \frac{dE}{dt} = v\frac{dE}{dx} = -\frac{Z^2 e^4 N_e}{4\pi\epsilon_0^2 v\, m_e} \ln\left(\frac{2m_e v^2}{\bar{I}}\right). \tag{5.40}$$

The value of $\bar{I}$ cannot be calculated explicitly (except for very simple models of atoms) and should thus be measured by experiment. Typical values for the logarithmic factor, called Coulomb logarithm, are around 20. We recognize in Eq. (5.40) the ratio of the maximum kinetic energy ($2m_e v^2$) on the binding energy, justifying the normalization of $\bar{I}$. A careful inspection of Eq. (5.40) reveals that the energy loss of an energetic moving particle depends on the charge and the velocity of the particle, but not on its mass (except for the Coulomb logarithm), and that low energy particles lose their energy more rapidly than high energy particles.

5.4.4 Thomson Scattering

The Thomson scattering is the process in which a free charge radiates in response to an incident electromagnetic wave. If the charge oscillates at non-relativistic velocities, $v \ll c$, then we may directly apply the Larmor's formula computed in the previous section. This effect arises in astroparticle typically when a photon radiated by annihilation of dark matter for instance interacts with cosmic rays. If one supposes an incident *linearly polarized wave*, the force of the electromagnetic wave $\mathbf{F}$ on the charged particle is

$$\mathbf{F} = e\,\boldsymbol{\epsilon}\,E_0 \sin\omega_0 t \Rightarrow m\dot{\mathbf{v}} = e\,E_0 \sin\omega_0 t \times \boldsymbol{\epsilon} \Rightarrow \langle P \rangle = \frac{e^4 E_0^2}{12\pi\epsilon_0\, m^2 c^3}, \tag{5.41}$$

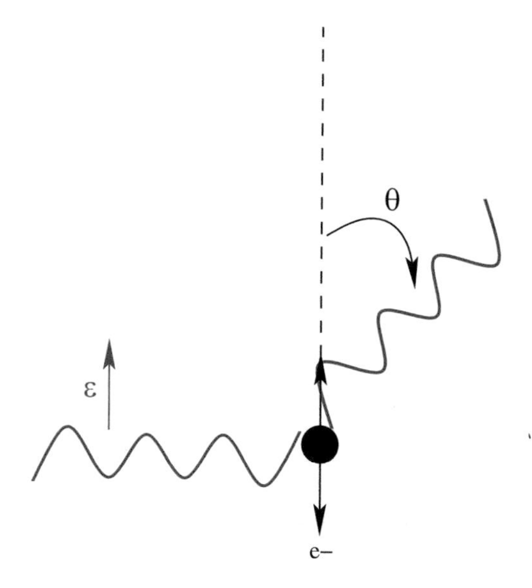

Fig. 5.11 Thomson scattering of a polarized radiation by a charged particle

where $\boldsymbol{\epsilon}$ is the normalized **E**-field direction (see Fig. 5.11). To obtain the above equation, we just implemented the value of the mean acceleration $\langle|\dot{\mathbf{v}}|^2\rangle = \frac{e^2}{2m^2}E_0^2$ into the Larmor's formula (5.31). Defining the Thomson cross section σ_T:

$$\langle P\rangle = \sigma_T \times \text{[initial flux of energy]} = \sigma_T \left(c\frac{\epsilon_0}{2}|E_0|^2\right) = \sigma_T\, cU_{rad}, \tag{5.42}$$

where $U_{rad} = \frac{\epsilon_0}{2}|E_0|^2$ is the energy density of radiation in which the electron is located,[5] we obtain

$$\sigma_T = \frac{8\pi}{3}\left(\frac{e^2}{4\pi\epsilon_0 mc^2}\right)^2 = \frac{8\pi}{3}r_0^2 = 6.65\times10^{-29}\,\mathrm{m}^2 = 1714\,\mathrm{GeV}^{-2}, \tag{5.43}$$

with $r_0 = \frac{e^2}{4\pi\epsilon_0\, mc^2}$ is the classical radius of the charged particle.[6] σ_T is the Thomson scattering cross section and represents the fraction of the initial wave scattered by the charged particle.

Thomson scattering is one of the most important processes that impedes the escape of photons from any region. Indeed, notice that there is no change in the energy of the photons in the process: the energy is first transferred to the electron, which then by its acceleration radiates back the energy it received. After the scattering, the photons follow a random walk, each step between a mean free path depending on

[5] Indeed, in the interstellar medium, the density of radiation depends on the localization and the morphology of the gazes.

[6] The radius r_0 of a particle of mass m is defined by $mc^2 = V(r_0) = \frac{e^2}{4\pi\epsilon_0 r_0}$.

the density of the electrons. The electrons then prevent the photons to escape from their own plasma. The Thomson free path is then $\lambda_T = (\sigma_T N_e)^{-1}$, N_e being the number of electrons per unit volume. This process is one of the more fundamental ones in particle physics as well as astroparticle or cosmology. It is for instance the dominant cross section to determine the last scattering surface, i.e. the decoupling time as it is described in Sect. 3.3.4.2.

5.4.5 Cyclotron Radiation

Cyclotron radiation corresponds to the radiation emitted by a non-relativistic charged particle moving in a magnetic field **B**. The Lorentz force acting perpendicular to the motion of the particle creates an acceleration that in turn transforms into loss of energy by radiation. Cyclotron effect becomes a synchrotron effect when the particle becomes relativistic (see Fig. 5.5) and is treated in Sect. 5.6.3. The fundamental principle of dynamics applied to the charged particle of mass m and charge e at a distance r from the source of the magnetic field gives us

$$\boxed{\begin{aligned} &\frac{mv^2}{r} = eB\,v \;\Rightarrow\; \omega_L = 2\pi \times \nu_L = 2\pi \times \frac{v}{2\pi r} = \frac{eB}{m} \\ &\Rightarrow\; \nu_L = \frac{eB}{2\pi m} \simeq 14\ \text{MHz}\left(\frac{e}{1.6\times 10^{-19}}\right)\left(\frac{B}{1\ \text{Gauss}}\right)\left(\frac{1\ \text{GeV}}{m}\right) \end{aligned}} \quad (5.44)$$

ω_L (ν_L) being the cyclotron (Larmor) pulsation (frequency). Notice that the frequency is independent of the position of the particle or its velocity. Moreover, as we saw in Eq. (5.33), the radiated field is monochromatic and has the same frequency as that of the moving particle itself. As all the particles in rotation around **B** possess the *same* frequency, independently of their rotation radius, they will all contribute to the electromagnetic field at the same frequency. For an electron in a typical 10 μG magnetic field from galactic center, we expect a frequency around the MHz, which is the range of measurement of several experiments. In the relativistic case treated below, we should add the Doppler effect given by Eq. (5.68) with a mean on $\cos\theta$, which gives a frequency multiplied by the factor $\gamma = \frac{1}{\sqrt{1-\frac{v^2}{c^2}}}$. In other words, a relativistic electron of 1 GeV would give a synchrotron radiation at a frequency around 2000 times larger than that in the non-relativistic case, so between 100 MHz and 1 GHz.

Combining Eq. (5.31) with Eq. (5.44), we can then compute the energy lost per second by a particle of mass m and charge $q \times e$ moving around a magnetic field of intensity B, noticing that in the non-relativistic case, $E = \frac{1}{2}mv^2$:

$$\frac{dE}{dt} = -\frac{e^4 B^2 E}{3\pi\epsilon_0 c^3 m^3} \Rightarrow E = E_0 e^{-t/\tau}, \text{ with } \tau = 3\pi\epsilon_0 \frac{c^3 m^3}{e^4 B^2}. \tag{5.45}$$

Exercise Noticing that $|\dot{v}| = \frac{v^2}{r}$, recover the expression (5.45).

Implementing some more natural units for particle physicists, we can express

$$\tau = 3\pi(3\times 10^8)^3 \left(\frac{1}{q\times 1.6\times 10^{-19}}\right)^4 \left(\frac{1\ \text{Tesla}}{B}\right)^2 \left(\frac{1.88\times 10^{-27} m}{1\ \text{GeV}}\right)^3$$
$$\simeq \frac{2\times 10^{18}}{q^4}\left(\frac{1\ \text{Gauss}}{B}\right)^2 \left(\frac{m}{1\ \text{GeV}}\right)^3 \text{ s.}$$

To give an example, an electron in typical local magnetic fields of a galaxy (around 1 μG, see Sect. 5.10.4.1 for a more detailed modelization), we obtain $\tau \simeq 8\times 10^{12}$ years, much larger than the age of the Universe. For a proton, the lifetime is even greater. In other words, cyclotron radiation is not a significant cooling mechanism for interstellar gas.

5.4.6 Bremsstrahlung Radiation

In 1927, Carl Anderson found that the ionization rate given by the Bethe–Bloch formulae (5.39) and (5.75) underestimates the energy loss rate for relativistic electrons. The additional loss was associated to the acceleration of the electron in the electromagnetic field. Indeed, we computed in Eq. (5.31) the energy lost by an accelerated particle. Then, a particle passing through the magnetic field generated by a gas of ionized particle will lose energy by exchanging momentum with constituents of the gas (Coulomb/ionization scattering) but also from its own acceleration under the impact of the surrounding electromagnetic field. This phenomena was in fact already observed by Nikola Tesla in 1880 in another context and was called *bremsstrahlung*. The process is identical to what is called *free–free emission* in the language of atomic physics, in the sense that the radiation corresponds to transitions between unbound states of the electron in the field of the nucleus. In 1934, computations of the spectrum of non-relativistic and relativistic bremsstrahlung were carried out by Bethe and Heitler.

The bremsstrahlung, from the German *bremsen* "to brake" and *Sthralung* "radiation," i.e. "braking radiation" or "deceleration radiation," is the electromagnetic radiation produced by the deceleration of a charged particle when deflected by another charged particle, typically an electron by an atomic nucleus (see Fig. 5.5). The moving particle loses kinetic energy, which is transformed into an outgoing photon because energy is conserved. Strictly speaking, braking radiation is any radiation due to the acceleration of charged particles. However, the term is frequently used in the more narrow sense of radiation from electrons slowing in matter.

5.5 Notions of Relativity

5.5.1 Main Idea

In a lot of astrophysical processes, the charged particle radiating electromagnetic field is produced with a relativistic velocity. It is for instance the case for GeV-like dark matter annihilating into leptons. We would like to apply our preceding results to such an energetic particle. The main idea is to move to an instantaneous rest frame K' such that the particle has zero velocity at a certain time and we then can apply all the previous results we obtained. During an infinitesimal lapse of time the particle moves non-relativistically and we can use the Larmor's formula.

5.5.2 Lorentz Transformations

We recall in this section the main results concerning the transformations induced by special relativity. Consider two frames K and K' that are moving with respect to each other with a relative velocity v along the x-axis. The origins are assumed to coincide at $t = 0$. If a pulse of light is emitted at $t = 0$, each observer will see an expanding sphere centered on his own origin:

$$x^2 + y^2 + z^2 - c^2t^2 = 0, \qquad x'^2 + y'^2 + z'^2 - c^2t'^2 = 0.$$

The actual relations between (x, y, z, t) and (x', y', z', t') are called the *Lorentz transformations*:

$$x' = \gamma(x - vt); \quad y' = y, \quad z' = z, \quad t' = \gamma\left(t - \frac{v}{c^2}x\right), \tag{5.46}$$

with $\gamma = \frac{1}{\sqrt{1-\frac{v^2}{c^2}}}$. The inverse relations $x_i = f(x_i')$ have obviously the same form with $v \to -v$. It is common to define quadrivectors $x_\mu = (ct, x, y, z)$, μ being the Lorentz index. Any vectors respecting the Lorentz transformation laws are then called quadrivectors. As we will see, it is also the case for the energy–momentum quadrivector $p_\mu = (\frac{E}{c}, p_x, p_y, p_z)$ for a particle of energy E and momentum $\mathbf{p}$.

There are multiple ways to demonstrate the Lorentz transformations (5.46). Some with arguments are based purely on physics of transformations, some others taking into account the ether and then defining a *virtual* time t' where the equations of Maxwell are invariant (this was the original method of Poincaré and Lorentz). I give an example based on physics argument in the following frame. However, the one I use to teach is (to my taste) more elegant at the price of being more mathematical. Here is how I do.

If we suppose a constant velocity of light, and we need to define an invariant of a transformation, we can, by default, set this invariant $cd\tau$ to 0 for a particle moving at the speed of light. There exist two possibilities, $x = ct$ or $x = -ct$. We can then define

$$c^2 d\tau^2 = (cdt - dx)(cdt + dx) = c^2 dt^2 - dx^2. \tag{5.47}$$

$d\tau$ is called *proper time*. Then, restricting ourselves to a translation along the x-direction (generalization is straightforward) we can write the transformation

$$\begin{aligned} x' &= \Lambda_0^1\, t + \Lambda_1^1\, x \\ t' &= \Lambda_0^0\, t + \Lambda_1^0\, x. \end{aligned} \tag{5.48}$$

Take an observer at O who does not move ($dx = 0$); the observer O' moving with a velocity v with respect to him will feel the change of coordinate

$$dx' = \Lambda_0^1\, dt \tag{5.49}$$

$$dt' = \Lambda_0^0\, dt \quad \Rightarrow \quad \Lambda_0^1 = \frac{dx'}{dt'} \Lambda_0^0 = v \Lambda_0^0, \tag{5.50}$$

and demanding for $d\tau$ to be constant, we have

$$c^2 dt^2 - dx^2 = c^2 dt'^2 - dx'^2 =$$
$$\left[c^2 (\Lambda_0^0)^2 - (\Lambda_0^1)^2\right] dt^2 + \left[c^2 (\Lambda_1^0)^2 - (\Lambda_1^1)^2\right] dx^2 + 2\left[c^2 \Lambda_0^0 \Lambda_1^0 - \Lambda_0^1 \Lambda_1^1\right] dxdt,$$

which gives

$$\begin{aligned} c^2 (\Lambda_0^0)^2 - (\Lambda_0^1)^2 &= c^2 \\ c^2 (\Lambda_1^0)^2 - (\Lambda_1^1)^2 &= -1 \\ c^2 \Lambda_0^0 \Lambda_1^0 - \Lambda_0^1 \Lambda_1^1 &= 0. \end{aligned} \tag{5.51}$$

Solving this set of equations combined with (5.50), we obtain easily

$$\Lambda_0^0 = \frac{1}{\sqrt{1 - \frac{v^2}{c^2}}}$$

$$\Lambda_0^1 = \frac{v}{\sqrt{1 - \frac{v^2}{c^2}}}$$

$$\Lambda_1^0 = \frac{\frac{v}{c^2}}{\sqrt{1-\frac{v^2}{c^2}}}$$

$$\Lambda_1^1 = \frac{1}{\sqrt{1-\frac{v^2}{c^2}}}, \tag{5.52}$$

which are the set of Eq. (5.46), up to a sign on v corresponding to the direction of the translation.

Deriving the Lorentz Transformations

A complete section is devoted to the Lorentz transformation in Appendix A.1.1. I will give here just some fundamentals. Suppose a light signal proceeding through the x-axis in the reference (lab) frame K, and one should have

$$x - ct = 0. \tag{5.53}$$

On the other hand, the hypothesis of constant velocity of light c in any frame should impose, in a system K' moving with a velocity v relative to K,

$$x' - ct' = 0. \tag{5.54}$$

Combining the two previous equations, and adding the situation of a light traveling toward negative x-axis, one can write

$$x' - ct' = \lambda(x - ct), \quad x' + ct' = \mu(x + ct), \tag{5.55}$$

which gives by combination

$$x' = ax - bct, \quad ct' = act - bx, \tag{5.56}$$

with $a = \frac{\lambda+\mu}{2}$ and $b = \frac{\lambda-\mu}{2}$. For the origin of the reference frame K', we always have $x' = 0$, which gives $x = \frac{b}{a}ct$. Remembering that K' travels with a velocity v relative to K, this gives $v = \frac{b}{a}c$. Now, if one makes a snapshot at $t = 0$ of a unit length $\Delta x = 1'$, the observatory in the lab frame will observe $\Delta x = \frac{\Delta x'}{a} = \frac{1}{a}$. On the other hand, the unit length $\Delta x = 1$ will be observed in the moving frame at $t' = 0$ as $\Delta x' = \Delta x \left(a - \frac{b^2}{a}\right) = a\Delta x \left(1 - \frac{v^2}{c^2}\right) = a\left(1 - \frac{v^2}{c^2}\right)$. Imposing the principle of relativity (the observed length is the

(continued)

same, independent of the reference frame, $\Delta x = \Delta x'$) we obtain

$$a = \frac{1}{\sqrt{1-\frac{v^2}{c^2}}}, \quad \text{and then} \quad b = \frac{\frac{v}{c}}{\sqrt{1-\frac{v^2}{c^2}}}. \tag{5.57}$$

The transformation (5.56) then becomes the Lorentz transformation

$$x' = \frac{x - vt}{\sqrt{1-\frac{v^2}{c^2}}}, \qquad ct' = \frac{ct - \frac{v}{c}x}{\sqrt{1-\frac{v^2}{c^2}}}. \tag{5.58}$$

We can then easily check that $x'^2 - ct'^2 = x^2 - ct^2$ confirming that the velocity of light is the same in both frames.

Writing the Lorentz transformation into a differential form, one can deduce the transformation laws for velocities:

$$dx = \gamma(dx' + vdt'); \quad dy = dy', \quad dz = dz', \quad dt = \gamma\left(dt' + \frac{v}{c^2}dx'\right) \quad \Rightarrow$$

$$u_x = \frac{dx}{dt} = \frac{\gamma(dx' + vdt')}{\gamma(dt' + \frac{v}{c^2}dx')} = \frac{u'_x + v}{1 + \frac{vu'_x}{c^2}},$$

$$u_y = \frac{u'_y}{\gamma(1 + \frac{vu'_x}{c^2})},$$

$$u_z = \frac{u'_z}{\gamma(1 + \frac{vu'_x}{c^2})}. \tag{5.59}$$

The generalization of such transformations to an axis not parallel to the x-axis can be written by decomposing u into its parallel ($u_\parallel$) and perpendicular ($u_\perp$) components with respect to $\mathbf{v}$:

$$u_\parallel = \frac{u'_\parallel + v}{1 + \frac{vu'_\parallel}{c^2}}, \quad u_\perp = \frac{u'_\perp}{\gamma(1 + \frac{vu'_\parallel}{c^2})}. \tag{5.60}$$

For $u' = c$ (light emitted in K') we can notice

$$\tan\theta = \frac{u_\perp}{u_\parallel} = \frac{c}{\gamma v}, \tag{5.61}$$

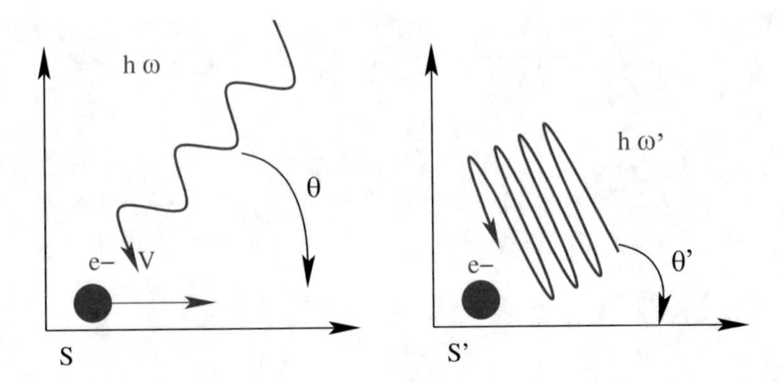

Fig. 5.12 Doppler effect and aberration (angular contraction)

which means that if the distribution of light is isotropic in the K' frame, the observer, for highly relativistic speeds ($\gamma \gg 1$), will see half of the photons lying within a cone of half angle $1/\gamma$. This is called the *beaming effect* that we have illustrated in Fig. 5.12.

We can also compute explicitly the angle (see Fig. 5.12) θ' under which is seen an incident radiation from a moving electron with a velocity **v** in its rest frame as a function of the incident angle θ measured in the galactic (laboratory) frame:

$$\sin\theta' = \frac{v'_y}{c} = \frac{\sin\theta}{\gamma\left(1 - \frac{v}{c}\cos\theta\right)}, \qquad \cos\theta' = \frac{v'_x}{c} = \frac{\cos\theta - \frac{v}{c}}{(1 - \frac{v}{c}\cos\theta)} \tag{5.62}$$

with $\sin\theta = v_y/c$, $\cos\theta = v_x/c$. Notice that $v \to -v$ in the formula compared to Eq. (5.59) because we express the measured value in S' as a function of the ones in S.

It becomes then straightforward to deduce the Lorentz transformations for the acceleration:

$$\begin{aligned}
a_x &= \frac{a'_x}{\gamma^3(1 + \frac{vu_x}{c^2})^3} \\
a_y &= \frac{a'_y}{\gamma^2(1 + \frac{vu_x}{c^2})^2} - \frac{u'_y v}{c^2}\frac{a'_x}{\gamma^2(1 + \frac{vu_x}{c^2})^3} \\
a_z &= \frac{a'_z}{\gamma^2(1 + \frac{vu_x}{c^2})^2} - \frac{u'_z v}{c^2}\frac{a'_x}{\gamma^2(1 + \frac{vu_x}{c^2})^3}.
\end{aligned} \tag{5.63}$$

If the particle is at rest in K', we obtain

$$a'_\parallel = \gamma^3 a_\parallel, \qquad a'_\perp = \gamma^2 a_\perp. \tag{5.64}$$

Exercise Recover the transformation laws for the acceleration.

We can complete the list with the Lorentz transformation for the electromagnetic fields

$$E'_{\parallel} = E_{\parallel}, \qquad B'_{\parallel} = B_{\parallel}$$
$$E'_{\perp} = \gamma\left(E_{\perp} + \frac{1}{c}\mathbf{v} \wedge \mathbf{B}\right), \qquad B'_{\perp} = \gamma\left(B_{\perp} - \frac{1}{c}\mathbf{v} \wedge \mathbf{E}\right). \tag{5.65}$$

Exercise Recover the Lorentz transformation for the electric and magnetic fields, and deduce the equations for **E** and **B** in the moving frame K' as a function of (x', y', z'). Compute the time t' necessary in K' to recover Maxwell equations "as if" the moving observer is at rest.[7] Give your conclusion (this was the first step to relativity proposed by Lorentz in 1894).

5.5.3 Relativistic Larmor's Formula

From Eq. (5.64) one can then deduce the loss of energy of a relativistic particle in the laboratory frame. We just need to apply the Larmor's formula (5.31) in the rest frame of the particle and then make a Lorentz boost. It gives

$$P = \frac{e^2\gamma^4}{6\pi\epsilon_0 c^3}(|a_{\perp}|^2 + \gamma^2|a_{\parallel}|^2), \tag{5.66}$$

where P is the power emitted by the radiating charge.

5.5.4 Doppler Effect

Any periodic phenomenon in the moving frame K' will appear to have a longer period by a factor γ when viewed by local observers in frame K. We also have to add the additional effect on the period due to the delay time for light propagation between the journey $1 \rightarrow$ observer and $2 \rightarrow$ observer. The joint effect is called Doppler effect. Suppose that the moving source emits one period of radiation as it moves from point 1 to point 2 at velocity v. If the frequency of the radiation in the rest frame of the source is ω', then the time to move from point 1 to point 2 in the

[7] To respect the Galilean principle.

observer's frame is given by the time dilatation effect

$$\Delta t = \frac{2\pi\gamma}{\omega'}. \tag{5.67}$$

Adding the classical Doppler effect taking into account the fact that the journey of the light is shorter[8] by a time d/c with $d = v\Delta t \cos\theta$, θ being the angle between the gamma ray and $\mathbf{v}$ in the observer referential at the time of *emission*, one obtains

$$\omega = \frac{\omega'}{\gamma\left(1 - \frac{v}{c}\cos\theta\right)}. \tag{5.68}$$

This is the relativistic Doppler formula, where ω is the frequency observed by the fixed observer and is illustrated by Fig. 5.12.

5.5.5 Transformations on the Energies

Another quadrivector that we can build and that respects the Lorentz transformation (5.46) is the energy–momentum one, $p_\mu = (E/c, \mathbf{p})$. In this case the transformation can be written as:

$$p'_x = \gamma\left(p_x + \frac{v}{c}\frac{E}{c}\right); \quad p'_y = p_y, \quad p'_z = p_z, \quad \frac{E'}{c} = \gamma\left(\frac{E}{c} + \frac{v}{c}p_x\right) \tag{5.69}$$

with for a particle at rest in K, $\mathbf{p} = 0$ and $E = mc^2$. Notice that comparing to (5.46), the transformation corresponds to a reference frame K' moving with a velocity $-v$ compared to the K frame where the particle is at rest ($p = 0$). We then deduce the energy and momentum of a particle of velocity $+v$ compared to the laboratory frame:

$$E' = \frac{mc^2}{\sqrt{1-\frac{v^2}{c^2}}} = \gamma mc^2, \quad p'_x = \frac{mv}{\sqrt{1-\frac{v^2}{c^2}}} = \gamma mv, \quad p'_y = p_y, \quad p'_z = p_z. \tag{5.70}$$

We remark that, by expanding this expression at low velocities $v \ll c$, one obtains

$$E' \sim mc^2 + \frac{1}{2}mv^2,$$

[8]This convention applies if the source at K' moves *toward* the observer at rest at K. If opposite, just replace v by $-v$.

and we find back the expression of the energy as potential (mass) energy plus kinetic energy. We can also rephrase (5.70) in a useful form

$$v = \frac{|p'|c^2}{E'} = \frac{|p'|}{\gamma m}. \tag{5.71}$$

Note also that if a particle of rest mass m moves with a velocity v and loses or gains the energy E_0 (through absorption or emission of a photon for instance), its energy in the laboratory frame is increased (decreased) by $\pm E_0/\sqrt{1-\frac{v^2}{c^2}}$, which means that the total energy in the laboratory frame is now $(m \pm \frac{E_0}{c^2})c^2/\sqrt{1-\frac{v^2}{c^2}}$. Its behavior is then equivalent of a particle of mass $m \pm E_0/c^2$. This is the famous formulae proposed by Einstein in addenda of his article published in October 1905: a particle at velocity v receiving (losing) an amount of energy E_0 without altering its velocity receives a mass contribution $\Delta m = \pm E_0/c^2$ (the famous formula "$E = mc^2$"). This way of "proving" relativity was the one preferred by Einstein himself, but one has to wait until the work of Cockcroft and Walton in 1932 who observed the first transmutation mechanism, shooting beam of protons at very high speeds. Firing protons like bullets, into metal targets, they converted some of the atoms' mass into energy.

5.5.6 Fizeau Experiment

One of the first proof of special relativity was made by Fizeau in 1851, even before Einstein was born,[9] when he measured the relative speed of light in moving water of velocity v. Fizeau used a special interferometer arrangement to measure the effect of movement of a medium upon the speed of light. For a light of velocity $w = c/n$, n being the refraction index of water, the law of composition should give the velocity in the lab frame u (5.59):

$$u = \frac{w+v}{1+\frac{w.v}{c^2}} \simeq (w+v)\left(1-\frac{w.v}{c^2}\right) \simeq w + v\left(1-\frac{w^2}{c^2}\right) = w + v\left(1-\frac{1}{n^2}\right), \tag{5.72}$$

which is exactly the experimental result measured by Fizeau in his 1851 experiment.

[9]Einstein himself used to tell that the Fizeau experiment was his first motivation to establish his own laws of transformation.

5.6 Radiative Processes in Astrophysics Part II: The Relativistic Case

5.6.1 Relativistic Coulomb Scattering or Ionization Losses

5.6.1.1 Ionization Loss

For the treatment of relativistic Coulomb loss we use Eqs. (5.59) and (5.65) to compute the transfer of momentum from the moving frame (rest frame of the moving particle where the treatment is non-relativistic) to the rest frame of the observer.[10] If we write $\gamma = E/mc^2$, we have in the rest frame

$$\Delta p = \Delta p_\perp = F_\perp dt = F'_\perp dt' = \Delta p'_\perp \tag{5.73}$$

where $F_\perp$ ($E_\perp$) represents the component of the force (electric field) perpendicular to the direction of motion, which is the only active one as we discussed in Sect. 5.4.3. The transferred momentum is the same in the relativistic frame than in the rest frame. It is logical because only the perpendicular component of the momentum is transferred in the Coulomb interaction, and perpendicular components are never affected by transformations in relativity. Equation (5.36) is then valid even for a relativistic moving particle. The difference is then due to the factors $b_{\min}$ and $b_{\max}$. $b_{\max}$ is greater by a factor γ because the duration of the impulse is shorter by this factor. In the case of $b_{\min}$, the transverse momentum of the electron is greater by a factor γ, and hence, because of the Heisenberg uncertainty principle,

$$\Delta x \simeq b_{\min} = \frac{\hbar}{\Delta p} \propto \gamma^{-1}, \tag{5.74}$$

where we used the Lorentz transformation for the momentum (velocity) obtained in Eq. (5.60). Thus, we expect the logarithmic term to have the form $\ln(2\gamma^2 m_e v^2/\bar{I})$. A more precise computation taking into account relativistic quantum effects is given by the Bethe–Bloch formula

$$\frac{dE}{dx} = -\frac{Z^2 e^4 N_e}{4\pi\epsilon_0^2 v^2 m_e}\left[\ln\left(\frac{2\gamma^2 m_e v^2}{\bar{I}}\right) - \frac{v^2}{c^2}\right].$$

We derived all the expressions above, except for the factor v^2/c^2, which is always practically very small compared to $\ln(2\gamma^2 m_e v^2/\bar{I})$, which is $\gtrsim 20$ in the majority of physical processes in the galactic medium. As discussed earlier, $\bar{I}$ is a parameter that should be obtained experimentally. We usually express the energy loss as a function

[10] It can be the solar system or even the galaxy itself that is considered at rest compared to the moving relativistic particles.

of the density of mass of the material crossed by the high energy particle using[11]

$$\frac{dE}{dt} = -\frac{4\pi Z^2 r_0^2 m_e c^3 Z N_A}{\beta A}\left[\ln\left(\frac{2\gamma^2 m_e c^2 \beta^2}{\bar{I}}\right) - \beta^2\right] \times \rho$$

with $r_0 = e^2/4\pi\epsilon_0 m_e c^2$ the classical radius of the electron, $\beta = \frac{v}{c}$, and we substitute $x \to vt$. The exact treatment of the ionization energy loss of heavy particles, taking into account that the interstellar medium is mainly composed of hydrogen and helium, leads to

$$\frac{dE}{dt} = -\frac{4\pi Z^2 r_0^2 m_e c^3 Z N_A}{\beta A}\sum_{i=H,He}\left[\frac{1}{2}\ln\left(\frac{2\gamma^2 m_e c^2 \beta^2 E_{\max}}{\bar{I}_i^2}\right) - \beta^2 - \frac{\delta}{2}\right] \times \rho$$

$$= -\frac{4\pi Z^2 r_0^2 m_e c^3 Z N_s}{\beta}\sum_{i=H,He}\left[\frac{1}{2}\ln\left(\frac{2\gamma^2 m_e c^2 \beta^2 E_{\max}}{\bar{I}_i^2}\right) - \beta^2 - \frac{\delta}{2}\right], \quad (5.75)$$

where δ represents a screening effect due to the density of atoms of the medium crossed by the particle and $E_{\max} = 2m_e c^2\beta^2\gamma^2/(1 + 2\gamma m_e/M + (m_e/M)^2)$, M being the nucleon mass, is the maximum kinetic energy transferred to one atom and N_s the number density of the medium. The density correction can be neglected in our case, and the values of $I_H = 19$ eV and $I_{He} = 44$ eV are commonly used. A more complete treatment in quantum field theory gives the Bethe–Bloch formula for the ionization loss

$$\left(\frac{dE}{dt}\right)_I = -2\pi r_0^2 m_e c^3 \frac{1}{\beta}\sum_{s=H,He} Z_s n_s\left[\ln\left(\frac{(\gamma-1)\beta^2 E^2}{2I_s^2}\right) + \frac{1}{8}\right]. \quad (5.76)$$

5.6.1.2 Coulomb Scattering

The Coulomb collisions in a completely ionized plasma are dominated by scattering off the thermal electrons. The corresponding energy losses are given by

$$\left(\frac{dE}{dt}\right)_{Coul} = -2\pi r_0^2 m_e c^3 \frac{1}{\beta} Zn\left[\ln\left(\frac{E m_e c^2}{4\pi r_0 \hbar^2 c^2 Zn}\right) - \frac{3}{4}\right] \quad (5.77)$$

$$\simeq -5.5\times 10^{-16}\left(\frac{Zn}{1\ \mathrm{cm}^{-3}}\right)\left(\frac{1}{1-(mc^2/E)^2}\right)\left[1 + \frac{1}{74}\ln\left(\frac{E/m_e c^2}{Zn/(1\ \mathrm{cm}^{-3})}\right)\right]\ \mathrm{GeV\ s^{-1}}$$

[11]Remembering that the Avogadro number N_A is by definition the number of carbon12 atoms in 12 grams of carbon, or equivalently, N_A/A represents the number of atoms of nuclei A in one gram, and thus $Z \times N_A/A$ represents the number of protons (and then electrons) per gram of material. We can then deduce $N_e = \rho Z N_A/A$, ρ being the mass density of material A.

n being the density of ions of charge Z in the ionized plasma and m the mass of the traveling particle. The values of the densities n_e, n_H, or n_{He} depend on the position in the galactic plane. As the Sun reference point (8 kpc) they are $n_e = 0.0024\ \text{cm}^{-3}$, $n_H = 0.6\ \text{cm}^{-3}$, and $n_{He} = 0.144\ \text{cm}^{-3}$.

5.6.2 Inverse Compton Scattering

The Inverse Compton interaction corresponds to the relativistic limit of the Thomson scattering (see Fig. 5.5). The strategy to compute it is quite straightforward: to place ourselves in the framework where the charged particle is at rest, and then to apply a Lorentz transformation to obtain the result in the galactic (observer) coordinate frame. The energy loss per unit of time being Lorentz invariant, $P' = dE'/dt' = dE/dt = P$, we need to compute $dE'/dt' = \sigma_T c U'_{rad}$ (Eq. 5.42). For that we need to compute $U'_{rad} = N\hbar\omega'$ as a function of $U_{rad} = N\hbar\omega$, N being the density of photons. The frequency $\omega' \to \omega$ follows the classical Doppler effect described through Eq. (5.68). However, there is another subtlety, which is equivalent to a second Doppler effect in a sense: cU'_{rad} is a *flux* of energy. One then needs to compute the interval of time $\Delta t'$ between the arrival of two pulses on the electron in S' as a function of the interval of the same event Δt measured in S. After a look at Fig. 5.13 we can write

$$\Delta t = t_2 - t_1 + \frac{x_2 - x_1}{c} \cos\theta = (t'_2 - t'_1)\gamma[1 + (v/c)\cos\theta]. \tag{5.78}$$

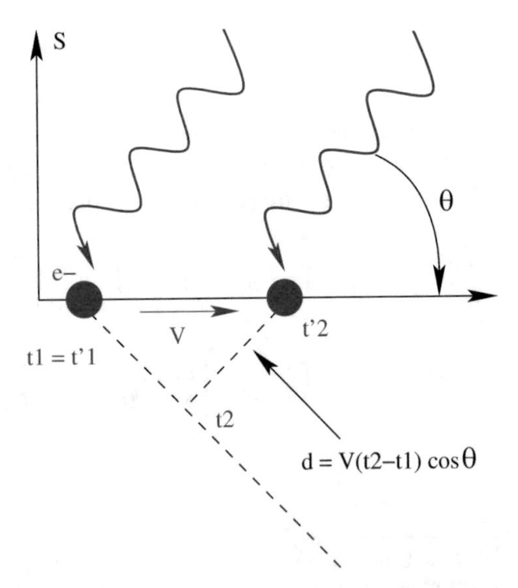

Fig. 5.13 Inverse Compton scattering geometry

One obtains, exactly as for the Doppler effect, $\Delta t = \Delta t' \times \gamma[1 + (v/c)\cos\theta]$, that is, the time interval between the arrival of photons from the direction θ is shorter by a factor $\gamma[1 + (v/c)\cos\theta]$ in S' than in S. Thus, the rate of arrival of photons, and correspondingly their number density, is greater by this factor. Then, as observed in S', the energy of the beam is therefore

$$U'_{rad} = \gamma^2[1 + (v/c)\cos\theta]^2 \times U_{rad}. \tag{5.79}$$

After a mean on the angle θ and using the relation $\gamma^2 - 1 = (v/c)^2\,\gamma^2$ we can write

$$U'_{rad} = \frac{1}{4\pi}\int_0^{\pi} \gamma^2[1 + (v/c)\cos\theta]^2\, 2\pi\,\sin\theta d\theta = \frac{4}{3}U_{rad}\left(\gamma^2 - \frac{1}{4}\right)$$

$$\Rightarrow \langle P\rangle_{IC} = \langle P'\rangle = \frac{4}{3}\sigma_T c U_{rad}\left(\frac{v^2}{c^2}\right)\gamma^2$$

$$\simeq 2.56 \times 10^{-17}\left(\frac{E}{1\text{ GeV}}\right)^2\left(\frac{511\text{ keV}}{m}\right)^2 \text{GeV s}^{-1}, \tag{5.80}$$

where we have used $v \simeq c$, σ_T form (5.43) and $U_{rad} = 2.4 \times 10^{-10}$ GeV cm^{-3} from Eq. (3.47).

5.6.3 Synchrotron Radiation

Synchrotron radiation is the relativistic limit of the cyclotron mechanism (see Fig. 5.5). We will compute the synchrotron emission with both methods: first directly in the observer (fixed) frame using the relativistic Larmor's formula. We will then recover this result computing the emission in the particle frame (particle will be at rest) but feeling the electric field generated by the relativistic boost.

5.6.3.1 From the Observer Point of View

To obtain the formulation for the cyclotron radiation produced by a relativistic particle (called synchrotron radiation) we first need to compute the acceleration of the particle. The relativistic dynamics principle gives us

$$\frac{d}{dt}\gamma m\mathbf{v} = e(\mathbf{v} \times \mathbf{B}). \tag{5.81}$$

We then should recover the same result than for the synchrotron replacing $m \to \gamma m$ and applying the relativistic Larmor's formula, noticing that there is no acceleration parallel to $\mathbf{B}$ and that the acceleration is perpendicular to the velocity vector (Eq. 5.81), and we deduce $|a_\parallel|^2 = 0$. We then obtain

$$P = \frac{e^4\gamma^2B^2v_\perp^2}{6\pi\epsilon_0 m^2c^3} = \frac{e^4\gamma^2B^2v^2\sin^2\theta}{6\pi\epsilon_0 m^2c^3}. \tag{5.82}$$

If one averages on the velocity angle:

$$\langle v_\perp^2 \rangle = \frac{v^2}{4\pi} \int \sin^2 \theta d\Omega = \frac{2}{3} v^2 \quad \Rightarrow \langle P \rangle = \langle P \rangle_{synch} = \frac{e^4 \gamma^2 B^2 v^2}{9\pi \epsilon_0 m^2 c^3} \tag{5.83}$$

If the particle is relativist, one can consider $E = pc$ with $p = \gamma m v$, which implies $v^2 = E^2/(\gamma^2 m^2 c^2)$ and

$$\langle P \rangle_{synch} = -\frac{dE}{dt} = \frac{e^4 B^2 E^2}{9\pi \epsilon_0 m^4 c^5} = \frac{2e^4 E^2}{9\pi \epsilon_0^2 m^4 c^7} \left(\frac{B^2}{2\mu_0} \right)$$

$$\simeq 2.51 \times 10^{-18} \left(\frac{E}{\text{GeV}} \right)^2 \left(\frac{B}{\mu\text{G}} \right) \left(\frac{511 \text{ keV}}{m} \right)^4 \text{GeV s}^{-1}. \tag{5.84}$$

5.6.3.2 From the Particle Point of View

In the rest frame of the particle, the fundamental principle of dynamic can be written as

$$\frac{d}{dt'} m\mathbf{v}' = e(\mathbf{E}' + \mathbf{v}' \times \mathbf{B}') = e\mathbf{E}' = \frac{e}{c} \gamma v B \sin\theta \frac{\mathbf{v}'}{v'} \tag{5.85}$$

because the particle is at rest, $\mathbf{v}' = \mathbf{0}$, θ being the angle between $\mathbf{v}'$ and $\mathbf{B}'$. We used Eq. (5.65) for the Lorentz transformation on $\mathbf{E}'$. We then obtain

$$P = -\left(\frac{dE}{dt} \right)' = \frac{e^2 |\dot{\mathbf{v}}'|^2}{6\pi \epsilon_0 c^3} = \frac{e^4 \gamma^2 B^2 v^2 \sin^2 \theta}{6\pi \epsilon_0 m^2 c^3} = 2\sigma_T c U_B \left(\frac{v}{c} \right)^2 \gamma^2 \sin^2 \theta$$

which implies

$$\boxed{\langle P \rangle_{synch} = \frac{4}{3} \sigma_T c U_B \left(\frac{v}{c} \right)^2 \gamma^2} \tag{5.86}$$

with $U_B = B^2/2\mu_0$, recovering Eq. (5.83).

5.6.4 Relativistic Bremsstrahlung

When a charged particle passes through an electric field, it can be accelerated or decelerated, depending on its charge. In both cases, the change in the velocity will generate a loss of energy, following the Larmor's formula. In the presence of magnetic field, this loss of energy corresponds to the synchrotron radiation, whereas in the electric field case we use to call it bremsstrahlung. In this case, we define a radiation length X by the thickness of the medium over which the energy of the incident particle is reduced by a factor e:

$$E = E_0 e^{-\frac{x}{X}}. \tag{5.87}$$

X can be expressed in cm, but it is more common to write it in $g.cm^{-2}$ dividing it by the mass density

$$X \to \frac{X}{nM} \tag{5.88}$$

with n the density of the ionized particles in the gas and M its atomic mass. We can then express the loss of energy by bremsstrahlung as

$$-\left(\frac{dE}{dt}\right)_{brem} = c\,E\,\frac{n\,M}{X}, \tag{5.89}$$

where c is the velocity of light.

5.6.5 Energy Losses: Summary

As a summary, we can write the total energy loss for an electron in a galactic medium from synchrotron (*synch*) (Eq. 5.86), Inverse Compton (*IC*) (Eq. 5.80), Coulomb (*Coul*), ionization (*I*), and bremsstrahlung (*brem*) radiation in the relativistic case ($v \simeq c$) summarized in Fig. 5.5:

$$\begin{aligned}
\langle P\rangle &= \langle P\rangle_{synch} + \langle P\rangle_{IC} + \langle P\rangle_{Coul} + \langle P\rangle_{I} + \langle P\rangle_{brem} \\
&= \frac{4}{3}\sigma_T c U_B + \frac{4}{3}\sigma_T c U_{rad} + 2\pi r_0^2 m_e c^3 \frac{1}{\beta} Zn\left[\ln\left(\frac{E m_e c^2}{4\pi r_0 \hbar^2 c^2 n Z}\right) - \frac{3}{4}\right] \\
&+ 2\pi r_0^2 m_e c^3 \frac{1}{\beta} \sum_{s=H,He} Z_s n_s \left[\ln\left(\frac{(\gamma-1)\beta^2 E^2}{2 I_s^2}\right) + \frac{1}{8}\right] + cE \sum_{s=H,He} \frac{n_s M_s}{X_s} \\
&= 2.51\times 10^{-18}\left(\frac{E}{\text{GeV}}\right)^2\left(\frac{B}{\mu\text{G}}\right)\left(\frac{511\text{ keV}}{m}\right)^4 \\
&+ 2.56\times 10^{-17}\left(\frac{E}{\text{GeV}}\right)^2\left(\frac{511\text{ keV}}{m}\right)^2 \\
&+ 5.5\times 10^{-16}\left(\frac{Zn}{1\text{ cm}^{-3}}\right)\left(\frac{1}{\beta}\right)\left[1 + \frac{1}{74}\ln\left(\frac{E/m_e c^2}{Zn/(1\text{ cm}^{-3})}\right)\right] \\
&+ 7.4\times 10^{-18}\left(\frac{1}{\beta}\right)\sum_{s=H,He}\left(\frac{Z_s n_s}{1\text{ cm}^{-3}}\right)\left[\ln\left(\frac{(\gamma-1)\beta^2 E^2}{2 I_s^2}\right) + \frac{1}{8}\right]
\end{aligned}$$

Table 5.3 Number densities of atomic hydrogen (H_I), molecular hydrogen (H_2), ionized hydrogen (H_{II}), and helium (He) as a function of the galactic radius R

R (kpc)	n_{HI} (atom cm^{-3})	$2n_{H2}$ (mol cm^{-3})	n_{HII} (atom cm^{-3})	n_{He} (atom cm^{-3})
0	0.16411	0.	0.0275	0.01805
1	0.16411	0.06211	0.04226	0.02488
2	0.20349	0.12422	0.08674	0.03605
3	0.24616	0.47352	0.15635	0.07917
4	0.36104	0.76053	0.19357	0.12337
5	0.51694	1.04366	0.15542	0.17167
6	0.51694	0.99151	0.08488	0.16593
7	0.54976	0.88814	0.03951	0.15817
8	0.60720	0.71078	0.02390	0.14498
9	0.50280	0.39278	0.02024	0.09851
10	0.65116	0.10128	0.01901	0.08277
11	0.40338	0.	0.01802	0.04437
12	0.57451	0.	0.01701	0.06320
13	0.48365	0.	0.01598	0.05320
14	0.22138	0.	0.01494	0.02435
15	0.14515	0.	0.01389	0.01597
16	0.09448	0.	0.01286	0.01039
17	0.06782	0.	0.01184	0.00746
18	0.04867	0.	0.01085	0.00535
19	0.03491	0.	0.00989	0.00384
20	0.02504	0.	0.00897	0.00275

$$+3 \times 10^{10} \left(\frac{E}{\text{GeV}}\right) \sum_{s=H,He} \frac{n_s M_s}{X_s} \text{ GeV s}^{-1} \tag{5.90}$$

M_s being the atomic mass of species s and X_s its radiation length ($X_H \simeq 62.8$ g cm^{-2}, $X_{He} \simeq 93.1$ g cm^{-2}). The local density of different species at a certain distance from the Milky Way center can be found in Table 5.3. The electron species can be identified with the ionized hydrogen HII species). We show in Fig. 5.14 the summary of the losses as a function of the charged particle energy.

5.7 Ultra-High Energetic (UHE) Processes

If a very high energetic particle like a proton for instance interacts with the photons of the Cosmological Microwave Background (CMB), pions or other particles can be created onshell. As a consequence, it exists a maximum energy (and thus free path) above which these particles (protons or photons) cannot propagate in the Universe

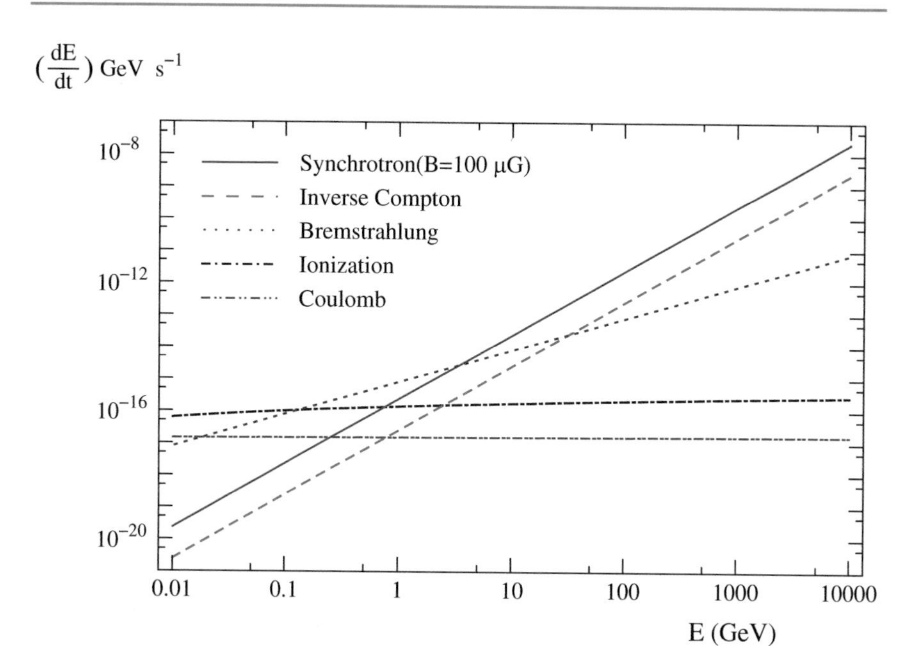

Fig. 5.14 Combining the energy loss mechanisms for $B = 100\mu G$

anymore. Their spectrum should then possess a typical cut that should be observable by dedicated experiences (called GZK cut for Greisen–Zatsepin–Kuzmin in the case of protons). In other words, if ultra-high energy events are observed by such experiments, the origin would be local (like dark matter origin for instance), or at least relatively near the Earth, corresponding to a short mean free path.

5.7.1 Cosmic Rays Case

Photons of the CMB act on accelerated protons through the Δ^+ resonance

$$\begin{aligned}
&\gamma + p \rightarrow \Delta^+ \rightarrow n + \pi^+ \rightarrow n + \mu^+ + \nu_\mu \\
&\gamma + p \rightarrow \Delta^+ \rightarrow p + \pi^0 \rightarrow p + \gamma + \gamma \\
&\gamma + p \rightarrow p + N\pi,
\end{aligned} \tag{5.91}$$

the charged pions decaying into ultra-high energy muons and muons neutrinos.[12] These high energy neutrinos can be detected by ground based detectors like Icecube. Computing the maximum energy allowed by the proton before the reaction occurs is equivalent to compute the energy necessary for the reaction to occur in the rest frame of the proton and then to make a Lorentz transformation back in the galactic (laboratory) frame.
In the proton rest frame R, the threshold energy E_γ^R of the photon is obviously the mass of the pion $E_\gamma^R > 200$ MeV. Using the formulae computed for the Doppler effect (Eq. 5.68) we deduce for the energy of the photon in the laboratory frame E_γ^0

$$200 \text{ MeV} \lesssim E_\gamma^R = E_\gamma^0 \gamma \left(1 + \frac{v}{c}\cos\theta\right) < 2\gamma E_\gamma^0 \tag{5.92}$$

with $E_\gamma^0 = 6 \times 10^{-4}$ eV (see Sect. E.1.1), which gives

$$\gamma \gtrsim 1.7 \times 10^{11} \quad \Rightarrow \quad E_p^0 = \gamma E_p^R = \gamma m_p c^2 \gtrsim 1.7 \times 10^{20} \text{ eV} = E_p^{\max}. \tag{5.93}$$

A more detailed treatment, taking into account the integration over the Planck spectrum of the CMB, decreases this value to $E_p^{\max} = 5 \times 10^{19}$ eV. Any proton of energy $E_p \gtrsim 5 \times 10^{19}$ eV reacts then with the CMB to produce pions and then muons and muon neutrinos. This limit is called the *GZK cutoff* computed for the first time by Greisen and independently by Kuzmin and Zatsepin in 1966. The mean free path λ for a single scattering is then given by

$$\lambda \times \sigma_{p\pi} n_\gamma^0 = 1 \quad \Rightarrow \quad \lambda = (\sigma_{p\pi} n_\gamma^0)^{-1} \simeq 10^{23}\text{m} \simeq 3 \text{ Mpc} \tag{5.94}$$

$\sigma_{p\pi} \simeq 2.5 \times 10^{-4}$ barns being the interaction cross section of the reactions (5.91) and $n_\gamma = 411$ cm^{-3} (Eq. 3.42) the density of photons of the CMB. After each reaction described in (5.91), the proton in the galactic frame loses the kinetic energy $\Delta E_p = \gamma m_\pi c^2$. Its initial energy being $E_p^0 = \gamma m_p c^2$, the relative loss of energy $\Delta E_p / E_p^0 \simeq 1/10$. As a consequence, if the highest energy cosmic rays are protons, they cannot have originated from farther than about 30 Mpc from our galaxy.

5.7.2 Photons and Neutrinos Cases

The phenomena of a GZK cutoff exist also for photons and neutrinos. Indeed, a very highly energetic photon can hit a photon from the CMB, producing a e^+e^-

[12] For the reader not used to the particle physics contents, the pions are composed of $\pi^0 = \frac{1}{\sqrt{2}}(\bar{u}u + \bar{d}d)$, $\pi^+ = \bar{d}u$ and $\pi^- = d\bar{u}$. The Delta baryons Δs also called *Delta resonances* are made of $\Delta^{++} = uuu$, $\Delta^+ = uud$, $\Delta^0 = udd$, and $\Delta^- = ddd$. Their masses are about 1.232 GeV and can be considered (in the case of Δ^+ and Δ^0) as excited states of the proton and neutron, respectively.

pair onshell via a t-channel exchange of a virtual electron. Taking into account that the CMB temperature is $T_0 \simeq 2.34 \times 10^{-4}$ eV,

$$(P_1 + P_2)^2 = 2P_1.P_2 \simeq 4E_{CMB}E_\gamma = 4m_e^2 \;\Rightarrow\; E_\gamma \simeq 400 \text{ TeV}. \qquad (5.95)$$

In fact, when taking into account the mean free path, a photon of 100 TeV will lose 90% of its energy in the Milky Way before crossing it. The same exercise can be done for neutrinos; considering the process on the CMB through the t-channel exchange of an electron with a photon + W final state, we obtain $E_\nu^{cut} \simeq \frac{M_W^2}{4E_{CMB}} \simeq 5 \times 10^{15}$ GeV.

Exercise Compute the mean free path for a photon and neutrino as a function of its energy in the Milky Way.

5.8 Indirect Detection of Gamma Ray

5.8.1 The Principle

The main principle of the indirect detection of gamma rays is to observe the sky, in different ranges of energy, depending on the satellites or telescopes. It can range from the keV scale for XMM Newton satellite to more than 10 TeV scale for the HESS telescope in Namibia. The main advantage of terrestrial detectors concerns its ability to measure higher energetic photons, whereas the size of the satellites limits drastically the sensitivity of the instruments. The FERMI satellite, however, is able to give limits on the flux up to 300 GeV.
For the continuum of gamma rays, the observed differential flux at the Earth coming from a direction forming an angle ψ with respect to the galactic center is

$$\frac{d\Phi_\gamma(E_\gamma, \psi)}{dE_\gamma} = \sum_i \frac{dN_\gamma^i}{dE_\gamma} \langle\sigma_i v\rangle \frac{1}{8\pi m_{\text{DM}}^2} \int_{\text{line of sight}} \rho^2 \, dl\,, \qquad (5.96)$$

where the discrete sum is over all dark matter annihilation channels, dN_γ^i/dE_γ is the differential gamma-ray yield, $\langle\sigma_i v\rangle$ is the annihilation cross section averaged over its velocity distribution, m_{DM} is the mass of the dark matter particle, and ρ is the dark matter density. Assuming a spherical halo, one has $\rho = \rho(r)$ with the galactocentric distance $r^2 = l^2 + R_0^2 - 2lR_0 \cos\psi$, where R_0 is the solar distance to the galactic center ($\simeq$ 8 kpc). It is customary to separate in the above equation the particle physics part from the halo model dependence introducing the (dimensionless) quantity $J(\psi)$ we already discussed in Eq. (5.3):

$$J(\psi) = \frac{1}{8.5\text{ kpc}} \left(\frac{1}{0.3\text{ GeV/cm}^3} \right)^2 \int_{\text{line of sight}} \rho^2(r(l, \psi))\, dl \, . \tag{5.97}$$

We can explain the origin of Eq. (5.96) in easily. The differential flux (number of particles arriving per unit of surface per unit of time and per unit of energy) is the cross section of a given final state i (σ_i) multiplied by the density of the target on its way in one second ($\frac{\rho}{m_{\text{DM}}} \times v$) multiplied by the number of particles per unit of surface ($\frac{\rho}{m_{\text{DM}}} dl$) giving a spectrum $\frac{dN_i}{dE}$. If one looks per solid angle, one also needs to divide by 4π, adding $\frac{1}{2}$ factors, because we can detect only one of the two final states, the second one flies in the opposite direction. One can then write

$$\frac{d\Phi_\gamma(E_\gamma, \psi)}{dE_\gamma} = 0.94 \times 10^{-13}\text{ cm}^{-2}\text{ s}^{-1}\text{ GeV}^{-1}\text{ sr}^{-1} \times \sum_i \frac{dN_\gamma^i}{dE_\gamma} \left(\frac{\langle \sigma_i v \rangle}{10^{-29}\text{cm}^3\text{s}^{-1}} \right) \left(\frac{100\text{ GeV}}{m_{\text{DM}}} \right)^2 J(\psi) \, . \tag{5.98}$$

Actually, when comparing to experimental data, one must consider the integral of $J(\psi)$ over the spherical region of solid angle $\Delta\Omega$ given by the angular acceptance of the detector that is pointing toward the galactic center, i.e. the quantity $\bar{J}(\Delta\Omega)$ with

$$\bar{J}(\Delta\Omega) \equiv \frac{1}{\Delta\Omega} \int_{\Delta\Omega} J(\psi)\, d\Omega \, , \tag{5.99}$$

must be used in the above equation. For example, for experiments like FERMI, MAGIC, or HESS $\Delta\Omega \simeq 10^{-5}$ sr. The gamma-ray flux can now be expressed as

$$\Phi_\gamma(E_{thr}) = 0.94 \times 10^{-13}\text{ cm}^{-2}\text{ s}^{-1} \times \sum_i \int_{E_{thr}}^{m_\chi} dE_\gamma \frac{dN_\gamma^i}{dE_\gamma} \left(\frac{\langle \sigma_i v \rangle}{10^{-29}\text{cm}^3\text{s}^{-1}} \right) \left(\frac{100\text{ GeV}}{m_{\text{DM}}} \right)^2 \bar{J}(\Delta\Omega)\Delta\Omega \, , \tag{5.100}$$

where E_{thr} is the lower threshold energy of the detector. Concerning the upper limit of the integral, note that the dark matter moves at galactic velocity and therefore its annihilation occurs (almost) at rest. We also notice that for typical values of $\bar{J}(\Delta\Omega)\Delta\Omega \sim 1 - 10^3$ (see Table 5.2), and WIMP cross section $\langle \sigma v \rangle \simeq 10^{-26}\text{ cm}^3\text{s}^{-1}$ (see Eq. 3.160) we expect a flux of 1 GeV photons around $10^{-2} - 10$ per m^2 per second. That is a very low flux, and the sensitivity of the experiments dedicated to indirect detection searches should be extremely high.

5.8.2 Galactic Halo

A crucial ingredient for the calculation of $\bar{J}$ in (5.99), and therefore for the calculation of the flux of gamma rays, is the dark matter density profile of our galaxy. As we discussed after Eq. (5.2), the different profiles that have been proposed in the literature can be parameterized as

$$\rho(r) = \frac{\rho_0[1 + (R_0/a)^\alpha]^{(\beta-\gamma)/\alpha}}{(r/R_0)^\gamma[1 + (r/a)^\alpha]^{(\beta-\gamma)/\alpha}}, \tag{5.101}$$

where ρ_0 is the local (solar neighborhood) halo density, R_0 the distance from the Sun to the galactic center ($\simeq 8.5$ kpc) and a is a characteristic length. The measured value $\rho_0 = 0.3$ GeV/cm^3 is commonly used in analysis, since this is just a scaling factor, and modifications to its value can be straightforwardly taken into account. As we discussed in Sect. 5.2, the values of α, β, and γ are determined by N-body simulations.

5.8.3 Adiabatic Compression Mechanism

Highly cusped profiles are deduced from N-body simulations. In particular, Navarro, Frenk, and White obtained in 1996 a profile with a behavior $\rho(r) \propto r^{-1}$ at small distances. A more singular behavior, $\rho(r) \propto r^{-1.5}$, was obtained by Moore et al. in 1999. To show how sensitive are these results to the ingredients you consider in the simulation, let us take the example of the adiabatic compression process. Indeed, these predictions are valid only for halos without baryons. One can improve simulations in a more realistic way by taking into account the effect of the normal gas (baryons). While they lose their energy through radiative processes falling to the central region of the galaxy, they modify slowly (adiabatically) the mass distribution. The resulting gravitational potential is then deeper, and the dark matter must move closer to the center increasing its density, in other words, and *adiabatic compression*.

The increase in the dark matter density is often treated using adiabatic invariants. We assume spherical symmetry, circular orbit for the particles, and conservation of the angular momentum $M(r)r$ = constant, where $M(r)$ is the total mass enclosed within radius r. The mass distributions in the initial (r) and final (r_f) configurations are therefore related by $[M(r) + M_b(r)]r = [M_b(r_f) + M(r)]r_f$, where $M(r)$ and $M_b(r)$ are, respectively, the mass profile of the galactic halo before the cooling of the baryons at a distance r (obtained through N-body simulations), and the baryonic composition of the Milky Way observed now. From this expression, one can deduce r_f as a function of r and then the new distribution of dark matter. A more precise approximation can be obtained including the possibility of elongated orbits. In this case, the mass inside the orbit, $M(r)$, is smaller than the real mass, the one the particle "feels" during its revolution around the galactic center. As a consequence,

Table 5.4 NFW and Moore et al. density profiles without and with adiabatic compression (NFW$_c$ and Moore$_c$, respectively) with the corresponding parameters, and values of $\bar{J}(\Delta\Omega)$

	a (kpc)	α	β	γ	$\bar{J}(10^{-3}\text{sr})$	$\bar{J}(10^{-5}\text{sr})$
NFW	20	1	3	1	1.214×10^3	1.264×10^4
NFW$_c$	20	0.8	2.7	1.45	1.755×10^5	1.205×10^7
Moore et al.	28	1.5	3	1.5	1.603×10^5	5.531×10^6
Moore$_c$	28	0.8	2.7	1.65	1.242×10^7	5.262×10^8

the modified compression model is based on the conservation of the product $M(\bar{r})r$, where $\bar{r}$ is the averaged radius of the orbit. The time average radii are given by $\bar{x} \sim 1.72x^{0.82}/(1+5x)^{0.085}$, with $x \equiv r/r_s$, and r_s the characteristic radius of the assumed approximation.

The models and constraints applied for the Milky Way are summarized in Table 5.4. There we label the resulting NFW and Moore et al. profiles with adiabatic compression by NFW$_c$ and Moore$_c$, respectively. As one can see, at small r, the dark matter density profile following the adiabatic cooling of the baryonic fraction is a steep power law $\rho \propto r^{-\gamma_c}$ with $\gamma_c \approx 1.45(1.65)$ for a NFW$_c$(Moore$_c$) compressed model. We observe for example that the effect of the adiabatic compression on an NFW profile is basically to transform it into a Moore et al. one. The important point we wanted to underline here was not specifically on the adiabatic compression mechanism, but just to give an idea to the reader of why so many simulations can give so many different results for the dark matter distribution in the Milky Way, generating orders of magnitude of differences in the theoretical predictions of fluxes.

Of course, these results have important implications for the computation of the gamma-ray fluxes from the galactic center. In particular, in Table 5.4 we see that for $\Delta\Omega = 10^{-3}$ sr one has $\bar{J}_{\text{NFW}_c}/\bar{J}_{\text{NFW}} \simeq 145$ and $\bar{J}_{\text{Moore}_c}/\bar{J}_{\text{Moore et al.}} \simeq 77$. Thus the effect of the adiabatic compression is very strong, increasing the gamma-ray fluxes about two orders of magnitude. Similarly, for $\Delta\Omega = 10^{-5}$ sr one obtains $\bar{J}_{\text{NFW}_c}/\bar{J}_{\text{NFW}} \simeq 953$ and $\bar{J}_{\text{Moore}_c}/\bar{J}_{\text{Moore et al.}} \simeq 95$.

Let us finally remark that the $\bar{J}$ calculation is usually regulated by assuming a constant density for $r < 10^{-5}$ kpc. This procedure has consequences essentially for divergent $\bar{J}$ when $\gamma \geq 1.5$. We use a cutoff in density, $\rho(r < R_s) = 0$, R_s being the Schwarzschild radius of a $3 \times 10^6 M_\odot$ black hole. In addition we also consider dark matter annihilation and estimate its effect by requiring an upper value for dark matter density, $\rho(r < r_{\max}) = \rho(r_{\max}) = m_\chi/(\langle\sigma v\rangle.t_{BH})$ ($t_{BH} \sim 10^{10}$ yr, being the age of the black hole). We also apply the same procedure with a more realistic cut-off value $\rho(r < 10^{-6}\ \text{pc}) = 0$ suggested by dark matter particle scatterings on stars. This kind of effects can be significant only for the Moore$_c$ halo model and increase the $\bar{J}$ value by less than an order of magnitude for $\langle\sigma v\rangle$.

5.9 The Tricky Case of the Galactic Center

5.9.1 The Idea

As we understood, astrophysical searches for dark matter are a fundamental part of the experimental efforts to explore the dark sector. The strategy is to search for dark matter annihilation products in preferred regions of the sky, i.e., those with the highest expected dark matter concentrations and still close enough to yield high induced fluxes at the Earth. For that reason, the galactic center, nearby dwarf spheroidal galaxy (dSphs) satellites of the Milky Way, as well as local galaxy clusters, are thought to be among the most promising objects for the searches. In particular, dSphs represent very attractive targets because they are highly dark matter dominated systems and are expected to be free from any other astrophysical gamma-ray emitters that might contaminate any potential dark matter signal. Although the expected signal cannot be as large as that from the galactic center, dSphs may produce a larger signal-to-noise (S/N) ratio. This fact allows us to place very competitive upper limits on the gamma-ray signal from dark matter annihilation, using data collected by the Large Area Telescope onboard the Fermi gamma-ray observatory for instance. These are often referred to as the most stringent limits on dark matter annihilation cross section obtained so far.

Despite these interesting limits derived from dSphs, the galactic center is still expected to be the brightest source of dark matter annihilations in the gamma-ray sky by several orders of magnitude. Although several astrophysical processes at work in this crowded region make it extremely difficult to disentangle the signal from conventional emissions, the induced gamma-ray emission is expected to be so large there that the search is still worthwhile. Furthermore, the dark matter density in the galactic center may be larger than what is typically obtained in N-body cosmological simulations. Ordinary matter (baryons) dominates the central region of our galaxy. Thus, baryons may significantly affect the distribution as we just saw when discussing the adiabatic compression phenomenon. It is also observed in many cosmological simulations that include hydrodynamics and stars formation. If this is the only effect of baryons, then the expected annihilation signal will substantially increase.

In this section, we analyze in detail the constraints that can be obtained for generic dark candidates from *Fermi*-LAT inner galaxy gamma-ray measurements assuming some specific (and well motivated) dark matter distributions. This should be seen as a concrete application of the theoretical approach we had in the previous sections. Our approach is conservative, requiring simply that the expected dark matter signal does not exceed the gamma-ray emission observed by the *Fermi*-LAT in an optimized region around the galactic center. The region is chosen in such a way that the S/N ratio is maximized. This kind of analysis is quite common in the field of indirect detection searches.

5.9.2 Dark Matter Density Profiles

As we discussed in the preceding section, cosmological N-body simulations provide important results regarding the expected dark matter density in the central region of our galaxy. Simulations suggest the existence of a universal dark matter density profile, valid for all masses and cosmological epochs. It is convenient to use the following parameterization for the dark matter halo density, which covers different approximations for dark matter density and is a condensed form of Eq. (5.101), easier to manipulate analytically:

$$\rho(r) = \frac{\rho_s}{\left(\frac{r}{r_s}\right)^{\gamma}\left[1+\left(\frac{r}{r_s}\right)^{\alpha}\right]^{\frac{\beta-\gamma}{\alpha}}}\,, \tag{5.102}$$

where ρ_s and r_s represent a characteristic density and a scale radius, respectively. The NFW density profile, with $(\alpha,\beta,\gamma) = (1,3,1)$, is by far the most widely used in the literature. Another approximation is the so-called Einasto profile

$$\rho_{\text{Ein}}(r) = \rho_s \exp\left\{-\frac{2}{\alpha}\left[\left(\frac{r}{r_s}\right)^{\alpha} - 1\right]\right\}\,, \tag{5.103}$$

which provides a better fit than NFW to numerical results. Finally, we will also consider dark matter density profiles that possess a core at the center, such as the purely phenomenologically motivated Burkert profile:

$$\rho_{\text{Burkert}}(r) = \frac{\rho_s\, r_s^3}{(r + r_s)\,(r^2 + r_s^2)}\,. \tag{5.104}$$

Early results on the central slopes of the dark matter profiles showed some significant disagreements between the estimates, with values ranging from $\gamma = 1.5$ to $\gamma = 1$. As the accuracy of the simulations improves, the disagreement became smaller. For the Via Lactea II (VLII) simulation, the slope was estimated to be $\gamma = 1.24$. A re-analysis of the VLII simulation and new simulations performed by the same group gives the slope $\gamma = 0.8 - 1.0$, which is consistent with the Aquarius simulation. Another improvement comes from the fact that the simulations now resolve the cusp down to a radius of ~ 100 pc, which means that less extrapolation is required for the density of the central region.

Yet, there is an additional ingredient that is expected to play a prominent role in the centers of dark matter halos: baryons. Although only a very small fraction of the total matter content in the Universe is due to baryons, they represent the dominant component at the very centers of galaxies like the Milky Way. Actually, the fact that current N-body simulations do not resolve the innermost regions of the halos is a minor consideration relative to the uncertainties due to the interplay between baryons and dark matter. As we pointed out, the effect of the baryonic adiabatic compression might be crucial for indirect dark matter searches, as it increases by several orders of magnitude the gamma-ray flux from dark matter annihilation in the inner regions, and therefore its detectability.

There is, however, another possible effect related to baryons that tends to *decrease* the dark matter density and flatten the central cusp The mechanism relies on numerous episodes of baryon infall followed by a strong burst of star formation, which expels the baryons. At the beginning of each episode the baryons dominate the gravitational potential. The dark matter contracts to respond to the changed potential. A sudden onset of star formation drives the baryons out. The dark matter also moves out because of the shallower potential. Each episode produces a relatively small effect on the dark matter, but a large number of them results in a significant decline of its density. Indeed, cosmological simulations that implement this process show a strong decline of the dark matter density. Whether the process happens in reality is still unclear. Simulations with the cycles of infall-burst-expansion process require that the gas during the burst stage does not lose energy through radiation, which is not realistic. Still, the strong energy release needed by the mechanism may be provided by other processes and the flattening of the central cusp may occur. If this happened to our galaxy, then the dark matter density within the central $\sim$ 500 pc may become constant. This would reduce the annihilation signal by orders of magnitude. We note that this mechanism would wipe out the dark matter cusp also in centers of dwarf galaxies. Yet, recent works that also include stellar feedback offer a much more complicated picture in which galaxies may retain or not their dark matter cusps depending on the ratio between their stellar-to-halo masses.

We show in Table 5.5 the values obtained after simulations for the different parameters entering in the profiles, as well as an illustration of the density profiles in Fig. 5.15.

The effect of baryonic adiabatic compression is clearly noticed in Table 5.5, where at small r as a steep power law $\rho \propto 1/r^{\gamma}$ with $\gamma = 1.37$ for NFW$_c$ [3], which is in contrast to the standard NFW value, $\gamma = 1$ [4].

A value of $\gamma = 1.37$ is indeed perfectly consistent with what has been found in recent hydrodynamic simulations, and it is also compatible with current observational constraints (mainly derived from microlensing and dynamics) on the slope of the DM density profile in the central regions of the Milky Way. Some studies actually allow for even steeper adiabatically contracted profiles. Finally, for the Burkert profile, we decided to choose a core radius of 2 kpc. This core size is in line with that suggested by recent hydrodynamic simulations of Milky Way size halos. For the normalization of the profile researchers choose the value of the local

Table 5.5 Dark matter density profiles, following the notation of Eqs. (5.102–5.104)

Profile	α	β	γ	ρ_s [GeV cm^{-3}]	r_s [kpc]
Burkert	–	–	–	37.76	2
Einasto	0.22	–	–	0.08	19.7
NFW	1	3	1	0.14	23.8
NFW$_c$	0.76	3.3	1.37	0.23	18.5

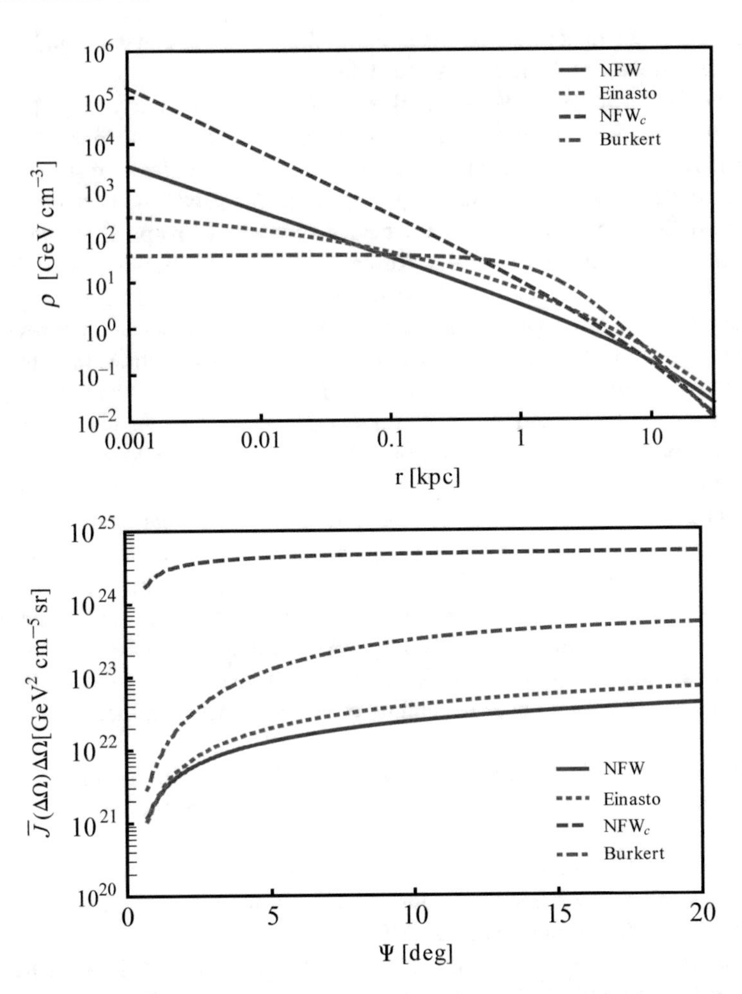

Fig. 5.15 Top panel: dark matter density profiles, with the parameters given in Table 5.5. Bottom panel: The $\bar{J}(\Delta\Omega)\Delta\Omega$ quantity integrated on a ring with an inner radius of 0.5° ($\sim$ 0.07 kpc) and an external radius of Ψ ($R_\odot \tan\Psi$) for the DM density profiles given in Table 5.5. Blue (solid), red (long-dashed), green (short-dashed) and yellow (dot-dashed) lines correspond to NFW, NFW$_c$, Einasto, and Burkert profiles, respectively. The four dark matter density profiles are compatible with current observational data

density suggested by analysis of local motions of stars, giving $\rho(r = 8.5 \text{ kpc}) \simeq 0.3$ GeV cm^{-3} for all the profiles. The resulting profiles are all compatible with current observational constraints. Note, however, that some recent work favors a substantially larger core radius and a slightly higher normalization for Burkert-like profiles. Let us finally point out that there are other possible effects driven by baryons that might steepen the dark matter density profiles in the centers of dark matter halos, such as central black holes, which we will not consider in this textbook.

5.9.3 Gamma-Ray Flux from Dark Matter Annihilation

The gamma-ray flux from dark matter annihilation in the galactic halo has two main contributions: prompt photons and photons induced via Inverse Compton scattering (see Sect. 5.6.2). The former are produced indirectly through hadronization, fragmentation, and decays of the annihilation products or by internal bremsstrahlung, or directly through one-loop processes (but these are typically suppressed in most dark matter models). The second contribution originates from electrons and positrons produced in dark matter annihilations, via Inverse Compton scattering of the ambient photon background. The other two possible contributions to the gamma-ray flux from dark matter annihilation can be neglected at the first approximation: radiation from bremsstrahlung is expected to be subdominant with respect to Inverse Compton scattering in the energy range considered (1–100 GeV) and a few degrees of the galactic plane (see Fig. 5.14), and synchrotron radiation is only relevant at radio frequencies, below the threshold of the main satellites or telescopes. Thus the gamma-ray differential flux from dark matter annihilation from a given observational region $\Delta\Omega$ in the galactic halo can be written as follows:

$$\frac{d\Phi_\gamma}{dE_\gamma}(E_\gamma, \Delta\Omega) = \left(\frac{d\Phi_\gamma}{dE_\gamma}\right)_{prompt} + \left(\frac{d\Phi_\gamma}{dE_\gamma}\right)_{ICS} . \tag{5.105}$$

We discuss in detail both components in the next subsections.

5.9.3.1 Prompt Gamma Rays

A continuous spectrum of gamma rays is produced mainly by the decays of π^0's generated in the cascading of annihilation products and also by internal bremsstrahlung. While the former process is completely determined for each given final state of annihilation ($b\bar{b}$, $\tau^+\tau^-$, $\mu^+\mu^-$, and W^+W^- channels), the latter depends in general on the details of the dark matter model such as the dark matter particle spin and the properties of the mediating particle. For the prompt contribution, we can directly take our Eq. (5.96)

$$\left(\frac{d\Phi_\gamma}{dE_\gamma}\right)_{prompt} = \sum_i \frac{dN_\gamma^i}{dE_\gamma} \frac{\langle\sigma_i v\rangle}{8\pi m_{DM}^2} \bar{J}(\Delta\Omega)\Delta\Omega . \tag{5.106}$$

This equation has to be multiplied by an additional factor of $1/2$ if the DM particle studied is its own antiparticle. But be careful, the Majorana case has a lot of subtleties as we explain in detail in Sect. B.2.2. The discrete sum is over all DM annihilation channels. As we underlined in the previous section, dN_γ^i/dE_γ is the differential gamma-ray yield, $\langle\sigma_i v\rangle$ is the annihilation cross section averaged over the velocity distribution of the dark matter, m_{DM} is the mass of the dark matter particle, and the quantity $\bar{J}(\Delta\Omega)$ (commonly known as the *J-factor*) was defined in Eq. (5.99):

$$\bar{J}(\Delta\Omega) \equiv \frac{1}{\Delta\Omega}\int d\Omega \int_{l.o.s.} \rho^2(r(l,\Psi))\, dl\,. \tag{5.107}$$

The J-factor accounts for both the dark matter distribution and the geometry of the system. The integral of the square of the dark matter density ρ^2 in the direction of observation Ψ is along the line of sight ($l.o.s$), and r and l represent the galactocentric distance and the distance to the Earth, respectively. Indeed, in Eq. (5.106), all the dependences on astrophysical parameters are encoded in the J-factor itself, whereas the rest of the terms encode the particle physics input.[13] The most crucial aspect in the calculation of $\bar{J}(\Delta\Omega)\Delta\Omega$ is related to the modeling of the DM distribution in the GC.

In the right panel of Fig. 5.15, the $\bar{J}(\Delta\Omega)\Delta\Omega$ quantity corresponding to each of the four profiles discussed in Sect. 5.9.2 is shown as a function of the angle Ψ from the GC. The associated observational regions $\Delta\Omega$ to each Ψ are taken around the GC. The angular integration is over a ring with an inner radius of 0.5° and an external radius of Ψ. We have assumed a $r = 0.1$ pc constant density core for both NFW and NFW$_c$, although the results are almost insensitive to any core size below $\sim$1 pc. Remarkably, the adiabatic compression increases the DM annihilation flux by several orders of magnitude in the inner regions. This effect can turn out to be especially relevant when deriving limits on the DM annihilation cross section. We also note that for the Burkert profile the value of $\bar{J}(\Delta\Omega)\Delta\Omega$ is larger than for the NFW and Einasto profiles. This is so because of the relative high normalization used for this profile compared to the others and, especially, due to the annular region around the GC where most studies are focused on, which excludes the GC itself (where such cored profiles would certainly give much less annihilation flux compared to cuspy profiles, see left panel of Fig. 5.15). We note, however, that the use of another Burkert-like profile with a larger DM core than the one used here may lead to substantially lower $\bar{J}(\Delta\Omega)\Delta\Omega$ values, and thus to weaker DM constraints. Notice finally that the NFW$_c$ profile reaches a constant value of $\bar{J}(\Delta\Omega)\Delta\Omega$ for a value of Ψ smaller than the other profiles. This is relevant for optimization of the region of interest for DM searches, since we see that for NFW$_c$ a larger region of analysis will not increase the DM flux significantly as for NFW, Einasto, and Burkert profiles. We show in Fig. 5.16 a typical region that is used to analyze data

[13]Strictly speaking, both terms are not completely independent of each other, as the minimum predicted mass for halos is set by the properties of the dark matter particle and is expected to play an important role also in the J-factor when substructures are taken into account. At a first approximation, we do not consider the effect of substructures on the annihilation flux, as large substructure boosts are only expected for the outskirts of dark matter halos, and thus they should have a very small impact on inner galaxy studies.

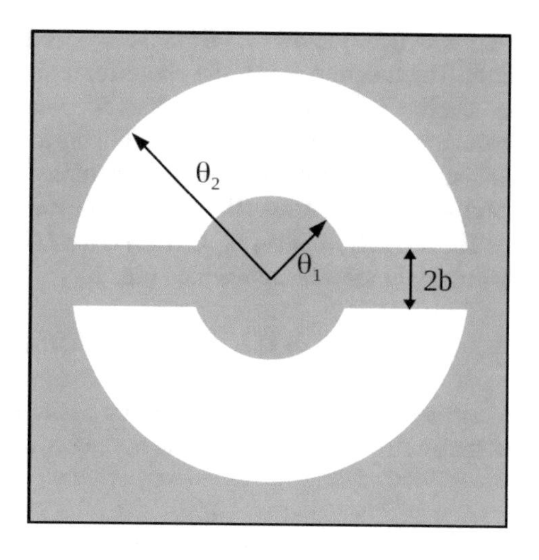

Fig. 5.16 Example of a region that we need to focus on around the galactic center (GC) to look for possible traces of dark matter annihilation

from the galactic center. Notice how we exclude the Milky Way bar and the pure galactic center to maximize the signal-to-noise ratio. The aim of indirect searches of dark matter is to see, in the white region, if any extra-photons can be observed, and then constrain the annihilation cross section (or even exclude some models) applying directly the theoretical prediction (5.106).

5.9.3.2 Gamma Rays from Inverse Compton Scattering

Electron and positron ($e^{\pm}$) fluxes are generated in dark matter annihilations mainly through the hadronization, fragmentation, and decays of the annihilation products, since direct production of e^+e^- is suppressed by small couplings in most dark matter models. These $e^{\pm}$ propagate in the galaxy and produce high energy gamma rays via ICS of the ambient photon background. The differential flux produced by ICS from a given observational region $\Delta\Omega$ in the galactic halo is given by

$$\frac{d\Phi_\gamma^{ICS}}{dE_\gamma} = \sum_i \frac{\langle\sigma_i v\rangle}{8\pi m_{DM}^2} \int_{m_e}^{m_{DM}} \frac{dE_I}{E_\gamma} \frac{dN_{e^\pm}^i}{dE_e}(E_I) \int d\Omega \, \frac{I_{IC}(E_\gamma, E_I; \Psi)}{E_\gamma} \,, \tag{5.108}$$

where E_I is the $e^{\pm}$ injection energy, Ψ corresponds to the angular position where the ICS gamma rays are produced, and the function $I_{IC}(E_\gamma, E_I; \Psi)$ is given by

$$I_{IC}(E_\gamma, E_I; \Psi) = 2E_\gamma \int_{l.o.s.} dl \int_{m_e}^{E_I} dE_e \, \frac{P_{IC}(E_\gamma, E_e; \mathbf{x})}{b_T(E_e; \mathbf{x})} \, \tilde{I}(E_e, E_I; \mathbf{x}) \,. \tag{5.109}$$

Here $\mathbf{x} = (l, \Psi)$ and $b_T \propto E^2$ is the energy loss rate of the electron in the Thomson limit. The function P_{IC} is the photon emission power for ICS, and it depends on the interstellar radiation (ISR) densities for each of the species composing the photon background. It is known that the ISR in the inner galactic region can be well modeled as a sum of separate black-body radiation components corresponding to star-light (SL), infrared radiation (IR), and cosmic microwave background (CMB).

The last ingredient in Eq. (5.109) is the $\tilde{I}(E_e, E_I; \mathbf{x})$ function, which can be given in terms of the well-known halo function,

$$I(E, E_I; \mathbf{x}) = \tilde{I}(E, E_I; \mathbf{x})[(b_T(E)/b(E, \mathbf{x}))(\rho(\mathbf{x})/\rho_\odot)^2]^{-1}, \tag{5.110}$$

where $\rho_\odot$ is the DM density at Sun's position and $b(E, \mathbf{x})$ encodes the energy loss of the $e^\pm$. The $\tilde{I}(E_e, E_I; \mathbf{x})$ function obeys the diffusion-loss equation

$$\nabla^2 \tilde{I}(E_e, E_I; \mathbf{x}) + \frac{E_e^2}{K(E_e; \mathbf{x})} \frac{\partial}{\partial E_e} \left[\frac{b(E_e; \mathbf{x})}{E_e^2} \tilde{I}(E_e, E_I; \mathbf{x}) \right] = 0 \,, \tag{5.111}$$

and is commonly solved by modeling the diffusion region as a cylinder with radius $R_{\max}$ =20 kpc, height z equal to $2L$, and vanishing boundary conditions. Also the diffusion coefficient $K(E; \mathbf{x})$ can be considered homogeneous inside the cylinder with an energy dependence following a power law $K(E) = K_0(E/1\mathrm{GeV})^\delta$. For these three parameters L, K_0, and δ, the so-called diffusion coefficients, one usually adopts three sets referred to as MIN, MED, and MAX models, which account for the degeneracy given by the local observations of the cosmic rays at the Earth including the boron to carbon ratio, B/C. The use of the different diffusion models, MIN, MED, MAX, does not introduce a large variation in the dark matter constraints.

Let us finally make some remarks concerning the importance of the energy loss function $b(E; \mathbf{x})$. We develop the phenomenon of energy loss in the section devoted to the synchrotron radiation (5.10) but we can say few words here. The two main energy loss mechanisms of $e^\pm$ in the galaxy are the ICS and synchrotron radiation produced by interaction with the galactic magnetic field. The former is the only contribution to the energy losses that is usually considered, since it is the most important one in studies of sources far from the GC. But when the $e^\pm$ energy reaches several hundreds of GeV, synchrotron radiation can dominate the energy loss rate due to the suppression factor in the ICS contribution in the Klein–Nishina regime. By contrast, synchrotron radiation losses do not have this suppression and are driven by the magnetic field energy density $u_B(\mathbf{x}) = B^2/2$. Although the strength and exact shape of the galactic magnetic field is not well known, in the literature it is broadly described by the from

$$B(r, z) = B_0 \exp\left(-\frac{r - 8.5\,\mathrm{kpc}}{10\,\mathrm{kpc}} - \frac{z}{2\,\mathrm{kpc}} \right) , \tag{5.112}$$

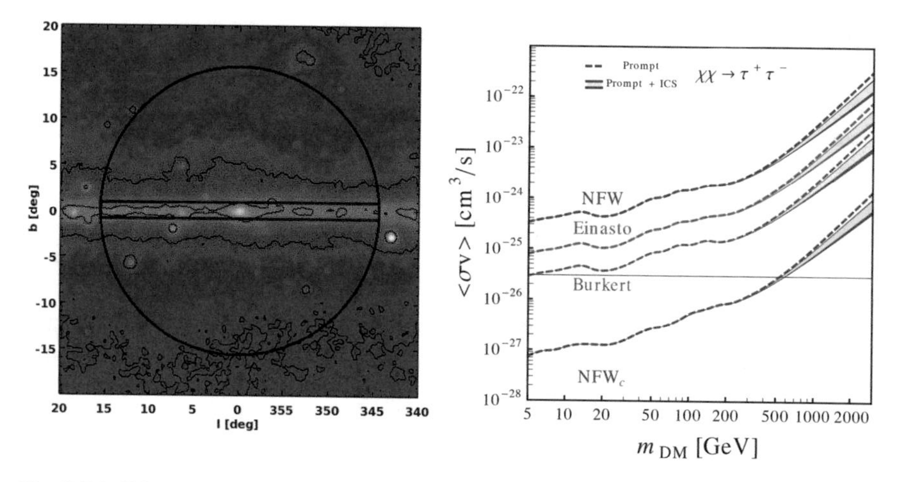

Fig. 5.17 Galactic plane as observed by the FERMI telescope, and our region of analysis (left); limit obtained on the annihilation cross section for different types of halo profiles (right) where we considered a $\tau^+\ \tau^-$ final state

normalized with the strength of the magnetic field around the solar system, B_0, which is known to be in the range of 1 to 10 μG. This field grows toward the GC, and therefore one should expect that the energy losses are dominated by synchrotron radiation in the inner part of the galaxy. On the other hand, we can expect that when the magnetic field is stronger, the energy of the injected $e^{\pm}$ is more efficiently liberated in the form of synchrotron emission resulting in a softer spectrum, and producing therefore smaller constraints on the dark matter annihilation cross section.

We present the result of a typical analysis in Fig. 5.17. On the left, we show the galactic plane as observed by the FERMI telescope. The brightest regions correspond to the highest fluxes of photons. We have superimposed on it the drawing of the region of interest (ROI) that was used for the analysis. We have excluded the galactic bar of the Milky Way. On the right side, the non-observation of gamma-ray excess over the background gives limit on the annihilation cross section of the dark matter (and, as a consequence, on the model) when comparing the data with Eq. (5.106). We find that cross sections above $\simeq 10^{-24}\text{cm}^3\text{s}^{-1}$ are excluded for a 100 GeV dark matter in the case of an NFW halo, whereas a compressed NFW halo excludes $\langle\sigma v\rangle \gtrsim 5\times 10^{-27}\text{cm}^3\text{s}^{-1}$ for the same mass. Notice that in the later case, thermal dark matter would be excluded because it predicts $\langle\sigma v\rangle \simeq 3\times 10^{-26}\text{cm}^3\text{s}^{-1}$ (3.160, and horizontal line in the figure). The final state was chosen to be $\tau^+\tau^-$ and results can slightly change for different final states, due to their different shapes in the spectrum. Indeed, some final states have tendency to radiate more photons than other ones and can then put stronger limits on the annihilation cross section $\langle\sigma v\rangle$. We also see that, in this specific final state, the ICS is clearly subdominant and begins to restrict the parameter space only for dark matter masses above the TeV scale.

5.10 Dark Matter and Synchrotron Radiation

As we discussed in great detail in Sect. 5.6.3, an electron moving in the magnetic field(s) generated inside our galaxy will lose energy by synchrotron (cyclotron in the non-relativistic case) radiation. We can then easily understand that, if dark matter particles annihilate into charged particles, the e^+e^- pair being the simplest of the possible final states, experiments can observe the effects of these charged particles that should produce synchrotron emissions. The task is far to be easy but reveals to be promising. For that, one needs to understand in detail the behavior of an electron in a magnetic field. Generically speaking, for an electron propagating in the interstellar medium, the diffusion-loss equation for its density N_e can be written as:

$$\frac{\partial}{\partial_t}\frac{\mathrm{d}N_e}{\mathrm{d}E_e} = \frac{\partial}{\partial_{E_e}}\left[b(E_e, \mathbf{r})\frac{\mathrm{d}N_e}{\mathrm{d}E_e}\right] + Q(E_e, \mathbf{r}), \tag{5.113}$$

where

$$Q(E_e, r) = \frac{1}{2}\left(\frac{\rho(r)}{m_\chi}\right)^2 \langle\sigma v\rangle \frac{\mathrm{d}N_e}{\mathrm{d}E_e} \tag{5.114}$$

is the electrons injected by dark matter annihilation at a constant rate (per second per GeV) and $b(E)$ represents the energy loss rate, i.e. the energy lost per electron in 1 s at the energy E. It is time to explain and analyze this equation step by step.

5.10.1 Neglecting Diffusion

The number of electrons with energies lower than E_e is given by

$$N_e(E_e) = \int_0^{E_e} \frac{dN_e}{dE_e} dE_e.$$

During 1 s, the electrons of energy $E(\gtrsim E_e)$ lose an amount of energy $\frac{dE}{dt}(E) = b(E)$. One can thus deduce that all electrons having an energy between E_e and $E_e + b(E_e)dt$ will have their energy *below* E_e after 1 s. This number is $\frac{dN_e}{dE} \times \frac{dE}{dt} = b(E_e)\frac{dN_e}{dE}$. In other words, $N(E_e) \rightarrow N(E_e) + b(E_e)dt\frac{dN_e}{dE_e}$ during the time dt. Per GeV and per second, the number of particles "lost" at energy E_e (*i.e.* having their energy dropped below E_e) is then $\partial_{E_e}\left[b(E_e)\frac{dN_e}{dE_e}\right]$, which represents the loss energy term in Eq. (5.113).

If we suppose that the transport timescale of the particle (which will fly away from its production point) is much longer than the cooling timescale responsible for the loss of energy, a steady state equation ("steady state" in the sense that we supposed that the galaxy has reached an equilibrium) can be written as:

Number of particles created between E_e and $E_{\max}$ + number of particles lost by cooling effects = 0,

or after integration of Eq. (5.113) and reminding that $dN_e/dE_e(\infty, \mathbf{r}) = 0$.

$$\int_{E_e}^{\infty} dE'_e Q(E'_e, \mathbf{r}) + \left[0 - b(E_e)\frac{dN_e}{dE_e}(E_e, \mathbf{r})\right]$$
$$= \int_{E_e}^{m_{\rm DM}} dE'_e Q(E'_e, \mathbf{r}) - b(E_e)\frac{dN_e}{dE_e}(E_e, \mathbf{r}) = 0$$

implying

$$\frac{dN_e}{dE_e}(E_e, \mathbf{r}) = \frac{\int_{E_e}^{m_{\rm DM}} dE'_e Q(E'_e, \mathbf{r})}{b(E_e)}. \tag{5.115}$$

$b(E)$ represents the loss energy rate and can have different origins: synchrotron radiation (discussed below), Inverse Compton, bremsstrahlung, or Coulomb. We will note them b_{syn}, b_{IC}, b_{brem}, and b_{com}, respectively: $b(E) = b_{syn}(E) + b_{IC}(E) + b_{brem}(E) + b_{com}(E)$.

5.10.2 Synchrotron Loss of Energy

As we just saw, among the indirect dark matter detection channels, gamma-ray emission represents one of the most promising opportunities due to the very low attenuation in the interstellar medium and to its high detection efficiency. Positrons and protons strongly interact with gas, radiation, and magnetic field in the galaxy and thus the expected signal sensibly depends on the assumed propagation model. However, during the process of thermalization in the galactic medium the high energies e^+ and e^- release secondary low energy radiation, especially in the radio and X-ray band, which, hence, can represent a chance to look for dark matter annihilation. A synchrotron emission is the relativistic limit of the cyclotron radiation given by the Larmor's formula (see Sect. 5.4.2 for more details concerning radiative loss of energy), which gives the total power radiated by a non-relativistic point charge as it accelerates (or decelerates):

$$P^{SI}_{Larmor} = \frac{e^2}{6\pi\epsilon_0 c^3}|\mathbf{a}|^2; \quad P^{CGS}_{Larmor} = \frac{2}{3}\frac{e^2}{c^3}|\mathbf{a}|^2, \tag{5.116}$$

where $\mathbf{a}$ stands for $\dot{\mathbf{v}}$ and SI or CGS (Centimeter–Gram–Second system) refers to the metric system (Longair versus Rybicki convention, see Eq. (5.31) for more details). We will give the results in both conventions as different authors use freely different types of conventions through their works. One can express the Larmor's formula as a function of the B-field after applying the equation of motion

$$m\frac{d\mathbf{v}}{dt} = e\mathbf{v}\times\mathbf{B} \;\Rightarrow\; \mathbf{v}_{\parallel} = \text{cste};\;\; \dot{\mathbf{v}}_{\perp} = \mathbf{a}_{\perp} = \frac{ev_{\perp}B}{m} \qquad [\text{SI}]$$
$$m\frac{\mathbf{v}}{dt} = \frac{e}{c}\mathbf{v}\times\mathbf{B} \;\Rightarrow\; \mathbf{v}_{\parallel} = \text{cste};\;\; \dot{\mathbf{v}}_{\perp} = \mathbf{a}_{\perp} = \frac{ev_{\perp}B}{mc} \qquad [\text{CGS}]. \quad (5.117)$$

Implementing Eq. (5.117) in Eq. (5.116) one obtains

$$P^{SI}_{Larmor} = \frac{e^4B^2v_{\perp}^2}{6\pi\epsilon_0 m^2c^3}; \quad P^{CGS}_{Larmor} = \frac{2}{3}\frac{e^4B^2v_{\perp}^2}{m^2c^5}. \quad (5.118)$$

The cyclotron Larmor's formula transforms into a synchrotron radiation for very relativistic electrons, with the replacement[14]

$$m \;\rightarrow \gamma m\;; \quad (|a_{\parallel}|^2 + |a_{\perp}|^2) \rightarrow \gamma^4(\gamma^2|a_{\parallel}|^2 + |a_{\perp}|^2). \quad (5.119)$$

Implementing in Eq. (5.118) one can write

$$P = \frac{e^4B^2\gamma^2v_{\perp}^2}{6\pi\epsilon_0 m^2c^3} = \sigma_T^{SI}\epsilon_0\gamma^2\beta_{\perp}^2B^2c^3 = 2\sigma_T^{SI}\gamma^2\beta_{\perp}^2c\left(\frac{B^2}{2\mu_0}\right) \qquad [\text{SI}]$$
$$\Rightarrow \langle P\rangle = \frac{4}{3}\sigma_T^{SI}c\gamma^2\beta^2U_B^{SI}, \quad (5.120)$$

where we used

$$\langle\beta_{\perp}^2\rangle = \beta^2\langle\sin^2\theta\rangle = \frac{1}{4\pi}\int\sin^2\theta d\phi d\cos\theta = \frac{2}{3}. \quad (5.121)$$

For the CGS system,

$$P = \frac{2e^4B^2\gamma^2v_{\perp}^2}{3m^2c^5} = \frac{1}{4\pi}\sigma_T^{CGS}c\gamma^2B^2v_{\perp}^2 = 2\sigma_T^{CGS}\gamma^2\beta_{\perp}^2c\left(\frac{B^2}{8\pi}\right) \qquad [\text{CGS}]$$
$$\Rightarrow \langle P\rangle = \frac{4}{3}\sigma_T^{CGS}c\gamma^2\beta^2U_B^{CGS}, \quad (5.122)$$

[14] Remember that in special relativity the Lorentz transformation on accelerations gives $a'_{\parallel} = \gamma^3 a_{\parallel}$ and $a'_{\perp} = \gamma^2 a_{\perp}$, see Eq. (5.63).

which at the end can be finally written universally as

$$\langle P \rangle = \frac{4}{3} \sigma_T^i c \gamma^2 \beta^2 U_B^i \tag{5.123}$$

with $\sigma_T^{SI} = \frac{8\pi}{3} \left(\frac{e^2}{4\pi\epsilon_0 mc^2} \right)^2$, $\sigma_T^{CGS} = \frac{8\pi}{3} \left(\frac{e^2}{mc^2} \right)^2$, $U_B^{SI} = B/2\mu_0$ and $U_B^{CGS} = B/8\pi$. σ_T is the Thompson cross section (see Sects. 3.3.4.2 and 5.4.4), equal to $0.665 \times 10^{-24}\text{cm}^2$ for electrons. In a third step, one needs to transform Eq. (5.120) as a function of the energy of the electron E_e. If it is considered as relativistic, this energy can be written as $E_e^2 = p^2c^2$, with $p = \gamma m v \simeq \gamma m c$, and one obtains $E_e^2 = \gamma^2 m^2 v^2 c^2 \simeq \gamma^2 m^2 c^4$, which gives us

$$P = \frac{e^4 B^2 E_\perp^2}{6\pi\epsilon_0 m^4 c^5} \Rightarrow \langle P \rangle = \frac{2}{3} \frac{e^4 B^2 E_e^2}{6\pi\epsilon_0 m^4 c^5} = P_0 \left(\frac{E_e}{\text{GeV}} \right)^2 \left(\frac{B}{\mu\text{G}} \right)^2 \tag{5.124}$$

with $P_0 = 2.51 \times 10^{-18}$ GeV s^{-1}. One can also write Eq. (5.124) with a more explicit dependence on the energy density of the magnetic field B:

$$\langle P \rangle = \frac{4}{3} \frac{e^4}{6\pi\epsilon_0^2} \frac{E_e^2}{m_e^4 c^7} \left(\frac{B^2}{2\mu_0} \right) = \frac{4}{3} \frac{e^4}{6\pi\epsilon_0^2} \frac{E_e^2}{m_e^4 c^7} U_B = \tilde{P}_0 \left(\frac{E_e}{\text{GeV}} \right)^2 \left(\frac{U_B}{\text{ev/cm}^3} \right) \tag{5.125}$$

with $\tilde{P}_0 = 1.02 \times 10^{-16}$ GeV s^{-1}. One can apply Eq. (5.125) to the Inverse Compton loss of energy process. Indeed, this process is generated by the interaction of an electron with a photon from the background of energy density U_{rad}. One can then write $b_{IC}(E_e) = \tilde{P}_0 \left(\frac{E_e}{\text{GeV}} \right)^2 \left(\frac{U_{rad}}{\text{ev/cm}^3} \right)$. The relevant radiation background for the Inverse Compton scattering is given by an extragalactic uniform contribution consisting of the CMB, with $U_{CMB} = 8\pi^5 (kT)^4 / 15(hc)^3 \simeq 0.26$ eV/cm^3, the optical/infrared extragalactic background, and the analogous spatially varying galactic contribution, the Interstellar Radiation Field (ISRF). The ISRF intensity near the solar position is about 5 eV/cm^3 and reaches values as large as 50 eV/cm^3 in the inner kpc's. Combining the synchrotron and Inverse Compton loss of energy processes, and forgetting in the first approximation the bremsstrahlung and Compton effects, one obtains

$$b(E) = b_{syn}(E) + b_{IC}(E) = \tag{5.126}$$

$$\left[2.51 \times 10^{-18} \left(\frac{E}{\text{GeV}} \right)^2 \left(\frac{B}{\mu\text{G}} \right)^2 + 1.02 \times 10^{-16} \left(\frac{E}{\text{GeV}} \right)^2 \left(\frac{U_{rad}}{\text{eV/cm}^3} \right) \right] \text{GeV s}^{-1}.$$

One can see from Eq. (5.126) that the Inverse Compton scattering dominates the loss energy process up to a magnetic field $B \lesssim 50\mu G$.

To compare our prediction with the experiment, one needs to convert the power spectrum, which is a function of the energy of the electron, into a spectrum in frequencies ν. We do so considering that the total energy emitted between ν and $\nu + d\nu$ is equal to the energy emitted between E and $E + dE$: $P(\nu)d\nu = n(E,r)P(E)dE$, or $P(\nu) = n(E,r)P(E)dE/d\nu$. $P(\nu)$ is called the emissivity and $n(E,r) = dN/dE$ is the density number of particles with energy comprised between E and $E + dE$. We thus see that we need to know the "conversion" factor $dE/d\nu$ to compute the emissivity; in other words, we need to express the frequency of the synchrotron radiation as a function of the electron energy. In fact, the synchrotron spectrum of an electron gyrating in a magnetic field has its prominent peak at the resonance frequency ν_c

$$\nu_c = \frac{3}{2}\gamma^2\nu_L = \frac{3eE^2B}{4\pi m_e^3c^4} = \nu_0\left(\frac{B}{\mu G}\right)\left(\frac{E}{\text{GeV}}\right)^2 \Rightarrow E(\nu) \simeq \sqrt{\frac{\nu}{\nu_0}}\left(\frac{B}{\mu G}\right)^{-1/2} \text{GeV}, \tag{5.127}$$

where ν_L is the Larmor's frequency (5.44) and $\nu_0 = 16.02 \times 10^6$ Hz.

Exercise Derive the expression above, taking the relativistic limit of the cyclotron frequency Eq. (5.44) and applying a mean on the solid angle.

This implies that, in practice, a $\delta-$approximation around the peaks $\nu \simeq \nu_c$ works extremely well. To compute the total fluxes $\mathcal{F}_\nu$ (Watt per meter square per second per Hertz), one needs to integrate the expression (5.115) along the line of sight (los) and the solid angle $d\Omega$:

$$\mathcal{F}_\nu = P(E) \times \frac{dE}{d\nu} \times \frac{1}{4\pi}\int d\Omega \int_{los} ds\, \frac{dN}{dE}(r). \tag{5.128}$$

Combining Eqs. (5.115), (5.124), (5.127), and (5.128) one obtains for the flux, expressed in Jansky (10^{-26} Wm^{-2}Hz^{-1}):

$$\begin{aligned}\mathcal{F}_\nu = 2.92 \times 10^7\text{Jy} \times &\left(\frac{1}{1 + \frac{\tilde{P}_0}{P_0}\left(\frac{B}{\mu G}\right)^2}\right) \times \left(\frac{\nu}{\text{GHz}}\right)^{-1/2} \times \left(\frac{B}{\mu G}\right)^{-1/2} \\ \times &\int_{E_\nu}^{m_{DM}} \frac{dN_{\text{source}}}{dE} dE \times \left(\frac{\sigma v}{3.10^{-26}\text{cm}^3\text{s}^{-1}}\right) \times \left(\frac{m_{DM}}{\text{GeV}}\right)^{-2} \\ \times &\left(\frac{\rho_0}{\text{GeV/cm}^3}\right)^2 \times \frac{1}{4\pi}\int d\Omega \int \frac{ds(r,\theta)}{\text{kpc}} f^2(r),\end{aligned} \tag{5.129}$$

where we have defined $\rho(r) = \rho_0 \times f(r)$, $\rho_0 = 0.3$ GeV cm^{-3} being the local dark matter density. The integral on the solid angle and line of sight can be developed using

$$\frac{1}{4\pi}\int d\Omega \int ds(r,\theta) = \frac{1}{2}\int_0^{\theta} \sin\theta' d\theta' \int_0^{D(1-\sin\theta')} dl, \tag{5.130}$$

where $D = 8.5$ kpc is the distance from the observer (on Earth) to the galactic center. The angle θ is the one formed between the direction of D and the line of sight, and l is related to r through the relation $r = \sqrt{D^2 + l^2 - 2Dl\cos\theta}$.

Concerning the loss of energy, we only considered the synchrotron and Inverse Compton process. If one takes into account Coulomb losses and Bremsstrahlung, one would obtain

$$b(E) = P_0 \left(\frac{E}{\text{GeV}}\right)^2 \left(\frac{B}{\mu\text{G}}\right)^2 + \tilde{P}_0 \left(\frac{E}{\text{GeV}}\right)^2 \tag{5.131}$$

$$+ b_0^{Coul} n(1 + \log(\gamma/n)/75) + b_0^{Brem} n(0.36 + \log(\gamma/n)), \tag{5.132}$$

where n is the mean number density of electrons in cm^{-3}, and the average over space gives about $n \simeq 1.3 \times 10^{-3}$, $\gamma = E/m_e$ and $b_0^{Coul} \simeq 6.13 \times 10^{-16}$, $b_0^{Brem} \simeq 1.51 \times 10^{-16}$ in GeVs^{-1} units.

5.10.3 Taking into Account Spatial Diffusion**

If we take into account the spatial diffusion, Eq. (5.113) becomes

$$\frac{\partial n(x,E)}{\partial t} = \nabla \cdot [K(x,E)\, \nabla n(x,E)] + \frac{\partial}{\partial E}[b(x,E) n(x,E)] = q(x,E)\,, \tag{5.133}$$

where $b(x, E)$ encodes the energy loss rate, $n(x, E) = dN/dE$ is the number density of electron per energy, and $q(x, E)$ is the source term. The problem is more complex by the dependence of the processes on the position x. As we discussed above, cosmic ray electrons lose energy mainly through synchrotron radiation and Inverse Compton scattering (IC), with a rate $b(x, E)$ that at the galactic medium is typically of the order 10^{-16} GeV s^{-1}. Additional bremsstrahlung losses of electron energies in the interstellar medium are neglected in our approach, although they may have a very small effect, since for the synchrotron frequency we work with, the electron energies of interest are around 1 GeV.

Assuming a steady state, Eq. (5.133) can be re-expressed as

$$\frac{\partial \tilde{n}(x,E)}{\partial \tilde{t}} - K_0\, \Delta \tilde{n}(x,E) = \tilde{q}(x,E) \tag{5.134}$$

with $K(x, E) = K_0 E^\delta$ and in which the derivative with respect to energy has been parameterized in terms of the parameter $\tilde{t} \equiv -\int dE(E^\delta/b(x, E))$. If at the level of propagation one considers that energy losses have an average value over all the diffusion regions, i.e. $b(x, E) \approx b(E) = E^2/\tau$ GeV s^{-1}, as in Eq. (5.126), then in (5.134) $\tilde{n}(x, E) = E^2 n(x, E)$ and $\tilde{q}(x, E) = E^{2-\delta} q(x, E)$. The solution of this equation can be found in the Green function formalism to be

$$n(x, E) = \frac{1}{b(E)} \int_E^\infty dE_s \times \int_{\mathrm{DZ}} d^3x G(x, E \leftarrow x_s, E_s) q(x_s, E_s), \tag{5.135}$$

where the volume integral is over the diffusion zone (DZ). The Green function $G(x, E \leftarrow x_s, E_s)$ gives the probability for an electron injected at x_s with energy E_s to reach x with energy $E < E_s$ and has a general solution of the form

$$G(x, E \leftarrow x_s, E_s) = \frac{\tau}{E^2} G(x, \tilde{t} \leftarrow x_s, \tilde{t}_s) \tag{5.136}$$

$$G(x, \tilde{t} \leftarrow x_s, \tilde{t}_s) = \left(\frac{1}{4\pi K_0 \Delta\tilde{t}} \right)^{3/2} e^{-\frac{(\Delta x)^2}{4K_0 \Delta\tilde{t}}}, \tag{5.137}$$

where $\Delta\tilde{t} = \tilde{t} - \tilde{t}_s$ and $(\Delta x)^2 = (\mathbf{x} - \mathbf{x}_s)^2$. The unique argument of the Green function is actually the diffusion length $\lambda = 4K_0\Delta\tilde{t}$ because the energy dependence enters only in this combination. This is the characteristic length of an electron traveling during its propagation. Here we are going to focus on diffusion models for which the half-thickness $L_h \sim 4$ kpc is small compared to the radius of the disk R_h, such that in practice the radial boundary has negligible effect on propagation. The Green function for which $n(x, E)$ vanishes at $z = \pm L_h$ may be expressed as

$$\tilde{G}(\lambda, L) = \frac{1}{(\sqrt{\pi}\lambda)^3} \sum_{n=-\infty}^{n=\infty} (-1)^n \exp^{-(2nL+(-1)^n z_s)^2/\lambda^2}. \tag{5.138}$$

As for the source term, in the case of production from DM annihilation with cross section $\langle \sigma v \rangle$, it can be expressed as

$$q(x, E) = \eta \, \langle \sigma v \rangle \left\{ \frac{\rho(x)}{m_{DM}} \right\}^2 \frac{dN}{dE}. \tag{5.139}$$

Here $\rho(x)$ is the DM profile, given in units of GeV/cm^3, and $\eta = 1/4$ or $\eta = 1/2$ depending on the Dirac or Majorana nature of DM. The injection spectrum of electrons is given by dN/dE.

With all these ingredients we can express the electron number density as

$$n(x, E) = \eta \langle \sigma v \rangle \left(\frac{\rho_{\odot}}{m_{DM}} \right)^2 \frac{1}{b(E)} \int_E^{m_{DM}} dE_s \frac{dN}{dE_s} I_{\text{halo}} . \tag{5.140}$$

Here $I(\lambda)$ is the so-called halo function, which is defined as

$$I_{\text{halo}} = \int_{\text{DZ}} d^3x_s \tilde{G}(x, E \leftarrow x_s, E_s) \left\{ \frac{\rho(x_s)}{\rho_{\odot}} \right\}^2 . \tag{5.141}$$

When electrons (and positrons) are created in the galaxy, they are accelerated by the local magnetic field and produce synchrotron radiation with an energy flux per unit frequency ν per solid angle (or spectral energy distribution) given by

$$F_\nu = \frac{1}{4\pi} \int_{\text{los}} dl \int_{m_e}^{m_{DM}} dE \; P(x, \nu, E)\, n(x, E), \tag{5.142}$$

where the integration is performed along the line of sight (los) and on the electron energies. In this relation $P(x, \nu, E)$ is the synchrotron power (per unit frequency) emitted at ν by an electron of energy E that, for an energy distribution of electrons $n(x, E)$, must be integrated over all the electron energies E that lead to synchrotron radiation at the same frequency ν. In practice we saw that the radiation power as a function of ν is strongly peaked near a so-called critical frequency ν_c, defined as

$$\nu_c = \frac{3eBE^2}{4\pi m_e^3} \equiv 16\text{MHz} \left(\frac{E}{\text{GeV}} \right)^2 \left(\frac{B}{\mu\text{G}} \right) \tag{5.143}$$

in natural units, so that there is a near one-to-one correspondence between ν and E that, we take to be such that

$$P(\nu, E) \approx \frac{4}{27} \frac{e^3 B}{m_e} \delta \left(\frac{\nu}{\nu_c(E)} - 0.29 \right) . \tag{5.144}$$

Indeed, the peak of the synchrotron radiation, when numerically solved, is not exactly at $\nu = \nu_c$ but more precisely at $\nu = 0.29\, \nu_c$.

Using this handy approximation, which, in one form or another, is often adopted in the literature, the flux of Eq. (5.142) takes a simple form,

$$F_\nu = \frac{1}{4\pi} \int_{\text{los}} dl \frac{E}{2\nu} P_{\text{sync}}(E)\, n(x, E), \tag{5.145}$$

where

$$E \equiv E(\nu) \approx \left(\frac{\nu}{4.7\text{MHz}} \right)^{1/2} \left(\frac{\mu\text{G}}{B} \right)^{1/2} \text{GeV}, \tag{5.146}$$

which stems from $\nu = 0.29\,\nu_c(E)$, and where $P_{\rm sync}$ is the total synchrotron energy loss rate of an electron of energy E,[15]

$$P_{\rm sync}(E) \hat{=} \int d\nu P(\nu, E) = \frac{e^4 B^2 E^2}{9\pi m_e^4}$$
$$\simeq 2.5 \cdot 10^{-18} \left(\frac{E}{\rm GeV}\right)^2 \left(\frac{B}{\mu\rm G}\right)^2 \,\rm GeV\, s^{-1}. \tag{5.147}$$

Finally, we may express the synchrotron energy flux as

$$F_\nu = 1.21 \times 10^8 \frac{\rm Jy}{\rm sr} \left\{ \frac{\eta}{2} \left(\frac{\langle \sigma v \rangle}{3.1 \times 10^{-26} {\rm cm}^3/s} \right) \right. \tag{5.148}$$
$$\times \left(\frac{1\rm GeV}{m_{DM}}\right)^2 \left(\frac{\rho_\odot}{\rm GeV/cm^3}\right)^2 \left(\frac{\mu G}{B}\right)^{1/2}$$
$$\times \left(\frac{\rm GHz}{\nu}\right)^{1/2} \frac{P_{\rm sync}(E)}{b(E)} \frac{1}{4\pi} \int \frac{dl}{\rm kpc} \int_E^{m_{DM}} dE_s \frac{dN}{dE_s} I_{\rm halo}\,,$$

using jansky flux units (Jy), 1 Jy = 10^{-26} W·m^{-2}·Hz^{-1}. The total energy loss assumed here is given by (see also Eq. (5.90))

$$b(E) = P_{\rm sync}(E)(1 + r_{\rm IC/sync}), \tag{5.149}$$

where

$$r_{\rm IC/sync} = \frac{2}{3}\frac{U_{\rm rad}}{B^2/2} \simeq 2 \left(\frac{U_{\rm rad}}{8\ \rm eV/cm^3}\right) \left(\frac{B}{10\ \mu\rm G}\right)^{-2} \tag{5.150}$$

is the ratio between IC and synchrotron energy loss and $U_{\rm rad}$ is the total radiation density.

5.10.4 General Astrophysical Setup*

5.10.4.1 Astrophysical Uncertainties

As we discussed when analyzing gamma ray from the galactic center in Sect. 5.9.2, studies coming from N-body simulations have led to popular expressions for the distribution of DM in the galactic halo, like the Navarro–Frenk–White (NFW)

[15]In these expressions, we are a bit loose regarding the definition of the magnetic field. In principle B should be the effective magnetic field felt by the electron, $B_\perp$. In practice, assuming an isotropic distribution of electron velocities, $\langle B_\perp^2 \rangle = 2/3 B^2$. This is explicit in Eq. (5.147). For convenience, this factor of 2/3 has been included in the definition of Eq. (5.144).

density profile

$$\rho_{\rm NFW}(r) = \frac{\rho_s}{\frac{r}{r_s}\left(1 + \frac{r}{r_s}\right)^2} . \tag{5.151}$$

That is a specific case of Eq. (5.102) or the Einasto profile (5.103)

$$\rho_{\rm EIN}(r) = \rho_s \exp\left\{-\frac{2}{\alpha}\left(\left(\frac{r}{r_s}\right)^\alpha - 1\right)\right\} , \tag{5.152}$$

where r is the radial distance from the center of the DM halo, and (ρ_s, r_s) are parameters that are fitted in the simulation to recover astronomical observables. On the other hand, observations of galactic rotation curves as well as some of the simulations that include also baryonic feedback on DM density find DM density profiles that are more cored toward the inner regions of the galaxy. One example of such profile is the isothermal profile

$$\rho_{\rm iso}(r) = \rho_s \frac{r_s^2}{r^2 + r_s^2} \tag{5.153}$$

or modified Einasto profile (in this case the parameter α is smaller when compared to the parameters found in simulations that contain only DM component). While the parameter α for the Einasto profile is fixed from a fit to the simulations, the values of parameters ρ_s, a typical scale density, and r_s, a typical scale radius for the Milky Way, are determined from astrophysical observations (e.g. local stellar surface brightness, stellar rotational curves, total Milky Way mass within a given distance…). There exist other alternatives like Burkert-like profile of Eq. (5.104) that we will, however, not consider in this section. As we will see in Sect. 5.10.4.2, the ratio of the synchrotron signal calculated with DM density of these three profiles gets smaller at higher latitudes, as at those distances (closer to the solar position) DM density is better constrained. As the rotational curve measurements are poor at distances smaller than 2 kpc (or $\sim$ 10°) from the galactic center and those regions of the galaxy are baryon dominated, the chosen region of interest (ROI) for analysis usually spans $|b| \in (10 \pm 3)^\circ$ in galactic latitudes, and $|l| \lesssim 3^\circ$ in galactic longitudes.

Together with DM density profile, the cosmic ray propagation parameters pose one of the main uncertainties in prediction of the synchrotron signal. This corresponds, for instance, to different values of the width of the Milky Way where the diffusion takes place (L_h) or the spatial diffusion parameter K_0 in Eq. (5.134). Three typical models used in the literature are called MIN/MED/MAX and featured in Table 5.6. They are used to probe the uncertainty in cosmic ray propagation parameters. Originally, those parameters were derived to produce the maximal, median, and minimal anti-proton flux from dark matter, while being compatible with the cosmic ray secondary to primary B/C ratio measurement. Therefore, by

Table 5.6 Upper three rows: parameter sets derived using a semi-analytical approach, to lead to MIN/MED/MAX anti-proton fluxes at Earth from an exotic galactic component. Lower four rows: parameters consistent with a fit to cosmic ray data, and shown to reproduce the gamma-ray diffuse data well. Last row: plain diffusion model, shown to be consistent with the radio data at 22 MHz–94 GHz frequencies

Model	L_h [kpc]	K_0 [cm^2 s^{-1}]	δ	v_a [km s^{-1}]
MIN	1	4.8 10^{26}	0.85	0
MED	4	3.4 10^{27}	0.70	0
MAX	15	2.3 10^{28}	0.46	0
1a	4	6.6 10^{28}	0.26	34.2
1b	4	6.6 10^{28}	0.35	42.7
2a	10	1.2 10^{29}	0.3	39.2
2b	10	1.05 10^{29}	0.3	39.2
PD	4	3.4 10^{28}	0.5	0

construction, they do not necessarily capture the uncertainty in the electron fluxes in the inner galaxy, which is of interest here. In Sect. 5.10.4.3 we will comment in more detail on the impact of a choice of cosmic ray parameters, exploring additional sets consistent with the cosmic ray data and which were i) derived numerically and ii) shown to reproduce the observed whole sky gamma ray or radio emission, therefore probing more directly the signals in the inner galaxy.

The galactic magnetic field (GMF) is considered possibly the most important ingredient when dealing with synchrotron radiation. In the diffuse interstellar medium it has a large scale regular component as well as a small scale random part, both having a strength of order micro-Gauss. The best available constraints in determining the *large scale* GMF are Faraday rotation measures and polarized synchrotron radiation, while *random* component is deduced mainly based on the synchrotron emission. Several 3D models of the large scale magnetic field have been developed the recent years. It is common to use a simple parameterization for a *total* magnetic field as customary in the literature:

$$B(\rho, z) \propto \exp\left(-\frac{r - r_\odot}{R_m} - \frac{|z|}{L_m}\right) . \tag{5.154}$$

The parameters R_m and L_m should in principle depend on the diffusion model assumed (or *vice versa*), since the propagation in the galactic medium is intimately related with the magnetic field. We took $L_m = \delta \cdot L_h$ and $R_m = \delta \cdot R_h$ in Table 5.6. An actual extent of L_m and R_m does not play a critical role (the difference between a constant magnetic field and the exponential form defined above is $\lesssim 30\%$), as long as the field extends into the region of interest (i.e. $L_m \gtrsim 1$ kpc). The normalization at Sun's position $B_\odot$ is more or less well constrained and is evaluated to $B_\odot \simeq 6\,\mu$G, consistent with the present measurements. However, the value of the field in the inner galaxy is considerably less known and we can rewrite the normalization of (5.154) as

$$B_0 \equiv B_\odot[1 + K\ \Theta(R_{IG} - r)], \tag{5.155}$$

where $\Theta(R_{IG} - r)$ is the unit step function as a function of $r = \sqrt{\rho^2 + z^2}$. With this change, while leaving $B_\odot$ unchanged *locally* we allow for the magnetic field to have a higher effective normalization $B_\odot\ (1+K)$ in the inner galaxy (IG). A typical value of R_{IG} is $R_{IG} = 2$ kpc. Now that we have fixed the spatial dependence of the B-field, we will explore the impact of overall normalization $B_\odot\ (1 + K)$ in the inner galaxy, on synchrotron fluxes in Sect. 5.10.4.4.

Contributions to the energy losses for electrons are assumed to come only from synchrotron and IC processes in our semi-analytical approach, as commented above. In principle, the radiation density $U_{\rm rad}$ (see Eq. (5.150)) has a spatial profile, which affects the synchrotron flux estimations. However, in the semi-analytical estimations used here we take $U_{\rm rad}$ to be constant, which turns out to be a good approximation for any analysis, as is justified in the next section. A value of $U_{\rm rad} \simeq 8$ eV cm^{-3} is usually considered in the literature.[16] Concerning the frequency of observation we focus on the data taken at 45 MHz. Indeed, in a wide range of frequencies (22 to 1420 MHz) the change in the measured synchrotron flux is very small, while going to lower frequencies maximizes a synchrotron signal of low-mass WIMPs.

5.10.4.2 Synchrotron Signal for Different Choices of DM Density Profile

In the remainder of the text we will focus on two DM profiles, NFW (Eq. 5.151) and isothermal (ISO) (Eq. 5.153), using the following values of parameters, consistent with observations: $\rho_s = 0.31$ GeV cm^{-3}, $r_s = 21$ kpc, for an NFW profile, and $\rho_s = 1.53$ GeV cm^{-3} and $r_s = 5$ kpc for isothermal profile.[17] However, in this section we also show the prediction for the synchrotron signal in the case of a modified Einasto profile. In particular, this profile is modified to have a shallower inner slope than the usual Einasto profile found in DM-only simulations and it describes better results of simulations that include baryonic feedback (parameters we use are $\alpha = 0.11$, $r_s = 35.24$ kpc, $\rho_s = 0.041$ GeV cm^{-3}). As this profile is "bulkier" at distances 1 kpc from the GC, the DM signals are generally higher than those of NFW in that region.

[16] Note, however, that the conditions in the inner galaxy might be quite different from a simple CR propagation setup assumed here. In particular, observations of the bubble-like structures centered at the galactic center and extending to 50 deg in latitudes in gamma rays and WMAP haze at microwave frequencies, observed by the WMAP satellite and Planck satellite, witness of possibly more complicated configuration of the magnetic fields and CR propagation parameters in that region. While bubble-like structures appear subdominant with respect to the standard components of the diffuse emission, we caution that before their origin is understood, the actual structure of magnetic fields or the CR propagation conditions cannot be reliably modeled.

[17] Note that we make a conservative choice by choosing a rather extended core. Smaller values of r_s would result in fluxes more similar to those obtained with the NFW profile.

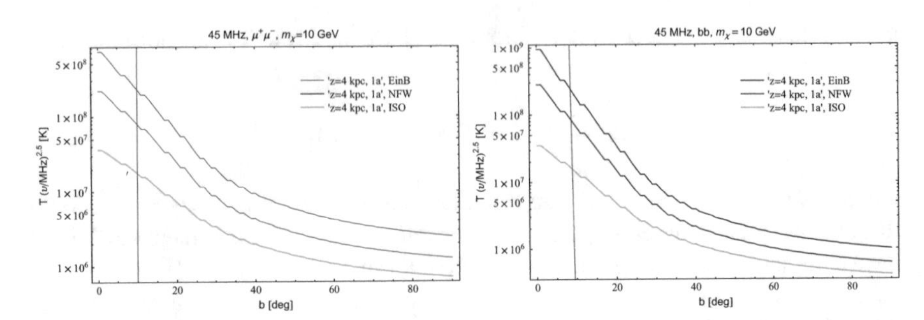

Fig. 5.18 Comparison between synchrotron signals for a DM mass of 10 GeV annihilating to muons (left) and b quarks (right) for three DM density profiles: modified Einasto, NFW, and ISO thermal profile. $\rho_\odot = 0.43$ GeV cm^{-3} is assumed for this plot (see text for the remaining parameters) and propagation of electrons is done using a CR propagation setup as shown in Table 5.6

The local value of DM density is set to $\rho_\odot = 0.43$ GeV cm^{-3}. One should also keep in mind that the overall normalization of DM distribution $\rho_\odot$ is uncertain, being in the $(0.43 \pm 0.113 \pm 0.096)$ GeV cm^{-3} range[18]. Therefore, in addition to the differences in the signal caused by the DM profile shape, and shown in Fig. 5.18, synchrotron signals scale with $\rho_\odot^2$. We usually express the flux F_ν in terms of the *brightness temperature* of the radiation, T, where $F_\nu = \frac{2h\nu^3}{c^2} \frac{1}{e^{\frac{h\nu}{kT}} - 1}$ is the usual black-body relation.

5.10.4.3 Synchrotron Signal for Different Choices of Cosmic Ray Parameters

As discussed above, MED/MIN/MAX sets of CR propagation parameters were derived using a semi-analytical description of CR propagation, and a fit to B/C measurement, with a requirement to produce minimal, medium, and maximal DM generated anti-proton fluxes, at a solar position. We show a comparison of synchrotron signal calculated with these choices of CR propagation parameters in Fig. 5.19. We see that the propagation parameters can generate one order of magnitude of uncertainties in the synchrotron flux produced by dark matter annihilation.

5.10.4.4 Synchrotron Signal for Different Choices of Magnetic Field**

In this section we study how dark matter limits change depending on the assumptions of the overall normalization of the magnetic field in our ROI. It has already been noticed that for a fixed electron injection spectrum there exists an optimum value for the magnetic field that maximizes the synchrotron flux at a given

[18]However, depending on the analysis, the uncertainty window on this value can vary in the 0.2–0.8 GeV cm^{-3} range.

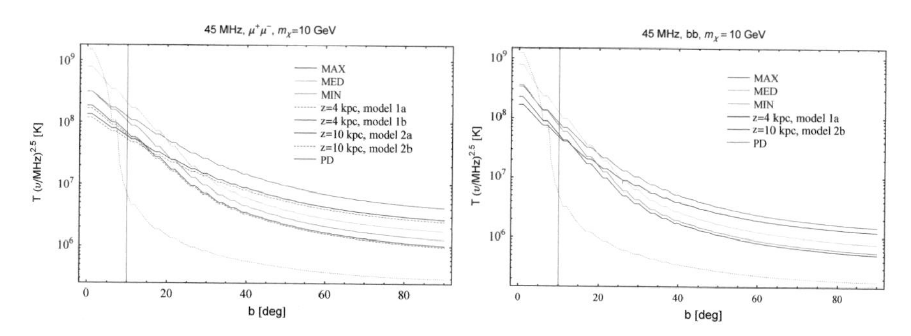

Fig. 5.19 Comparison between synchrotron signals for a DM mass of 10 GeV annihilating to muons (top) and b quarks (bottom figure) for different CR propagation setups, detailed in Table 5.6. NFW DM profile is assumed here

synchrotron frequency. However, in this section we want to understand this fact in more detail. As can be seen in Eq. (5.148), the magnetic field influences the flux through the energy losses, and through the electron energy. Indeed, given a specific frequency ν, and a given annihilation channel, it can be expressed in the following form:

$$F(\nu, B) \propto \left(\frac{B^2}{\alpha + B^2}\right) \frac{1}{\sqrt{B}}, \tag{5.156}$$

where α represents here the rest of energy losses, here assumed to be only IC. Note that since both synchrotron and IC losses scale with energy as E^2, the energy dependence cancels in this particular analysis. From (5.156) one observes two extremal cases: one in which synchrotron loss is negligible with respect to the rest of energy losses, for which the flux scales with B as $F \sim B^{3/2}$, thus increasing as B increases, and the other, in which synchrotron is actually the dominant energy loss, after which the flux scales as $F \sim 1/\sqrt{B}$, thus decreasing as B increases. In other words, there will be an intermediate value of the magnetic field for which synchrotron becomes the dominant energy loss, and this value is actually the one maximizing the flux.

Figure 5.20 shows the shape of synchrotron flux as a function of the value of magnetic field at GC. Taking into account only IC (apart from synchrotron of course), one can have an idea about the maximum of the flux already by direct differentiation of (5.156), assuming the values of α correspondent to this case. The value of B for which the flux is maximal scales as $B_{\rm GC}^{\rm max} \propto \sqrt{U_{\rm rad}}$. For $U_{\rm rad} = 8$ eV/cm^3, $B_{\rm GC}^{\rm max} \simeq 26\,\mu$G.

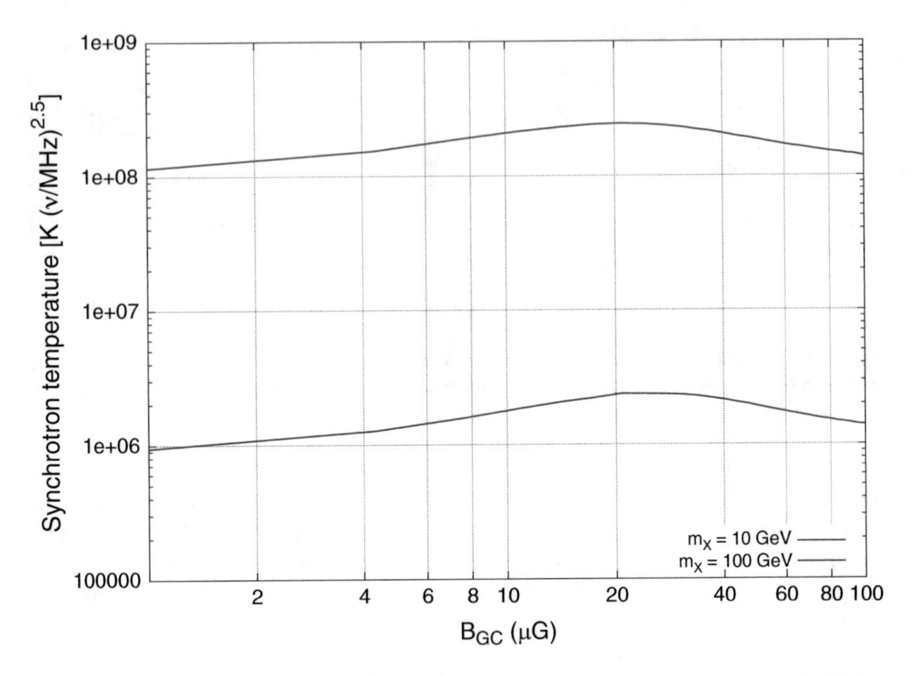

Fig. 5.20 Flux predicted at 10 deg off the GC, as a function of the value of magnetic field at the GC (roughly, at our ROI, we have $B_{\mathrm{ROI}} \approx 0.5 B_{\mathrm{GC}}$). Lines represent the results in the semi-analytical approach. We assume a DM annihilating directly to electrons, using the NFW profile and the MED diffusion model

5.11 Sommerfeld Enhancement

5.11.1 Generalities

The limit on the annihilation cross section we just computed, from γ-ray, anti-matter, or synchrotron radiation, can be largely modified by an effect called the "Sommerfeld enhancement." The Sommerfeld enhancement is an elementary effect in non-relativistic quantum mechanics, which accounts for the effect of a potential on the interaction cross section. This enhancement is named after Sommerfeld who proposed it in 1931. More technical details are developed in Sect. B.6. The application of this effect has been used extensively in the dark matter sector to explain (for instance) positron excesses observed by a satellite named PAMELA some years ago (and not confirmed by other experiments since).

We can find a gravitational classical analogy of this quantum effect. Indeed, if a meteor of mass m approaches the Earth of mass M with a velocity v, the lower is the velocity, the more chance it has to enter into collision with it as the gravitational attraction will compensate the kinetic energy of the meteor as we can see in Fig. 5.21. Without any force of gravity, the cross section is simply $\sigma_0 = \pi R$,

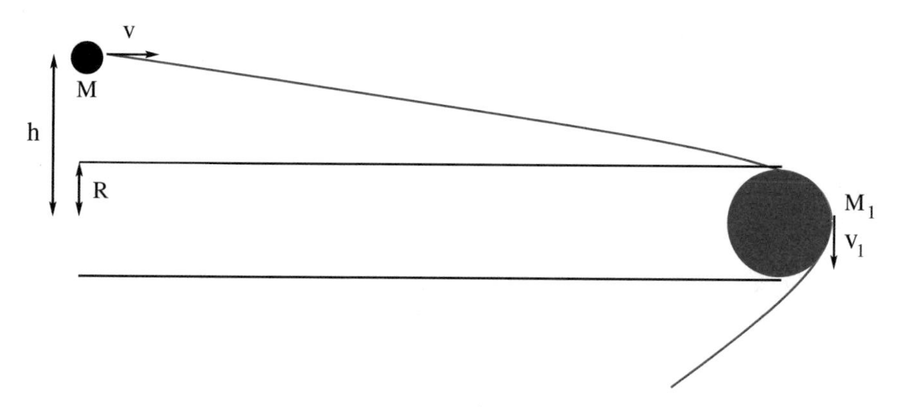

Fig. 5.21 Classical manifestation of the Sommerfeld enhancement in the case of a meteor hitting the Earth

R being the radius of the Earth. By conservation of angular momentum we can write $L_i = L_f$ with

$$L_i = |\mathbf{OM} \times m\mathbf{v}| = mhv, \quad |\mathbf{OM_1} \times m\mathbf{v_1}| = mRv_1 \quad \Rightarrow \quad v_1 = \frac{vh}{R} \tag{5.157}$$

where h is the impact parameter and R is the radius of the Earth (see the figure). The conservation of energy gives

$$\frac{1}{2}mv^2 = \frac{1}{2}mv_1^2 - \frac{G_N Mm}{R} \quad \Rightarrow \quad v_1^2 = v^2 + \frac{2G_N M}{R} \tag{5.158}$$

G_N being the Newton gravitational constant. Combining Eqs. (5.157) and (5.158) we obtain

$$h^2 = R^2\left(1 + \frac{2G_N M}{v^2 R}\right) \quad \Rightarrow \quad \sigma = \pi h^2 = \sigma_0\left(1 + \frac{2G_N M}{v^2 R}\right). \tag{5.159}$$

For very small values of v, there is a large enhancement of the cross section due to the gravity. Even if the correction vanishes as gravity is switched off ($G_N \to 0$), the expansion parameter is $2G_N M/(Rv^2)$, which can become large at small velocity.

5.11.2 Solving the Schrodinger's Equation

Before reading this paragraph, we advice the reader to read Sect. B.6 to remind some basic structures of the solutions of Schrodinger's equation. We are considering the solution to Schrodinger's equation for scattering of an incoming plane wave in the z-direction $e^{ikz} = e^{ikr\cos\theta}$ by a potential localized in a region near the origin, so that the total wave function beyond the range of the potential has the form[19]

$$\psi_k = e^{ikr\cos\theta} + f(\theta)\frac{e^{ikr}}{r}. \tag{5.160}$$

The overall normalization is of no concern, as we are only interested in the *fraction* of the ingoing wave that is scattered. Since the interaction or the annihilation is taking place locally, near $r = 0$, the only effect of a perturbation V to the Schrodinger's equation described in Sect. B.6 is to change the value of the modulus of the wave function at the origin relative to its unperturbed value. Then we can write

$$\sigma = \sigma_0 S_k, \tag{5.161}$$

where the Sommerfeld enhancement factor S is simply

$$S_k = \frac{|\psi_k(0)|^2}{|\psi_k^{(0)}(0)|^2} = |\psi_k(0)|^2. \tag{5.162}$$

One can wonder why using ingoing plane wave following the z-axis and not directly a spherical plane wave. Indeed, many potentials in nature are spherically symmetric, or nearly so, and from a theorist point of view it would be nice if the experimentalists could exploit this symmetry by arranging to send in spherical waves corresponding to different angular momenta rather than breaking the symmetry by choosing a particular direction. Unfortunately, this is difficult to arrange, and we must be satisfied with the remaining azimuthal symmetry of rotations about the ingoing beam direction. In fact, though, a full analysis of the outgoing scattered waves from an ingoing plane wave yields the same information as would spherical wave scattering. This is because a generic wave can actually be written as a *sum over spherical waves*:

$$\psi_k = \sum_l A_l \; P_l(\cos\theta) \; R_{kl}(r). \tag{5.163}$$

Note that the A_l coefficients are independent on ϕ as they correspond to the $m = 0$ solution of the $Y_{lm}(\theta, \phi)$ eigenvector of the $\hat{l}$ operator (see Sect. B.6 for more details). To determine the coefficients A_l, we use the fact that the combination $A_l \times P_l(\cos\theta)$ is independent of the radius r. Thus, we can compute them for large values of r where the influence of the potential $V(r)$ is negligible and $\psi_k \simeq e^{ikz}$ can be approximated by a plane wave. We will then look at the influence of $V(r)$ for small values of the radius r, keeping the values of A_l obtained for $r \to \infty$. We use the asymptotic expansion of e^{ikz} and identify it with the asymptotic Schrodinger

[19]The symmetry around the azimuthal axis imposes $f(\theta, \phi) = f(\theta)$.

solution for ψ_k with a null potential[20] for $R_{kl}(r) = R^0_{kl}(r)$ obtained[21] in Eq. (B.208) and using the fact that for large r, $\delta_l \ll kr$

$$e^{ikz} = \frac{1}{2ikr}\sum_l (2l+1)P_l(\cos\theta)[e^{ikr} - (-1)^l e^{-ikr}] \tag{5.164}$$

$$\simeq \sum_l A_l\, P_l(\cos\theta)\, \frac{1}{r}\sin\left(kr - l\frac{\pi}{2} + \delta_l\right) \Rightarrow \quad A_l = \frac{1}{k} i^l (2l+1) e^{-i\delta_l}.$$

It is now very simple to determine $\psi_k(0)$ and then the Sommerfeld factor (5.162). If the potential $V(r)$ does not blow faster than $1/r$ near $r \simeq 0$, then we can ignore it relative to the kinetic terms, and we have $R_{kl}(r) \propto r^l \rightarrow 0$ as we showed in Eq. (B.206). As a consequence, all but the term with a null momentum $l = 0$ vanish at the origin.

$$S_k = \left|\frac{R_{k,0}(0)}{k}\right|^2 \tag{5.165}$$

with $R_{k0}(r) = \chi_k(r)/r$ solution of Eq. (B.209) with $l = 0$

$$\frac{d^2\chi_k(r)}{dr^2} + \left[k^2 - 2MV(r)\right]\chi_k(r) = 0 \tag{5.166}$$

with the normalization condition $\chi_k(r) \rightarrow \sin(kr + \delta_0) \simeq \sin(kr)$ when $r \rightarrow \infty$, corresponding to the solution of a free particle obtained in Eq. (B.204) as $V(r)$ has no influence anymore when $r \rightarrow \infty$. As we noticed, $R_{k0} \rightarrow$ constant when $r \rightarrow 0$, one deduces $\chi_k(r) \rightarrow 0$ when $r \rightarrow 0$, implying $\chi_k(r) \simeq r\frac{d\chi_k(r)}{dr}$ when $r \simeq 0$. We then can express the Sommerfeld factor

$$S_k = \left|\frac{\frac{d\chi_k}{dr}(0)}{k}\right|^2 \tag{5.167}$$

with the boundary conditions $\chi(r) \simeq \sin kr$, $r \rightarrow \infty$ and $\chi(0) = 0$. We can check that for a null potential, $\chi^0_k = A\sin(kr)$, with $A = 1$ by boundary condition. Then $(\chi^0_k)' = k$, and $S_k = 1$ as we expected.

[20] Suppose the potential $V(r) \rightarrow 0$ when $r \rightarrow \infty$.

[21] Note the change or normalization by multiplying R^0_{kl} by the momentum k.

5.11.3 The Coulomb Potential

Let us begin to compute the Sommerfeld enhancement for the attractive Coulomb potential $V(r) = -\frac{\alpha}{r}$. Equation (5.166) can then be written as

$$\frac{d^2\chi}{dx^2} + \frac{v^2}{\alpha^2}\chi + \frac{2}{x}\chi - \frac{l(l+1)}{x^2}\chi = 0 \tag{5.168}$$

with $x = r \times \alpha M$. The solutions for this equation are usually expressed in terms of confluent hypergeometric functions. Its solution is the regular Coulomb Wave function that plays an important role in various problems in quantum mechanics and can be expressed, for $l = 0$:

$$\chi(x) = \frac{-2\pi\frac{\alpha}{v}}{(e^{-2\pi\frac{\alpha}{v}} - 1)}\frac{v}{\alpha}x\left(1 + \sum_i A_i x^i\right) \tag{5.169}$$

A_i being coefficients of the order of unity, which does not have influence for the Sommerfeld enhancement as $x,\ r \to 0$. Applying the solution of (5.169) to Eq. (5.167) we obtain

$$S_k = \left|\frac{2\pi\frac{\alpha}{v}}{1 - e^{-2\pi\frac{\alpha}{v}}}\right|^2. \tag{5.170}$$

Note that as $v \to \infty$, $S_k \to 1$, which means that the high velocity regime does not affect the interaction cross section. However, as $v \to 0$ $S_k \to 2\pi\frac{\alpha}{v}$ and what is named Sommerfeld enhancement. We show in Fig. 5.22 the evolution of the Sommerfeld enhancement in the case of the attractive Coulomb interaction for different values of the coupling α. We clearly see that the effect can enhance the interaction or annihilation cross section of orders of magnitude. We also notice that the enhancement is independent on the mass of the interacting particle.

5.11.4 The Yukawa Interaction

In the preceding Coulomb potential, generated by electromagnetic force, the particles exchanged are massless photons. In general, the mediating particles ϕ can be massive. This interaction is described by a Yukawa potential

$$V(r) = -\frac{\alpha}{r}e^{-m_\phi r}, \tag{5.171}$$

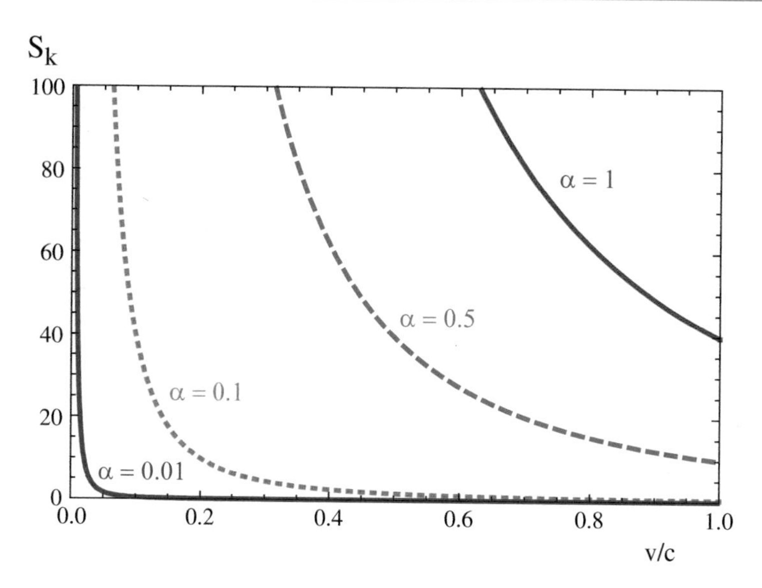

Fig. 5.22 Sommerfeld enhancement as a function of v/c in the case of an attractive Coulomb potential $V(r) = -\frac{\alpha}{r}$ for different values of α

where m_ϕ is the mass of the exchanged particle. The Yukawa potential is clearly a generalization of the Coulomb potential that we recover for $m_\phi = 0$. The Schrodinger equation then becomes

$$\frac{d^2\chi}{dx^2} + \frac{v^2}{\alpha^2}\chi + \frac{2e^{-\frac{m_\phi x}{\alpha M}}}{x}\chi - \frac{l(l+1)}{x^2}\chi = 0. \tag{5.172}$$

The above Eq. (5.172) does not possess analytical solution. A numerical solution exists and has been studied extensively in the literature. However, it is possible to give an approximate analytical expression if one approximates the Yukawa potential by the Hulthen potential:

$$V(r) \simeq V_H(r) = \frac{\alpha\delta e^{-\delta r}}{1 - e^{-\delta r}}. \tag{5.173}$$

One can find the best choice of δ to best reproduce the Yukawa potential,

$$\delta = \frac{\pi^2 m_\phi}{6}. \tag{5.174}$$

Exercise Recover analytically this value for δ.

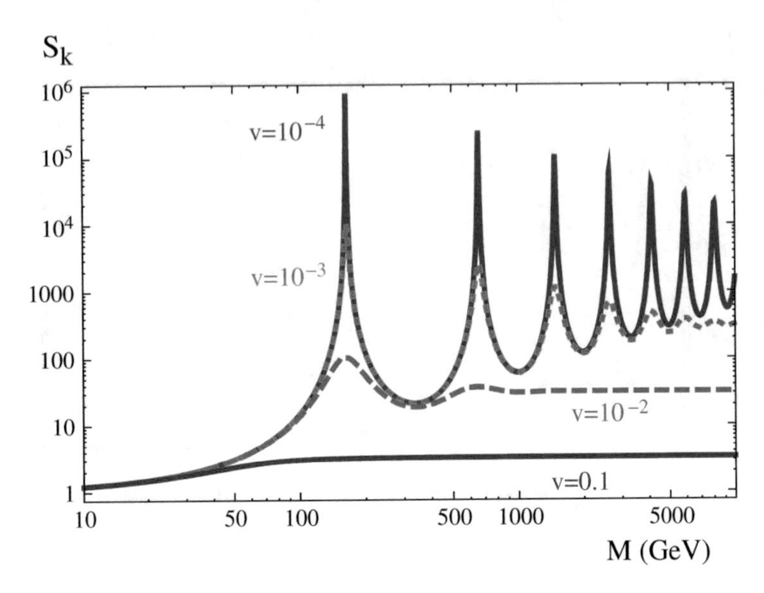

Fig. 5.23 Sommerfeld enhancement as a function of the mass of the dark matter M in the case of an attractive Yukawa potential $V(r) = -\frac{\alpha e^{-m_\phi r}}{r}$ for $\alpha = 0.1$ and $m_\phi = 10$ GeV for different values of v/c

Approximating also $\frac{l(l+1)}{Mr^2}$ by $\frac{l(l+1)}{M}\frac{\delta^2 e^{-\delta r}}{(1-e^{-\delta r})^2}$ we can derivate an analytical solution of Eq. (5.172):

$$S_k \simeq \left(\frac{\pi\alpha}{v}\right)\frac{\sinh\left(\frac{12vM}{\pi m_\phi}\right)}{\cosh\left(\frac{12vM}{\pi m_\phi}\right) - \cos\left(2\pi\sqrt{\frac{6\alpha M}{\pi^2 m_\phi} - \frac{36v^2M^2}{\pi^4 m_\phi^2}}\right)}. \qquad (5.175)$$

We want to underline that the obtention of such result is far from being straightforward and need a solid algebraic work. We propose to the reader to do it as an exercise. The advantage of having expressed Eq. (5.175) analytically is that its utilization is straightforward and avoids any computing consuming time. We can notice that we recover the Coulomb limit in the case $m_\phi \to 0$ (Fig. 5.23).

5.12 Structure Formation Constraints

Other constraints can be extracted from the observation of the sky, not only related to annihilation processes, as it is the case for the observation of γ-rays, anti-matter, or synchrotron radiation. For instance, the formation of large scale structures (LSS) is a fundamental keypoint in any strategy to unveil the nature of the dark matter. Indeed, like clouds of smokes in a room, energetic movements of the hands will destroy any formation of small structures; the dark matter, if too hot, will prevent the formation of the first structures. This constraint on the temperature of the dark matter is for instance one of the main arguments that excludes light neutrinos as the main component of the Universe. It is often said that "a dark matter mass below 1 keV is excluded by LSS," and we will recover this result in the following section, before looking in more detail the mechanism that leads to galactic scale structure and the influence of the dark matter on it.

5.12.1 Free Streaming

If dark matter is composed of collisionless particles (like neutrino for instance), after they are kinematically decoupled from the bath (see Sect. 3.2), they are subject to Landau damping , also known as collisionless phase mixing or free streaming. Until perturbations become Jeans unstable (see Sect. 5.12.2) and begin to grow after the decoupling time, collisionless particles can stream out of overdense regions and into underdense regions, smoothing out inhomogeneities. We will estimate the scale of collisionless damping in this section.[22]
Once a species decouples from the plasma, it simply travels in free fall in the expanding Universe. We thus may choose the particle motion to be along $d\phi = d\theta = 0$ so that the motion of a freely propagating particle is simply given by $a(t)dr = v(t)dt$. The distance λ_{FS} traversed by a free streaming particle at a time t can then be written

$$\lambda_{FS}(t) = \int dr = \int_0^t \frac{v(t')}{a(t')}dt'. \tag{5.176}$$

While the particles are relativistic, $v = c$, and later, $v \propto a^{-1}$ because the momentum is redshifted. We thus need to introduce a new scale, the scale factor a_{nr} where the dark matter particles become non-relativistic. It is defined by the condition $kT_{nr} = a_{nr}^{-1}kT_0$. Following Eq. (A.25) we have the link between temperature and scaling factor: $T \propto a^{-1}$. We thus can deduce the temperature when a particle χ becomes non- relativistic: $m_\chi \simeq T_{nr} \Rightarrow a_{nr} \simeq \frac{T_0}{m_\chi}$. In the case of the neutrino $\chi = \nu$, Eq. (3.89) gave us $m_\nu \approx \Omega_\nu h^2 \times 92$ eV implying

$$a_{nr} = \frac{T_0}{m_\nu} = \frac{2.35 \times 10^{-4}}{92}\left(\Omega_\nu h^2\right)^{-1} = 2.6 \times 10^{-6}\left(\Omega_\nu h^2\right)^{-1}. \tag{5.177}$$

[22] In order to take this effect into account properly, one must integrate the Boltzmann equation that describes the collisionless component.

We will then integrate the free streaming equation in the 3 regimes independently: t between 0 and t_{nr}, between t_{nr} and t_{EQ}, and finally between t_{EQ} and t_0:

- Between $t = 0$ and $t = t_{nr}$, the neutrino is relativistic and thus $v = c$ and the expansion factor is proportional to $t^{1/2}$ [Eq. (A.27)], $a(t) = \alpha\, t^{1/2}$

$$\lambda_{FS}^{0-nr}(t) = \int_0^t \frac{dt'}{a(t')} = \frac{2}{\alpha} t^{1/2} = 2\frac{a(t)t_{nr}}{a_{nr}^2} \qquad [0 < \mathrm{t} < \mathrm{t_{nr}}]. \tag{5.178}$$

- Between t_{nr} and t_{EQ}, the velocity of the particle $v(t)$ is proportional to $a^{-1}(t)$ by redshift, which means $v(t) = c\frac{a_{nr}}{a(t)} = \frac{a_{nr}}{a(t)}$, which gives

$$\lambda_{FS}^{nr-EQ}(t) - \lambda_{FS}^{nr-EQ}(t_{nr}) = \lambda_{FS}^{nr-EQ}(t) - \lambda_{FS}^{0-nr}(t_{nr}) \tag{5.179}$$

$$= \int_{t_{nr}}^t \frac{a_{nr} dt'}{a^2(t')} = \frac{t_{nr}}{a_{nr}} \ln\left(\frac{t}{t_{nr}}\right)$$

$$\Rightarrow \lambda_{FS}^{nr-EQ}(t) = 2\frac{t_{nr}}{a_{nr}} + 2\frac{t_{nr}}{a_{nr}} \ln\left(\frac{a(t)}{a_{nr}}\right)$$

$$= 2\frac{t_{nr}}{a_{nr}}\left[1 + \ln\left(\frac{a(t)}{a_{nr}}\right)\right] \qquad [\mathrm{t_{nr}} < \mathrm{t} < \mathrm{t_{EQ}}].$$

- Between $t = t_{EQ}$ and $t = t_0$, the dependence on the velocity is of course still redshifted, but the relation between time and scale factor from Eq. (A.27) is $a(t) = \alpha\, t^{2/3}$. Remarking $t_{nr} = t_{EQ}(a_{nr}/a_{EQ})^2$ we then obtain

$$\lambda_{FS}^{EQ-t_0}(t) = \lambda_{FS}^{nr-EQ}(t_{EQ}) + a_{nr} \int_{t_{EQ}}^t \frac{dt'}{a^2(t')} \tag{5.180}$$

$$= 2\frac{t_{nr}}{a_{nr}}\left[1 + \ln\left(\frac{a_{EQ}}{a_{nr}}\right)\right] + \frac{3t_{nr}}{a_{nr}}\left[1 - \left(\frac{a_{EQ}}{a(t)}\right)^{1/2}\right]$$

$$\Rightarrow \lambda_{FS}^{EQ-t_0}(t) = \frac{2t_{nr}}{a_{nr}}\left[\frac{5}{2} + \ln\left(\frac{a_{EQ}}{a_{nr}}\right) - \frac{3}{2}\left(\frac{a_{EQ}}{a(t)}\right)^{1/2}\right] \qquad [\mathrm{t_{EQ}} < \mathrm{t} < \mathrm{t_0}].$$

We can see that $\lambda_{FS}(t)$ quickly reaches and asymptotic value λ_{FS}^{∞}. With the values obtained in Eq. (3.52) and Eqs. (A.22–A.27) we can deduce t_{nr} and a_{EQ} and thus compute the asymptotic limit λ_{FS}^{∞}:

$$\lambda_{FS}^{\infty} = \frac{M_P}{T_0 T_{nr}} \sqrt{\frac{45}{4\pi^3 g_\rho}} \left[\frac{5}{2} + \ln\left(\frac{\rho_0^R T_{nr}}{\rho_0^M T_0}\right)\right] \approx 70\ \mathrm{Mpc}\frac{1\ \mathrm{eV}}{T_{nr}} \approx 210\ \mathrm{Mpc}\frac{1\ \mathrm{eV}}{m_\nu} \tag{5.181}$$

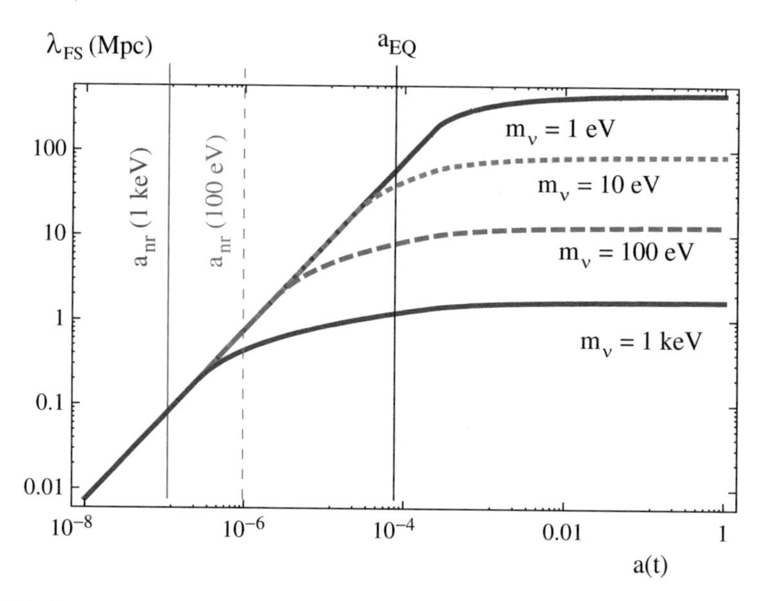

Fig. 5.24 Free streaming distance (in Mpc) as a function of the scaling parameter $a(t)$ for different masses of neutrino (from 1 eV to 1 keV)

if we suppose $T_{nr} \simeq m_\nu/3$ (an exact solution of the Boltzmann solution is needed here). It means that structures of smaller scales than 70 Mpc should have been destroyed by neutrino of masses $m_\nu < 1$ eV. If they were the main matter constituent of the density of the Universe. We plotted a numerical solution of λ_{FS} in Fig. 5.24. The limit $m_\nu = 1$ keV is known as the warm dark matter limit. Indeed, the free streaming of such particle is around 1 Mpc, which is the typical size of protogalaxies. Hot dark matter are particles still relativistic at the recombination time ($m_\nu \lesssim 10$ eV as we can see from Fig. 5.24). In fact, more generically speaking, candidate particles can be grouped into three categories on the basis of their effect on the fluctuation spectrum. If the dark matter is composed of abundant light particles that remain relativistic until shortly before recombination, then it may be called "hot." The best candidate for hot dark matter is a neutrino. A second possibility is for the dark matter particles to interact more weakly than neutrinos, to be less abundant, and to have a mass of order 1 keV. Such particles are dubbed "warm dark matter" because they have lower thermal velocities than massive neutrinos. There are few candidate particles that fit this description. Particles from supersymmetric frameworks like gravitinos and photinos or even sterile neutrinos have been suggested. Any particles that became non-relativistic very early, and so were able to diffuse a negligible distance, are termed "cold" dark matter (CDM) .

5.12.2 Jeans Radius and Mass

The structure formation through gravitational collapses is a highly non-trivial science, treated in numerous textbooks. We will try in this very short notice to give the main ideas and results following a (false but approximate) linear treatment. In two words, if one contracts a volume V of a gas of density ρ in equilibrium, an increase of pressure will counterbalance the contraction. However, if the sound speed in the gas c_s is not sufficient, the time for the information to cross the radius R ($t_s = \frac{R}{c_s}$) will be too long, and the gain of the gravitational potential generated by smaller distances between bodies can dominate before the pressure has time to act. In other words, if

$$t_s = \frac{R}{c_s} \gtrsim t_G = \frac{1}{\sqrt{G\rho}}, \tag{5.182}$$

where t_G is the gravitational free-fall characteristic time (obtained from the virial theorem $v^2 \sim R^2 G\rho$) the collapse occurs. We can then extract from Eq. (5.182) the Jeans radius R_J (Jeans mass M_J) corresponding to the maximal radius (mass) of a gas of constant density ρ that can keep its equilibrium state:

$$R_J = \frac{c_s}{\sqrt{G\rho}}; \quad M_J \sim R_J^3 \rho = \frac{c_s^2}{G^{\frac{3}{2}}\sqrt{\rho}}, \tag{5.183}$$

which gives

$$M_J = 2 \times 10^{12} M_\odot \left(\frac{c_s}{148}\right) \left(\frac{10^{-3}\ \text{GeV/cm}^3}{\rho}\right)^{\frac{1}{2}}, \tag{5.184}$$

where we used the sound velocity $c_s = 1480\,\text{km/s}\sqrt{\frac{T_g}{10^8\ \text{K}}}$, T_g being the temperature of the gas. Above M_J, the volume of gas begins to contract under gravity. These numbers ($T_g \simeq 10^6\ K$, $M = 2 \times 10^{12} M_\odot$) correspond roughly to the characteristics of the Milky Way, the mean mass density, ρ, being mainly from dark matter source, being less well measured. We then understand that a too small deviation from a mean density $\delta\rho$ will not induce a sufficiently efficient gravitational collapse to generate galactic- or cluster-sized objects.

The Speed of Sound

The speed of sound c_s appearing in Eq. (5.182) is a generalization of the distance travelled by a sound wave while propagating through an elastic medium. This fundamental notion appears in structure formations, as well as primordial baryonic oscillation (BAO) or even aspects of dark matter production. In an ideal gas, it depends only on its temperature and its composition. Sound wave is then generalized to any kind of fluid, dark one, or the primordial plasma. Consider the sound wave propagating at the velocity c_s in an elastic medium of density ρ. Per unit of time, the mass passing through a surface A along the direction z is

$$\dot{m} = \rho \times \frac{dz}{dt} \times A = \rho \times c_s \times A. \tag{5.185}$$

The flux being the same at the entrance and the exit of the volume, we should have

$$\frac{d}{dt}(\rho c_s) = 0 \quad \Rightarrow \quad c_s \frac{d\rho}{dt} = -\rho \frac{dc_s}{dt}. \tag{5.186}$$

From Newton second's law

$$\rho \times A \times dz \times \frac{dc_s}{dt} = -dP \times A, \tag{5.187}$$

where we introduced the pressure, as the force acting per unit of surface on A; combining Eqs. (5.186) and (5.187) we obtain

$$c_s^2 = \frac{dP}{d\rho}, \tag{5.188}$$

where we used $\frac{dz}{dt} = c_s$. Considering the equation of state $P = w\rho c^2$, we obtain $c_s = 0$ for a dust-type gas, and $c_s = \frac{c}{\sqrt{3}}$ for a relativistic gas. As one can see, the speed of sound in the plasma is *different* from the speed of light.

If we are in the presence of a plasma made of non-relativistic baryons and photons, noticing that $\frac{\delta\rho_b}{\rho_b} = 3\frac{\delta T}{T}$ and $\frac{\delta\rho_R}{\rho_R} = 4\frac{\delta T}{T}$, we can write

$$\frac{\delta\rho_b}{\delta\rho_R} = \frac{3}{4}\frac{\rho_b}{\rho_R} \quad \Rightarrow \quad \delta\rho = \delta\rho_R\left(1 + \frac{3}{4}\frac{\rho_b}{\rho_R}\right) \tag{5.189}$$

(continued)

and

$$c_s^2 = \frac{\delta P}{\delta \rho} = \frac{\delta P_R}{\delta \rho} = \frac{1}{3}\frac{\delta \rho_R}{\delta \rho} = \frac{c^2}{3\left(1 + \frac{3}{4}R_b\right)}. \tag{5.190}$$

Exercise Show that for a plasma made of an admixture of radiation (ρ_R) and non-relativistic baryons (ρ_b), $w = \frac{P}{\rho c^2} = \frac{1}{3(1+R_b)}$, with $R_b = \frac{\rho_b}{\rho_R}$. Decomposing $\frac{dP}{d\rho} = \frac{dP}{da}\frac{da}{d\rho}$, recover Eq. (5.190).

The influence of the baryon is then to "slow down" the photons in the plasma. In other words, the measurement of the horizon at CMB time gives a hint concerning R_b, the baryonic content of the plasma, see Sect. 3.7.2.4. Notice that we supposed an isolated system with constant entropy. If not, the real expression should be read

$$c_s^2 = \left(\frac{\partial P}{\partial \rho}\right)_S + \left(\frac{\partial P}{\partial S}\right)_P. \tag{5.191}$$

It is also interesting to ask what is the timescale needed for collapsing a structure of density ρ that generates itself a gravitational potential Φ. Using the Poisson law (5.23),

$$\Delta\Phi = 4\pi G\rho, \tag{5.192}$$

and the conservation of mass passing through a surface S, $\delta M \propto \rho S dr$:

$$\frac{d\delta M}{dt} \propto \frac{\partial \rho}{\partial t}\mathbf{S}.\mathbf{dr} + \mathbf{dv}.\mathbf{S}\,\rho = 0 \;\Rightarrow\; \frac{\partial \rho}{\partial t} = -\rho \times \nabla.\mathbf{v}. \tag{5.193}$$

If ρ_0 is the mean density, we can write $\rho = \rho_0(1 + \delta(t))$, and Eq. (5.193) becomes at first order in δ

$$\dot{\delta} = -\nabla.\mathbf{v} \;\Rightarrow\; \nabla.\dot{\mathbf{v}} = -\ddot{\delta} = -\Delta\Phi, \tag{5.194}$$

where, for the last expression, we used the Newtonian equation of motion

$$\dot{\mathbf{v}} = -\nabla\Phi, \tag{5.195}$$

and supposed to simplify a spherical symmetry. Solving Eq. (5.194) gives then

$$\ddot{\delta} = 4\pi\rho_0 G\delta \;\Rightarrow\; \delta = \delta_0 e^{\sqrt{4\pi\rho_0 G}t} \propto e^{t/t_G}, \tag{5.196}$$

where δ_0 is the original seed of the perturbation. The exponential growth is the characteristic of a Newtonian static Universe. A more complex analysis, taking into account the dilution effect due to an expanding Universe, was performed by Evgenii Lifshitz in 1946, showing that the growth is much softer and follows a power law:

$$\delta \propto t^{2/3}. \tag{5.197}$$

It was noticed very early after the discovering of the CMB that structure formation was in tension with the data. Indeed, if one supposes that the decoupling time happens at a redshift of $\sim$ 1100, and that the matter density nowadays is roughly $\rho_m^0 \simeq 2.5\times10^{-30}$ g cm^{-3}, the mass included in a volume $V = \frac{4\pi}{3}(R_H^{CMB})^3$, with $R_H^{CMB} = ct_{CMB}$ the horizon size at decoupling time, is:

$$M_H = \frac{4\pi}{3}(ct_{CMB})^3\rho_m^0(1+z_{CMB})^3 \tag{5.198}$$

with t_{CMB} =380,000 years and $z_{CMB} = 1100$; we obtain

$$M_H \simeq 3.3\times10^{17}M_\odot, \tag{5.199}$$

which is much larger than the galactic structures we observe nowadays. In a top-down scenario, where the galactic substructures would have been formed from fragmentation of huge gravitationally bound systems that would be acceptable. This scenario was the one preferred by the Soviet cosmologist Yakov Zeldovich in 1977. However, even the first serious simulations running in the 80s immediately revealed that in this case, the galaxy-mass objects form rather late, at a redshift $z \sim 2$, whereas we know that galaxies existed long before, from their direct observation at much higher redshift. The bottom-up scenario is then largely favored by the data, which necessitates the presence of structures below the galactic scale $\simeq 10^{12}M_\odot$ shortly after the decoupling epoch. The presence of a dark matter component can solve this problem.

5.12.3 The Influence of Dark Matter

As we discussed above, the large mass trapped inside the horizon size at the decoupling epoch would have generated too large structures. However, if we introduce a massive component m_{dm} that decoupled from the thermal bath at a temperature $T_{dm} \sim m_{dm}$ (corresponding to a time $t_{dm} \sim \frac{1}{H} \sim \frac{M_P}{T_{dm}^2}$), we deduce that the mass included in the dark matter horizon from the origin to t_{dm} is

$$M_{dm} \sim \frac{4\pi}{3}(ct_{dm})^3\rho(T_{dm}=m_{dm}) \sim 10^9 M_\odot\left(\frac{1\text{ keV}}{m_{dm}}\right)^2. \tag{5.200}$$

Dark matter masses above the keV scale can then generate structures of the size of dwarf galaxies. Notice how this simple argument is in agreement with the more complex free-streaming constraints of Fig. 5.24 or even the Tremaine–Gunn bound (3.98). However, the fact that inhomogeneous systems can be formed thanks to the presence of a dark component is far to be sufficient to explain the structure of the Universe observed today at the cluster scale.

When the physics community left the steady state paradigm (proposed by Bondi, Gold and Hoyle) to adopt the Big Bang option, it became obvious that the baryonic fraction η_b deduced from measurements in the galaxies and cluster of galaxies (see Fig. 1.5) is too small to be compatible with an Einstein–de Sitter model where $\Omega_m = 1$. That implies the need for dark "something" that could be dark matter, dark energy, or both. James Peebles, one of the first post-CMB cosmologists, noticed it already in 1965. But he was also one of the first to look at the influence of the dark matter in the structure formation. Indeed, whereas in the steady state paradigm, the Universe being here forever, the time needed to produce the first galaxies does not really matter. However, once Big Bang appears as the dominant paradigm, quickly James Peebles understood that the time to produce the first complex structures is quite short because it should begin after the recombination time, around 380,000 years after the Big Bang, when the baryons have decoupled from the thermal plasma and are allowed to collapse gravitationally. This was even more problematic if one considers a Universe only dominated by a baryonic component Ω_b, in which density is strongly limited by the measured light elements abundance (see Fig. 1.5).

What is the level of inhomogeneity in our present Universe? We observe that at scales below $R \lesssim 8\ h^{-1} \sim 12$ Mpc, roughly the cluster scale, the inhomogeneity began to be larger than the mean mass density, in other words,

$$\frac{\delta\rho}{\rho}(R \lesssim 12\ \text{Mpc}) \gtrsim 1. \tag{5.201}$$

From the Lifshitz relation (5.197), one deduces that

$$\left.\frac{\delta\rho}{\rho}\right|_z = \left.\frac{\delta\rho}{\rho}\right|_0 \frac{1}{1+z} \sim \frac{1}{1+z} \tag{5.202}$$

in a matter dominated Universe where $a \propto t^{2/3}$. We should then expect, at the decoupling time t_{CMB}, where $z_{CMB} = 1100$, an amplitude of fluctuation

$$\left.\frac{\delta\rho}{\rho}\right|_{CMB} \sim 10^{-3} \quad \Rightarrow \quad \left.\frac{\delta T}{T}\right|_{CMB} \simeq 2.5 \times 10^{-4}, \tag{5.203}$$

where we have supposed, in the last equality, that the Universe is *only* composed of baryons, and that, at decoupling time, $\frac{\delta\rho}{\rho} \sim \frac{\delta T^4}{T^4} \sim 4\frac{\delta T}{T}$. This level of fluctuations should appear at the decoupling time at a scale corresponding to $\sim$ 12 Mpc presently, which means $\sim$ 10 kpc at $z = 1100$.

Exercise Show that a present cluster scale $R = 12$ Mpc corresponds to an angle of the order of arcminute in the CMB map.

However, already in 1984, experimental limits on the CMB anisotropies were $\frac{\delta T}{T} \lesssim 10^{-4}$ on an angular scale corresponding to regions containing the masses of galaxies or clusters of galaxies at $z = 1100$. Nothing being observed, researchers like James Peebles or Joseph Silk proposed that a massive hypothetical particle could resolve this contradiction. Indeed, if a dark matter fluid has decoupled from the plasma at $z \sim 10{,}000$, the dark matter fluctuations have grown to $\frac{\delta\rho}{\rho} \sim 10^{-4}$ at the decoupling time, while the photon–baryon sound-wave fluctuations are stuck to 10^{-5}, consistent with observations.

One can then ask what is the value of $\frac{\delta T}{T}$ expected at the decoupling time if one supposes $\frac{\delta\rho}{\rho} = 1$ at a cluster scale of 8 h^{-1} Mpc. Locally, fluctuations produce a gravitational potential $\delta\Phi$ such as

$$\delta\Phi = \frac{G\delta M}{R} \sim G\delta\rho\, R^2 \sim \frac{R^2\rho}{M_P^2}\left(\frac{\delta\rho}{\rho}\right)_R \sim 3R^2H^2\left(\frac{\delta\rho}{\rho}\right)_R \sim 3\left(\frac{Rh}{3000\ \text{Mpc}}\right)^2\left(\frac{\delta\rho}{\rho}\right)_R .$$

The observation of a cutoff at $R \sim 8h^{-1}$Mpc implies

$$\delta\Phi \simeq 2.1 \times 10^{-5}\left(\frac{\delta\rho}{\rho}\right)_{8\ \text{Mpc}} = 2.1 \times 10^{-5}. \tag{5.204}$$

From (5.197), we see that $\delta\Phi \propto \delta\rho R^2$ is scale invariant for a matter dominated Universe. We then deduce

$$\delta\Phi|_{CMB} = 10^{-5}. \tag{5.205}$$

To find the relation between $\delta\Phi$ and $\frac{\delta T}{T}$ one should use the Sachs–Wolfe method . From General Relativity, we know that the invariant measure in the presence of a gravitational perturbation can be written in the Newtonian gauge, in the weak field limit,

$$ds^2 = (1 + 2\delta\Phi)dt^2 - R^2(t)(1 - 2\delta\Phi)d\chi^2, \tag{5.206}$$

where χ is the comoving coordinate (see Eq. 2.20). The effect of the potential is to redshift the energy of photons emitted from the regions where $\delta\Phi \neq 0$.

$$\frac{p'}{p} = \frac{\sqrt{1 - 2\delta\Phi}}{\sqrt{1 + 2\delta\Phi}} \simeq 1 - 2\delta\Phi, \tag{5.207}$$

where p' (p) is the momentum[23] with (without) $\delta\Phi$. This corresponds to a photon redshift

$$z = \frac{E' - E}{E} = \frac{\delta E}{E} = \frac{p' - p}{p} = \frac{\delta T}{T} = 2\delta\Phi \quad \Rightarrow \quad \left.\frac{\delta T}{T}\right|_{CMB} \sim 2 \times 10^{-5}. \tag{5.208}$$

We have seen that a present amplitude of mass fluctuation of the order $\frac{\delta\rho}{\rho} = 1$ at a scale $R = 8\ h^{-1}$ Mpc implies, at the decoupling time, a variation in the CMB temperature fluctuation of the order $\frac{\delta T}{T} \simeq 2 \times 10^{-5}$, which is effectively what has been observed by the satellites WMAP and PLANCK. That is one of the greatest achievements of the CDM model.

5.12.4 Correlation Function*

Two-point correlation function of galaxies is the simplest estimator of the galaxies distribution at large scale and defines the amplitude of fluctuations. The two-point correlation function quantifies the excess of probability to find a galaxy 2 at a distance r from a galaxy 1 selected randomly, compared to a mean uniform distribution. For instance, if the galaxies are distributed following a Poisson distribution with a mean density ρ_0, the probability $dN(r)$ to find a galaxy in a volume dV_2 at a distance r from the galaxy 1 in the volume dV_1 is the fraction of galaxies in $dV_1 \times$ the fraction of galaxies in dV_2:

$$dN(r) = \rho_0^2 dV_1 dV_2. \tag{5.209}$$

When we write *Poissonian* distribution, we mean random stationary process (identical mean properties whatever is the size or the place of the sample considered), i.e. the galaxies are placed by chance and independently one with respect to each other, with an homogeneous spatial probability. The spectrum of such a distribution is thus flat: this is the distribution of a "white noise." If the distribution of the galaxies differs slightly from a Poissonian distribution, the probability will be

$$dN(r) = \rho_0^2(1 + \zeta(r))dV_1 dV_2, \tag{5.210}$$

where $\zeta(r)$ is the two-point correlation function of the galaxies. We used r instead of $\mathbf{r}$ with the hypothesis of the isotropy of the distribution. $\zeta(r) > 1$ indicates a surdensity, $\zeta(r) = 0$ indicates a galaxy density equals to the mean density, and

[23] To see it in the non-relativistic limit, just consider $p' = mv'$, and $v'^2 = \frac{1-2\delta\Phi}{1+2\delta\Phi}v^2$.

$\zeta(r) < 0$ indicates an underdensity. The probability of finding a galaxy at a distance r from another one can thus be written with Eq. (5.209)

$$dN(x, r) = \rho(x)dV1\ \rho(x + r)dV_2. \tag{5.211}$$

We can then introduce the contrast of density δ_ρ

$$\delta_\rho(x) = \frac{\rho(x) - \rho_0}{\rho_0} \quad \Rightarrow \quad \rho(x) = \rho_0(1 + \delta_\rho(x))$$
$$\Rightarrow \quad dN(r) = \langle dN(x, r)\rangle = \rho_0^2(1 + \langle\delta_\rho(x)\delta_\rho(x + r)\rangle)dV_1 dV_2$$
$$= \rho_0^2(1 + \zeta(r))dV_1 dV_2$$

with $\zeta(r) = \langle\delta_\rho(x)\delta_\rho(x + r)\rangle$ is called the function of *autocorrelation of the matter distribution*.

5.12.5 Power Spectrum $P(k)$*

We can define a power spectrum function $P(k)$ by

$$\zeta(r) = \frac{V}{(2\pi)^3}\int e^{ik.r}P(k)d^3k. \tag{5.212}$$

$P(k)$ is then the Fourier transform of the two-point correlation function $\zeta(r)$. In other words, $P(k)$ represents the probability of having distribution of matter periodically distributed in the close Universe with a wave vector $k \simeq 1/\lambda$. In the isotrope case, Eq. (5.212) can be written as

$$\zeta(r) = \frac{V}{2\pi^2}\int_0^\infty P(k)\frac{\sin(kr)}{kr}k^2dk. \tag{5.213}$$

r represents a comoving scale (or wavelength) corresponding to the wave number $k = \frac{2\pi}{r}$. The function $\frac{\sin(kr)}{kr}$ acts like a windows letting only the modes with $k < \frac{2\pi}{r}$ to contribute to the fluctuation at the scale r. For these reasons, fluctuations with $k > \frac{2\pi}{r}$ (i.e. small scales compared to r) do not contribute at the scale r. As a conclusion the correlation function at a scale r will be sensible only to power spectrum corresponding to scales $> r$.
To have an idea, recent measurements with some simulation hypotheses give

$$\zeta(r) \simeq \left(\frac{r}{r_0}\right)^{-\gamma} \tag{5.214}$$

with $\gamma \simeq 1.77$ and $r_0 \simeq 5h^{-1}$ Mpc for scales from 0.1 to 10 h^{-1} Mpc. Higher is r_0 larger is $\zeta(r)$ and so the structure features. r_0 quantifies the clustering.

References

1. J.D. Jackson, *Classical Electrodynamics* (Wiley, New York). ISBN 978-0-471-30932-1. OCLC 925677836
2. M.S. Longair, *High Energy Astrophysics: Volume 1, Particles, Photons and their Detection* (Cambridge University Press, Cambridge, 2011) (ISBN 0521756189), 1992, 440pp., (ISBN 0521387736)
3. O.Y. Gnedin, A.V. Kravtsov, A.A. Klypin, D. Nagai, Astrophys. J. **616**, 16–26 (2004). https://doi.org/10.1086/424914. [arXiv:astro-ph/0406247 [astro-ph]]
4. J.F. Navarro, C.S. Frenk, S.D.M. White, Astrophys. J. **462**, 563–575 (1996). https://doi.org/10.1086/177173. [arXiv:astro-ph/9508025 [astro-ph]]

A Cosmology and Astrophysics

A.1 Useful Cosmology

A.1.1 Lorentz Transformation

In this section, I will review very briefly the Lorentz transformations, more from an historical perspective that one can read in classical textbook on the subject. I urge the reader to have a look at the original Einstein's article of 1905 which is very clear and pedagogical. The first step made by Einstein is to consider a space of what he called himself "*events*." Let us suppose the *event* as being a ray of light leaving one observer in movement, reaching a mirror and going back to him. The fact that the light comes back to the observer is fundamental, as the observer can define (and then measure) the time that the light takes to make a round trip. If one considers only a travel from the observer A to another observer B, it would not be possible to define a traveling time (or an interval of time), independent of the *unknown* distance between A and B. This is deeply linked with the concept of *simultaneity*, i.e. "what does it mean to A that the light arrives to B and how does he know it if it exists a maximal velocity in Nature". In a Newtonian world where *instantaneous* signals are possible, this question makes no sense. However, this is not the case in the Einstein world, that is the reason why the first 5 pages of his original article are uniquely dedicated to the definition of simultaneity.

As we saw, it is important for A to being able to measure (and thus define) a time delay, which means, he has to see for himself the ray of light coming back. That will be two events in space-time : (x'_A, t'_1) when the light leaves A and (x'_A, t'_2) when the light reaches A again after being reflected by the mirror. If one supposes that A moves with a velocity v [in its own referential of events (x',t')] with respect to a static[1] reference frame (x, t) where an observer B stays. Let us suppose that the

[1]Obviously, the notion of *static* and *moving* reference frames has no meaning in the Einstein approach where absolute space does not exist (contrarily to the Newtonian approach). These

Y. Mambrini, *Particles in the Dark Universe*,
https://doi.org/10.1007/978-3-030-78139-2

distance between the two mirrors is l' in the moving referential (x', t') and l in the static referential (x, t). The round trip time will be for A

$$t' = \frac{2l'}{c}, \tag{A.1}$$

where c is the velocity of light, whereas for the static observer B, it will be

$$t = \frac{l}{v - c} + \frac{l}{v + c} = \frac{2l}{c} \frac{1}{1 - \frac{v^2}{c^2}}. \tag{A.2}$$

Being conscious of the Lorentz proposition of lengths contraction (following the null result of the Michelson–Morley experiment[2]), $l' = \frac{l}{\sqrt{1-\frac{v^2}{c^2}}}$ and combining Eqs. (A.1) and (A.2), we obtain

$$t = \frac{t'}{\sqrt{1 - \frac{v^2}{c^2}}}. \tag{A.3}$$

It is interesting to notice that in the articles of Lorentz (and Fitzgerald), the authors supposed a real physical contraction of objects (and not of space) due to atoms being closer in the direction of motion induced by the effects of the electromagnetic forces.

We can then use this result to compute the transformation laws and express the events in the moving frame with respect to events in the static frame. l can be written as Δx, distance between two points in the static frame, that can be translated, in the moving frame using Lorentz contraction by $\Delta x' = \frac{\Delta x}{\sqrt{1-\frac{v^2}{c^2}}}$, which allows us to write

$$x' = \frac{x}{\sqrt{1 - \frac{v^2}{c^2}}} + \alpha t \tag{A.4}$$

giving for the point $x = vt$ $(x' = 0)$, $\alpha = -\frac{v}{\sqrt{1-\frac{v^2}{c^2}}}$, or in other words

$$x' = \frac{x - vt}{\sqrt{1 - \frac{v^2}{c^2}}}. \tag{A.5}$$

denominations are just here to help visualizing the situation of the relative motion of two observers. Any static frame is a moving frame in another referential.

[2]To be more precise, the first proposition of lengths contraction was made by Fitzgerald in 1889, whereas the Lorentz paper was published in 1892.

If we apply the same techniques to find the transformation law for the time, we can write

$$t' = at + bx. \tag{A.6}$$

Considering the point $x = 0$, it travels at velocity $-v$ with respect to the moving frame, so one can write, for this point, $t(x = 0) = t'\sqrt{1 - \frac{v^2}{c^2}}$. Indeed, the time of the static referential is diluted in this case because the "static" frame is the moving frame view from the observer B. Implementing this value of t and $x = 0$ in Eq. (A.6) we have

$$a = \frac{1}{\sqrt{1 - \frac{v^2}{c^2}}}, \tag{A.7}$$

whereas if one considers the point $x' = 0$ $(x = vt)$, applying this time $t'(x = vt) = \sqrt{1 - \frac{v^2}{c^2}}\, t$ we obtain

$$\left(\sqrt{1 - \frac{v^2}{c^2}}\right) t = \frac{1}{\sqrt{1 - \frac{v^2}{c^2}}}\, t + b\, x (= vt) \;\; \Rightarrow \;\; b = \frac{\frac{v}{c^2}}{\sqrt{1 - \frac{v^2}{c^2}}}. \tag{A.8}$$

Finally, combining all the preceding equations, we can write

$$x' = \frac{x - vt}{\sqrt{1 - \frac{v^2}{c^2}}} \tag{A.9}$$

$$t' = \frac{t - \frac{v}{c^2}x}{\sqrt{1 - \frac{v^2}{c^2}}} \tag{A.10}$$

which is the set of equations usually called Lorentz transformations.[3]

We let the reader check that the relations (A.9) and (A.10) keep constant the quantity $c^2t^2 - x^2$ (or $c^2t^2 - x^2 - y^2 - z^2$ in the 3 dimensional space). At a classical Galilean level, where the time does not transform, it is *both* the quantities ct^2 and $x^2 + y^2 + z^2$ that are conserved independently. This cannot be the case in relativity for the obvious reason that time and position are tightly related. However, as noticed

[3]Even if it is Poincaré who wrote them for the first time two weeks before the Einstein paper in 1905, Lorentz having not seen a group structure and the symmetry $x \leftrightarrow t$ in the transformations had a different asymmetric solution for t'. It is Poincaré, with respect to the Dutch physicist who insisted to call them itself "Lorentz transformations".

by Poincaré, the structure of group theory still preserves one quantity in the Lorentz transformations. It is a length, completely independent of the referential of study, such that all the observers will agree on, moving or not, and it can be written

$$ds^2 = c^2dt^2 - dx^2 - dy^2 - dz^2 = c^2dt^2 - dl^2. \tag{A.11}$$

In fact, as it is presented in some textbook, demanding for ds to be an invariant of the theory leads to the Lorentz transformations (A.9)–(A.10).

A.1.2 Friedmann Equation

The scale factor of the Universe (his size) is governed by Friedmann's equation,

$$\left(\frac{\dot{a}}{a}\right)^2 = \frac{8\pi G}{3}\rho - \frac{K}{a^2} = \frac{8\pi}{3M_{\text{Pl}}^2}\rho - \frac{K}{a^2}, \tag{A.12}$$

where we have ignored the cosmological constant for simplicity. Neglecting in the first step the curvature term (as one can do at early time), and approximating $\dot{a} \sim a/t$, Eq. (A.12) implies the familiar result that the expansion timescale $t \propto \rho^{-1/2}$. We choose the boundary condition $a(t_0) = 1$ at present time and $a(0) = 0$ at the early stage of the Universe. The Hubble constant H_0 is the present time value of the Hubble parameter $H_0 = (\dot{a}/a)_{t_0}$. K is the value of hypersurfaces of space–time, and ρ is the total matter density. We define the quantity

$$\rho_{\text{cr}} = \frac{3H_0^2}{8\pi G} \approx 2 \times 10^{-29}\, h^2 \text{g cm}^{-3} = 1.1 \times 10^{-5}\ \text{GeV cm}^{-3} \tag{A.13}$$

which is called the critical density for reasons that will become clear below. The density Ω_0 is the current total matter density $\rho(t_0) = \rho_0$ in unit of ρ_{cr},

$$\Omega_0 = \frac{\rho_0}{\rho_{\text{cr}}}. \tag{A.14}$$

The Hubble constant is commonly written as

$$H_0 = 100\, h \text{ km s}^{-1}\ \text{Mpc}^{-1} \approx 3.2 \times 10^{-18} h\ \text{s}^{-1}, \tag{A.15}$$

where $0.5 < h < 1$ express our ignorance of H_0. We can then extract from the Friedmann's equation (A.12) remembering that $a(t_0) = 1$,

$$K = H_0^2(\Omega_0 - 1) \tag{A.16}$$

In other words, a Universe which contains the critical matter density is flat.

The first law of thermodynamics with conservation of entropy, $dS = dU + pdV$ can be written as

$$\frac{d(\rho a^3)}{dt} + p\frac{da^3}{dt} = 0 \tag{A.17}$$

For ordinary matter, $\rho \gg p \sim 0$, hence $\rho \propto a^{-3}$. For relativistic matter, $p = \rho/3$, hence $\rho \propto a^{-4}$. Therefore, the matter density changes with a as

$$\rho(a) = \rho_0 a^{-n} = \rho_{\rm cr}\Omega_0 a^{-n}, \tag{A.18}$$

where $n = 3$ for ordinary matter ("dust"), and $n = 4$ for relativistic matter ("radiation").

Summarizing, we can rewrite Friedmann's equation into the form

$$H^2(a) = H_0^2[\Omega_0 a^{-n} - (\Omega_0 - 1)a^{-2}]. \tag{A.19}$$

Since the Universe expands, $a < 1$ for $t < t_0$, and so the expansion rate $H(a)$ was larger in the past. At very early times, $a \ll 1$, the first term in Eq. (A.19) dominates because $n \geq 3$ and we can write

$$H(a) = H_0\Omega_0^{1/2}a^{-n/2}. \tag{A.20}$$

This is called the Einstein–de Sitter limit of Friedmann's equation.

A.1.3 The Horizon

The size of casually connected regions of the Universe is called the *horizon size*. It is given by the distance a photon can travel in the time since the Big Bang. Since the appropriate timescale is provided by the inverse Hubble parameter, the horizon size is $d'_H = cH^{-1}(a)$, and the *comoving* horizon size is (reminding that the velocity of light $c = 1$)

$$d_H(a) = \frac{c}{aH(a)} = \frac{1}{aH(a)} = \frac{1}{H_0}\Omega_0^{-1/2}a^{n/2-1}, \tag{A.21}$$

where we have inserted the Einstein–de Sitter limit of Friedmann's equation. $cH_0^{-1} = 3\,h^{-1}$ Gpc is called the *Hubble radius.*

The previous calculations show that the matter density today is completely dominated by ordinary rather than relativistic matter. But since the relativistic matter density grows faster with decreasing scale factor a than the ordinary matter density, there had to be a time $a_{eq} \ll 1$ before which relativistic matter dominated. Using

Eq. (3.50), the condition

$$\Omega_0^R a_{eq}^{-4} = \Omega_0^M a_{eq}^{-3} \tag{A.22}$$

yields[4]

$$a_{eq} = \frac{3.7 \times 10^{-5}}{0.15} \simeq 2.7 \times 10^{-4} \tag{A.23}$$

which means that the Universe was ten thousands time smaller at the epoch when radiation and matter were in equilibrium. We have seen in Sect. 5.12 that a_{eq} plays a big role in structure formation constraints. If we suppose that matter dominated completely for all $a > a_{eq}$ (i.e. ignoring the contribution from radiation to the matter density), Eq. (A.21) yields

$$d_H(a_{eq}) = \frac{\Omega_0^{-1/2}}{H_0} a_{eq}^{1/2} \approx 90 \text{ Mpc}. \tag{A.24}$$

As we noticed in Eq. (A.18) the photon temperature evolves as a^{-4}. We also have demonstrated in Eq. (3.27) that the energy density of radiative particles evolves as T_γ^4 which implies that $T_\gamma \propto a^{-1}$. We can then compute that temperature of the photon at the time when $a = a_{eq}$:

$$T_\gamma(a_{eq}) = T_\gamma^0 \times a_{eq}^{-1} = \frac{2.4 \times 10^{-4}}{2.7 \times 10^{-4}} \simeq 0.9 \text{ eV}. \tag{A.25}$$

We recover the results obtained in the Sect. 3.1.6.

We can also compute the time dependance of the evolution of the scale factor. Indeed, from $\dot{a}/a = H \propto \rho_0^{1/2} a^{-n/2}$ which gives after integration

$$t \propto a^{(n-2)/2} \tag{A.26}$$

which means

$$a \propto t^{1/2} \text{ [radiation dominated]} \qquad a \propto t^{2/3} \text{ [matter dominated]} \tag{A.27}$$
$$\Rightarrow H = \frac{\dot{a}}{a} = \frac{1}{2t} \text{ [radiation dominated]} \quad \frac{2}{3t} \text{ [matter dominated]}.$$

From the previous section, we computed the energy density of relativistic particle $\rho_{\gamma\nu} = \frac{3g}{\pi^2} e^{\mu/T} T^4$, which means $\rho_{\gamma\nu} \propto T^4$ for a null chemical potential μ.

[4]For another way of computing the equilibrium time or temperature, see Sect. 3.1.6.

Moreover, if we write $p = adr/dt$, and knowing that t scales as a^{-2} in radiation time, one can deduce that p scales as a^{-1} in radiation dominated epoch.

$$p \propto a^{-1}, \tag{A.28}$$

which means that for a non-relativistic particle $p = mv$ implies $v \propto a^{-1}$.
Reminding that the TOTAL entropy of the Universe $S = sa^3 = g_{*S}T^3a^3$ is constant, one can deduce

$$a(T) \propto g_{*S}^{1/3} T^{-1} \quad \text{[in matter AND radiation dominated epoch]} \tag{A.29}$$

implying

$$a(T) = a_0 \frac{2.4 \times 10^{-4}\ \text{eV}}{T} = a_{DEC} \frac{0.3\ \text{eV}}{T}. \tag{A.30}$$

Combining with the results from the thermodynamics equilibrium of the previous section and supposing that during the early radiation dominated epoch $\rho \sim \rho_R$, Eq. (3.27) combined with Eq. (A.12), one obtains

$$H = \sqrt{\frac{8\pi G}{3}} \rho_R^{1/2} = 1.66 g_\rho^{1/2} \frac{T^2}{M_{Pl}} = 0.32 g_\rho^{1/2} \frac{T^2}{M_P}. \tag{A.31}$$

Moreover, if $a \propto t^n$, one observes that $H = \dot{a}/a = nt^{-1}$ implying

$$t = 0.301 g_*^{-1/2} \frac{M_{Pl}}{T^2} \sim \left(\frac{T}{\text{MeV}}\right)^{-2} \quad \text{[radiation dominated : n=1/2]} \tag{A.32}$$

$$t \propto T^{-3/2} \ \text{[matter dominated: n = 2/3]}, \tag{A.33}$$

where $g_\rho = g_{boson} + \frac{7}{8} g_{fermion}$ counts the total number of massless degrees of freedom ($m \ll T$).

A.2 Basics of General Relativity

A.2.1 The Context

"$G_{\mu\nu} = \kappa T_{\mu\nu}$". This equation was on my wall, next to the photo of the 1927 Solvay congress, during a large part of my study. As we discussed in the Sect. A.1.1, the special relativity transformations and physics can be reconstructed from the basic concept of the invariance of the length scale $ds^2 = c^2dt^2 - dl^2$ between referentials in relative movement with constant velocity. This is obviously not the case in *classical* Galilean transformations which impose dt^2 *and* dl^2 independent

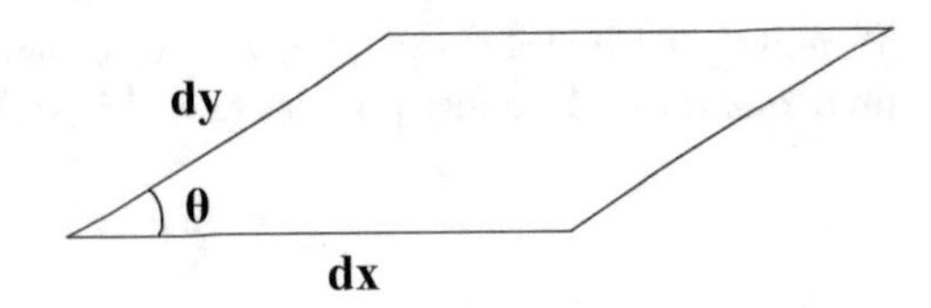

Fig. A.1 Illustration of a 2d surface

on the reference frame, separately. In the same manner, the Einstein transformation laws of General Relativity concern the independence of the physics on the referential frame, in any metric, including curved space-time. It can be derived from the conservation of a unit of volume, as the Lorentz transformation can be deduced from the conservation of ds^2.

A.2.2 Measuring a Length, a Surface, or a Volume

Before entering in detail, dealing with volume in the 4-dimensional space-time means dealing with volume. We remind that under a change of variable, $x \to x' = f(x)$, the unit of length dx transforms to

$$dx' = \frac{\partial f(x)}{\partial x} dx. \tag{A.34}$$

This can be generalized in a 2-dimensional space to compute an element of surface. Indeed, the area $\mathcal{A}$ of a parallelepiped with sides $d\mathbf{x}$ and $d\mathbf{y}$ is the base $|d\mathbf{x}|$ multiplied by the height $h = |d\mathbf{y}| \times \sin\theta$, θ being the angle between $d\mathbf{x}$ and $d\mathbf{y}$, see Fig. A.1, or

$$\mathcal{A} = |d\mathbf{x}| \times |d\mathbf{y}| \times \sin\theta = |d\mathbf{x} \wedge d\mathbf{y}|. \tag{A.35}$$

Under a change of variables $x = f(u, v)$ and $y = g(u, v)$, we can write

$$d\mathbf{x} = \frac{\partial}{\partial u} f(x, y)\, d\mathbf{u} + \frac{\partial}{\partial v} f(u, v)\, d\mathbf{v}\,; \quad d\mathbf{y} = \frac{\partial}{\partial u} g(x, y)\, d\mathbf{u} + \frac{\partial}{\partial v} g(u, v)\, d\mathbf{v}$$

$$\Rightarrow \mathcal{A} = |d\mathbf{x} \wedge d\mathbf{y}| = \left| \frac{\partial f}{\partial u}\frac{\partial g}{\partial v} - \frac{\partial f}{\partial v}\frac{\partial g}{\partial u} \right| |d\mathbf{u} \wedge d\mathbf{v}| = J \times |d\mathbf{u} \wedge d\mathbf{v}|, \tag{A.36}$$

J being the Jacobian of the transformation.[5] We can generalize Eq. (A.36) to any manifold and dimensions, defining the transformation $(x^\mu \to x'^\mu(x^\nu))$ of an

[5] A Jacobian of a transformation $x^\mu \to x'^\mu$ with $x'^\mu = x'^\mu(x^\nu)$ is given by $J = \det\left|\frac{\partial x^\mu}{\partial x'^\nu}\right|$.

element of volume $d\tau$ into $d\tau'$ by

$$d\tau = J \times d\tau' \; ; \quad d\tau' = J^{-1} \times d\tau \tag{A.37}$$

with $J = |\frac{\partial x'^{\mu}}{\partial x^{\nu}}|$ and $J^{-1} = |\frac{\partial x^{\nu}}{\partial x'^{\mu}}|$.
I hope this little demonstration will answer the typical question I had in my lectures on the subject: " why the element of surface $dxdy$ does not simply transform to $\left(\frac{\partial}{\partial u} f(u, v) + \frac{\partial}{\partial v} f(u, v)\right) \times \left(\frac{\partial}{\partial u} g(u, v) + \frac{\partial}{\partial v} g(u, v)\right) dudv$". The negative sign generated by the wedge product translates the obvious fact that a surface under two parallel vectors is null.

A.2.3 The Einstein–Hilbert Action (I)

We remind the reader that a contravariant vector A^{μ} and a covariant vector A_{μ} are defined by their transformation laws under a change of coordinates $x^{\mu} \to x'^{\mu}$:

$$A'^{\mu} = \frac{\partial x'^{\mu}}{\partial x^{\alpha}} A^{\alpha} \; ; \quad \text{and} \;\; A'_{\mu} = \frac{\partial x^{\alpha}}{\partial x'^{\mu}} A_{\alpha}. \tag{A.38}$$

This definition ensures the conservation of the norm $A^{\mu} A_{\mu}$, especially the square of the invariant distance $ds^2 = dx^{\mu} dx_{\mu}$.

Exercise Prove this statement.[6]

More, precisely, writing the element of length

$$ds^2 = g_{\mu\nu}(x) dx^{\mu} dx^{\nu} \tag{A.39}$$

we let the reader check that if one defines

$$dx_{\mu} = g_{\mu\nu}(x) dx^{\nu},$$

dx_{μ} possesses the transformation properties of a covariant vector ($dx'_{\mu} = \frac{\partial x^{\alpha}}{\partial x'^{\mu}} dx_{\alpha}$), and we can rewrite Eq. (A.39)

$$ds^2 = dx_{\mu} dx^{\mu}.$$

Notice that, if one needs to deal with fermionic fields, it is more convenient to work in the tetrad (*vierbein* in German) framework. This allows to place ourselves in the Minkowski flat space X^a, with the metric η^{ab}, where the γ^a matrices take their Lorentz form (B.16), and then "bend" them in the physical metric $g^{\mu\nu}$. In other

[6] You will need for this to notice that $\delta^{\beta}_{\alpha} = \frac{\partial x^{\beta}}{\partial x^{\alpha}} = \frac{\partial x^{\beta}(x')}{\partial x^{\alpha}} = \frac{\partial x'^{\mu}}{\partial x^{\alpha}} \times \frac{\partial x^{\beta}}{\partial x'^{\mu}}$.

words, one introduces *locally* the flat metric η^{ab}, where the vector basis includes four orthonormal vectors $\mathbf{e}_a$ such that $\mathbf{e}_a.\mathbf{e}_b = \eta_{ab}$. Here, the $\mathbf{e}_a$ are 4-dimensional vectors in the tangent Minkowski space. With respect to the physical space x^μ, we can write $\mathbf{e}_a = e_a^\mu \mathbf{e}_\mu$, where $\mathbf{e}_\mu$ is a corresponding local basis, and e_a^μ serves as the intermediate coefficients between the two basis, which means

$$e_a^\mu = \frac{\partial x^\mu}{\partial X^a}, \quad e_\mu^a = \frac{\partial X^a}{\partial x^\mu}. \tag{A.40}$$

and from

$$\eta_{ab} dX^a dX^b = \eta_{ab} \frac{\partial X^a}{\partial x^\mu} \frac{\partial X^b}{\partial x^\nu} dx^\mu dx^\nu, \tag{A.41}$$

one deduces

$$g_{\mu\nu} = e_\mu^a e_\nu^b \eta_{ab}. \tag{A.42}$$

and

$$\gamma^\mu = e_a^\mu \gamma^a. \tag{A.43}$$

With the same philosophy, one can define a contravariant and covariant tensor by

$$T'^{\mu\nu} = \frac{\partial x'^\mu}{\partial x^\alpha} \frac{\partial x'^\nu}{\partial x^\beta} T^{\alpha\beta} \; ; \quad \text{and} \quad T'_{\mu\nu} = \frac{\partial x^\alpha}{\partial x'^\mu} \frac{\partial x^\beta}{\partial x'^\nu} T_{\alpha\beta}. \tag{A.44}$$

If one defines a space-time dependant metric $g_{\mu\nu}(x)$, it is straightforward to prove that the condition

$$ds^2 = g_{\mu\nu}(x) dx^\mu dx^\nu = \text{constant} \tag{A.45}$$

under a change of coordinates $x^\mu \to x'^\mu(x)$, or $dx^\mu = \frac{\partial x^\mu}{\partial x'^\alpha} dx'^\alpha$ implies

$$g'_{\mu\nu}(x') = \frac{\partial x^\alpha}{\partial x'^\mu} \frac{\partial x^\beta}{\partial x'^\nu} g_{\alpha\beta}(x). \tag{A.46}$$

It then follows that metric $g_{\mu\nu}(x)$ is a covariant tensor of second rank. Another very interesting consequence of the relation (A.46) is that we can write, in the matrix form

$$\mathbf{g}'(x) = \mathbf{J} \times \mathbf{g} \times \mathbf{J}^{\mathbf{T}} \quad \Rightarrow \quad \det[\mathbf{g}'] = (\det[\mathbf{J}])^2 . \det[\mathbf{g}], \tag{A.47}$$

in other words, combining Eqs. (A.37) and (A.47), we notice that under coordinate transformations $d\tau \rightarrow = J^{-1} d\tau$ and $\sqrt{|g|} \rightarrow J\sqrt{|g|}$, with $g = \det g_{\mu\nu}$, which means that the product

$$\sqrt{|g|} d^4x = \sqrt{-g} d^4x \tag{A.48}$$

defines the invariant 4-volume element of the theory. The $|g| = -g$ relation arises from the fact that the matrix $g_{\mu\nu}$ has three negatives and one positive eigenvalues,[7] the determinant g is then negative, and $\sqrt{-g}$ is real. One can then define a first invariant gravitational action, which should be written in terms of covariant quantities and invariant under the general change of coordinates. We can express it as function of a scalar Lagrangian, integrated over an invariant 4-volume. The simplest Scalar Lagrangian is obviously a constant Λ. S_Λ is then given by

$$S_\Lambda = -\frac{1}{8\pi G} \Lambda \int d^4x \sqrt{-g}, \tag{A.49}$$

G being the Newton constant of gravity. This action will play a fundamental role in cosmology, because of describing the cosmological constant action. The $8\pi G$ factor is a convention that will simplify the Einstein equations of fields as we will see later on.

A.2.4 Tooling with the Metric

When dealing with fundamental fields, one needs to deal with derivatives of the field with respect to the coordinate system, especially derivative of vector fields. Whereas it is obvious that for a scalar field ϕ, $\partial^\mu \phi$ $(\partial_\mu \phi)$ behaves like a contravariant (covariant) vector fields, it is not the case for $\partial_\mu A^\nu$. Indeed, under a change of coordinate

$$\partial_\mu A^\nu \quad \rightarrow \quad \partial'_\mu A'^\nu = \frac{\partial x'^\nu}{\partial x^\rho} \frac{\partial x^\sigma}{\partial x'^\mu} \frac{\partial A^\rho}{\partial x^\sigma} + \frac{\partial x^\sigma}{\partial x'^\mu} \frac{\partial^2 x'^\nu}{\partial x^\rho \partial x^\sigma} A^\rho. \tag{A.50}$$

Whereas the first term corresponds indeed to a tensor transformation of the type Eq. (A.44), the second part should be eliminated, the same way covariant derivatives are defined in gauge theory, defining

$$D_\mu A^\nu = \partial_\mu A^\nu + \Gamma^\nu_{\alpha\mu} A^\alpha, \tag{A.51}$$

[7] Also called a Lorentzian signature.

where $\Gamma^{\nu}_{\alpha\mu}$, the analog of the gauge field, is called a Christoffel symbol or spin connection. Imposing that $D_\mu A^\nu$ transforms as a tensor,

$$D'_\mu A'^\nu = \frac{\partial x^\alpha}{\partial x'^\mu}\frac{\partial x'^\nu}{\partial x^\beta} D_\alpha A^\beta,$$

one obtains

$$\Gamma'^{\nu}_{\alpha\mu} = \frac{\partial x'^\nu}{\partial x^\beta}\frac{\partial x^\rho}{\partial x'^\alpha}\frac{\partial x^\sigma}{\partial x'^\mu}\Gamma^{\beta}_{\rho\sigma} + \frac{\partial x'^\nu}{\partial x^\rho}\frac{\partial^2 x^\rho}{\partial x'^\alpha \partial x'^\mu}. \tag{A.52}$$

Exercise Demonstrate this relation.

As we can notice, the first term in the right-hand side of (A.52) corresponds to a classical covariant-contravariant transformation for the rank-3 tensor $\Gamma^{\nu}_{\alpha\mu}$. The second term is the one needed to cancel the extra-contribution generated by the derivative in the right-hand side of (A.50). Notice that we can also derive the expression of the derivative for a covariant vector by solving

$$D'_\nu A'_\mu = \frac{\partial x^\alpha}{\partial x'^\nu}\frac{\partial x^\beta}{\partial x'^\mu} D_\alpha A_\beta. \tag{A.53}$$

Another possibility is to use the Leibniz formula[8]

$$D_\mu(A_\nu A^\nu) = A_\nu D_\mu A^\nu + (D_\mu A_\nu)A^\nu = D_\mu |A|^2 = \partial_\mu |A|^2 \tag{A.54}$$

which gives

$$D_\mu A_\nu = \partial_\mu A_\nu - \Gamma^{\alpha}_{\nu\mu} A_\alpha$$

At first sight, we can imagine thousands (and much more) of tensors of rank-3 satisfying the condition (A.52). However, one needs two extra constraints. First of all, to preserve the metric in the differentiation process, one needs the covariant derivative to commute with the metric tensor, i.e.

$$g_{\mu\nu}(x).D_\alpha A^\nu = D_\alpha\left(g_{\mu\nu}(x).A^\nu\right) \tag{A.55}$$

which gives, using Leibniz relation

$$D_\alpha g_{\mu\nu} = 0 = \partial_\alpha g_{\mu\nu} - \Gamma^{\beta}_{\mu\alpha}\, g_{\beta\nu} - \Gamma^{\beta}_{\nu\alpha}\, g_{\mu\beta} \tag{A.56}$$

[8] As an exercise, show that $D_\mu(AB) = A(D_\mu B) + (D_\mu A)B$, whatever are the tensors A and B.

that we can reduce to

$$\partial_\alpha g_{\mu\nu} = \Gamma^\beta_{\mu\alpha}\, g_{\beta\nu} + \Gamma^\beta_{\nu\alpha}\, g_{\mu\beta}. \tag{A.57}$$

Adding the symmetry condition $\Gamma^\alpha_{\mu\nu} = \Gamma^\alpha_{\nu\mu}$, one can derive uniquely the connection tensor

$$\Gamma^\alpha_{\mu\nu} = \frac{1}{2} g^{\alpha\rho}\left[\partial_\mu g_{\alpha\nu} + \partial_\nu g_{\alpha\mu} - \partial_\alpha g_{\mu\nu}\right]. \tag{A.58}$$

The Covariant Derivative for Spinor Fields
The expression (A.51) is valid for the derivative of a vector field. For a scalar, it is straightforward to see that $D_\mu\phi = \partial_\mu\phi$. For a spinor field ψ, the construction of a covariant derivative is a little bit more involved. If we define

$$D_\mu\psi = \partial_\mu\psi - \frac{1}{4}\omega^{ab}_\mu[\gamma_a\gamma_b - \gamma_b\gamma_a]\psi, \tag{A.59}$$

in order to obtain the equation for spinor connections (A.51),

$$D_\mu(\bar{\psi}\gamma^\nu\psi) = \partial_\mu(\bar{\psi}\gamma^\nu\psi) + \Gamma^\nu_{\alpha\mu}\bar{\psi}\gamma^\alpha\psi, \tag{A.60}$$

using the *vierbein* conventions (A.40) and (A.43), we let the reader prove that

$$\omega^{ab}_\mu = \frac{1}{4}(e^b_\alpha e^{\beta a} - e^a_\alpha e^{\beta b})\Gamma^\alpha_{\beta\mu} + \frac{1}{4}(e^{\alpha b}\partial_\mu e^a_\alpha - e^{\alpha a}\partial_\mu e^b_\alpha). \tag{A.61}$$

Just a remark, when developing $D_\mu(\bar{\psi}\gamma^\alpha\psi)$, do not forget to include the term $\bar{\psi}(D_\mu\gamma^\nu)\psi$ with $\gamma^\nu = e^\nu_a\gamma^a$.

A.2.5 A Geometrical Approach

We can also recover the expression (A.52) with a pure geometrical approach, as it is done in several very good textbooks. The idea is quite elegant. Instead of imposing a covariant constraint (A.53), the Christoffel symbol is in fact defined by the change induced on a vector A^μ when traveling from the point x to $\tilde{x}$ with

$$\tilde{x}^\mu = x^\mu + dx^\mu \ ; \quad A^\mu(x) \to \tilde{A}(\tilde{x}) = A^\mu(x) - \Gamma^\mu_{\alpha\beta} A^\alpha dx^\beta. \tag{A.62}$$

We will recover the transformation law on $\Gamma^\mu_{\alpha\beta}$ (A.52) (which was obtained by asking $D_\mu A^\nu$ to behave as a tensor) demanding this time that the variation of the vector field A^μ *at any position* transforms as a vector. For that, one needs to compute $A'^\mu(\tilde{x}')$ and $\tilde{A}'^\mu(\tilde{x}')$.[9] We then can write from (A.62)

$$\tilde{A}'^\mu(\tilde{x}') = A'^\mu(x) - \Gamma'^\mu_{\alpha\beta} A'^\alpha(x) dx'^\beta = \frac{\partial x'^\mu}{\partial x^\alpha} A^\alpha(x) - \Gamma'^\mu_{\alpha\beta} \frac{\partial x'^\alpha}{\partial x^\rho} A^\rho(x) \frac{\partial x'^\beta}{\partial x^\sigma} dx^\sigma \tag{A.63}$$

and from the definition of $\tilde{A}'^\mu$ as a vector

$$\tilde{A}'^\mu(\tilde{x}') = \frac{\partial \tilde{x}'^\mu}{\partial x^\alpha} \tilde{A}^\alpha(\tilde{x}) = \left[\frac{\partial x'^\mu}{\partial x^\alpha} + \frac{\partial^2 x'^\mu}{\partial x^\alpha \partial x^\sigma} dx^\sigma \right] \times \left[A^\alpha(x) - \Gamma^\alpha_{\rho\sigma} A^\rho dx^\sigma \right], \tag{A.64}$$

where we used $\tilde{x}'^\mu = x'^\mu + dx'^\mu$. Comparing Eqs. (A.63) and (A.64) and developing at the first order in dx^μ, we obtain

$$\Gamma'^\mu_{\alpha\beta} = \frac{\partial x'^\mu}{\partial x^\nu} \frac{\partial x^\rho}{\partial x'^\alpha} \frac{\partial x^\sigma}{\partial x'^\beta} \Gamma^\nu_{\rho\sigma} - \frac{\partial x^\rho}{\partial x'^\alpha} \frac{\partial x^\sigma}{\partial x'^\beta} \frac{\partial^2 x'^\mu}{\partial x^\rho \partial x^\sigma} \tag{A.65}$$

which is equivalent to Eq. (A.52) (*exercise*). Notice also that the definition of $D_\nu A^\mu$ (A.51) has also a geometrical interpretation. Indeed, it is straightforward to check that

$$A^\mu(\tilde{x}) - \tilde{A}^\mu(\tilde{x}) = D_\nu A^\mu . dx^\nu. \tag{A.66}$$

$D_\nu A^\mu$ can then be interpreted as the differential vector at the point $\tilde{x}$, or in other words, between the vector at $\tilde{x}$ and the *transported* vector from x to $\tilde{x}$ (see the box below). The last (but not the least) tensor one should take into account while looking for invariant in general space-time transformations is the Riemann tensor.

[9] Notice that both A and $\tilde{A}$ should be computed in the same point $\tilde{x}'$ to be able to compare their difference. That a subtle but fundamental point.

The Parallel Transport

The parallel transport is a fundamental notion in General Relativity, especially when comes the time to measure movement in gravitational potential and definitions of geodesics. It is in fact the essence of the General Relativity principle. It answers to the question : "how do the coordinates of a tensor evolve, staying parallel to itself, during its motion on a trajectory in space-time?". We illustrate the situation in Fig. A.2, where we plotted the tangential vector along a path s, $\frac{dx^\mu}{ds}$ (in blue) and the parallel transported vector (in red). From Eq. (A.66) the transported vector along ds should follow $\tilde{A}^\mu(\tilde{x}) = A^\mu(\tilde{x})$, or

$$D_\nu A^\mu dx^\nu = 0 \;\; \Rightarrow \;\; \frac{dA^\mu}{ds} + \Gamma^\mu_{\alpha\nu}\frac{dx^\nu}{ds}A^\alpha = 0. \tag{A.67}$$

Notice that if one takes $A^\mu = \frac{dx^\mu}{ds}$, the tangential vector itself, we obtain the equation

$$\frac{d^2x^\mu}{ds^2} + \Gamma^\mu_{\alpha\nu}\frac{dx^\alpha}{ds}\frac{dx^\nu}{ds} = 0, \tag{A.68}$$

which is the equation of a geodesic, or as we can feel after a look at Fig. A.2, the equation of a trajectory that keeps constant its tangential vector, i.e. the shorter way from a point A to a point B (straight line in the case of flat space, great radius in the case of a sphere). These are the trajectories that follow a particle under a space-time deformed by gravity.

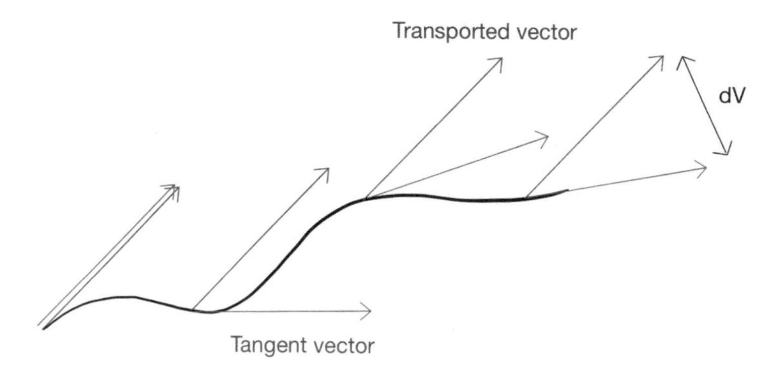

Fig. A.2 Illustration of a parallel transported vector along a path s

A.2.6 The Riemann Tensor

The geometry of a manifold is not determined by how a tensor evolves under the change of coordinates, but how a vector evolves along a closed curve on the manifold. In other way, how the covariant derivative ($D_\nu A^\mu \propto dA^\mu$) behaves in one direction of the curve, and then in the other direction. This behavior is determined by the commutation relation $D_\nu D_\mu - D_\mu D_\nu$, defining the Riemann tensor $R^\lambda_{\sigma\mu\nu}$ by

$$D_\mu D_\nu A^\rho - D_\nu D_\mu A^\rho = R^\rho_{\alpha\mu\nu} A^\alpha. \tag{A.69}$$

This is reminiscent to the commutation relation in gauge theories, where one can think $R^\rho_{\alpha\mu\nu}$ as the structure function of the gauge transformation. Applying (A.51) twice, it is straightforward to check that

$$R^\rho_{\alpha\mu\nu} = \partial_\mu \Gamma^\rho_{\alpha\nu} - \partial_\nu \Gamma^\rho_{\alpha\mu} + \Gamma^\rho_{\sigma\mu}\Gamma^\sigma_{\alpha\nu} - \Gamma^\rho_{\sigma\nu}\Gamma^\sigma_{\alpha\mu}. \tag{A.70}$$

Exercise Check the relation (A.70). Then, take a vector A^μ at x^ν and parallel transport it to $\tilde{x}^\nu = x^\nu + dy^\nu + dz^\nu$ following two different ways: moving along dy and then dz, or along dz and then dy. Show that

$$\tilde{A}^\rho(dy \to dz) - \tilde{A}^\rho(dz \to dy) = A^\alpha R^\rho_{\alpha\mu\nu} dz^\mu dy^\nu, \tag{A.71}$$

expressing the fact that the Riemannian tensor is an intrinsic characteristic of the curvature.

The interesting point is to be able to define a scalar from the Riemann tensor. Indeed, as the proper length $ds^2 = c^2dt^2 - dl^2$ defines the invariant of any coordinates transformation in the Minkowski space-time, asking for the invariance of the scalar which is related to the curvature will naturally lead to the Einstein equation of General Relativity. Defining as a first step the Ricci tensor $R_{\mu\nu}$

$$R_{\mu\nu} = R^\sigma_{\mu\sigma\nu} = \partial_\sigma \Gamma^\sigma_{\mu\nu} - \partial_\mu \Gamma^\sigma_{\sigma\nu} + \Gamma^\sigma_{\alpha\sigma}\Gamma^\alpha_{\mu\nu} - \Gamma^\sigma_{\alpha\mu}\Gamma^\alpha_{\nu\sigma}, \tag{A.72}$$

we obtain the scalar curvature R by contracting the Ricci indices,

$$R = g^{\mu\nu} R_{\mu\nu}. \tag{A.73}$$

We can construct an invariant volume from R, as we did for the cosmological constant, defining this way the Einstein–Hilbert action S_{EH}

$$S_{EH} = -\frac{1}{16\pi G}\int d^4x\sqrt{-g}R = -\frac{M_P^2}{2}\int d^4x\sqrt{-g}R. \tag{A.74}$$

A.2.7 The Einstein Equation of Fields in Vacuum

We have now everything in the hand to derive the Einstein equation of fields. Let us derive them in vacuum first. Adding to the cosmological constant action (A.49) the curvature scalar, one can write the total action S

$$S = S_\Lambda + S_{EH} = -\frac{1}{16\pi G}\int d^4x\sqrt{-g}\,(R+2\Lambda). \tag{A.75}$$

To obtain the equations of fields, one needs to minimize the action, or in other words, to satisfy the condition $\delta S = 0$. To do that, we will first minimize the cosmological constant part S_Λ part in Eq. (A.75):

$$\delta S_\Lambda = -\frac{\Lambda}{16\pi G}\int d^4x\frac{\delta(-g)}{\sqrt{-g}} = -\frac{\Lambda}{16\pi G}\int d^4x\sqrt{-g}g^{\mu\nu}\delta g_{\mu\nu}, \tag{A.76}$$

where we used (A.80)

$$\delta g = g g^{\mu\nu}\delta g_{\mu\nu}. \tag{A.77}$$

How to Differentiate a Determinant

We recall here the basics of matrix computation. The determinant of a matrix a_{ij} can be written as $\det a = \sum_{i,j} a_{ij}A^{ji}$ where A^{ij} is the comatrix of a_{ij}. A^{ij} is obtained by taking first the matrix of the Minors (a Minor $m_{i,j}$ is the determinant of the matrix where the row i and column j have been suppressed), then multiplying each Minor by $(-1)^{i+j}$. The resulting matrix is the matrix of the cofactors A^{ij}. The inverse of the matrix a_{ij} can then be written as (to demonstrate)

$$b^{ij} = \frac{1}{\det a}(A^{ij})^\top = \frac{1}{\det a}A^{ji}. \tag{A.78}$$

(continued)

Then, from

$$\det a = \sum a_{ij} A^{ji}, \tag{A.79}$$

we can deduce,

$$\frac{\partial \det a}{\partial a_{ij}} = A^{ji} = \det a \times b^{ij}. \tag{A.80}$$

For the Einstein–Hilbert part, we can write

$$\begin{aligned} \delta S_{EH} &= -\frac{1}{16\pi G} \int d^4x \left[\delta(\sqrt{-g}) R + \sqrt{-g} \delta g^{\mu\nu} R_{\mu\nu} + \sqrt{-g} g^{\mu\nu} \delta R_{\mu\nu} \right] \\ &= \delta S_1 + \delta S_2 + \delta S_3. \end{aligned} \tag{A.81}$$

The δS_1 part is straightforward as it corresponds exactly to the same kind of variation we computed for the cosmological constant part δS_Λ :

$$\delta S_1 = -\frac{1}{32\pi G} \int d^4x \sqrt{-g} R g^{\mu\nu} \delta g_{\mu\nu}. \tag{A.82}$$

To compute δS_2, we notice that

$$g^{\mu\alpha} g_{\alpha\mu} = \delta^\mu_\nu \;\Rightarrow\; \delta(g^{\mu\alpha} g_{\alpha\nu}) = 0 \;\Rightarrow\; \delta g^{\mu\nu} = -g^{\mu\alpha} \delta g_{\alpha\beta} g^{\beta\nu} \tag{A.83}$$

we obtain

$$\delta S_2 = \frac{1}{16\pi G} \int d^4x \sqrt{-g} g^{\mu\alpha} \delta g_{\alpha\beta} g^{\beta\nu} R_{\mu\nu} = \frac{1}{16\pi G} \int d^4x \sqrt{-g} R^{\mu\nu} \delta g_{\mu\nu}. \tag{A.84}$$

For δS_3, we let the reader check that, using $\delta R_{\mu\nu} = D_\alpha \delta \Gamma^\alpha_{\mu\nu} - D_\nu \Gamma^\alpha_{\mu\alpha}$, we can compute

$$\begin{aligned} \delta S_3 &= -\frac{1}{16\pi G} \int d^4x \sqrt{-g} D_\alpha [g^{\mu\nu} \delta \Gamma^\alpha_{\mu\nu} - g^{\mu\alpha} \delta \Gamma^\alpha_{\mu\alpha}] \\ &= -\frac{1}{16\pi G} \int d^4x \partial_\alpha \left(\sqrt{-g} [g^{\mu\nu} \delta \Gamma^\alpha_{\mu\nu} - g^{\mu\alpha} \delta \Gamma^\alpha_{\mu\alpha}] \right) = 0 \end{aligned}$$

because δS_3 is the integral of a total derivative. At the end, combining (A.76), (A.82), and (A.84) and asking for $\delta S_\Lambda + \delta S_1 + \delta S_2 = 0$, we obtain

$$G^{\mu\nu} = R^{\mu\nu} - \frac{1}{2}Rg^{\mu\nu} = \Lambda g^{\mu\nu} \tag{A.85}$$

which are the Einstein equation of fields in the vacuum.

A.2.8 Adding Matter Fields

Obviously, the Universe is far to be empty. One then needs to introduce a matter Lagrangian $\mathcal{L}_m$ and the matter action S_m

$$S_m = \sum_{\text{fields}} \int d^4x\sqrt{-g}\mathcal{L}_{\text{fields}} = \int d^4x\sqrt{-g}\mathcal{L}_m, \tag{A.86}$$

where the sum runs over all the fields of the theory under consideration. We can then define the stress–energy–momentum tensor $T_{\mu\nu}$ by

$$\delta S_m = \frac{1}{2}\int d^4x\sqrt{-g}T_{\mu\nu}\delta g^{\mu\nu} = -\frac{1}{2}\int d^4x\sqrt{-g}T^{\mu\nu}\delta g_{\mu\nu}, \tag{A.87}$$

where we used (A.83) for the negative sign. The Einstein equations then become

$$G^{\mu\nu} = R^{\mu\nu} - \frac{1}{2}Rg^{\mu\nu} = 8\pi G\, T^{\mu\nu} + \Lambda g^{\mu\nu}, \tag{A.88}$$

where $T^{\mu\nu}$ is the energy tensor and $G = 1/M_{Pl}^2$ the gravitational coupling (M_{Pl} is the Planck mass = 1.22×10^{19} GeV). Different $T^{\mu\nu}$ are computed in the main text (see Sect. 2.2.3 for a scalar field for instance). In the following, we compute explicitly the stress–energy tensor for a perfect fluid.

A.2.9 The Perfect Fluid Stress–Energy–Momentum Tensor

To solve the set of equations (A.85), one needs to look into more detail the stress–energy–momentum tensor $T^{\mu\nu}$. In Sect. 2.2.3 we computed it for a scalar field, which is very useful when dealing with inflation if we consider a scalar inflation. However, after inflation occurred, the Universe is filled with a relativistic gas and then dominated by non-relativistic matter (dust). To understand the evolution of the

scale parameter R during the reheating (and post reheating) epoch, one needs to find the energy content of a set of N particles in a given volume V. We will neglect the effects of gravity on these particles at a first approximation.
When we consider a set of N particles (or equivalently a set of N charges in the electromagnetic theory of fields), it is useful to define a quadrivector describing the inflow and outflow of particles (or charges) in a given volume V, or more precisely the flow of the density of particles. If a volume V contains N particles, we can define in the rest frame a *number density* $n = \frac{N}{V}$. For an observer moving at a velocity v with respect to the rest frame (under any direction x^1, x^2, or x^3), the contraction of length (A.9) tells us that $n \to \frac{n}{\sqrt{1-\frac{v^2}{c^2}}}$. The dilatation of density in a moving referential compared to a rest frame can be interpreted as the behavior of the time component of a quadrivector. If one writes $n = n_0 \frac{dt}{d\tau}$ with $d\tau = \frac{\sqrt{ds}}{c} = \sqrt{1 - \frac{v^2}{c^2}}$ (from A.11), n_0 being a constant, we can build a four-vector

$$n^\mu = n_0 \frac{dx^\mu}{d\tau}, \tag{A.89}$$

where n_0 and $d\tau$ are constant, invariant under Lorentz transformations. It is then obvious that n^μ has the properties of a vector, whose time component is the density of particles. How should we then interpret $n^i = n_0 \frac{dx^i}{d\tau}$? If we develop (A.89) as function of the local time t we obtain

$$n^\mu = \frac{n_0}{\sqrt{1 - \frac{v^2}{c^2}}} \begin{pmatrix} 1 \\ \frac{dx^1}{dt} \\ \frac{dx^2}{dt} \\ \frac{dx^3}{dt} \end{pmatrix} = \frac{n_0}{\sqrt{1 - \frac{v^2}{c^2}}} \begin{pmatrix} 1 \\ v^1 \\ v^2 \\ v^3 \end{pmatrix}, \tag{A.90}$$

where $\mathbf{n} = \frac{n_0}{\sqrt{1-\frac{v^2}{c^2}}} \mathbf{v}$ is called the *number density current* and is the equivalent of the current J^μ in electrodynamics. In a sense, it corresponds to the number of particles (or charges) that penetrate the volume due to its movement in space.

How to build a tensor from the number density vector? The easiest way is to suppose that the energy (momentum) is proportional to the density (current density) of the particles in the volume ΔV, or in other words

$$p^\mu = T^{\mu\nu} n_\nu \Delta V. \tag{A.91}$$

It is indeed easy to check that in the rest frame, the components are

$$T^{00} = \frac{p^0}{n_0 \Delta V} = \frac{\epsilon}{n_0} \quad \text{and} \quad T^{i0} = \frac{p^i}{n_0 \Delta V} = \pi^i \tag{A.92}$$

if we set $n_0 = 1$ by simplicity,[10] $T^{00} = \epsilon$ represents then the energy density whereas $T^{i0} = \pi^i$ is the momentum density as it would be measured by an observer at rest in the inertial frame under discussion.

The computation (and interpretation) of the components T^{0i} can be done directly from T^{i0}. Indeed, from (A.91) we can write along the direction x^1 (direction perpendicular to the plane (x^2, x^3)):

$$\Delta p^0 = T^{01} \Delta x^2 \Delta x^3 \Delta x^0 \quad \Rightarrow \quad T^{01} = \frac{\Delta E}{\Delta S \Delta t} \tag{A.93}$$

with $\Delta S = \Delta x^2 \Delta x^3$. T^{0i} can then be interpreted by the flux of energy passing through a surface perpendicular to the ith direction, which is equivalent to the flux of momenta T^{i0}:

$$T^{0i} = T^{i0}.$$

Finally, for the spatial–spatial component, by the same reasoning

$$T^{ij} = \frac{\Delta p^i}{\Delta S_j \Delta t} = \frac{\Delta F^i}{\Delta S_j} = P^{ij} \tag{A.94}$$

with ΔS_j the element of surface perpendicular to the jth direction and F^i the force exerted on the element of surface ΔS_j, in other words, $P^{ij} = T^{ij}$ is the pressure oriented in the ith direction, exerted on the surface perpendicular to the jth direction, which is also called the stress tensor. In summary,

$$T^{\mu\nu} = \begin{pmatrix} \textbf{Energy density} & | \ \textbf{Energy flow} \\ \textbf{Momentum density} & | \ \textbf{Stress tensor} \end{pmatrix} \tag{A.95}$$

which can be written, in the rest frame and considering a perfect fluid (homogeneous and isotrope), as

$$T_0^{\mu\nu} = \begin{pmatrix} \rho & 0 & 0 & 0 \\ 0 & P & 0 & 0 \\ 0 & 0 & P & 0 \\ 0 & 0 & 0 & P \end{pmatrix} \tag{A.96}$$

To generalize $T^{\mu\nu}$ in *any* referential frame, it should be in a tensor form with $T_0^{\mu\nu}$ as a limit in the rest frame. The only covariant form can be written $T^{\mu\nu} = Au^\mu u^\nu + B\eta^{\mu\nu}$, with $u^\mu = \frac{dx^\mu}{d\tau}$. Demanding to recover (A.96) in the rest frame, we

[10] The tensor $T^{\mu\nu}$ can always been defined up to a constant.

obtain

$$T^{\mu\nu} = (\rho + P)\frac{dx^\mu}{d\tau}\frac{dx^\nu}{d\tau} - \eta^{\mu\nu}P = (\rho + P)\frac{dx^\mu}{d\tau}\frac{dx^\nu}{d\tau} - g^{\mu\nu}P, \qquad \text{(A.97)}$$

where, to generalize $\eta^{\mu\nu} \to g^{\mu\nu}$, we have used the fact that any curved metric can be viewed as a Minkowski metric locally.

A.2.10 Deflection Angle

As it is well known, the general theory of relativity has been confirmed in 1919 by Eddington when he measured a deviation in the light emitted by stars and passing near the Sun during the 1919 eclipse. It is obviously beyond the scope of this book (and this appendix) to give a complete mathematical description of the theory of General Relativity (I let the reader jump to [2] for a nice introduction in the subject). The basics of deflected light, or even the computation of the perihelion of Mercury is based on the fact that light as particles move on a timeline geodesic, and so we classically study some of the geodesics of the Schwarzschild vacuum solution:

$$2K = (1 - 2m/r)\dot{t}^2 - (1 - 2m/r)^{-1}\dot{r}^2 - r^2\dot{\theta}^2 - r^2\sin^2\theta\dot{\phi}^2 \qquad \text{(A.98)}$$

with $2K = 0, +1$ or -1 depending on whether the tangent vector is null, or has positive or negative length, respectively. In other words, when computing the Mercury perihelion, $2K = 1$ whereas the light travels on a null geodesic and thus $2K_\gamma = 0$. For the light, the Eq. (A.98) then becomes (you can check it as an exercise)

$$\frac{d^2u}{d\phi^2} + u = 3mu^2 \qquad \text{(A.99)}$$

with $u = 1/r$. In the limit of special relativity, m vanishes and the equation becomes

$$\frac{d^2u}{d\phi^2} + u = 0, \qquad \text{(A.100)}$$

the general solution of which can be written in the form

$$u = \frac{1}{b}\sin(\phi - \phi_0), \qquad \text{(A.101)}$$

where b is the impact parameter, the distance of closest approach to the origin (see Fig. A.3 for illustration). This is the equation of a straight line as ϕ goes from ϕ_0 to $\phi_0 + \pi$. The straight line motion is the same as is predicted by Newtonian theory.

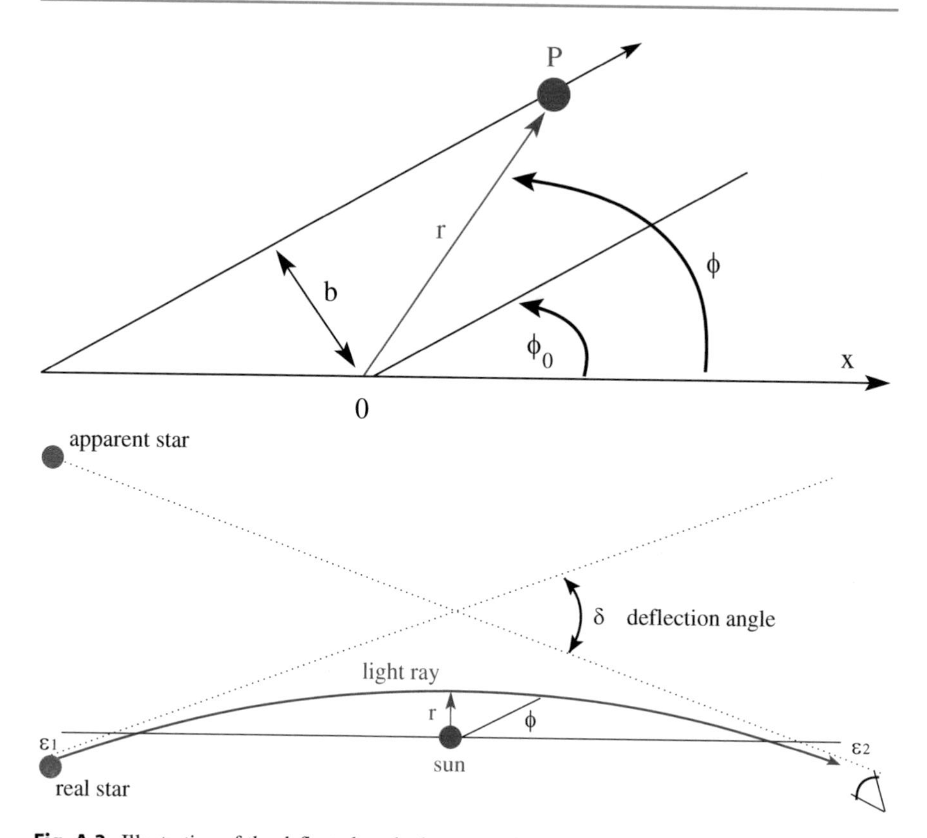

Fig. A.3 Illustration of the deflected angle due to gravitational field in General Relativity

The equation of a light ray in Schwarzschild space-time can be thought of as a perturbation of the classical equation, treating m/r as small. We therefore look for a solution of the form

$$u = u_0 + 3mu_1, \tag{A.102}$$

where u_0 is the solution of (A.101). Taking $\phi_0 = 0$ for convenience, we obtain

$$u_1'' + u_1 = u_0^2 = \frac{\sin^2\phi}{b^2} \quad \Rightarrow \quad u \simeq \frac{\sin\phi}{b} + \frac{m(1 + C\cos\phi + \cos^2\phi)}{b^2}, \tag{A.103}$$

where C is an arbitrary constant of integration for u_1. Since m/b is small, this is clearly a perturbation from straight line motion. We are interested in determining the angle of deflection δ for a light ray in the presence of a spherically symmetric source, such as the Sun. A long way from the source $r \to \infty$ and hence $u \to 0$, which requires the right-hand side of the Eq. (A.103) to vanish. Let us take the values of ϕ for which $r \to \infty$, that is the angles of the asymptote, to be $-\epsilon_1$ and

$\pi + \epsilon_2$, respectively, as shown in Fig. A.3. Using the small angle approximation for ϵ_1 and ϵ_2, we get

$$-\frac{\epsilon_1}{b} + \frac{m}{b^2}(2 + C) = 0, \quad -\frac{\epsilon_2}{b} + \frac{m}{b^2}(2 - C) = 0. \tag{A.104}$$

Adding, we find

$$\delta = \epsilon_1 + \epsilon_2 = \frac{4m}{b} \tag{A.105}$$

or, in non-relativistic units,

$$\delta = \frac{4GM}{c^2 b} \tag{A.106}$$

A.3 Matter/Radiation Domination

As we have already seen , the radiation[11] ($\gamma\nu$) and matter (M) densities are directly linked to the scale R of the Universe by (see the preceding paragraph)

$$\rho_\gamma \propto a^{-4} \qquad \rho_\nu \propto a^{-4} \qquad \rho_M \propto a^{-3}. \tag{A.107}$$

From this relation, one can understand that with time, and the expansion of the Universe, the radiation density decreases faster than the matter density. Nowadays, the last measurements of the present densities give $\rho^0_{\gamma\nu} \simeq 7.8 \times 10^{-34}(T_0/2.725)^4$ g/cm^3 and $\rho^0_M \simeq 1.88 \times 10^{-29}\Omega_M h^2$ g/cm^3. From Eq. (A.107) we obtain $\rho_{\gamma\nu}(t)/\rho_M(t) \propto a_0/a(t) = 1 + z$ one can deduce the radius/redshift/temperature of equal matter and radiation densities (a_{EQ}, z_{EQ}, T_{EQ}) with $T_0 = 2.725$ K and $\Omega_M h^2 \simeq 0.146$:

$$1 + z_{EQ} = a_0/a_{EQ} = 2.3 \times 10^3 \tag{A.108}$$

$$T_{EQ} = T_0(1 + z_{EQ}) = 0.77 \text{ eV} = 9000K \tag{A.109}$$

$$t_{EQ} = 1.4 \times 10^5 \text{ years.} \tag{A.110}$$

One can also calculate directly T_{EQ} from $\frac{T_{EQ}}{T_0} = \frac{\Omega_m}{\Omega_{\gamma\nu}}$ with $T_0 = 2.75$ K, $\Omega_{\gamma\nu}h^2 = 4.15 \times 10^{-5}$ and $\Omega_m h^2 = 0.146$ gives $T_{EQ} = 9000$ K/ 0.81 eV/80,000 years.

[11] We will note $\rho_{\gamma\nu}$ the energy density of relativistic particle (CMB + ν) and ρ_γ the energy density of the CMB.

A.4 Thermodynamical Fundamental Relations

$$\begin{aligned}
g_\rho(today) &= 3.36; \quad g_s(today) = 3.91 \\
\rho_R &= \frac{\pi^2}{30} g_* T^4 = 8.09 \times 10^{-34} \text{g cm}^{-3} \text{ today} \\
s &= \frac{2\pi^2}{45} g_{*s} T^3 \simeq 2909 \text{ cm}^{-3} \text{ today} \\
n_\gamma &= \frac{2\zeta(3)}{\pi^2} T^3 = 411 \text{ cm}^{-3} \text{ today}
\end{aligned} \tag{A.111}$$

Average energy per particles:

$$\langle E \rangle_{E \gg m} \equiv \frac{\rho}{n} = \frac{\pi^2}{30\zeta(3)} T \simeq 2.701\, T \quad \text{(BOSE)} \tag{A.112}$$

$$\langle E \rangle_{E \gg m} = \frac{7\pi^4}{180\zeta(3)} T \simeq 3.151\, T \quad \text{(FERMI)}$$

$$\langle E \rangle_{E \ll m} = m + \frac{3}{2} T$$

with $\zeta(3) \simeq 1.20206$.

$$\frac{n_B}{n_\gamma} = \eta = 2.68 \times 10^{-8}\ \Omega_B h^2. \tag{A.113}$$

A.5 Classical Thermodynamic: The Laplace's Law

The laws that are used in thermodynamics of the early Universe are the ones corresponding to a gas of particle transforming adiabatically. We demonstrate in this section the Laplace law PV^γ = constant, valid in such transformations. The first principle of thermodynamics affirms that " the change in the internal energy dU of a closed system is equal to the amount of heat δQ supplied to the system, plus the amount of work δW received by the system from its surroundings" (or equivalently "minus the amount of work done by the system on its surroundings"). In the case of a thermodynamical system, only the internal energy varies

$$dU = \delta W + \delta Q.$$

The mechanical work is the product of the external pressure P multiplied by the variation of volume dV, or, if we consider the work *received* by the system, one should write

$$\delta W = -PdV.$$

Moreover, if the process is adiabatic it is generated without exchange of heat, $\delta Q = 0$. One then obtains

$$dU = -PdV.$$

If one considers the enthalpy of the system, $H = U + PV$, one can write for an adiabatic transformation

$$dH = VdP.$$

Supposing that the gas is perfect, the enthalpy and internal energy depend only on the temperature T

$$dU = C_v dT, \quad dH = C_p dT,$$

where C_v and C_p (in units of Joule per Kelvin) are, respectively, the specific heat at constant volume and pressure.[12] We can then deduce

$$C_p dT = VdP$$
$$C_v dT = -PdV$$

which gives after combining both equations

$$\gamma \frac{dV}{V} + \frac{dP}{P} = 0, \quad \Rightarrow \quad PV^{\gamma} = \text{constant} \tag{A.114}$$

with $\gamma = \frac{C_p}{C_v}$. This is the Laplace's law. Another interesting relation is between internal energy and pressure, called the *equation of state* of the gas. If one defines α by $P = \alpha u$, u being the density of internal energy (kinetic one in the absence of other sources and in adiabatic processes), one can write $PV = \alpha U$. Combining with Eq. (A.114) and remembering that $dU = -PdV$ one obtains

$$\alpha dU = PdV + VdP = PdV - \gamma PdV = PdV(1-\gamma) = -\alpha PdV \Rightarrow \alpha = \gamma - 1. \tag{A.115}$$

[12] A specific heat, or heat capacity is the heat required to raise a unit mass of the material by one degree. This can be done at constant volume or at constant pressure and the corresponding symbols are C_v and C_p.

A.6 Tooling with Math

A.6.1 Function $\Gamma(z)$

A.6.1.1 Definition, Propriety

$$\begin{aligned}
&\Gamma(z) = \int_0^\infty t^{z-1} e^{-t} dt \\
&\Gamma(z+1) = z\Gamma(z) \\
&\Gamma(n) = (n-1)!
\end{aligned} \tag{A.116}$$

A.6.1.2 Some Values

$$\begin{aligned}
&\Gamma(-3/2) = \frac{4}{3}\sqrt{\pi} \approx 2.36 \\
&\Gamma(-1/2) = -2\sqrt{\pi} \approx -3.54 \\
&\Gamma(1/2) = \sqrt{\pi} \approx 1.77 \\
&\Gamma(1) = 0! = 1 \\
&\Gamma(3/2) = \frac{1}{2}\sqrt{\pi} \approx 0.89 \\
&\Gamma(2) = 1! = 1 \\
&\Gamma(5/2) = \frac{3}{4}\sqrt{\pi} \approx 1.33 \\
&\Gamma(3) = 2! = 2 \\
&\Gamma(7/2) = \frac{15}{8}\sqrt{\pi} \approx 3.32 \\
&\Gamma(4) = 3! = 6.
\end{aligned} \tag{A.117}$$

A.6.2 The Riemann Zeta Function $\zeta(z)$

A.6.2.1 Definition, Propriety

$$\zeta(z) = \sum_1^\infty n^{-z} = \frac{1}{1^z} + \frac{1}{2^z} + \frac{1}{3^z} + \frac{1}{4^z} + \frac{1}{5^z} + \dots \tag{A.118}$$

A.6.2.2 Some Values

$$\begin{aligned}
&\zeta(0) = -\frac{1}{2} \\
&\zeta(1/2) \approx -1.46
\end{aligned}$$

$$
\begin{aligned}
\zeta(1) &= \infty \\
\zeta(3/2) &\approx 2.61 \\
\zeta(2) &\approx 1.64 = \frac{\pi^2}{6} \\
\zeta(3) &\approx 1.2 \\
\zeta(4) &\approx 1.08 = \frac{\pi^4}{90} \\
\zeta(6) &= \frac{\pi^6}{945} \\
\zeta(8) &= \frac{\pi^8}{9450} \\
\zeta(10) &= \frac{\pi^{10}}{93555}.
\end{aligned}
\tag{A.119}
$$

A.6.3 Modified Bessel Function of the 2nd Kind $K_n(z)$

A.6.3.1 Definition, Propriety

$$
\begin{aligned}
& z^2 K_n''(z) + z K_n'(z) - (z^2 + n^2) K_n(z) = 0 \\
& K_n(z) = \frac{\sqrt{\pi}\, z^n}{2^n \Gamma(n + \frac{1}{2})} \int_1^\infty e^{-zt} (t^2 - 1)^{n - \frac{1}{2}} dt = \frac{\Gamma(n + \frac{1}{2})(2z)^n}{\sqrt{\pi}} \int_0^\infty \frac{\cos t dt}{(t^2 + z^2)^{n + \frac{1}{2}}} \\
& \frac{n}{z} K_n(z) - K_n'(z) = K_{n+1}(z) \\
& K_0(z) \sim -\log z \\
& K_1(z) = z \int_1^\infty e^{-zt} \sqrt{t^2 - 1} dt \\
& K_2(z) = \frac{z^2}{3} \int_1^\infty e^{-zt} (t^2 - 1)^{3/2} dt = K_1'(z) - \frac{1}{z} K_1(z) = z \int_1^\infty t e^{-zt} \sqrt{t^2 - 1} dt \\
& z \to \infty, \quad K_n(z) \simeq \sqrt{\frac{\pi}{2z}} e^{-z}.
\end{aligned}
\tag{A.120}
$$

A.6.3.2 Some Values

$$
\begin{gathered}
K_1(0.01) = 99.97; \quad K_1(0.1) = 9.84; \quad K_1(1) = 0.60; \\
K_1(10) = 1.9 \times 10^{-5}; \quad K_1(100) = 4.7 \times 10^{-45}
\end{gathered}
$$

$$K_2(0.01) = 20000; \quad K_2(0.1) = 200; \quad K_2(1) = 1.63;$$
$$K_2(10) = 2. \times 10^{-5}; \quad K_2(100) = 4.7 \times 10^{-45}$$
$$\int_0^{\infty} z^2 K_1(z) dz = 2; \quad \int_0^{\infty} z^4 K_1(z) dz = 16;$$
$$\int_0^{\infty} z^3 K_2(z) dz = 8; \quad \int_0^{\infty} z^5 K_2(z) dz = 96;$$

A.6.3.3 Some Approximations

For $x \gtrsim 10$, we can write

$$K_1(x) \simeq \left(\frac{\pi}{2x}\right)^{\frac{1}{2}} e^{-x}\left[1 + \frac{3}{8x}\left(1 - \frac{5}{16x}\left(1 - \frac{21}{24x}\right)\right)\right]. \tag{A.121}$$

A.6.4 Useful Integrals

$$\int e^{-cx^2} = \sqrt{\frac{\pi}{4c}}\text{erf}(\sqrt{c}x) \quad \text{erf being the error function}$$
$$\int xe^{-cx^2} = -\frac{1}{2c}e^{-cx^2}$$
$$\int x^2 e^{-cx^2} = \sqrt{\frac{\pi}{16c^3}}\text{erf}(\sqrt{c}x) - \frac{1}{2c}xe^{-cx^2}$$
$$\int x^3 e^{-cx^2} = -\frac{x^2}{2c}e^{-cx^2} - \frac{1}{2c^2}e^{-cx^2}$$
$$\int x^4 e^{-cx^2} = \frac{3}{8}\sqrt{\frac{\pi}{c^5}}\text{erf}(\sqrt{c}x) - \frac{e^{-cx^2}}{4c^2}[3x + 2cx^3]$$
$$\int x^5 e^{-cx^2} = -\frac{e^{-cx^2}}{c^3}\left[1 + cx^2 + \frac{c^2}{2}x^4\right]$$
$$\int x^6 e^{-cx^2} = \frac{15}{16}\sqrt{\frac{\pi}{c^7}}\text{erf}(\sqrt{c}x) - e^{-cx^2}\left[\frac{15}{8c^3}x + \frac{5}{4c^2}x^3 + \frac{1}{2c}x^5\right]$$
$$\int \frac{e^{cx}}{x}dx = \ln|x| + \sum_{n=1}^{\infty}\frac{(cx)^n}{n.n!}$$
$$\int \frac{e^{cx}}{x^n} = \frac{1}{n-1}\left(-\frac{e^{cx}}{x^{n-1}} + c\int \frac{e^{cx}}{x^{n-1}}\right) \quad (\text{for n} \neq 1)$$
$$\int_0^{+\infty} e^{-ax^2}dx = \frac{1}{2}\sqrt{\frac{\pi}{a}}$$

$$\int_0^{+\infty} x^2 e^{-ax^2} dx = \frac{1}{4}\sqrt{\frac{\pi}{a^3}}$$

$$\int_0^{+\infty} x^4 e^{-ax^2} dx = \frac{3}{8a^2}\sqrt{\frac{\pi}{a}}$$

$$\int_0^{+\infty} x^n e^{-ax^2} dx = \frac{(2k-1)!!}{2^{k+1}a^k}\sqrt{\frac{\pi}{a}} \quad \text{(n = 2k, k integer)}$$

$$\int_1^{\infty} t(t^2-1)^{1/2} e^{-zt} dt = \frac{1}{z}K_2(z)$$

$$\int_0^{\infty} dx \left(\frac{x^n}{e^x - \delta}\right) = \Gamma(n+1)\zeta(n+1)y(\delta) \tag{A.122}$$

with $y(\delta) = 1$ if $\delta = 1$ and $1 - \frac{1}{2^n}$ if $\delta = -1$. $n!!$ is the double factorial defined as $n!! = n \times (n-2) \times (n-4) \times ..$

$$\int_0^x t^{a-1} e^{-t} dt = \gamma(a, x), \tag{A.123}$$

$\gamma(a, x)$ being the incomplete gamma function that can be approximate

$$\gamma(a, x) \simeq \frac{x^a}{a}, \quad \text{for } x \ll 1. \tag{A.124}$$

And finally,

$$\int_0^{\infty} \frac{dx}{e^x + 1} = \log(2)\ ; \quad \int_0^{\infty} \frac{dx}{e^x - 1} = \infty \tag{A.125}$$

$$\int \frac{1}{ae^{\lambda x} + b} dx = \frac{x}{b} - \frac{1}{b\lambda} \ln\left(ae^{\lambda x} + b\right). \tag{A.126}$$

A.6.4.1 Euler–Masheroni Constant γ

$$\gamma = 0.577 \tag{A.127}$$

A.6.4.2 Gauss Error Function erf

$$\text{erf}(x) = \frac{2}{\sqrt{\pi}} \int_0^x e^{-t^2} dt. \tag{A.128}$$

A.6.4.3 Delta Dirac δ

One useful definition of the Dirac-δ function is

$$\delta(x-\alpha) = \frac{1}{2\pi}\int_{-\infty}^{+\infty} e^{ip(x-\alpha)}dp \tag{A.129}$$

$$\int_{-\infty}^{+\infty} \delta(x)dx = 1, \quad \int_{-\infty}^{+\infty} f(x)\delta(x-\alpha)dx = f(\alpha) \tag{A.130}$$

$$\delta(g(x)) = \frac{\delta(x-x_0)}{|g'(x_0)|}, \tag{A.131}$$

where x_0 is the root of the function $g(x)$. If $g(x)$ possesses several roots, one has to sum over them.

A.6.5 Laplace Operator

In mathematics, the Laplace operator Δ, or Laplacian, is given by the divergence of the gradient of a function f:

$$\Delta f = \nabla^2 f = \nabla.\nabla f, \tag{A.132}$$

where, in the n Euclidian space ∇ is defined by

$$\nabla = \left(\frac{\partial}{\partial x_1}, \ldots, \frac{\partial}{\partial x_n}\right), \tag{A.133}$$

which gives, in the Cartesian coordinates x_i :

$$\Delta f = \sum_i^n \frac{\partial^2 f}{\partial x_i^2}. \tag{A.134}$$

In (r, θ) polar coordinates, we can write

$$\Delta f = \frac{1}{r}\frac{\partial}{\partial r}\left(r\frac{\partial f}{\partial r}\right) + \frac{1}{r^2}\frac{\partial^2 f}{\partial \theta^2} = \frac{\partial^2 f}{\partial r^2} + \frac{1}{r}\frac{\partial f}{\partial r} + \frac{1}{r^2}\frac{\partial^2 f}{\partial \theta^2}, \tag{A.135}$$

whereas in (r, θ, z) cylindrical coordinates, we obtain

$$\Delta f = \frac{1}{\rho}\frac{\partial}{\partial \rho}\left(\rho\frac{\partial f}{\partial \rho}\right) + \frac{1}{\rho^2}\frac{\partial^2 f}{\partial \phi^2} + \frac{\partial^2 f}{\partial z^2}, \tag{A.136}$$

and in (r, θ, ϕ) spherical coordinates

$$\Delta f = \frac{1}{r^2}\frac{\partial}{\partial r}\left(r^2 \frac{\partial f}{\partial r}\right) + \frac{1}{r^2 \sin\theta}\frac{\partial}{\partial \theta}\left(\sin\theta \frac{\partial f}{\partial \theta}\right) + \frac{1}{r^2 \sin^2\theta}\frac{\partial^2 f}{\partial \phi^2}. \tag{A.137}$$

Particle Physics

B

B.1 Feynman Rules

In this section[1] we summarize point-by-point how to compute a decay rate and a cross section, from the amplitude of the process $\mathcal{M}$. We then make explicit the rules to compute the amplitude $\mathcal{M}$ in any microscopic model.

B.1.1 Decay Rates and Cross Sections

To compute a differential rate, with n particles in the final state, one should follow similar rules if you consider a decay rate or a scattering cross section. In the former case, for a decaying particle of energy E, the differential width $d\Gamma_n$ is obtained by multiplying the following factors:

- A factor of $(2\pi)^4\delta^4(P_f - P_i)$ where P_f is the total four-momentum of the n decay products and P_i is the four-momentum of the decaying particle, representing the condition of energy–momentum conservation;

[1]This section is freely adapted from the excellent book by Franz Gross "Relativistic Quantum Mechanics and Field Theory", Wiley Eds.

Y. Mambrini, *Particles in the Dark Universe*,
https://doi.org/10.1007/978-3-030-78139-2

- A factor

$$\frac{d^3 p_i}{(2\pi)^3 E_i}$$

for each particle in the final state where p_i and E_i are the momentum and energy of the ith particle;
- A factor $\frac{1}{2E}$ for the initial particle which is decaying;
- And the absolute square of the $\mathcal{M}$-matrix.

The differential width is then (we let the reader go through the Sect. B.4.1 for a more detailed analysis of the cross section computation, inspired by the Fermi's golden rule)

$$d\Gamma_n = (2\pi)^4 \delta^4(P_f - P_i) \frac{1}{2E} \prod_{i=1}^{i=n} \frac{d^3 p_i}{(2\pi)^3 2E_i} |\mathcal{M}|^2. \tag{B.1}$$

The total decay rate is obtained by integrating Eq. (B.1) over all outgoing momenta, *summing* over all final spins and *averaging* over the initial spin, S, of the decaying particle.

$$\Gamma = \int \frac{1}{2S+1} \sum_{\text{spins}} d\Gamma_n. \tag{B.2}$$

We usually note $|\bar{\mathcal{M}}|^2$ the mean of the amplitude square, summed over the final spins, averaged over the initial spin.

The differential cross section for the production of n particles is obtained multiplying the following factors:

- A factor of $(2\pi)^4 \delta^4(P_f - P_i)$where P_f is the total four-momentum of the n final states and P_i is the four-momentum of the two initial particles, representing the condition of energy–momentum conservation;
- A factor

$$\frac{d^3 p_i}{(2\pi)^3 E_i}$$

for each particle in the final state where p_i and E_i are the momentum and energy of the ith particle;

- A factor $\frac{1}{4EE'}$ where E and E' are the energies of the two particles in the initial states;
- A factor of $1/v$ where v is the flux, or relative velocity of the two (colinear) colliding particles, equal to

$$v = \frac{p}{E} + \frac{p'}{E'} \tag{B.3}$$

where p and p' are the magnitudes of their momenta, and
- The absolute square of the $\mathcal{M}$-matrix.

The differential cross section can then be written as

$$d\sigma = (2\pi)^4 \delta^4 (P_f - P_i) \frac{1}{4EE'v} \prod_{i=1}^{i=n} \frac{d^3 p_i}{(2\pi)^3 2E_i} |\mathcal{M}|^2 \ . \tag{B.4}$$

The unpolarized cross section for scattering into some final state in the phase volume $\Delta\Omega$ is therefore obtained by integrating Eq. (B.4) over all outgoing momenta in $\Delta\Omega$, *summing* over all final spins and *averaging* over initial spins:

$$\Delta\sigma = \int_{\Delta\Omega} \frac{1}{(2S+1)(2S'+1)} \sum_{\text{spins}} d\sigma \ , \tag{B.5}$$

where S and S' are the spins of the initial particles. Finally, in calculating both decay rates and cross sections, for *each* set of m identical particles in the final states, the integral over momenta must *either* be divided by $m!$ or limited to the restricted cone $\theta_1 < \theta_2 < \ldots < \theta_m$.

B.2 Feynman Rules

B.2.1 General Rules

We give in this section the generic rules to compute the amplitude $\mathcal{M}$.

- The diagrams consist of *lines* and *vertices*.
- Each internal line represents the propagation of a virtual particle, between two vertices which represent points in space-time. Each vertex is determined by an interaction term in the Lagrangian, where particles can be destroyed and/or created.

- Energy conservation fixes the four-momentum of each internal lines. If diagrams contain loops, there exists at least one momentum for each loop which cannot be determined, and should be integrated over.

Each diagram should be associated with a number with the following rules:

Rule 0

Multiplication by an overall factor of i, which is reminiscent of the form of the $\mathcal{S}$-matrix $\mathcal{S} = e^{i\mathcal{T}}$, see Eq. (B.83).

Rule 1

An operator for each vertex, determined by the form of the Lagrangian.

Rule 2

A propagator for each internal line with four-momentum k , the precise form of which depends on the particle propagating. For spin zero bosons with isospin indices (i,j), for fermions with Dirac indices (α,β), and for photons or massive vector bosons with polarization indices (μ,ν), they are written as

$$\text{Spin } 0 : \Delta_{ij}(k) = \frac{i\delta_{ij}}{m^2 - k^2 - i\epsilon}$$

$$\text{Spin } 1/2 : \Delta_{\alpha\beta}(k) = \frac{i(m + \not{k})_{\alpha\beta}}{m^2 - k^2 - i\epsilon}$$

$$\text{photon or gluon} : \Delta_{\mu\nu}(k) = \frac{-i}{-k^2 - i\epsilon}\left[\eta_{\mu\nu} - \frac{k_\mu k_\nu}{k^2}(1-\alpha)\right]$$

$$\text{vector boson} : \Delta_{\mu\nu}(k) = \frac{-i[\eta_{\mu\nu} - k_\mu k_\nu/m^2]}{m^2 - k^2 - i\epsilon}$$

energy momentum conservation determines k as function of the external momenta, whereas α appearing in the case of massless gauge boson (photon and gluon) depends on the chosen gauge : $\alpha = 0$ in the Landau Gauge whereas $\alpha = 1$ in the Feynman gauge for instance.

Rule 3

- Multiply from the *left* by $\bar{u}(p_3, s_3)$ for each *outgoing fermion* with momentum p_3 and spin s_3.
- Multiply from the *right* by $u(p_1, s_1)$ for each *incoming fermion* with momentum p_1 and spin s_1.
- Multiply from the *right* by $v(p_4, s_4)$ for each *outgoing antifermions* with momentum p_4 and spin s_4.
- Multiply from the *left* by $\bar{v}(p_2, s_2)$ for each *incoming antifermion* with momentum p_2 and spin s_2.
- Multiplying by ϵ^*_μ for each *outgoing* vector with polarization index μ.
- Multiplying by ϵ_μ for each *incoming* vector with polarization index μ.

These last polarization vectors should be contracted with the γ^μ matrices emerging from the vertex implying fermions.

Rule 4

Symmetrize between identical bosons in the initial or final state, *antisymmetrize* between identical fermions in the initial or final state. Concretely speaking, it means that if we have 2 identical bosons in the final state, one should divide the amplitude square by 2, as well as for Majorana final states (see below), the antisymmetrization cancelling the minus sign arising from the exchange of two spinors. In other words, $|\mathcal{M}|^2 = \frac{1}{2}|\mathcal{M}_1 + \mathcal{M}_2|^2$, where $\mathcal{M}_i$ are the amplitudes of the 2 identical bosons (or fermions with a minus sign).

Rule 5

Multiply by the momentum conservation factor $(2\pi)^4\delta^{(4)}(\sum_i p_i)$, p_i being the external momenta and integrate over each internal four-momentum k left undetermined

$$\int \frac{d^4k}{(2\pi)^4}.$$

Rule 6

Add a minus sign to each closed fermion loop.

Rule 7

Multiply by $\frac{1}{n!}$ for bubbles with $n!$ identical neutral bosons.

Rule 8

Include the field renormalization factors $\sqrt{Z_i}$ for each external particle i.

Rule 9

$\mathcal{M}$ is the sum over all topologically distinct diagrams with equal weight except for a relative (-1) if two diagrams differ by the permutation of two fermion operators. This gives for instance a relative sign between the two diagrams contributing to e^-e^- scattering and similarly the two contributing to Bhabha scattering.

Rule 10

For each particle with a mass which could be shifted by self-interactions, $m \to m + \delta m$, a mass counter-term $i\delta m$ is added to remove the mass shift.

B.2.2 Majorana Rules

In the case of Majorana fermions, the Feynman rules have to be adapted. Indeed, Majorana flows violate the fermion number, and we cannot compute the amplitude exactly the same way. We compute explicit examples in Sect. B.4.8. Please, read also the Sect. B.3.8 before this one to recall the clear definition of a Majorana fermion and its Lagrangian (which differs from a Dirac one by a factor 1/2). There are so many factors 2 in the calculation, that it is easy to be lost, and computation should be made carefully to avoid double counting.

Our point of departure will be the Lagrangian (B.62). We will take the specific example of the Majorana coupling to a Higgs-like field ϕ. The Majorana part is

$$\frac{1}{2} y_\lambda \phi \bar{\lambda} \lambda, \tag{B.6}$$

where we took $\Gamma = \mathbb{1}$, but the reasoning is valid for any kind of coupling structure. While treating Majorana fermions, the Feynman rules for identical particles in the final states apply: one should multiply by 2 the coupling which appears in the lagrangian (so a factor 4 in the amplitude square) and divide by 2 the physical process (antisymmetrisation because 2 identical fermions in the final state). We will explain the former factor "2" in detail in this section.
Indeed, to compute an amplitude, one needs to develop λ into the creation and destruction operators[2] $b^\dagger$ and b.

$$\lambda = \int \frac{d^3k}{(2\pi)^3 2E} (b(k) u e^{-ikx} + b(k)^\dagger v e^{ikx}) \tag{B.7}$$

with u and v the Dirac spinors studied in Sect. B.3.2. If one looks for any process, the fact of having two "b's" in the development of λ will generate a factor 2 of symmetry. This fundamentally comes from the fact that one cannot trace a fermion flow with Majorana fermion and "entering" or "exiting" flows can combine together.

Let us take a concrete example. A particle ϕ decaying into two λ's

$$\begin{aligned} \langle \lambda(p_1)\lambda(p_2)|\mathcal{L}|\phi\rangle &= \langle 0|b(p_1)b(p_2)\ \frac{1}{2} y_\lambda \phi \lambda\lambda\ a_\phi^\dagger|0\rangle \\ &= \frac{y_\lambda}{2}[\bar{u}(p_1)v(p_2) + \bar{u}(p_2)v(p_1)] = y_\lambda \bar{u}(p_1)v(p_2), \end{aligned}$$

where we used the identity for Majorana particles in the last equality, $u = C\bar{v}^T$ and $v = C\bar{u}^T$. As one notices, the factor $\frac{1}{2}$ in front of y_λ disappeared, and we recover *the same* Feynman rule than in the case of the Dirac fermion. That is the fundamental point, and main result of this section. For a concrete example, have a look at Sect. B.4.8 where we computed the decay of a scalar into Majorana and Dirac particles.

B.2.2.1 Another Interpretation

One can find another interpretation in the literature in [1]. This interpretation released on the fermionic fluxes. Indeed, whereas for general fermion fields χ, a

---

[2] In the case of a Dirac fermions, $b^\dagger$ should be replaced by $d^\dagger$, because a Dirac fermion ψ does not follow $\psi^C = \psi$.

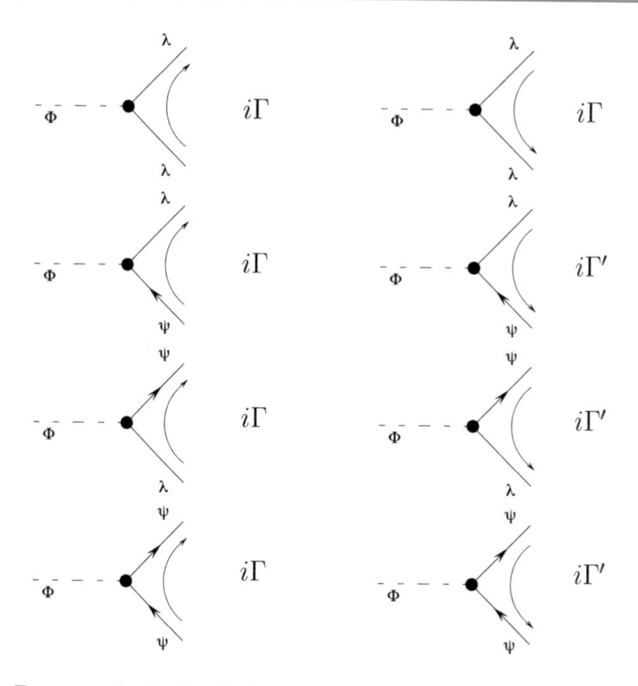

Fig. B.1 The Feynman rules for fermionic vertices with orientation (thin arrow)

Lagrangian can be written as

$$\bar{\chi}\Gamma\chi = \bar{\tilde{\chi}}\Gamma'\tilde{\chi}, \tag{B.8}$$

where Γ is the coupling matrix, $\tilde{\chi} = C\bar{\chi}^T$ (C being the charge conjugate operator (see Sect. B.3.7)) and $\Gamma' = C\Gamma^T C^{-1}$. This means that the *same term in the Lagrangian* will generate a process with *particle*(p_1)–*antiparticle*(p_2) in the final state, and the symmetric one, *antiparticle*(p_1)–*particle*(p_2), p_1 and p_2 being their respective momentum. Both final states are of course *different*. For a Majorana final state $\tilde{\chi} = \chi$. Therefore, if both $\chi's$ are Majorana fermions, $\Gamma = \Gamma'$. In other way, there is no flow of fermions, and one should compute matrix elements with Majorana states without distinguishing the possible flows of particles/antiparticles. This is illustrated more clearly in Fig. B.1 where we show the Feynman rules for fermionic vertices with orientation. As we can see, the two figures on top are the same, and one should not add them in a computation as it will be double counting. On the other hand, the two last figures represent the Dirac case, and two different states of particle/antiparticle with exchanged momentum. Only one of this diagram should also contribute to the process. In conclusion, for the computation of the matrix element, in the Majorana as in the Dirac case, one should count *only* one flow, as in one case (Dirac) the other flow correspond to another final state with exchanged momentum, whereas in Majorana case, that would be double counting.

Let us be more concrete. In the computation in a rate, implying two Majorana fermions in the final state, it is important to add a symmetry factor in the final result. Let us consider for instance the simplest example of the decay of a scalar ϕ into two Majorana fermions λ compared to a (Dirac) pair of electron positron final state of momentum p_1 and p_2. The Lagrangian

$$\mathcal{L} = y^e_\phi \, \phi \bar{e} e + \frac{1}{2} y^\lambda_\phi \phi \bar{\lambda}\lambda \tag{B.9}$$

gives, respectively, the width (see Eq. B.165) $\Gamma^i_\phi = \frac{|\mathcal{M}|^2}{16\pi \times sym \; m_\phi}$

$$\Gamma^{e^+e^-}_\phi = \frac{(y^e_\phi)^2 \, m_\phi}{8\pi}, \quad \Gamma^{\lambda\lambda}_\phi = \frac{(y^\lambda_\phi)^2 \, m_\phi}{16\pi}, \tag{B.10}$$

where we neglected the final state masses. Indeed, in the Dirac (electron) case, the amplitude is $\mathcal{M} = -i y^e_\phi \bar{u}(p_1) v(p_2)$ which implies $\sum_{spin} |\mathcal{M}|^2 = 4(y^e_\phi)^2 p_1.p_2 = 2(y^e_\phi)^2 m^2_\phi$, and no symmetry factor sym is present for Dirac final state because both particles are different. In the Majorana case λ, as we explained above, only one flow should be considered to avoid double counting. Then, $\sum_{spin} |\mathcal{M}|^2 = 4(y^\lambda_\phi)^2 p_1.p_2 = 2(y^\lambda_\phi)^2 m^2_\phi$, exactly as in the Dirac case. The $\left(\frac{1}{2}\right)$ symmetry factor sym between the two width can be understood by a look at Fig. B.1. For a given final state angle, the e^+e^- system is definite, whereas there are two possibilities for the $\lambda\lambda$ final state : both can have p_1 and/or p_2. There is then a double counting once we integrate on the total solid angle, and one should symmetrize it to take it into account, multiplying the final result by $\left(\frac{1}{2}\right)$.

B.2.3 Standard Model Couplings

We show in the following diagrams the Feynman rules in the Standard Model context and its simplest extension (Feynman rules = $i\mathcal{L}\times$ symmetry factors). In the case of the anomalies mediated trivectorial couplings, we are based on the Lagrangian

$$\mathcal{L}_{CS} = \alpha_1 \, \epsilon^{\mu\nu\rho\sigma} Z'_\mu Z_\nu F^Y_{\rho\sigma} + \alpha_2 \, \epsilon^{\mu\nu\rho\sigma} Z'_\mu Z_\nu F'_{\rho\sigma} \,, \tag{B.11}$$

which gives the interactions

$$\begin{aligned}
\Gamma^{\mu\nu\sigma}_{Z'ZZ}(p_3; p_1, p_2) &= 2\alpha_1 s_W \, \epsilon^{\mu\nu\rho\sigma} (p_1 - p_2)_\rho \,, \\
\Gamma^{\mu\nu\sigma}_{Z'Z\gamma}(p_3; p_1, p_2) &= 2\alpha_1 c_W \, \epsilon^{\mu\nu\rho\sigma} (p_2)_\rho \,, \\
\Gamma^{\mu\nu\sigma}_{Z'ZZ'}(p_3; p_1, p_2) &= 2\alpha_2 \, \epsilon^{\mu\nu\rho\sigma} (p_2 - p_3)_\rho \,,
\end{aligned} \tag{B.12}$$

with the obvious notation $s_W = \sin\theta_W$ and $c_W = \cos\theta_W$. Notice the decomposition $Y_\mu = \cos\theta_W A_\mu - \sin\theta_W Z_\mu$ and the symmetrization for the identical final state $Z'_\mu Z_\nu(p_1) Z_\sigma(p_2) \epsilon^{\mu\nu\rho\sigma}(p_2)_\rho = \frac{1}{2} Z'_\mu Z_\nu(p_1) Z_\sigma(p_2) \epsilon^{\mu\nu\rho\sigma}(p_2 - p_1)_\rho$ multiplied by the symmetry factor (2).

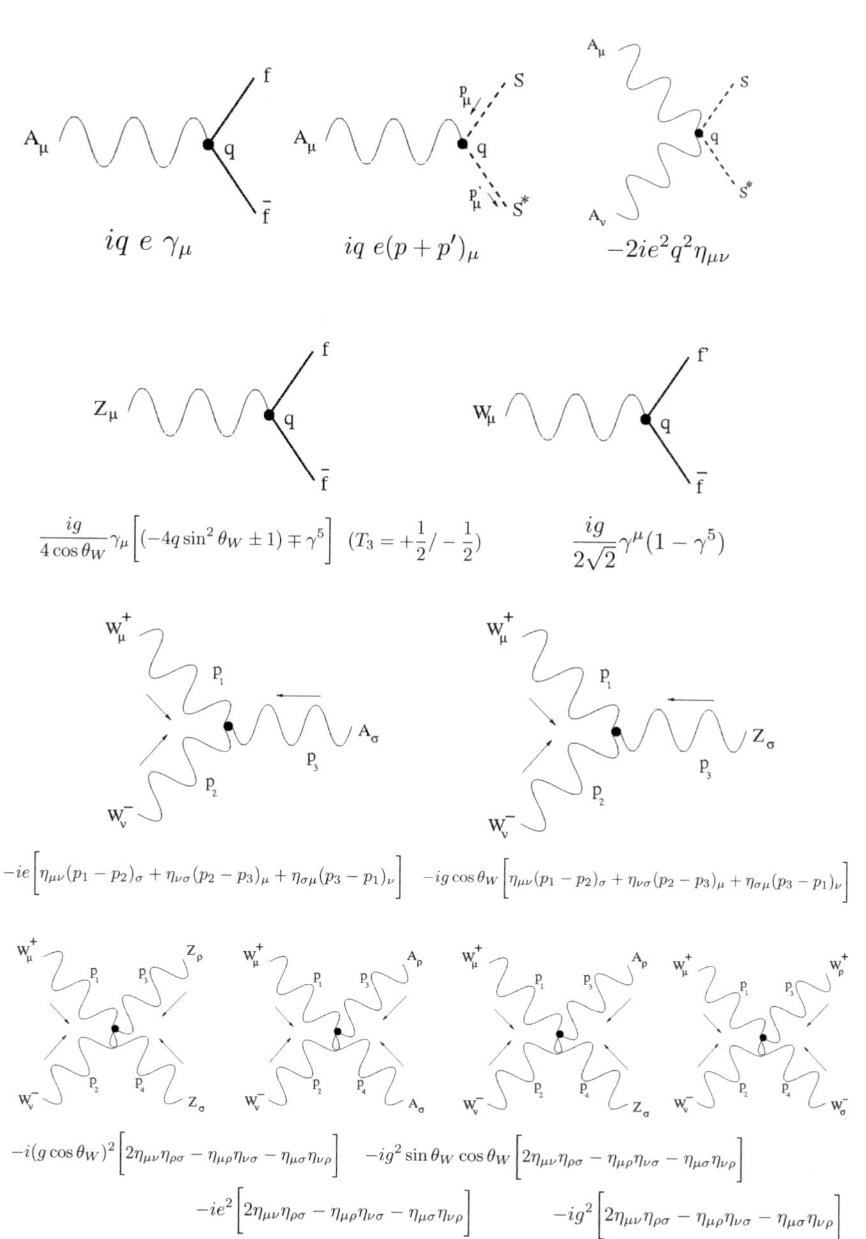

$$ig\, M_W\, \eta_{\mu\nu} \qquad \frac{ig}{\cos^2\theta_W} M_W\, \eta_{\mu\nu} \qquad \frac{i}{2}\left(\frac{g}{\cos\theta_W}\right)^2 \eta_{\mu\nu} \qquad \frac{i}{2} g^2 \eta_{\mu\nu}$$

$$-\frac{3i}{2}\left(\frac{gM_H}{M_W}\right)^2 \qquad -\frac{3i}{4}\left(\frac{gM_H}{M_W}\right)^2 \qquad -i\frac{Y_f}{\sqrt{2}} = -i\frac{m_f}{v} = -i\frac{gm_f}{2M_W}$$

$$m_f = Y_f\langle H\rangle = Y_f\frac{v}{\sqrt{2}}$$

$$2i\epsilon^{\mu\nu\rho\sigma}(p_2)_\rho = 2i\cos\theta_W(p_2^A)_\rho - 2i\sin\theta_W(p_2^Z - p_1^Z)_\rho \qquad 2i\epsilon^{\mu\nu\rho\sigma}(p_2 - p_3)_\rho$$

B.3 Diracology

B.3.1 Matrices

We present in this appendix the necessary tools to compute high energy processes, in the framework of quantum field theory.

Pauli Matrices

$$\sigma_1 = \begin{pmatrix} 0 & 1 \\ 1 & 0 \end{pmatrix} \quad \sigma_2 = \begin{pmatrix} 0 & -i \\ i & 0 \end{pmatrix} \quad \sigma_3 = \begin{pmatrix} 1 & 0 \\ 0 & -1 \end{pmatrix} \tag{B.13}$$

Dirac Matrices

$$\{\gamma^\mu, \gamma^\nu\} = \gamma^\mu\gamma^\nu + \gamma^\nu\gamma^\mu = 2\eta^{\mu\nu} I_4; \quad \gamma_\mu = \eta_{\mu\nu}\gamma^\nu \quad \{\gamma^5, \gamma^\mu\} = 0, \quad \gamma^0\gamma_\mu^\dagger\gamma^0 = \gamma_\mu \tag{B.14}$$

we name the operator $\sigma^{\mu\nu} = \frac{i}{2}[\gamma^\mu, \gamma^\nu] = \frac{i}{2}(\gamma^\mu\gamma^\nu - \gamma^\nu\gamma^\mu)$ with $\eta^{\mu\nu}$ the Minkowski metric with the signature (+,−,−,−) and I_4 the 4×4 identity matrix and

$$\gamma^i = \begin{pmatrix} 0 & \sigma_i \\ -\sigma_i & 0 \end{pmatrix} \tag{B.15}$$

which gives in explicit form,

$$\gamma^0 = \begin{pmatrix} 0&0&1&0 \\ 0&0&0&1 \\ 1&0&0&0 \\ 0&1&0&0 \end{pmatrix} \quad \gamma^1 = \begin{pmatrix} 0&0&0&1 \\ 0&0&1&0 \\ 0&-1&0&0 \\ -1&0&0&0 \end{pmatrix} \quad \gamma^2 = \begin{pmatrix} 0&0&0&-i \\ 0&0&i&0 \\ 0&i&0&0 \\ -i&0&0&0 \end{pmatrix}$$

$$\gamma^3 = \begin{pmatrix} 0&0&1&0 \\ 0&0&0&-1 \\ -1&0&0&0 \\ 0&1&0&0 \end{pmatrix} \quad \gamma^5 \equiv i\gamma^0\gamma^1\gamma^2\gamma^3 = \begin{pmatrix} -1&0&0&0 \\ 0&-1&0&0 \\ 0&0&1&0 \\ 0&0&0&1 \end{pmatrix}. \tag{B.16}$$

Another possible representation for γ^0 and γ^5, respecting the same commutation/anticommutation laws is

$$\gamma^0 = \begin{pmatrix} 1&0&0&0 \\ 0&1&0&0 \\ 0&0&-1&0 \\ 0&0&0&-1 \end{pmatrix} \quad \gamma^5 \equiv i\gamma^0\gamma^1\gamma^2\gamma^3 = \begin{pmatrix} 0&0&1&0 \\ 0&0&0&1 \\ 1&0&0&0 \\ 0&1&0&0 \end{pmatrix}. \tag{B.17}$$

Depending on the problem, it is sometimes easier to use conventions (B.16). This is the case when treating processes with interactions distinguishing left and right helicities for fermions, as in the Standard Model. Indeed, in this case, the eigenvectors are eigenvectors of γ^5, which is diagonal, rendering the calculations easier. On the other hand, when trying to find solutions of the Dirac equations, the convention (B.17), being diagonal in time derivative, gives simpler solutions. We will usually work respecting the definitions (B.16), explicitly indicating if we use convention (B.17).

B.3.2 Dirac Equation

The Dirac equation for a spinor ψ of mass m and energy $E^2 = p^2 + m^2$ should be written as

$$(i\gamma^\mu \partial_\mu - m)|\Psi\rangle = (i\gamma^0 \partial_t - i\gamma^i \partial_{x_i} - m)\,|\Psi\rangle = 0 \tag{B.18}$$

Using convention (B.17), the solutions of the equation are then

$$|\Psi_1\rangle = \sqrt{\frac{E+m}{2m}} \begin{pmatrix} 1 \\ 0 \\ \frac{p_3}{E+m} \\ \frac{p_1+ip_2}{E+m} \end{pmatrix} e^{-ipx} = u_1 e^{-ipx};$$

$$|\Psi_2\rangle = \sqrt{\frac{E+m}{2m}} \begin{pmatrix} 0 \\ 1 \\ \frac{p_1-ip_2}{E+m} \\ \frac{-p_3}{E+m} \end{pmatrix} e^{-ipx} = u_2 e^{-ipx};$$

$$|\Psi_3\rangle = \sqrt{\frac{E+m}{2m}} \begin{pmatrix} \frac{p_3}{E+m} \\ \frac{p_1+ip_2}{E+m} \\ 1 \\ 0 \end{pmatrix} e^{ipx} = v_2 e^{ipx};$$

$$|\Psi_4\rangle = \sqrt{\frac{E+m}{2m}} \begin{pmatrix} \frac{p_1-ip_2}{E+m} \\ \frac{-p_3}{E+m} \\ 0 \\ 1 \end{pmatrix} e^{ipx} = v_1 e^{ipx}. \tag{B.19}$$

It is often easier to work in the rest frame, and then applying a Lorentz transformation. The solutions for $\mathbf{p} = \mathbf{0}$ are then

$$|\Psi_1\rangle = \begin{pmatrix} 1 \\ 0 \\ 0 \\ 0 \end{pmatrix} e^{-imt} = u_1^0 e^{-imt}; \quad |\Psi_2\rangle = \begin{pmatrix} 0 \\ 1 \\ 0 \\ 0 \end{pmatrix} e^{-imt} = u_2^0 e^{-imt};$$

$$|\Psi_3\rangle = \begin{pmatrix} 0 \\ 0 \\ 1 \\ 0 \end{pmatrix} e^{imt} = v_2^0 e^{imt}; \quad |\Psi_4\rangle = \begin{pmatrix} 0 \\ 0 \\ 0 \\ 1 \end{pmatrix} e^{imt} = v_1^0 e^{imt}. \tag{B.20}$$

Exercise Find the eigenvectors of the Dirac equation, for a particle at rest, using the convention (B.16).

B.3.3 The Spin Matrix

The spin matrix Σ_i can be written as

$$\Sigma_i = \frac{\hbar}{2}\begin{pmatrix} \sigma_i & 0 \\ 0 & \sigma_i \end{pmatrix} \rightarrow \Sigma_1 = \frac{\hbar}{2}\begin{pmatrix} 0&1&0&0 \\ 1&0&0&0 \\ 0&0&0&1 \\ 0&0&1&0 \end{pmatrix}; \quad \Sigma_2 = \frac{\hbar}{2}\begin{pmatrix} 0&-i&0&0 \\ i&0&0&0 \\ 0&0&0&-i \\ 0&0&i&0 \end{pmatrix};$$

$$\Sigma_3 = \frac{\hbar}{2}\begin{pmatrix} 1&0&0&0 \\ 0&-1&0&0 \\ 0&0&1&0 \\ 0&0&0&-1 \end{pmatrix}. \tag{B.21}$$

Exercise Check that, in the case of a particle at rest, $\Sigma_3|\Psi_{1,3}\rangle = \frac{\hbar}{2}|\Psi_{1,3}\rangle$ and $\Sigma_3|\Psi_{2,4}\rangle = -\frac{\hbar}{2}|\Psi_{2,4}\rangle$.

For $p_3 \neq 0$ the result is still valid as a translation in z axis does not change the projection of the spin momentum on the z axis. However, after a translation on the x or y axis, a p_1 and p_2 component appears (relativistic boost). One can also give another useful expression for the spin matrix:

$$\langle\Psi^\dagger|\begin{pmatrix} \sigma_i & 0 \\ 0 & \sigma_i \end{pmatrix}|\Psi\rangle = \langle\Psi^\dagger|\gamma^0\gamma^i\gamma^5|\Psi\rangle = \langle\bar{\Psi}|\gamma^i\gamma^5|\Psi\rangle, \tag{B.22}$$

where we have used

$$\bar{\psi} = \psi^\dagger\gamma^0. \tag{B.23}$$

Some useful relations are sometimes used to simplify the calculation of amplitudes. From the Dirac equation we can write

$$/p_1 u(p_1) = m u(p_1); \quad /p_1 v(p_1) = -m v(p_1); \quad \bar{u}(p_2)/p_2 = m\bar{u}(p_2); \quad \bar{v}(p_2)/p_2 = -m\bar{v}(p_2). \tag{B.24}$$

The helicity operators $P_L = \frac{1-\gamma^5}{2}$ and $P_R = \frac{1+\gamma^5}{2}$ can then be written in the convention (B.16)

$$P_R = \begin{pmatrix} 0&0&0&0 \\ 0&0&0&0 \\ 0&0&1&0 \\ 0&0&0&1 \end{pmatrix} \qquad P_L = \begin{pmatrix} 1&0&0&0 \\ 0&1&0&0 \\ 0&0&0&0 \\ 0&0&0&0 \end{pmatrix}. \tag{B.25}$$

We can then decompose a Dirac spinor into its helicity (eigenvalues) component,

$$\Psi = \begin{pmatrix} \Psi_L \\ \Psi_R \end{pmatrix}. \tag{B.26}$$

B.3.4 Proca Equation

The Proca equation is the equation of movement of spin-1 fields. It can be deduced from the massive spin-1 Lagrangian

$$\mathcal{L} = -\frac{1}{4} F^{\mu\nu} F_{\mu\nu} + \frac{1}{2} M^2 A^\mu A_\mu \tag{B.27}$$

with

$$F_{\mu\nu} = \partial_\mu A_\nu - \partial_\nu A_\mu. \tag{B.28}$$

The Euler–Lagrange equation gives

$$\frac{\partial \mathcal{L}}{\partial A_\nu} - \partial_\mu \frac{\partial \mathcal{L}}{\partial(\partial_\mu A_\nu)} = 0 \;\Rightarrow\; \partial_\mu F^{\mu\nu} + M^2 A^\nu = 0. \tag{B.29}$$

Acting ∂_ν on (B.29), we obtain

$$\partial_\nu A^\nu = 0 \tag{B.30}$$

which transforms (B.29) into

$$\boxed{(\Box + M^2) A^\nu = 0,} \tag{B.31}$$

which is the Proca equation. We have 3 degrees of freedom, one being eliminated by (B.30). We then obtain 3 polarizations that we can write, supposing a particle moving along the z-axis:

$$\epsilon_{+1} = \frac{1}{\sqrt{2}} \begin{pmatrix} 0 \\ 1 \\ i \\ 0 \end{pmatrix}; \quad \epsilon_{-1} = \frac{1}{\sqrt{2}} \begin{pmatrix} 0 \\ 1 \\ -i \\ 0 \end{pmatrix}; \quad \epsilon_0 = \frac{1}{M} \begin{pmatrix} |\mathbf{p}| \\ 0 \\ 0 \\ E \end{pmatrix}; \tag{B.32}$$

where ϵ_λ represents the polarization vector with the spin projection λ along the z-axis.

Exercise Check that ϵ_λ are unitary vectors respecting the spin-1 relation (B.30), $p_\mu \epsilon_\lambda^\mu = 0$.

B.3.5 Rarita–Schwinger Equation

Even if the Nature showed us spin-0, spin-$\frac{1}{2}$, and spin-1 fields, some models exhibit spin-$\frac{3}{2}$ particles. The gravitino, partner of the graviton in supergravity is one example. In 1941, Rarita and Schwinger, based on a 1939 work by Fierz and Pauli, constructed the first spin-$\frac{3}{2}$ Lagrangian. The basic idea is quite simple. Such a field Ψ_μ should verify the Dirac *and* the Proca equation. The resulting wave function should then be a direct product of the Dirac solution (B.20) and Proca solution (B.32). In other words, Ψ_μ should satisfy

$$(i\gamma^\rho \partial_\rho - m)\Psi_\mu = 0; \quad \partial^\mu \Psi_\mu = 0. \tag{B.33}$$

The spin projection on the z-axis can take the values $+\frac{3}{2}$, $+\frac{1}{2}$, $-\frac{1}{2}$, and $-\frac{3}{2}$. We can then write, following the Clebsch–Gordan decomposition

$$\begin{aligned}
\Psi_\mu^{+\frac{3}{2}} &= \Psi^{+\frac{1}{2}} \epsilon_\mu^{+1} \\
\Psi_\mu^{+\frac{1}{2}} &= \frac{1}{\sqrt{3}} \Psi^{-\frac{1}{2}} \epsilon_\mu^{+1} + \sqrt{\frac{2}{3}} \Psi^{+\frac{1}{2}} \epsilon_\mu^{0} \\
\Psi_\mu^{-\frac{1}{2}} &= \frac{1}{\sqrt{3}} \Psi^{+\frac{1}{2}} \epsilon_\mu^{-1} + \sqrt{\frac{2}{3}} \Psi^{-\frac{1}{2}} \epsilon_\mu^{0} \\
\Psi_\mu^{-\frac{3}{2}} &= \Psi^{-\frac{1}{2}} \epsilon_\mu^{-1}.
\end{aligned}$$

Exercise In the rest frame, using for $\Psi^{+\frac{1}{2}}$, $|\Psi_1\rangle$ (or $|\Psi_3\rangle$), and for $\Psi^{-\frac{1}{2}}$, $|\Psi_2\rangle$ (or $\Psi_4\rangle$) of Eq. (B.20), and ϵ^λ of Eq. (B.32), show that

$$\gamma^\mu \Psi_\mu = 0. \tag{B.34}$$

This relation being Lorentz invariant, if it is valid in the rest frame, it is also valid in any Lorentz-transformed frame. With the three relations we have in (B.33) and (B.34), we see that we can generalize the Dirac Lagrangian, introducing terms of the type $\gamma^\mu \Psi_\mu$, $\partial^\mu \Psi_\mu$ and all types of combination of these terms. We then obtain

$$\mathcal{L}_{3/2} = \bar{\Psi}_\mu \left(i\eta^{\mu\nu}\gamma^\rho \partial_\rho - m\eta^{\mu\nu} - i\gamma^\mu \partial^\nu - i\gamma^\nu \partial^\mu + i\gamma^\mu \gamma^\rho \gamma^\nu \partial_\rho + m\gamma^\mu \gamma^\nu \right) \Psi_\nu. \tag{B.35}$$

Noticing that

$$\gamma^\mu \gamma^\nu = -\gamma^\nu \gamma^\mu + 2\eta^{\mu\nu}, \quad \gamma^{\mu\nu} = \frac{1}{2}[\gamma^\mu, \gamma^\nu] = \gamma^\mu \gamma^\nu - \eta^{\mu\nu} \tag{B.36}$$

and defining

$$\begin{aligned} \gamma^{\mu\nu\rho} &= \frac{1}{3!} \left[\gamma^\mu \gamma^\nu \gamma^\rho + \gamma^\nu \gamma^\rho \gamma^\mu + \gamma^\rho \gamma^\mu \gamma^\nu - \gamma^\nu \gamma^\mu \gamma^\rho - \gamma^\mu \gamma^\rho \gamma^\nu - \gamma^\rho \gamma^\nu \gamma^\mu \right] \\ &= \gamma^\mu \gamma^\nu \gamma^\rho - \eta^{\nu\rho}\gamma^\mu - \eta^{\mu\nu}\gamma^\rho + \eta^{\mu\rho}\gamma^\nu, \end{aligned} \tag{B.37}$$

we can then simplify (B.35)

$$\mathcal{L}_{3/2} = \bar{\Psi}_\mu \left(i\gamma^{\mu\rho\nu}\partial_\rho + m\gamma^{\mu\nu} \right) \Psi_\nu \tag{B.38}$$

which is the Rarita–Schwinger Lagrangian. Notice the combinations of $\gamma^\mu \partial^\nu$ factors that we added with a certain coefficient. In the original paper of 1941, it was a factor $\frac{1}{3}$ that was chosen by the authors.

B.3.6 Parity Operator

The parity operator, $\mathcal{P}$, inverts all space coordinates used in the description of a physical process. Consider for instance a scalar wave function $\psi(x, y, z, t)$. Performing the parity operation on this wave function will transform it to $\psi(-x, -y, -z, t)$, or

$$\mathcal{P}\psi(x, y, z, t) = \psi(-x, -y, -z, t). \tag{B.39}$$

The parity transformation can be viewed as a mirroring with respect to a plane (for instance $z \to -z$) followed by a rotation around an axis perpendicular to the plane (the z-axis). As angular momentum is conserved, physics will be invariant under the rotation and so the parity operation tests for invariance to mirroring with respect to a plane of arbitrary orientation. Parity conservation or $\mathcal{P}$-symmetry implies that any physical process will proceed identically, when viewed in mirror image. This sounds rather natural. After all, we would not expect a dice for instance to produce a different distribution of numbers if one swaps the position of the one and the six of the dice.

It is important to underline that we discuss in this section the *intrinsic* parity associated with each particle, and not the one associated with the orbital wave function of the particle. In other words, *the parity of the wave function of a particle or a system of particles is the product of the orbital parity times the product of the intrinsic parities of the particles involved.*
In interactions where parity is conserved (electromagnetic and strong), we can use this to compute selection rules for the various reactions. We do this aided by the following two ideas :

- Under parity, the orbital wave function will change by $(-1)^L$ where L is the angular momentum quantum number. This is because the parity operator $x_i \rightarrow -x_i$, changes $\theta \rightarrow \theta - \pi$ and $\phi \rightarrow \phi + \pi$. Due to this change in the angle, the orbital wave function, which always involves spherical harmonics (central potential) changes as $(-1)^L$. This is actually the reason that the photon has negative parity and is referred to as a 1^- state. Photons are emitted via atomic dipole transitions where $\Delta L = \pm 1$. Hence, the atomic parity changes by (-1) during these transitions, and for the overall parity of the system (atom + photon) to be conserved (electromagnetic interactions), we must have that the photon has negative parity.
- The overall wave function must be symmetric for systems of bosons and antisymmetric for systems of fermions.

In summary, the overall parity of the wave function of a set of particles is given by

$$P = (-1)^L \times P_1 \times P_2 \times \ldots \times P_s, \tag{B.40}$$

where L is the total angular momentum and P_1, P_2, .., P_s are the intrinsic parities of the particles involved.

B.3.6.1 Fermion Case

For fermions fields respecting the Dirac equation, imposing that the laws of physics are invariant under the change of space coordinates : $x_i \rightarrow \tilde{x}_i = -x_i$, Eq. (B.18) transforms as

$$\begin{aligned}&(i\gamma^0\partial_t + i\gamma^i\partial_{x_i} - m)|\Psi(-x)\rangle = 0 \;\Rightarrow\; \gamma^0(i\gamma^0\partial_t + i\gamma^i\partial_{x_i} - m)|\Psi(-x)\rangle = 0\\ &\Rightarrow (i\gamma^0\partial_t - i\gamma^i\partial_{x_i} - m)\gamma^0|\Psi(-x)\rangle = 0\end{aligned} \tag{B.41}$$

which means that under parity $|\Psi(x)\rangle \rightarrow |\Psi(-x)\rangle = e^{i\phi}\,\gamma^0|\Psi\rangle = \mathcal{P}|\Psi\rangle$, $e^{i\phi}$ being present as it is a general unitarity transformation:

$$|\Psi(x)\rangle \;\rightarrow_{\mathcal{P}}\; |\Psi(-x)\rangle = \mathcal{P}|\,\Psi(x)\rangle \;=\; e^{i\phi}\,\gamma^0|\Psi(x)\rangle; \quad \mathcal{P} = e^{i\phi}\,\gamma^0. \tag{B.42}$$

Note that under the parity $\mathcal{P} = \gamma^0$, the spinor $\Psi = \begin{pmatrix} \Psi_L \\ \Psi_R \end{pmatrix}$ transforms into $\Psi = \begin{pmatrix} \Psi_R \\ \Psi_L \end{pmatrix}$. This is easily understandable as the helicity (projection of the spin on the velocity vector) changes orientation in space under the parity operation: a left-handed particle becomes a right-handed particle. This was in fact the way taken by Wu for her experiment in 1956 [3] to show the violation of parity. Indeed, finding a process which produces a particle with a preferred helicity also proves that $\mathcal{P}$-symmetry is violated. Wu did it measuring the products of ^{60}Co decay.
From the definition of the parity operator, we can compute the parity of different bilinear forms which are useful to compute some processes. We can then question their existence :

$$\begin{aligned}
&\text{Scalar}: \quad \bar{\Psi}\Psi \rightarrow \Psi^\dagger(\gamma^0)^\dagger\gamma^0\gamma^0\Psi = +\bar{\Psi}\Psi, \quad \mathcal{P}_{\bar{\Psi}\Psi} = +1 \\
&\text{Pseudoscalar}: \quad \bar{\Psi}\gamma^5\Psi \rightarrow \Psi^\dagger(\gamma^0)^\dagger\gamma^0\gamma^5\gamma^0\Psi = -\bar{\Psi}\gamma^5\Psi, \quad \mathcal{P}_{\bar{\Psi}\gamma^5\Psi} = -1 \\
&\text{Vector}: \quad \bar{\Psi}\gamma^i\Psi \rightarrow \Psi^\dagger(\gamma^0)^\dagger\gamma^0\gamma^i\gamma^0\Psi = -\bar{\Psi}\gamma^i\Psi, \quad \mathcal{P}_{\bar{\Psi}\gamma^i\Psi} = -1; \ \mathcal{P}_{\bar{\Psi}\gamma^0\Psi} = +1 \\
&\text{Axial}: \quad \bar{\Psi}\gamma^i\gamma^5\Psi \rightarrow \Psi^\dagger(\gamma^0)^\dagger\gamma^0\gamma^i\gamma^5\gamma^0\Psi = +\bar{\Psi}\gamma^i\gamma^5\Psi, \quad \mathcal{P}_{\bar{\Psi}\gamma^i\gamma^5\Psi} = +1; \ \mathcal{P}_{\bar{\Psi}\gamma^0\gamma^5\Psi} = -1 \\
&\text{Tensor}: \quad \bar{\Psi}\sigma_{ij}\Psi \rightarrow \Psi^\dagger(\gamma^0)^\dagger\gamma^0\sigma_{ij}\gamma^0\Psi = +\bar{\Psi}\sigma_{ij}\Psi, \quad \mathcal{P}_{\bar{\Psi}\sigma_{ij}\Psi} = +1; \ \mathcal{P}_{\bar{\Psi}\sigma_{0j}\Psi} = -1
\end{aligned} \tag{B.43}$$

with $\sigma_{\mu\nu} = [\gamma_\mu, \gamma_\nu] = \gamma_\mu\gamma_\nu - \gamma_\nu\gamma_\mu$. Equation (B.43) summarize all the nature of couplings one can find in any specific microscopic models. Moreover, in view of the form of the matrix γ^0, we can notice that $\mathcal{P}|\Psi_{1,2}\rangle = +|\Psi_{1,2}\rangle$ and $\mathcal{P}|\Psi_{3,4}\rangle = -|\Psi_{3,4}\rangle$. In other words, the antiparticles states have opposite parity than the particle states.

B.3.6.2 Boson Case

In the case of a bosonic particle, one names *scalar* the particle ϕ with an even parity, $[\phi(-x_i, t) = \phi(x_i, t)]$ and *pseudo-scalar* the particle $\tilde{\phi}$ with an odd parity $[\tilde{\phi}(-x_i, t) = -\tilde{\phi}(x_i, t)]$. In the Standard Model, the two kinds of bosons exist: the Higgs h, which is a scalar particle, and the neutral pion π^0, a pseudo-scalar one. But how we can measure the parity of a particle?

B.3.7 The Charge Conjugate Operator

The charge conjugation operator C transforms a particle ψ into its antiparticle ψ^c. We can find this expression as we did for the parity operator, directly from the Dirac equation. Indeed, in the presence of an electromagnetic field, one should add the interaction of the photon field A_μ with the fermion of charge e. The Dirac equation for the field ψ then becomes

$$i\gamma^\mu(\partial_\mu - ieA_\mu)\psi - m\psi = 0 \tag{B.44}$$

which imply after taking the complex conjugate of the equation and multiplying by γ^2

$$-i\gamma^2(\gamma^\mu)^*(\partial_\mu + ieA_\mu)\psi^* - \gamma^2 m\psi^* = 0 \;\Rightarrow\; i\gamma^\mu(\partial_\mu + ieA_\mu)(\gamma^2\psi^*) - m(\gamma^2\psi^*) = 0, \tag{B.45}$$

where we used $(\gamma^{0,1,3})^* = \gamma^{0,1,3}$ and $(\gamma^2)^* = -\gamma^2$. Defining ψ^c as $\psi^c = i\gamma^2\psi^* = C\bar{\psi}^T = i\gamma^2\gamma^0\bar{\psi}^T$ (the factor i is added to integrate a phase and to keep C as a real operator), we notice that the Dirac equation for ψ^c is exactly the same than for ψ, but with an opposite charge: ψ^c is the antiparticle of ψ, and C the charge conjugate operator. We can then find the expression of C in the space of 4-dimensional spinors $\psi = (\chi, \eta)^T$. The operation can be written as

$$\psi^c = C\bar{\psi}^T = i\gamma^2\gamma^0\bar{\psi}^T = C\gamma_0^T\psi^*, \tag{B.46}$$

where

$$C = i\gamma^2\gamma^0 = \begin{pmatrix} 0 & 1 & 0 & 0 \\ -1 & 0 & 0 & 0 \\ 0 & 0 & 0 & -1 \\ 0 & 0 & 1 & 0 \end{pmatrix} \;\Rightarrow\; \psi = \begin{pmatrix} \chi \\ \eta \end{pmatrix} \rightarrow_C \psi^c = C\bar{\psi}^T = \begin{pmatrix} i\sigma_2\eta^* \\ -i\sigma_2\chi^* \end{pmatrix}; \tag{B.47}$$

we can then prove easily the following useful relations:

$$C^{-1} = -C^*;\;\; C^\top = -C;\;\; C^{-1}\gamma_\mu C = -\gamma_\mu^\top;\;\; C^\dagger = C^{-1}. \tag{B.48}$$

A Majorana particle is a particle which charge conjugate is equal to itself, $\psi^c = \psi$. We can then write

$$\psi = \begin{pmatrix} \chi \\ \eta \end{pmatrix} \;\Rightarrow\; \psi^c = i\gamma^2\gamma^0\bar{\psi}^T \begin{pmatrix} i\sigma_2\eta^* \\ -i\sigma_2\chi^* \end{pmatrix},\;\; \psi = \psi^c \;\Rightarrow\; \psi = \begin{pmatrix} \chi \\ -i\sigma_2\chi^* \end{pmatrix} \tag{B.49}$$

which is the generic form for a Majorana particle. Have a look at Sect. B.3.8 for a more detailed definition of a Majorana fermion in the context of Dirac equation and Sect. B.2.2 for the Feynman rules with Majorana particles. Some operators are suppressed for Majorana particles: For instance, if $\psi^c = \psi$ we can write

$$\bar{\psi^c}\gamma^\mu\psi^c = \psi^\top C^\dagger\gamma^\mu\gamma^0 C\psi^* = \psi^\top C^{-1}\gamma^\mu\gamma^0 C\psi^* = \psi^\top(\gamma^\mu)^\top(\gamma^0)^\top\psi^* \tag{B.50}$$

$$= (\psi^\top(\gamma^\mu)^\top(\gamma^0)^\top\psi^*)^\top = -\bar{\psi}\gamma^\mu\psi \;\Rightarrow\; \bar{\psi}\gamma^\mu\psi = \frac{1}{2}\left(\bar{\psi}\gamma^\mu\psi + \bar{\psi^c}\gamma^\mu\psi^c\right) = 0,$$

the negative sign is coming for the exchange of two fermions. As a consequence, we can deduce that a Majorana dark matter does not couple to a Z_μ (as the dark matter has no T^3 charges, no coupling to γ^5) or Z'_μ boson without axial coupling as the sum of both contributions will cancel. On the other hand, if a Majorana dark matter has an axial coupling to Z or Z' the annihilation rate is not null. We can also easily show with the same method that

$$\bar{\psi}^c \psi^c = \bar{\psi}\psi \tag{B.51}$$

showing that a Majorana dark matter can have a coupling to the Higgs boson. Another useful relation (used in the Sect. C.3) concerning the neutrino is (we let the reader to prove it)

$$\bar{\psi}_L \psi_R = \bar{\psi}_R^c \psi_L^c. \tag{B.52}$$

B.3.8 The Majorana Case

B.3.8.1 Definition

The Feynman rules for a Majorana fermion differ from Dirac fermion, the same manner Feynman rules for real scalar differ from imaginary scalars. All can be, at the end, summarized by factors 2 or 1/2, but it is important (and interesting) to understand how to treat the rules accurately. First of all, one should remember the basic facts of a Dirac equation. If one decomposes a (4-dimension) Dirac spinor ψ into its (2-dimension) Weyl spinors χ and η, we can write $\psi = \begin{pmatrix} \chi \\ \eta \end{pmatrix}$ which generates the Dirac equation (B.18) if one writes the Lagrangian

$$\mathcal{L}_{Dirac} = i\chi^\dagger \bar{\sigma}^\mu \partial_\mu \chi + i\eta^\dagger \sigma^\mu \partial_\nu \eta - M\eta^\dagger \chi - M^* \chi^\dagger \eta = \bar{\psi}(i\gamma^\mu \partial_\mu - M)\psi, \tag{B.53}$$

with $\bar{\sigma}^0 = \sigma^0$ and $\bar{\sigma}^i = -\sigma^i$. It is important to keep in mind that the 4-component Dirac spinor Ψ has been built from Eq. (B.53) which is the Lagrangian leading to the set of 2 Dirac equations for the 2-component spinors χ and η. We will try now, to build the same Lagrangian for a Majorana spinor. Before going into the details, one can have a guess of the result. Indeed, ψ is a 4-component spinors but have only 2 degrees of freedom in a Majorana case. A Majorana spinor to a Dirac one has the same correspondence than a real field to a scalar one. One should understand then that there should be a factor 1/2 entering into the game, $\mathcal{L}_{Majorana} = \frac{1}{2}\mathcal{L}_{Dirac}$. This is what we will demonstrate more rigorously.

B.3.8.2 Dirac-Like Majorana Equation

As we saw in the previous section, a (4-component) Majorana spinor λ is defined as $\lambda^C = C\bar{\lambda}^T = \lambda$. We can write, by analogy with the Dirac spinor $\lambda = \begin{pmatrix} \chi \\ \eta \end{pmatrix}$, and using Eq. (B.46)

$$\lambda^C = C\bar{\lambda}^T = i\gamma^2\gamma^0\bar{\lambda}^T = \begin{pmatrix} i\sigma_2\eta^* \\ -i\sigma_2\chi^* \end{pmatrix} = \begin{pmatrix} \epsilon^{ab}\eta_b^* \\ -\epsilon^{ab}\chi_b^* \end{pmatrix} \tag{B.54}$$

where

$$\epsilon^{ab} = \begin{pmatrix} 0 & 1 \\ -1 & 0 \end{pmatrix} \tag{B.55}$$

is the Kronecker symbol. The Kronecker symbol is in fact fundamental when dealing with Grassmann (anti-commuting) variables. One can consider it as a metric in this case, allowing to write a scalar product for anti-commuting variables. Indeed, if one defines $\eta.\chi = \eta_a\epsilon^{ab}\chi_b$, it is straightforward to check that $\eta.\chi = \chi.\eta$ using the property[3] of ϵ^{ab} and $\chi_a\eta_b = -\eta_b\chi_a$.

The condition $\lambda^C = \lambda$ leads to

$$\lambda = \begin{pmatrix} \chi \\ -\epsilon\ \chi^* \end{pmatrix} \;\Rightarrow\; \bar{\lambda} = \lambda^\dagger\gamma^0 = \left(-\chi^T\epsilon^T\ ;\ \chi^\dagger\right) \tag{B.56}$$

which implies

$$\bar{\lambda}\lambda = \chi^T\epsilon\chi - \chi^\dagger\epsilon\chi^*. \tag{B.57}$$

B.3.8.3 In the "Left–Right" Representation

With the Weyl representation in which we made all the calculation (defining our $\gamma's$ matrices) it is easy to rewrite the mass term using the "left" and "right" eigenvectors of the projection operators P_L and P_R (Eq. B.25). If one defines

$$\lambda_L = P_L\lambda = \begin{pmatrix} \chi \\ 0 \end{pmatrix} \text{ and } \lambda_R = (\lambda_L)^C = \begin{pmatrix} 0 \\ -\epsilon\chi^* \end{pmatrix}, \quad \lambda = \lambda_L + \lambda_R, \tag{B.58}$$

we can then write

$$\bar{\lambda}\lambda = \bar{\lambda}_L\lambda_R + \bar{\lambda}_R\lambda_L = \chi^T\epsilon\chi - \chi^\dagger\epsilon\chi^* \tag{B.59}$$

[3] In the case of the complex conjugate, the conventional notation is $\eta^*.\chi^* = (\eta^*)^a\epsilon_{ab}(\chi^*)^b$ with $\epsilon_{ab} = -\epsilon^{ab}$. This convention makes the scalar product easier to manipulate as one can see from Eq. (B.60) and Eq. (B.61).

(notice the change of sign comparing Eq. (B.57) and Eq. (B.59)). This reminds of course the definition of a Majorana mass term $M_M \overline{(\lambda_L)^C} \lambda_L$ which is the mass term we can write from a 4-component Majorana spinor. For more details on the Majorana mass, and its application to see-saw mechanism, see Sect. C.4.

However, the Dirac equation[4] $i\bar{\sigma}_\mu \partial_\mu \chi = -M^* \epsilon \chi^*$ is derived from the Lagrangian

$$\mathcal{L} = i\chi^\dagger \bar{\sigma}_\mu \partial_\mu \chi - \frac{M}{2} \chi . \chi + \frac{M^*}{2} \chi^* . \chi^*, \tag{B.60}$$

where we used the definition of the scalar product defined above. Notice the factor 1/2 in front of the mass term. It corresponds to the same factor for the mass term of real scalar fields (with respect to complex scalar fields), and appears naturally when one derives the Dirac equation from the equation of motion $\frac{\partial \mathcal{L}}{\partial \chi} = \partial_\mu \frac{\partial \mathcal{L}}{\partial(\partial_\mu \chi)}$. In 4-component notation, the Lagrangian (B.60) is then given by (for real mass term)

$$\mathcal{L}_{Majorana} = \frac{1}{2} i \bar{\lambda} \gamma^\mu \partial_\mu \lambda - \frac{M}{2} \bar{\lambda} \lambda \tag{B.61}$$

which corresponds indeed to our first guess we made at the beginning of the section $\mathcal{L}_{Majorana} = \frac{1}{2} \mathcal{L}_{Dirac}$. If moreover, one supposes that the mass term is coming from Yukawa-like couplings with a field ϕ, as in the Standard Model, one should write for generic couplings Γ (which contains $\gamma's$ and combinations of $\gamma's$ matrices),

$$\begin{aligned} \mathcal{L} &= \mathcal{L}_{Dirac} + \mathcal{L}_{Majorana} \\ &= \bar{\psi}(i\gamma^\mu \partial_\mu - M_\psi)\psi + \frac{1}{2}\bar{\lambda}(i\gamma^\mu \partial_\mu - M_\lambda)\lambda + y_\psi \phi \, \bar{\psi}\Gamma\psi + \frac{1}{2} y_\lambda \phi \, \bar{\lambda}\Gamma\lambda. \end{aligned} \tag{B.62}$$

The factor $\frac{1}{2}$ in front of the Yukawa coupling y_λ is fundamental when one needs to compute processes through the Feynman rules. Indeed, if ϕ is a scalar field developing a *vev* $\phi = \frac{1}{\sqrt{2}}(v + a + ib)$, the mass is then defined with the same convention, $M_\psi = \frac{y_\psi}{\sqrt{2}} v$, $M_\lambda = \frac{y_\lambda}{\sqrt{2}} v$. That point is very important once we made any computation with Majorana spinors: to check that the convention of the couplings fits with the mass term in the Lagrangian.

B.3.8.4 Furry's Theorem

In quantum electrodynamics, the Furry theorem tells that the contribution of a Feynman diagram, consisting of a closed polygon of fermion lines connected to

[4] Remember that the fundamental Dirac equation should be written in 2-component notation. The 4-component one, especially in the Weyl ("Left–Right") representation is mainly practical in the Standard Model because the helicities have different quantum number under the SU(2) gauge group.

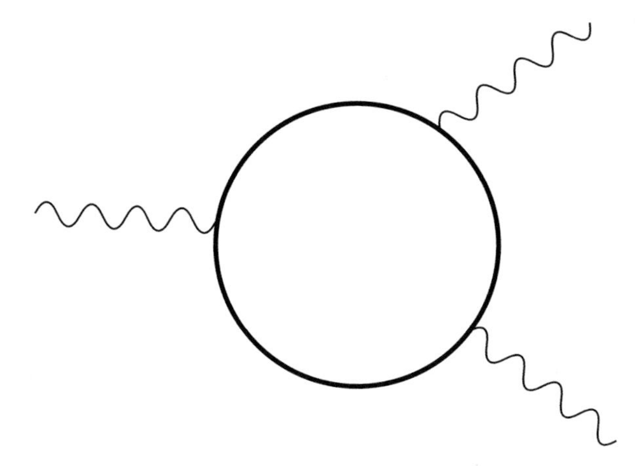

Fig. B.2 Illustration of the Furry's theorem, giving a null amplitude for an odd number of external electromagnetic legs

an odd number of photon lines, vanishes. The demonstration is quite easy if one considers a closed fermion loop of diagram with 3 external legs illustrated in Fig. B.2. Remembering that the electromagnetic current changes sign under charge conjugation *i.e.*

$$Cj^\mu C^\dagger = -j^\mu, \tag{B.63}$$

where we make use of Eq. (B.48). Remembering that the vacuum should be charge invariant ($C|0\rangle = 0$), we obtain

$$\langle 0|j^\mu|0\rangle = \langle 0|C^\dagger C j^\mu C^\dagger C|0\rangle = -\langle 0|j^\mu|0\rangle \;\Rightarrow\; \langle 0|j^\mu|0\rangle = 0. \tag{B.64}$$

The same argument applies to any vacuum expectation value of an odd number of electromagnetic currents, and by consequence, the vacuum expectation of all diagrams with odd number of legs vanishes. This property is known as Furry's theorem.

B.3.9 Traces

$$\eta^{\mu\nu}\eta_{\mu\nu} = 4$$

$$Tr[\gamma^\mu] = Tr[\gamma^5] = 0$$

$$Tr[\gamma^5\gamma^\mu] = Tr[\gamma^5\gamma^\mu\gamma^\nu] = Tr[\gamma^5\gamma^\mu\gamma^\nu\gamma^\rho] = 0$$

$$Tr[1] = 4$$

$$Tr[\gamma^\mu\gamma^\nu] = 4\eta^{\mu\nu}$$

$$
\begin{aligned}
&Tr[\not p_1 . \not p_2] = 4 p_1 . p_2 \\
&Tr[(\not p_1 - m_1)(\not p_2 - m_2)] = 4 p_1 . p_2 + 4 m_1 m_2 \\
&Tr[\gamma^\mu \gamma^\nu \gamma^\rho] = 0 \\
&Tr[\gamma^\mu \gamma^\nu \gamma^\rho \gamma^\sigma] = 4\left(\eta^{\mu\nu}\eta^{\rho\sigma} - \eta^{\mu\rho}\eta^{\nu\sigma} + \eta^{\mu\sigma}\eta^{\nu\rho}\right) \\
&Tr[\gamma^\mu \gamma^\nu \gamma^\rho \gamma^\sigma \gamma^5] = -4i\epsilon^{\mu\nu\rho\sigma} \\
&Tr[\not p_1 \gamma^\mu \not p_2 \gamma^\nu] = 4\left(p_1^\mu p_2^\nu - \eta^{\mu\nu} p_1 . p_2 + p_1^\nu p_2^\mu\right) \\
&Tr[\not p_1 \not p_2 \not p_3 \not p_4] = 4\left[(p_1 . p_2)(p_3 . p_4) - (p_2 . p_4)(p_1 . p_3) + (p_2 . p_3)(p_1 . p_4)\right] \\
&Tr[\not p_1 \gamma^\mu \not p_2 \gamma^\nu] Tr[\not p_3 \gamma_\mu \not p_4 \gamma_\nu] = 32\left[(p_1 . p_3)(p_2 . p_4) + (p_1 . p_4)(p_2 . p_3)\right] \\
&Tr[\not p_1 \gamma^\mu \not p_2 \gamma^\nu] Tr[\gamma_\mu \gamma_\nu] = -32 p_1 . p_2 \\
&Tr[(\not p_1 - M_\chi)\gamma^\mu(\not p_2 + M_\chi)\gamma^\nu] Tr[(\not p_3 - m_f)\gamma_\mu(\not p_4 + m_f)\gamma_\nu] \\
&= 32\left[(p_1 . p_3)(p_2 . p_4) + (p_1 . p_4)(p_2 . p_3) + 2M_\chi^2 m_f^2 + (p_1 . p_2) m_f^2 + (p_3 . p_4) M_\chi^2\right] \\
&Tr[(\not p_1 + M_\chi)\gamma^\mu(\not p_2 + M_\chi)\gamma^\nu] Tr[(\not p_3 + m_f)\gamma_\mu(\not p_4 + m_f)\gamma_\nu] \\
&= 32\left[(p_1 . p_3)(p_2 . p_4) + (p_1 . p_4)(p_2 . p_3) + 2M_\chi^2 m_f^2 - (p_1 . p_2) m_f^2 - (p_3 . p_4) M_\chi^2\right] \\
&\mathrm{Tr}[\gamma^\alpha \gamma^\beta \gamma^\mu \gamma^\nu \gamma^\rho \gamma^\sigma] \\
&= 4\eta^{\alpha\beta}(\eta^{\mu\nu}\eta^{\rho\sigma} - \eta^{\mu\rho}\eta^{\nu\sigma} + \eta^{\mu\sigma}\eta^{\nu\rho}) - 4\eta^{\alpha\mu}(\eta^{\beta\nu}\eta^{\rho\sigma} - \eta^{\beta\rho}\eta^{\nu\sigma} + \eta^{\beta\sigma}\eta^{\nu\rho}) \\
&+4\eta^{\alpha\nu}(\eta^{\beta\mu}\eta^{\rho\sigma} - \eta^{\beta\rho}\eta^{\mu\sigma} + \eta^{\beta\sigma}\eta^{\mu\rho}) - 4\eta^{\alpha\rho}(\eta^{\beta\mu}\eta^{\nu\sigma} - \eta^{\beta\nu}\eta^{\mu\sigma} + \eta^{\sigma\beta}\eta^{\mu\nu}) \\
&+4\eta^{\alpha\sigma}(\eta^{\beta\mu}\eta^{\nu\rho} - \eta^{\beta\nu}\eta^{\mu\rho} + \eta^{\rho\beta}\eta^{\mu\nu})
\end{aligned} \tag{B.65}
$$

Throughout the book, we will use the notation P_μ for a quadrivector and p_i for a 3-dimensional vector.

B.3.10 Mandelstam Variables

In a process $1 + 2 \to 3 + 4$, one usually defines the Mandelstam variables by

$$
\begin{aligned}
s &= (p_1 + p_2)^2 = (p_3 + p_4)^2 \\
t &= (p_3 - p_1)^2 = (p_4 - p_2)^2 \\
u &= (p_3 - p_2)^2 = (p_4 - p_1)^2.
\end{aligned} \tag{B.66}
$$

Developing the expression above, we can express all the scalar products as function of these Lorentz invariant variables and the masses of the scattering particles. Once we obtain an expression for an amplitude, or amplitude square as function of s, t

and u, we can transpose it in any reference frame. We have

$$p_1.p_2 = \frac{s - m_1^2 - m_2^2}{2}; \qquad p_3.p_4 = \frac{s - m_3^2 - m_4^2}{2}$$

$$p_1.p_3 = \frac{t - m_1^2 - m_3^2}{2}; \qquad p_2.p_4 = \frac{t - m_2^4 - m_4^2}{2}$$

$$p_1.p_4 = \frac{u - m_1^2 - m_4^2}{2}; \qquad p_2.p_3 = \frac{u - m_2^4 - m_3^2}{2}$$

and

$$s + t + u = m_1^2 + m_2^2 + m_3^2 + m_4^2. \tag{B.67}$$

Exercise Recover the expressions above.

B.3.11 The Generators T_i^a

The generators of the non-Abelian gauge groups $SU(2)$ and $SU(3)$ are given by

For $SU(2)$, $T_2^i = \frac{\sigma_i}{2}$, σ_i Being the Pauli Matrices

$$T_2^1 = \begin{pmatrix} 0 & 1/2 \\ 1/2 & 0 \end{pmatrix} \quad T_2^2 = \begin{pmatrix} 0 & -i/2 \\ i/2 & 0 \end{pmatrix} \quad T_2^3 = \begin{pmatrix} 1/2 & 0 \\ 0 & -1/2 \end{pmatrix}$$

For $SU(3)$, the Gell-Mann Matrices Are

$$\lambda_1 = \begin{pmatrix} 0 & 1 & 0 \\ 1 & 0 & 0 \\ 0 & 0 & 0 \end{pmatrix} \quad \lambda_2 = \begin{pmatrix} 0 & -i & 0 \\ i & 0 & 0 \\ 0 & 0 & 0 \end{pmatrix} \quad \lambda_3 = \begin{pmatrix} 1 & 0 & 0 \\ 0 & -1 & 0 \\ 0 & 0 & 0 \end{pmatrix} \quad \lambda_4 = \begin{pmatrix} 0 & 0 & 1 \\ 0 & 0 & 0 \\ 1 & 0 & 0 \end{pmatrix}$$

$$\lambda_5 = \begin{pmatrix} 0 & 0 & -i \\ 0 & 0 & 0 \\ i & 0 & 0 \end{pmatrix} \quad \lambda_6 = \begin{pmatrix} 0 & 0 & 0 \\ 0 & 0 & 1 \\ 0 & 1 & 0 \end{pmatrix} \quad \lambda_7 = \begin{pmatrix} 0 & 0 & 0 \\ 0 & 0 & -i \\ 0 & i & 0 \end{pmatrix} \quad \lambda_8 = \frac{1}{\sqrt{3}} \begin{pmatrix} 1 & 0 & 0 \\ 0 & 1 & 0 \\ 0 & 0 & -2 \end{pmatrix}.$$

B.4 Lorentz Invariant Scattering Cross Section and Phase Space

B.4.1 The FERMI's Golden Rule

B.4.1.1 The Non-relativistic Case

The golden rule, which in fact was first proposed by P. Dirac is in fact a natural consequence of the Schrodinger equation. Supposing a system of n eigenstates of a Hamiltonian H_0

$$H_0|n\rangle = E_n|n\rangle, \tag{B.68}$$

we can expand in the plane wave solutions $|n\rangle$ the solution of a perturbed system $H = H_0 + \delta H$, at a time t

$$|\phi(t)\rangle = \sum_n a_n(t)e^{-iE_nt}|n\rangle. \tag{B.69}$$

Projecting on $\langle f|$ the solution of the equation

$$\begin{aligned} H|\phi(t)\rangle &= i\frac{\partial}{\partial t}|\phi(t)\rangle \\ \Rightarrow\ i\frac{da_f(t)}{dt} &= \sum_n \langle f|\delta H|n\rangle a_n(t)e^{i(E_f-E_n)t}. \end{aligned} \tag{B.70}$$

Supposing an initial state $|\phi_i\rangle = e^{-iE_it}$, we obtain

$$\begin{aligned} ia_k(t) &= 2\langle f|\delta H|i\rangle e^{i\frac{(E_f-E_i)}{2}t}\frac{\sin\left(\frac{E_f-E_i}{2}\right)t}{E_f-E_i} \\ |a_k(t)|^2 &= 4|\langle f|\delta H|i\rangle|^2\frac{\sin^2\left(\frac{E_f-E_i}{2}\right)t}{|E_f-E_i|^2}. \end{aligned} \tag{B.71}$$

If one looks an interval of final energies, with a density of state $\rho(E_f)$ per unit energy interval, the probability of transition P_{fi} should be written as

$$P_{fi} = 4\int dE_f\rho(E_f)|\langle f|\delta H|i\rangle|^2\frac{\sin^2\left(\frac{E_f-E_i}{2}\right)t}{|E_f-E_i|^2}. \tag{B.72}$$

If one supposes that $\rho(E)$ varies slowly with E (by *slow* we mean that the observation time t is relatively long) and that $\langle f|\delta H|i\rangle$ is invariant across the final states, P_{fi} becomes

$$P_{fi} = 4\rho(E_f)|\langle f|\delta H|i\rangle|^2 \int_{-\infty}^{\infty} dE_f \frac{\sin^2\left(\frac{E_f - E_i}{2}\right)t}{|E_f - E_i|^2} = 2\pi\rho(E_f)|\langle f|\delta H|i\rangle|^2 t, \tag{B.73}$$

where we used

$$\int_{-\infty}^{\infty} dx \frac{\sin^2 \alpha x}{x^2} = \alpha\pi. \tag{B.74}$$

The golden rule can then be written, for the transition rate $i \to f$, $\Gamma_{fi} = \frac{dP_{fi}}{dt}$

$$\Gamma_{fi} = 2\pi\rho(E_f)|\mathcal{M}_{fi}|^2 \tag{B.75}$$

with $\mathcal{M}_{fi} = \langle f|\delta H|i\rangle$.

B.4.1.2 Normalization

In classical quantum mechanics, the wave function ϕ is normalized such as

$$\int |\phi|^2 d^3x = \int |\phi|^2 dV = 1. \tag{B.76}$$

This means that the particle has a probability of presence of 1 in the entire Universe. However, in field theory, there exists the possibility to create or annihilate particles from the void. The normalization is then a free parameter and a question of taste, as long as one stays coherent. Moreover, it is clear that the expression (B.76) is not Lorentz invariant due to the presence of the volume element of $dV = d^3x$. In order to restore the invariance, it is natural to normalise with a parameter with a dilution factor $\sqrt{\gamma}$. The usual convention is to $2E$ particle per unit of volume, E being the energy of the particle:

$$\int |\psi|^2 dV = 2E, \quad \phi = \frac{1}{\sqrt{2E}}\psi.$$

Once developed in the momentum phase space, as function of its Fourier transformation,

$$\psi(\mathbf{x}, t) = \int \frac{d^3\mathbf{p}}{\sqrt{(2\pi)^3 2E}} \left[\mathbf{a}(\mathbf{p})e^{iEt - i\mathbf{p}.\mathbf{x}} + \mathbf{a}^\dagger(\mathbf{p})e^{-iEt + i\mathbf{p}.\mathbf{x}}\right], \tag{B.77}$$

a and $\mathbf{a}^\dagger$ being expansion coefficients called destruction and creation operator, respectively, which commutation relation

$$[\mathbf{a}(\mathbf{p}), \mathbf{a}(\mathbf{p}')] = [\mathbf{a}^\dagger(\mathbf{p}), \mathbf{a}^\dagger(\mathbf{p}')] = 0 \tag{B.78}$$

and

$$[\mathbf{a}^\dagger(\mathbf{p}), \mathbf{a}(\mathbf{p}')] = [\mathbf{a}(\mathbf{p}), \mathbf{a}^\dagger(\mathbf{p}')] = \delta^{(3)}(\mathbf{p} - \mathbf{p}') \tag{B.79}$$

are deduced from the quantum relations

$$[\partial_t \psi(\mathbf{x}, t), \psi(\mathbf{x}', t)] = -i\delta^{(3)}(\mathbf{x} - \mathbf{x}') \tag{B.80}$$

and

$$[\partial_t \psi(\mathbf{x}, t), \partial_t \psi(\mathbf{x}', t)] = [\psi(\mathbf{x}, t), \psi(\mathbf{x}', t)] = 0. \tag{B.81}$$

The $\frac{1}{\sqrt{(2\pi)^3}}$ is the classical coefficient to normalize the wave function. Indeed, the plane wave function $\phi(x) = \frac{1}{\sqrt{2\pi}} \int e^{ipx} dp$ is normalized such as $|\phi(x)|^2 = 1$. We let the reader check it, knowing that the delta function $\delta(x)$ can be written as $\delta(x) = \frac{1}{2\pi} \int e^{ipx} dp$, see Eq. (A.129).

B.4.1.3 The $\mathcal{S}$-Operator

Let us denote $|i\rangle$ and $|f\rangle$ as the initial and final state, respectively, in a Fock space. We can define the transition between $|i\rangle$ and $|f\rangle$ by a $\mathcal{S}$ matrix which represents the dynamics of the process and whose amplitude squared is the probability of transition:

$$\mathcal{S}_{fi} = \langle f|\mathcal{S}|i\rangle.$$

Hence, the probability for the process $|i\rangle \to |f\rangle$ is

$$P(|i\rangle \to |f\rangle) = |\mathcal{S}_{fi}|^2. \tag{B.82}$$

In general, we can write

$$\mathcal{S}_{fi} = \delta_{fi} + i(2\pi)^4 \delta^{(4)}(p_f - p_i) \,.\, \mathcal{T}_{fi}$$

or

$$\mathcal{S} = \mathbb{I} + i\mathcal{T}, \tag{B.83}$$

the first part representing a non-interacting particle, while the second part is the dynamical part with the condition of conservation of energy–momentum. Moreover, put this way, it is a reminiscent of an exponential development, $\mathcal{S} = e^{i\mathcal{T}}$. What is physically of interest for us is the transition probability per unit time

$$w_{fi} = \frac{|\mathcal{S}_{fi}|^2}{T}.$$

From Eq. (B.82), we see that we must address the issue of defining the value of a squared Dirac δ-function. To do this we use a rather pragmatic approach due to Fermi himself:

$$[2\pi\delta(p_f^0 - p_i^0)]^2 = \int dt e^{i(p_f^0 - p_i^0)t} 2\pi\delta(p_f^0 - p_i^0) = T 2\pi\delta(p_f^0 - p_i^0)$$

$$[(2\pi)^3\delta^{(3)}(\mathbf{p}_f - \mathbf{p}_i)]^2 = \int d^3x e^{i(\mathbf{p}_f - \mathbf{p}_i).\mathbf{x}}(2\pi)^3\delta^{(3)}(\mathbf{p}_f - \mathbf{p}_i) = V(2\pi)^3\delta^{(3)}(\mathbf{p}_f - \mathbf{p}_i)$$

$$\Rightarrow w_{fi} = \frac{|\mathcal{S}_{fi}|^2}{T} = V(2\pi)^4\delta^{(4)}(p_f - p_i).|\mathcal{T}_{fi}|^2. \tag{B.84}$$

To talk about the transition rate, we look at Fock-space with a fixed number of particles. Experimentally, the angle and energy–momentum is only accessible up to a given accuracy. We therefore use differential cross section in angle $d\Omega$ and energy–momentum dp near Ω, p, respectively.

B.4.1.4 Computing the Rate

In a cubic box of volume $V = L^3$ with infinitely high potential wells, the authorized momentum-values are discretely distributed.

$$p = \frac{2\pi}{L}n \;\Rightarrow\; dn = \frac{L}{2\pi}dp \;\Rightarrow\; d^3n = \left(\frac{L}{2\pi}\right)^3 d^3p \tag{B.85}$$

and hence

$$dw_{fi} = V\,(2\pi)^4\delta^{(4)}(p_f - p_i)\,|\mathcal{T}_{fi}|^2 \prod_{f=1}^{n_f} \frac{V}{(2\pi)^3} d^3p_f \tag{B.86}$$

where n_f stands for the number of particles in the final state. In order to get rid of the normalization factors, we define a new matrix element $\mathcal{M}_{fi}$ by,

$$\mathcal{T}_{fi} = \left(\prod_{i=1}^{n_i} \frac{1}{\sqrt{2E_iV}}\right)\left(\prod_{f=1}^{n_f} \frac{1}{\sqrt{2E_fV}}\right)\mathcal{M}_{fi}. \tag{B.87}$$

At first sight, the apparition of the energies of both the initial and final states might be surprising. It is however needed in order to compensate the non-invariance of the volume, so that EV is a Lorentz invariant quantity as we saw in the previous section. This corresponds to the $2E$ normalization. Substituting the definition (B.87) in Eq. (B.86), we get the fundamentally important expression for the rate $R = dw_{fi}$

$$R = dw_{fi} = \frac{V^{1-n_i}}{(2\pi)^{3n_f}}(2\pi)^4\delta^{(4)}(p_f - p_i)|\mathcal{M}_{fi}|^2 \prod_{i=1}^{n_i}\frac{1}{2E_i}\prod_{f=1}^{n_f}\frac{d^3p_f}{2E_f}. \tag{B.88}$$

B.4.1.5 Application

We can apply the expression (B.88) to decay rate and scattering. For the decay of a particle a into n_f particle

$$a \rightarrow 1 + 2 + \ldots + n_f.$$

We have for the **total decay width**,

$$\Gamma_a = \frac{1}{2E_a}\frac{1}{(2\pi)^{3n_f}}\int \frac{d^3p_1}{2E_1}..\frac{d^3p_{n_f}}{2E_{n_f}}(2\pi)^4\delta^{(4)}(p_f - p_i)|\mathcal{M}_{fi}|^2. \tag{B.89}$$

We remark that since E_a is not a Lorentz invariant quantity, Γ_a also depends on the reference frame.

For the scattering cross section, the case of two particles interacting via the reaction

$$a + b \rightarrow 1 + 2 + \ldots + n_f \tag{B.90}$$

thus getting the **scattering cross section** $\sigma(a + b \rightarrow 1 + 2 + \ldots + n_f)$ defined by

$$\sigma = \frac{\#\text{ of transition } \; \mathrm{a+b} \rightarrow 1+2+\ldots+\mathrm{n_f} \text{ per unit time}}{\#\text{ of incoming particles per unit surface and time}} = \frac{w_{fi}}{\text{incoming flux}}. \tag{B.91}$$

The denominator can also be stated as

$$\text{incoming flux} = (\text{number density})\,.\,(\text{relative velocity}) \;=\; \frac{v_{ab}}{V}$$

we then find

$$\sigma_{i\to n_f} = \frac{1}{4F}\frac{1}{(2\pi)^{3n_f}}\int\left(\prod_{f=1}^{n_f}\frac{d^3p_f}{2E_f}\right)(2\pi)^4\delta^{(4)}(\sum_{f=1}^{n_f}p_f - p_a - p_b)|\mathcal{M}_{fi}|^2 \tag{B.92}$$

in which we see once more the Lorentz invariant Møller flux factor

$$\begin{aligned}F &= E_aE_bv_{ab} = \sqrt{(p_a.p_b)^2 - m_a^2m_b^2} \\ &= \frac{1}{2}\sqrt{(s-(m_a+m_b)^2)(s-(m_a-m_b)^2)}.\end{aligned} \tag{B.93}$$

We see that the total cross section is manifestly a Lorentz invariant quantity, since it only depends on Lorentz invariants.

B.4.2 Special Case

In the early Universe, the initial momentum of the scattering particles 1 and 2 is not uniquely determined as they live in a thermal plasma, with a statistical distribution of their momentum and energies f_1 and f_2. Moreover, it is not possible to define the relative velocity between the particles as they are all relativistic. One then needs to use the fundamental expression for the rate $R = n_1n_2\langle\sigma v\rangle$:

$$\begin{aligned}R &= \int f_1f_2d\Pi_1d\Pi_2(2\pi)^4\delta^{(4)}(P_1+P_2-P_3-P_4)d\Pi_3d\Pi_4|\overline{\mathcal{M}}_{1,2\to3,4}|^2 \\ &= \int f_1f_2\frac{d^3p_1}{(2\pi)^3}\frac{d^3p_2}{(2\pi)^3}(2\pi)^4|\overline{\mathcal{M}}_{1,2\to3,4}|^2\frac{d\Phi_2}{2E_12E_2}\end{aligned} \tag{B.94}$$

with $d\Pi_i = \frac{d^3p_i}{(2\pi)^32E_i}$ and where $d\Phi_2$ is the two body phase space, computed in the next section and $d\Omega$ the solid angle between particle 1 and 3 in the center of mass frame of (3,4). In the massless case ($m_3 = m_4 = 0$), one obtains $d\Phi_2 = \frac{d\Omega}{512\pi^6}$, see Eq. (B.101), which gives

$$R = \int f_1f_2\frac{d^3p_1}{(2\pi)^3}\frac{d^3p_2}{(2\pi)^3}\frac{|\overline{\mathcal{M}}_{1,2\to3,4}|^2}{128E_1E_2\pi^2}d\Omega. \tag{B.95}$$

I let the reader go to the Sect. 2.3.4.3 to have a more detailed study of this specific case.

B.4.3 Computing the Phase Space

We can show that $d\Pi = \frac{d^3p}{(2\pi)^3 2E}$ is a Lorentz invariant quantity. Indeed, if one considers a Lorentz boost of parameter[5] β in the z direction,

$$p'_x = p_x, \quad p'_y = p_y, \quad p'_z = \gamma(p_z - \beta E), \quad E' = \gamma(E - \beta p_z). \tag{B.96}$$

One then needs to check that $d\Pi = \frac{d^3p}{(2\pi)^3 2E} = \frac{dp'}{(2\pi)^3 2E'}$, or more simply $\frac{dp_z}{E} = \frac{dp'_z}{E'}$. For that, one can write

$$\frac{dp'_z}{E'} = \frac{\gamma dp_z - \beta\gamma dE}{\gamma(E - \beta p_z)} = \frac{dp_z(1 - \beta\frac{p_z}{E})}{E(1 - \beta\frac{p_z}{E})} = \frac{dp_z}{E}, \tag{B.97}$$

where we have used the fact that $E = \sqrt{p^2 + m^2} \Rightarrow \frac{dE}{dp_z} = \frac{p_z}{E}$. That proves the Lorentz invariant character of $d\Pi$.

B.4.3.1 2-Body Phase Space

One can then compute the two-body phase space $d\Phi_2$ in a reaction with center of mass energy $\sqrt{s}$ and quadri-momentum P_{com}

$$\begin{aligned} d\Phi_2 &= d\Pi_1 d\Pi_2\, \delta^{(4)}(P_{com} - P_1 - P_2) = \frac{d^3p_1}{(2\pi)^3 2E_1}\frac{d^3p_2}{(2\pi)^3 2E_2}\, \delta^{(4)}(P_{com} - P_1 - P_2) \\ &= \frac{p_2^2 dp_2 d\Omega}{4(2\pi)^6 E_1 E_2}\delta(\sqrt{s} - E_1 - E_2) \end{aligned} \tag{B.98}$$

with $E_1 = \sqrt{p_2^2 + m_1^2}$ and $E_2 = \sqrt{p_2^2 + m_2^2}$ in the center of mass frame. We can indeed compute the phase space in the center of mass frame as we showed in the previous section that it is Lorentz invariant. Using the relation[6] $\delta(g[x]) = \frac{\delta(x)}{g'(x)}$, one can write

$$\delta(\sqrt{s} - E_1 - E_2) = \delta(\sqrt{s} - \sqrt{p_2^2 + m_1^2} - \sqrt{p_2^2 + m_2^2}) = \frac{\delta(p_2 - p_2^*)}{\frac{p_2}{\sqrt{p_2^2+m_1^2}} + \frac{p_2}{\sqrt{p_2^2+m_2^2}}} \tag{B.99}$$

[5] Where, as usual, $\gamma = 1/\sqrt{1-\beta^2}$.

[6] From now on when using this formula we will note by simplification $\delta(x)$ as $\delta(x - x^*)$, where x^* is a zero of $g(x)$.

which gives

$$d\Phi_2 = \frac{p_2^* \, d\Omega}{4(2\pi)^6\sqrt{s}} = \frac{[s-(m_1-m_2)^2]^{1/2}[s-(m_1+m_2)^2]^{1/2}}{512\pi^6 s} d\Omega, \qquad \text{(B.100)}$$

where we used

$$p_2^* = \frac{[s-(m_1-m_2)^2]^{1/2}[s-(m_1+m_2)^2]^{1/2}}{2\sqrt{s}}.$$

In the massless states, one obtains

$$d\Phi_2(m_i = 0) = \frac{d\Omega}{512\pi^6}. \qquad \text{(B.101)}$$

B.4.3.2 N-Body Phase Space

One can generalize the previous computation in the case of N-body particles system. We can rewrite n-body phase space in the center of mass frame

$$d\Phi_n(\sqrt{s}; p_1, .., p_n) = \prod_{i=1}^{n} \frac{d^3 p_i}{(2\pi)^3 2E_i} \times \delta^{(3)}\Big(\sum_{i=1}^{n} p_i\Big)\, \delta\Big(\sum_{i=1}^{n} E_i - \sqrt{s}\Big) \qquad \text{(B.102)}$$

$$= \frac{d^3 p_n}{(2\pi)^3 2E_n} \prod_{i=1}^{n-1} \frac{d^3 p_i}{(2\pi)^3 2E_i} \delta^{(3)}\Big(\sum_{i=1}^{n-1} p_i - (-p_n)\Big)\delta\Big(\sum_{i=1}^{n-1} E_i - \sqrt{s} - (-E_n)\Big)$$

$$= \frac{d^3 p_n}{(2\pi)^3 2E_n} d\Phi_{n-1}(\epsilon; p_1, .., p_{n-1})$$

with $\epsilon^2 = (\sqrt{s} - E_n)^2 - p_n^2$, where we used the Lorentz invariance of the phase space to compute $d\Phi_{n-1}$ in the center of mass frame of the system $(p_1, .., p_{n-1})$ passing from $(\sqrt{s} - E_n; -p_n)$ to $(\sqrt{(\sqrt{s} - E_n)^2 - p_n^2}; 0)$ after a boost of momentum $+p_n$.

As an example, one can compute the three-body phase space, from the two-body one we computed already in Eq. (B.100).

$$d\Phi_3 = \frac{d^3 p_3}{(2\pi)^3 2E_3} \times \frac{p_2(\epsilon_3) d\Omega}{4(2\pi)^6 \epsilon_3}$$

with $\epsilon_3^2 = (\sqrt{s} - E_3)^2 - p_3^2$, which gives

$$d\Phi_3 = \frac{d^3 p_3 d\Omega}{8 \times (2\pi)^9 E_3} \frac{[(\sqrt{s}-E_3)^2 - p_3^2 - (m_1-m_2)^2]^{1/2}[(\sqrt{s}-E_3)^2 - p_3^2 - (m_1+m_2)^2]^{1/2}}{2\sqrt{(\sqrt{s}-E_3)^2 - p_3^2}}$$

which gives in the massless case:

$$d\Phi_3^{m_i=0} = \frac{p_3 dp_3 d\Omega_{13} d\Omega_{12}}{16(2\pi)^9}\sqrt{s - 2\sqrt{s}\,p_3}. \tag{B.103}$$

Another trick to compute a 3-body phase space is shown in Sect. B.4.6.2.

B.4.3.3 Summary in the Massless Case

We can then summarize our result for 2-, 3-, and 4-body final states in massless cases:

$$d\Phi_2^{m_i=0} = \frac{d\Omega_{12}}{8(2\pi)^6}$$

$$d\Phi_3^{m_i=0} = \frac{p_3 dp_3 d\Omega_{13} d\Omega_{12}}{16(2\pi)^9}.$$

B.4.3.4 Examples

We will compute in the next section, specific examples of decay or annihilation rates. They are given by

$$d\Gamma = \frac{|\mathcal{M}|^2}{2M}(2\pi)^4 d\Phi_n(P; p_1, \ldots, p_n) \tag{B.104}$$

with

$$d\Phi_n(P; p_1 \ldots, p_n) = \delta^4\big(P - \sum_{i=1}^{n} p_i\big)\prod_{i=1}^{n}\frac{d^3 p_i}{(2\pi)^3 2E_i}$$

for the width of a particle of mass M and amplitude of decay $\mathcal{M}$ decaying into n particles. In the case of annihilation between particles 1 and 2 into n final particles, the cross section $d\sigma$ can be written as

$$d\sigma = \frac{(2\pi)^4|\mathcal{M}|^2}{4\sqrt{(P_1.P_2)^2 - m_1^2 m_2^2}} \times d\Phi_n(p_1 + p_2; p_3, \ldots, p_{n+2}), \tag{B.105}$$

where we used Eq. (B.93) for the Lorentz invariant flux.

B.4.4 Annihilation

B.4.4.1 General Formulae

As we just saw in the previous section, the annihilation cross section between two-body states $(1, 2) \rightarrow (3, 4)$ can be written as

$$d\sigma = \frac{(2\pi)^4 |\mathcal{M}|^2}{4\sqrt{(P_1.P_2)^2 - m_1^2 m_2^2}} \delta^4(P_1 + P_2 - P_3 - P_4) \frac{d^3 p_3}{(2\pi)^3 2E_3} \frac{d^3 p_4}{(2\pi)^3 2E_4}. \tag{B.106}$$

Using $s = (P_1 + P_2)^2 = m_1^2 + m_2^2 - 2P_1.P_2$ and $m_1 = m_2 = M_\chi$, one has $\sqrt{(P_1.P_2)^2 - m_1^2 m_2^2} = \sqrt{(\frac{s}{2} - M_\chi^2)^2 - M_\chi^4} = \sqrt{(\frac{s}{2} - 2M_\chi^2)\frac{s}{2}}$. After integration on $d^3 p_3 \delta^3(p_1 + p_2 - p_3 - p_4)$, one can then rewrite Eq. (B.106)

$$d\sigma = \frac{|\mathcal{M}|^2}{64\pi^2 \sqrt{(\frac{s}{2} - 2M_\chi^2)\frac{s}{2}}} \frac{|p_4|^2 d\phi d\cos\theta d|p_4|}{\sqrt{|p_4|^2 + m_3^2}\sqrt{|p_4|^2 + m_4^2}} \delta \times \left(\sqrt{s} - \sqrt{|p_4|^2 + m_3^2} - \sqrt{|p_4|^2 + m_4^2}\right), \tag{B.107}$$

where we place ourself in the center of mass frame ($\mathbf{p_3} = -\mathbf{p_4}$, $|p_3| = |p_4|$) and $E_4 = \sqrt{|p_4|^2 + m_4^2}$. Reminding that the function delta respects $\delta[f(x)] = \frac{\delta(x)}{|f'(x)|}$,

$$\delta\left(\sqrt{s} - \sqrt{|p_4|^2 + m_3^2} - \sqrt{|p_4|^2 + m_4^2}\right) = \frac{\delta(|p_4|)}{\frac{|p_4|}{\sqrt{|p_4|^2 + m_3^2}} + \frac{|p_4|}{\sqrt{|p_4|^2 + m_4^2}}}, \tag{B.108}$$

one can write Eq. (B.107).

$$\begin{aligned} d\sigma &= \frac{|\mathcal{M}|^2}{64\pi^2 \sqrt{(\frac{s}{2} - 2M_\chi^2)\frac{s}{2}}} \frac{|p_4| d\phi d\cos\theta}{\sqrt{|p_4|^2 + m_3^2} + \sqrt{|p_4|^2 + m_4^2}} \\ &= \frac{|\mathcal{M}|^2}{64\pi^2 \sqrt{(\frac{s}{2} - 2M_\chi^2)\frac{s}{2}}} \frac{|p_4| d\phi d\cos\theta}{\sqrt{s}}, \end{aligned} \tag{B.109}$$

where we used $\sqrt{|p_4|^2 + m_3^2} + \sqrt{|p_4|^2 + m_4^2} = E_3 + E_4 = E_1 + E_2 = \sqrt{s}$. Moreover, $|p_4|^2 = E_4^2 - m_4^2$ and $E_4^2 = \frac{(s + m_4^2 - m_3^2)^2}{4s}$ implies $|p_4|^2 = \frac{1}{4s}(s^2 + m_4^4 + m_3^4 - 2sm_4^2 - 2sm_3^2 - 2m_3^2 m_4^2)$. Replacing this expression for $|p_4|$ in Eq. (B.109), we obtain, with $d\Omega = d\phi d\cos\theta$

$$\frac{d\sigma}{d\Omega} = \frac{|\mathcal{M}|^2}{128\pi^2 s}\frac{\sqrt{s^2 - 2m_3^2 s - 2m_4^2 s + (m_3^2 - m_4^2)^2}}{\sqrt{(P_1.P_2)^2 - m_1^2 m_2^2}} \tag{B.110}$$

$$= \frac{|\mathcal{M}|^2}{64\pi^2 s}\sqrt{\frac{[s-(m_3-m_4)^2][s-(m_3+m_4)^2]}{[s-(m_1-m_2)^2][s-(m_1+m_2)^2]}}$$

$$= \frac{|\mathcal{M}|^2}{64\pi^2 s}\frac{\sqrt{s - 2m_3^2 - 2m_4^2 + \frac{(m_3^2-m_4^2)^2}{s}}}{\sqrt{s - 4M_\chi^2}}.$$

But it is important to include the symmetric/statistic factors : 1/4 (1/(2S+1)* 1/(2S+1)) for a fermionic dark matter and 1/Sym for a symmetry factor if the particles 3 and 4 are identical ($Z^0 Z^0$ for instance but not W^+W^- or $f\bar{f}$). One then obtains

$$\frac{d\sigma}{d\Omega} = \frac{1}{(2S_1+1)(2S_2+1)Sym}\frac{|\mathcal{M}|^2}{64\pi^2 s}\frac{\sqrt{s - 2m_3^2 - 2m_4^2 + \frac{(m_3^2-m_4^2)^2}{s}}}{\sqrt{s - 4M_\chi^2}}$$

$$\simeq \frac{1}{(2S_1+1)(2S_2+1)Sym}\frac{|\mathcal{M}|^2}{32\pi^2 v s}\sqrt{1 - \frac{2m_3^2 + 2m_4^2}{4M_\chi^2} + \frac{(m_3^2 - m_4^2)^2}{16M_\chi^4}}, \tag{B.111}$$

where we have developed $s \simeq 4M_\chi^2 + M_\chi^2 v^2$, v being the relative velocity between the two colliding particles (also called Möller velocity). The simplified formulae for massless final state and amplitude which does not depend on the diffusion angle (scalar-like interactions) after integration on the solid angle is

$$\sigma v \simeq \frac{1}{(2S_1+1)(2S_2+1)Sym}\frac{\int |\mathcal{M}|^2 d\cos\theta}{16\pi s}. \tag{B.112}$$

B.4.4.2 A Shorter Formulation

In the literature, especially when discussing dark matter processes, one sometimes prefers to write another way Eq. (B.106) :

$$d\sigma = \frac{1}{2E_1 2E_2|v_2 - v_1|}\left(\prod_{final}\frac{d^3p_f}{(2\pi)^3}\frac{1}{2E_f}\right)(2\pi)^4\delta^{(4)}\left(P_1 + P_2 - \sum_{final} P_f\right)|\bar{\mathcal{M}}|^2, \tag{B.113}$$

where $|\bar{\mathcal{M}}|$ should be understood to mean the spin averaged squared amplitude. The two-body phase space is

$$dPS_2(p_3, p_4) = \left(\prod_{final} \frac{d^3 p_f}{(2\pi)^3} \frac{1}{2E_f}\right) \delta^{(4)}\left(P_1 + P_2 - \sum_{final} P_f\right) = \frac{d\Omega_{CM}}{256\pi^6}\left(\frac{|p_3|}{E_{CM}}\right). \tag{B.114}$$

I want to add an important note here. One must be careful when using "v." In Eq. (B.111) v represents the *relative* velocity of particle 1 with respect to particle 2, also noted $|v_2 - v_1|$, and *not* the absolute velocity or the velocity in the center of mass (CoM) frame. It is usual in dark matter literature to name v the *relative* velocity (in expressions of $\langle \sigma v \rangle$ for instance). However, when one computes cross section in quantum field theory, we usually do it in the center of mass frame, and $v_{CoM} = v/2$. In other words, $v_{dark\ matter} = v = |v_2 - v_1| = 2\ v_{CoM}$. This $v_{dark\ matter} = |v_2 - v_1|$ is the one used in Eq. (3.176) for instance. In other words, one can make all the computation in CoM frame, and then, when implementing velocities dependance we can substitute v_{CoM} by $v/2$, v being the relative velocity, which is the velocity which interests us because it is the one "measured" by the rotation of the dark matter halo.
Let us illustrate it concretely. When one has to compute $s = (P_1 + P_2)^2$, supposing $m_1 = m_2 = m$, one needs to develop

$$s = (P_1 + P_2)^2 = m_1^2 + m_2^2 + 2E_1E_1 - 2\mathbf{p}_1.\mathbf{p}_2 = 4m^2 + 4|p_{CoM}|^2. \tag{B.115}$$

Remembering that $\mathbf{p}_{CoM} = \gamma m v_{CoM}$, one deduces

$$s = 4m^2(1 + \gamma^2 v_{CoM}^2) = \frac{4m^2}{1 - v_{CoM}^2} = \frac{4m^2}{1 - v^2/4} \simeq 4m^2 + m^2v^2 \tag{B.116}$$

justifying the approximation used in Eq. (B.111).

B.4.4.3 A Note on the Symmetry Factor

We want to add a little remark concerning the symmetry factor $1/Sym$ in Eq. (B.112). If there are n_f identical particles in the final states, the symmetry factor is $Sym = \Pi_f(n_f!)$ and $\sigma = \frac{1}{Sym}\int d\sigma$. We need the symmetry factor because merely integrating over all the outgoing momenta in the phase space treats the final state as being labeled by an *ordered* list of momentum. But if some outgoing particles are identical, this is not correct; the momentum of the *identical particles* should be specified by an *unordered* list, because for example the state $a_1^\dagger a_2^\dagger|0\rangle$ is identical to the state $a_2^\dagger a_1^\dagger|0\rangle$. The symmetry factor provides the appropriate correction.

B.4.4.4 The Specific Case of Majorana or Identical Initial Particle

The counting of states in annihilation processes for Majorana fermions is non-trivial, and has led to a factor-of-two ambiguity in the literature which also propagated into different codes present in the market. In short, the $\langle \sigma v \rangle$ used when computing a relic abundance should not be confounded by the σv used when one needs to deal with indirect detection rates. In the Dirac case, or complex scalar dark matter, there are no differences, whereas a factor $1/2$ appears when dealing with Majorana/identical particles, the halo annihilation being twice less than the Boltzmann one.

The clearest way to see the origin of the factor of $1/2$ is probably to go back to the Boltzmann equation. In essence, one can view $\langle \sigma v \rangle$ as the thermal average (averaged over momentum and angles) of the cross section times velocity in the zero momentum limit; in this average one integrates over all possible angles. For identical particles in the initial state, one includes each possible initial state twice, therefore one needs to compensate by dividing by a factor of 2; the prefactor in the zero momentum limit becomes then $\langle \sigma v \rangle/2$. In the Boltzmann equation describing the time evolution of the dark matter candidate number density, the $1/2$ does not appear as it is compensated by the factor of 2 one has to include because 2 dark matter particles are depleted per annihilation, but we need to include the factor of $1/2$ explicitly in other cases where we need the annihilation rate (like for annihilation in the halo). Indeed, when integrating on phase space, we count twice $\chi(p_1)\chi(p_2)$ annihilation as they will exchange impulsion in the process of integrating on θ and ϕ.

B.4.4.5 Unitarity Limit

One interesting feature of the scattering theory is the possibility to extract constraints from theoretical bounds, like the unitarity limit. Indeed, asking for the $\mathcal{S}$ matrix (B.83) to respect

$$\mathcal{S}^\dagger \mathcal{S} = \mathbb{I}, \tag{B.117}$$

we obtain the condition

$$2\,\text{Im}\mathcal{T} = \mathcal{T}^\dagger \mathcal{T}. \tag{B.118}$$

The expression above is called the optical theorem. To appreciate its physical implications, defining the scattering amplitude of a process $p_1\ p_2 \to p_3\ p_4$ by (B.87)

$$\mathcal{M}_{p_1 p_2 \to p_3 p_4} = \langle p_3 p_4 | \mathcal{T} | p_1 p_2 \rangle,$$

the optical theorem becomes

$$-i\left[\mathcal{M}_{p_1p_2\to p_3p_4}-\mathcal{M}^*_{p_3p_4\to p_1p_2}\right] \tag{B.119}$$
$$=\sum_n\int\prod_{k=1}^{n}\frac{d^3q_k}{(2\pi)^3 2E_k}\mathcal{M}_{p_1p_2\to q_k}\mathcal{M}^*_{p_3p_4\to q_k}(2\pi)^4\delta^{(4)}(p_3+p_4-\sum_k q_k).$$

If one looks at the forward scattering amplitude, we can write

$$2\ \text{Im}\ \mathcal{M}_{p_1p_2}=\sum_n\int\prod_{k=1}^{n}\frac{d^3q_k}{(2\pi)^3 2E_k}(2\pi)^4\delta^{(4)}(p_1+p_2-\sum_k q_k)\times|\mathcal{M}_{p_1p_2\to q_k}|^2. \tag{B.120}$$

The optical theorem relates the forward scattering amplitude (on the left) to the *total* cross section (on the right). Indeed, we can recognize on the right-hand side of the equation the sum on the all possible final states *onshell* (described by the momentum q_k) for a given initial state (p_1, p_2). Using Eq. (B.106) or (B.113), we then obtain

$$\sigma_{tot}=\frac{2\times\text{Im}\ \mathcal{M}_{\text{p}_1\text{p}_2}}{4\ F}=\frac{2\times\text{Im}\ \mathcal{M}_{\text{p}_1\text{p}_2}}{2E_1 2E_2|v_2-v_1|}=\frac{\text{Im}\ \mathcal{M}_{p_1p_2}}{2p\sqrt{s}} \tag{B.121}$$

with $p=p_{com}$ and $F=\frac{1}{2}\sqrt{s(s-4M_\chi^2)}$. It is common to develop amplitudes $\mathcal{M}$ around Legendre polynomials

$$\mathcal{M}=\sum_l a_l(2l+1)P_l(\cos\theta) \tag{B.122}$$

with

$$a_l=8\pi\frac{\sqrt{s}}{p}\sin\delta_l e^{i\delta_l}, \tag{B.123}$$

where δ_l denotes the scattering-phase for the *l-th* partial wave. In a following exercise, we will justify this form for the a_l. In the forward direction, we then obtain for a non-relativistic particle of mass M_χ

$$\sigma_{tot}=\frac{4\pi(2l+1)\sin^2\delta_l}{p^2}=\frac{16\pi(2l+1)\sin^2\delta_l}{M_\chi^2 v^2},$$

where $v=|v_2-v_1|$.

Exercise Show that asking for $\Omega h^2 \lesssim 0.1$ while still respecting the unitarity constraint, in the case of a WIMP candidate, implies $M_\chi \lesssim 400$ TeV.

Exercise Considering the development

$$\mathcal{M}(s,\theta) = \sum_l a_l(2l+1)P_l(\cos\theta) \tag{B.124}$$

and using the property

$$\int_{-1}^{1} P_l(x)P_{l'}(x)dx = \frac{2}{2l+1}\delta_{ll'}, \tag{B.125}$$

with the help of Eq. (B.109) show that

$$\begin{aligned}&\int \frac{d^3p_3}{(2\pi)^3 2E_3}\frac{d^3p_4}{(2\pi)^3 2E_4}(2\pi)^2\delta^4(p_3+p_4-p_1-p_2)|\mathcal{M}(s,\theta)|^2 d\Omega\\ &= \frac{1}{4\pi}\frac{p}{\sqrt{s}}\sum_l |a_l|^2(2l+1),\end{aligned} \tag{B.126}$$

where $d\Omega = d\phi d\cos\theta$. From the optical theorem (B.120), applied to the extreme case where $\sigma_{tot} = \sigma_{elastic}$ (in other words, $q_1 = p_3$, $q_2 = p_4$, and $q_{k>2} = 0$), deduce that

$$\text{Im}\, a_l = \frac{p}{8\pi\sqrt{s}}|a_l|^2 \tag{B.127}$$

and then Eq. (B.123). We know that these a_l are the maximum values allowed by the optical theorem, meaning that the elastic scattering cross section computed with these values is the maximum cross section allowed by the unitarity constraint $S^\dagger S = 1$, or in other words,

$$\sigma_{scattering} < \frac{4\pi(2l+1)\sin^2\delta_l}{|p|^2}. \tag{B.128}$$

B.4.4.6 Scalar Fermi-Like Interaction

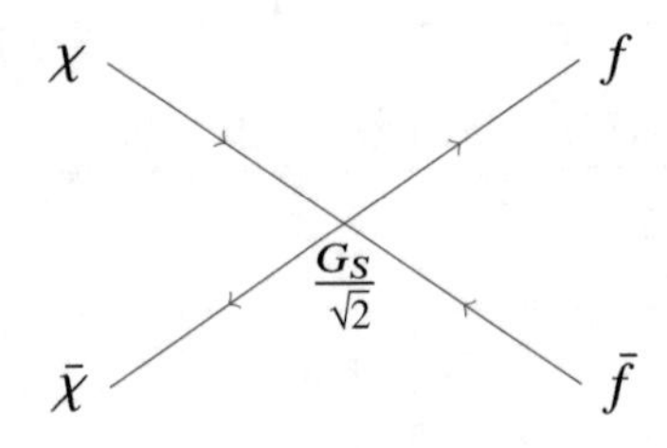

To illustrate this result, we can compute the annihilation cross section for a scalar–scalar type interaction of the form :

$$\mathcal{L}_S = \frac{G_S}{\sqrt{2}}\overline{\chi}\chi\overline{f}f. \tag{B.129}$$

The amplitude may be written as

$$\begin{aligned}
\mathcal{M} &= \frac{G_S}{\sqrt{2}}\overline{v}(p_{\bar{\chi}})u(p_\chi)\overline{u}(p_f)v(p_{\bar{f}}) \Rightarrow |\mathcal{M}|^2 \\
&= \frac{G_S^2}{2}Tr(\not{p}_{\bar{\chi}} - M_\chi)(\not{p}_\chi + M_\chi) \times Tr(\not{p}_{\bar{f}} - m_f)(\not{p}_f + m_f) \\
&= \frac{G_S^2}{2}(4P_{\bar{\chi}}.P_\chi - 4M_\chi^2)(4P_{\bar{f}}.P_f - 4m_f^2) \\
&= \frac{G_S^2}{2}(2s - 8M_\chi^2)(2s - 8m_f^2)
\end{aligned} \tag{B.130}$$

which imply

$$\sigma_S = \frac{G_S^2}{32\pi}c_f\frac{\sqrt{s - 4m_f^2}}{\sqrt{s - 4M_\chi^2}}\left(\frac{(s - 4M_\chi^2)(s - 4m_f^2)}{s}\right) \tag{B.131}$$

with c_f the color factor, equal to 3 for quarks and 1 for leptons. The annihilation cross section becomes after developing around v

$$\sigma_S v \simeq \frac{G_S^2 M_\chi^2}{16\pi}c_f\left(1 - \left(\frac{m_f}{M_\chi}\right)^2\right)^{\frac{3}{2}} v^2. \tag{B.132}$$

B.4.4.7 Vector Fermi-Like Interaction

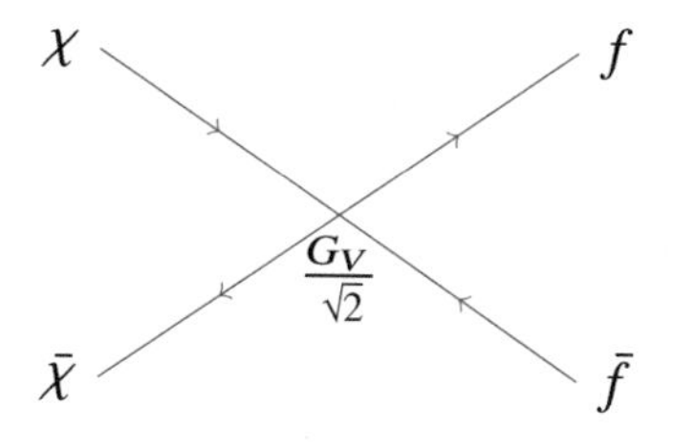

Another example is a vector–vector-like interaction:

$$\mathcal{L}_V = \frac{G_V}{\sqrt{2}}\overline{\chi}\gamma^\mu\chi\overline{f}\gamma_\mu f. \tag{B.133}$$

One can compute directly $|\mathcal{M}|^2$:

$$|\mathcal{M}|^2 = \frac{G_V^2}{2} Tr[(\not{p}_1 - M_\chi)\gamma^\mu(\not{p}_2 + M_\chi)\gamma^\nu] Tr[(\not{p}_3 - m_f)\gamma_\mu(\not{p}_4 + m_f)\gamma_\nu]$$

$$= \frac{G_V^2}{2} 32 \left[(p_1.p_3)(p_2.p_4) + (p_1.p_4)(p_2.p_3) + 2M_\chi^2 m_f^2 + (p_1.p_2)m_f^2 + (p_3.p_4)M_\chi^2\right]$$

$$= \frac{G_V^2}{2} 32 \left[(p_1.p_3)(p_2.p_4) + (p_1.p_4)(p_2.p_3) + \frac{s}{2}\left(M_\chi^2 + m_f^2\right)\right]. \tag{B.134}$$

Developing in the center of mass frame, where $\mathbf{p_1} = \mathbf{p_2} \Rightarrow E_1 = E_2 = \sqrt{s}/2 = E_3 = E_4$, one deduces $|p_1| = |p_2| = \sqrt{s/4 - M_\chi^2}$, $|p_3| = |p_4| = \sqrt{s/4 - m_f^2}$ and $p_1.p_3 = p_2.p_4 = E_1 E_3 - |p_1||p_3|\cos\theta \Rightarrow (p_1.p_3)(p_2.p_4) + (p_1.p_4)(p_2.p_3) = (p_1.p_3)^2 + (p_1.p_4)^2 = (E^2 - |p_1||p_3|\cos\theta)^2 + (E^2 + |p_1||p_3|\cos\theta)^2 = 2E^4 + 2|p_1|^2|p_2|^2\cos^2\theta$, with $E = E_1 = E_2 = E_3 = E_4$.
After integrating on the solid angle $d\cos\theta$, one obtains

$$\int d\cos\theta \left[(p_1.p_3)^2 + (p_2.p_4)^2\right] = 4E^4 + \frac{4}{3}|p_1|^2|p_3|^2 = \frac{s^2}{4} + \frac{4}{3}(\frac{s}{4} - M_\chi^2)(\frac{s}{4} - m_f^2) \tag{B.135}$$

which gives finally

$$\sigma v = \frac{G_V^2}{2}\frac{2\pi}{4}\frac{1}{64\pi^2}c_f \frac{\sqrt{s - 4m_f^2}}{\sqrt{s - 4M_\chi^2}} 32\left(\frac{s}{4} + \frac{4}{3s}(\frac{s}{4} - M_\chi^2)(\frac{s}{4} - m_f^2) + M_\chi^2 + m_f^2\right)$$

$$= \frac{G_V^2}{32\pi}c_f \frac{\sqrt{s - 4m_f^2}}{\sqrt{s - 4M_\chi^2}}\left(s + \frac{1}{3s}(s - 4M_\chi^2)(s - 4m_f^2) + 4(M_\chi^2 + m_f^2)\right). \tag{B.136}$$

Usually, when you will compute a cross section, you will use the easiest units of masses and energy which will be the GeV. However, fluxes are usually measured in more common units for experimentalists i.e. cm^{-2} or seconds. The conversion factor will be : 1 GeV^{-2} = 1.2×10^{-17}cm^3 s^{-1}.

B.4.4.8 Neutrino Interaction

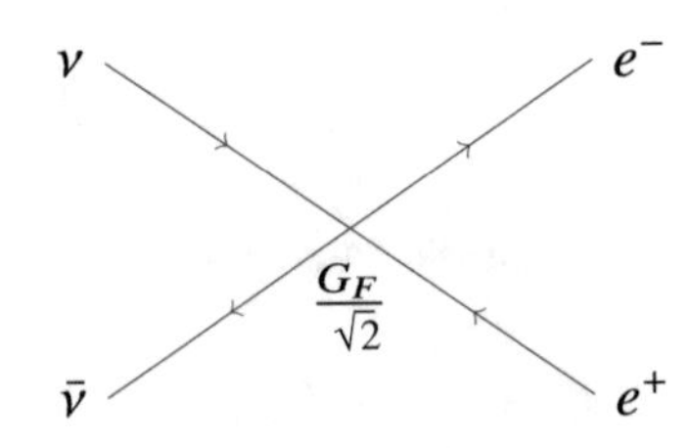

$$\mathcal{L} = \frac{G_F}{\sqrt{2}}\bar{\nu}\nu\bar{e}e \Rightarrow |\mathcal{M}|^2 = \frac{G_F^2}{2}2s(2s - 8m_e^2) \tag{B.137}$$

with $s = (p_\nu + p_{\bar{\nu}})^2 = 4E_\nu^2$ in the rest frame. Applying directly Eq. (B.111) to our case (being careful to use a massless neutrino approximation), one obtains

$$\sigma_{\nu\nu} = \frac{G_F^2}{32\pi}\frac{(s - 4m_e^2)^{3/2}}{\sqrt{s}} = \frac{G_F^2}{8\pi}\frac{(E_\nu^2 - m_e^2)^{3/2}}{E_\nu} \tag{B.138}$$

which finally gives, using the thermal average of relativistic species Eq. (A.112) :

$$\sigma_{\nu\nu} \simeq \frac{G_F^2}{8\pi}\left(\frac{7\pi^4}{180\zeta(3)}\right)^2 T^2 \simeq \frac{9.93G_F^2}{8\pi} \times T^2. \tag{B.139}$$

B.4.4.9 Annihilation into Monochromatic Photons

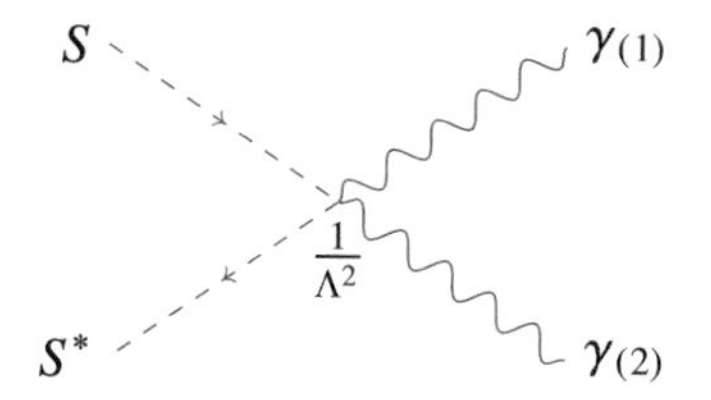

For a complex scalar dark matter, this process is effectively generated through the Lagrangian[7]

$$\mathcal{L} = \frac{1}{\Lambda^2}SS^*F_{\mu\nu}F^{\mu\nu} \tag{B.140}$$

with Λ a typical mass scale of particles running in the loop generating the effective interaction. These particles being charged under the electromagnetic forces they should be above $\simeq$ 500 GeV as no such particles have been found at LHC yet. This type of process is useful when searching for monochromatic lines in the sky, as it is still considered as a smoking gun signal for dark matter detection (see Sect. 5.145 for more details). Following the Feynman rule, the coupling can then be written as

$$C_{SS\gamma\gamma} = \frac{2i}{\Lambda^2}(\partial_\mu A_\nu^1 - \partial_\nu A_\mu^1)(\partial^\mu A_2^\nu - \partial^\nu A_2^\mu) \tag{B.141}$$

A_1 and A_2 being the photons fields 1 and 2, respectively, the factor 2 arising from the symmetry in the exchange of identical particles $\gamma_{(1)} \leftrightarrow \gamma_{(2)}$. We then need to

[7]Other dimension-6 point-like interactions of the form $SS^*F_{\mu\nu}\tilde{F}^{\mu\nu}$ can also be generated, see Sect. (B.3.7) for some examples.

develop the coupling, contracting it with the polarization of the photons to compute the amplitude $\mathcal{M}$:

$$\mathcal{M} = -\frac{2}{\Lambda^2}(p^1_\mu \epsilon^1_\nu - p^1_\nu \epsilon^1_\mu)(p^\mu_2 \epsilon^2_\nu - p^\nu_2 \epsilon^\mu_2) = -\frac{4}{\Lambda^2}\Big[(p_1.p_2)(\epsilon_1.\epsilon_2) - (p_2.\epsilon_1)(p_1.\epsilon_2)\Big]. \tag{B.142}$$

From $\mathcal{M}$, one can deduce $|\mathcal{M}|^2$

$$|\mathcal{M}|^2 = \frac{16}{\Lambda^4}\sum_{\mu,\nu,\mu',\nu'}\Big[(p_1.p_2)\epsilon^1_\mu \epsilon^\mu_2 - (p^2_\mu \epsilon^\mu_1)(p^1_\nu \epsilon^\nu_2)\Big]\Big[(p_1.p_2)\epsilon^1_{\mu'}\epsilon^{\mu'}_2 - (p^2_{\mu'}\epsilon^{\mu'}_1)(p^1_{\nu'}\epsilon^{\nu'}_2)\Big] \tag{B.143}$$

after developing and using $\sum_{\mu\mu'} \epsilon^i_\mu \epsilon^i_{\mu'} = -\eta_{\mu\mu'}$ one obtains

$$|\mathcal{M}|^2 = \frac{32}{\Lambda^4}(p_1.p_2)^2 = \frac{8}{\Lambda^4}s^2 \;\Rightarrow\; \sigma v = \frac{|\mathcal{M}|^2}{2\times 8\pi s} = \frac{s}{2\pi\Lambda^4} \simeq \frac{2m_s^2}{\pi\Lambda^4}, \tag{B.144}$$

where we have developed around $s \simeq 4m_s^2$ for low velocity[8] and divided by the symmetry factor 2 for the photons in the final state.

B.4.4.10 Annihilation in the Case of Real Scalar Dark Matter to Pairs of Fermions

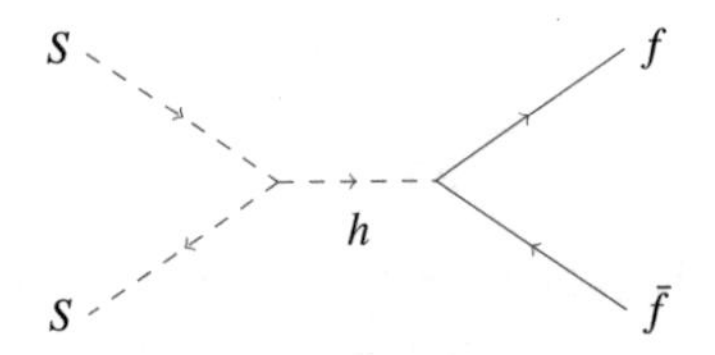

The two parts of the Lagrangian one needs to compute the scalar annihilation of Dark Matter $SS \to h \to \bar{f}f$ are (see B.236)[9]

$$\mathcal{L}_{HSS} = -\lambda_{HS}\frac{M_W}{2g}hSS \;\to C_{HSS} = -i\frac{\lambda_{HS}M_W}{g}$$
$$\text{and}\;\; \mathcal{L}_{Hff} = -\frac{gm_f}{2M_W}h\bar{f}f \;\to C_{H}ff = -i\frac{gm_f}{2M_W} \tag{B.145}$$

[8]Do not forget that v is the relative velocity of the colliding particles.

[9]Notice the factor 2 between $\mathcal{L}_{HSS}$ and C_{HSS} coming from the fact that S is real: $S = S^*$ (it corresponds to the 2 possible contractions).

which gives

$$|\mathcal{M}|^2 = \frac{\lambda_{HS}^2 m_f^2 (s/2 - 2m_f^2)}{(s - M_H^2)^2 + \Gamma_H^2 M_H^2} \tag{B.146}$$

Γ_H being the width of the Higgs boson (including its own decay into SS, see next section). When one implements this value of $|\mathcal{M}|^2$ into Eq. (B.111) one obtains after simplification

$$\langle \sigma v \rangle_{f\overline{f}}^S = \frac{|\mathcal{M}|^2}{8\pi s}\sqrt{1 - \frac{m_f^2}{M_S^2}} = \frac{\lambda_{HS}^2 (M_S^2 - m_f^2) m_f^2}{16\pi M_S^2 (4M_S^2 - M_H^2)^2}\sqrt{1 - \frac{m_f^2}{M_S^2}}. \tag{B.147}$$

B.4.4.11 Annihilation in the Case of Vectorial Dark Matter to Pairs of Fermions

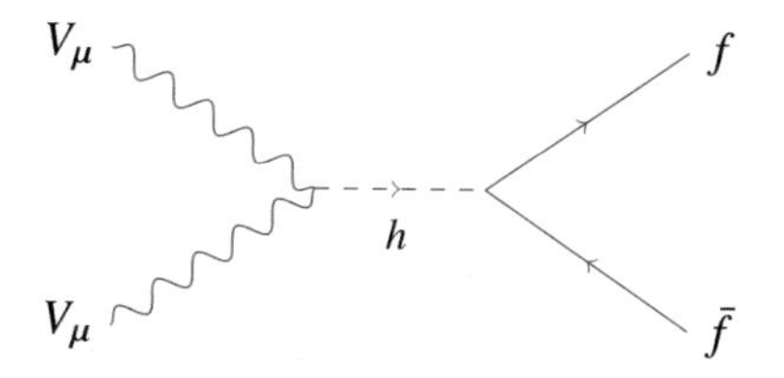

One can compute this annihilation cross section by the normal procedure or noticing that a neutral vectorial dark matter of spin 1 corresponds to 3 degrees of freedom. After averaging on the spin, one can then write $\langle \sigma v \rangle^V = \frac{3}{3\times 3}\langle \sigma v \rangle^S = \frac{1}{3}\langle \sigma v \rangle^S$. The academical computation for $V_\mu(p_1)V_\mu(p_2) \to f\overline{f}$ gives

$$\mathcal{L} \supset -\frac{\lambda_{hV}}{4}\eta_{\mu\nu} H^\dagger H V^\mu(p_1) V^\nu(p_2) \Rightarrow C_{hVV} = -\frac{i}{g}\lambda_{hV} M_W \eta_{\mu\nu}\epsilon_1^\mu \epsilon_2^\nu \tag{B.148}$$

$$\text{and} \quad C_{hf\overline{f}} = -\frac{i}{2}\frac{g m_f}{M_W}\overline{u}(p_f) v(p_{\overline{f}}) \quad \text{implying}$$

$$\mathcal{M} = -\epsilon_{1\lambda}^\mu \epsilon_{2\rho}^\nu \eta_{\mu\nu} \frac{\lambda_{hV} m_f}{2}\overline{u}(p_f)\frac{i}{(s - m_h^2) + i\Gamma_h m_h} v(p_{\overline{f}}),$$

$$\mathcal{M}^* = \epsilon_{1\lambda}^{\alpha *}\epsilon_{2\rho}^{\beta *}\eta_{\alpha\beta}\frac{\lambda_{hV} m_f}{2}\overline{v}(p_{\overline{f}})\frac{i}{(s - m_f^2) - i\Gamma_h m_h} u(p_f).$$

Using $\sum_\lambda \epsilon_{1\lambda}^\mu \epsilon_{1\lambda}^{\alpha *} = -\eta^{\mu\alpha} + \frac{p_1^\mu p_1^\alpha}{m_V^2}$, one obtains

$$|\mathcal{M}|^2 = \frac{2\lambda_{hV}^2 m_f^2 (m_V^2 - m_f^2)}{(s - m_h^2)^2 + \Gamma_h^2 m_h^2}\left(4 + \frac{(p_1 \cdot p_2)^2}{m_V^4} - \frac{p_1 \cdot p_1}{m_V^2} - \frac{p_2 \cdot p_2}{m_V^2}\right) = \frac{6\lambda_{hV}^2 m_f^2 (m_V^2 - m_f^2)}{(s - m_H^2)^2 + \Gamma_h^2 m_h^2} \tag{B.149}$$

implying, after averaging on the spin 1 initial state particles

$$\langle\sigma v\rangle^{V}_{f\bar{f}} = \frac{\lambda_{hV}^2 (m_V^2 - m_f^2) m_f^2}{48\pi m_V^2 [(4m_V^2 - m_h^2)^2 + \Gamma_h^2 m_h^2]} \sqrt{1 - \frac{m_f^2}{m_V^2}} \quad \text{(B.150)}$$

B.4.4.12 Exchange of a Vector: The Case of the Vectorial Coupling

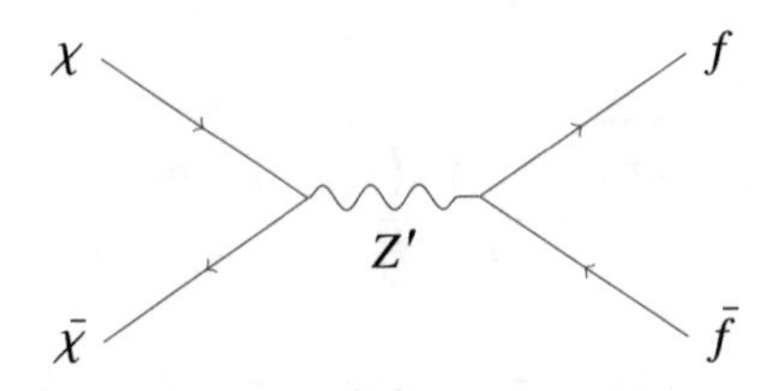

We compute in this section the process $\chi\bar{\chi} \to Z' \to f\bar{f}$.
From the Lagrangian $i\bar{\chi}\gamma^{\mu}D_{\mu}\chi + i\bar{f}\gamma^{\mu}D_{\mu}f$, we extract the interaction $g_D\left[q_{\chi}\bar{\chi}\gamma^{\mu}Z'_{\mu}\chi + q_f\bar{f}\gamma^{\mu}Z'_{\mu}f\right]$ which gives

$$\mathcal{M} = -i\frac{g_D^2 q_{\chi} q_f}{s - M_{Z'}^2}\bar{u}(p_f)\gamma^{\mu}v(p_{\bar{f}})\left[\eta_{\mu\nu} - \frac{p_{\mu}^{Z'}p_{\nu}^{Z'}}{M_{Z'}^2}\right]\bar{v}(p_{\bar{\chi}})\gamma^{\nu}u(p_{\chi}) \quad \text{(B.151)}$$

$$\Rightarrow |\mathcal{M}|^2 = 32\frac{g_D^4 q_{\chi}^2 q_f^2}{(s - M_{Z'}^2)^2}[(p_f.p_{\bar{\chi}})(p_{\bar{f}}.p_{\chi}) + (p_f.p_{\chi})(p_{\bar{f}}.p_{\bar{\chi}})$$
$$+m_f^2(p_{\chi}p_{\bar{\chi}}) + m_{\chi}^2(p_f.p_{\bar{f}}) + 2m_{\chi}^2 m_f^2]$$

$$\Rightarrow |\mathcal{M}|^2 = 64\frac{g_D^4 q_{\chi}^2 q_f^2}{(s - M_{Z'}^2)^2}[E^4 + (E^2 - m_f^2)(E^2 - m_{\chi}^2)\cos^2\theta + E^2(m_f^2 + m_{\chi}^2)].$$

After the average on the entering spins (1/4), we obtain

$$d\sigma = C_f\frac{|\bar{\mathcal{M}}|^2}{64\pi^2 s}\frac{\sqrt{s - 4m_f^2}}{\sqrt{s - 4m_{\chi}^2}}2\pi d(\cos\theta)$$

$$\Rightarrow \sigma = \frac{4\pi C_f\,\alpha_D^2\,q_{\chi}^2 q_f^2}{3E^2(4E^2 - M_{Z'}^2)^2}\frac{\sqrt{4E^2 - 4m_f^2}}{\sqrt{4E^2 - 4m_{\chi}^2}}(2E^2 + m_{\chi}^2)(2E^2 + m_f^2)$$

$$= \frac{4\pi C_f\,\alpha_D^2\,q_{\chi}^2 q_f^2}{3s(s - M_{Z'}^2)^2}\frac{\sqrt{s - 4m_f^2}}{\sqrt{s - 4m_{\chi}^2}}(s + 2m_{\chi}^2)(s + 2m_f^2) \quad \text{(B.152)}$$

C_f being the color index of the fermions in the final state and $\alpha_D = \frac{g_D^2}{4\pi}$, which gives

$$\langle \sigma v \rangle^{\chi}_{f\bar{f}} = \frac{8\pi C_f \, \alpha_D^2 \, q_\chi^2 q_f^2}{(4m_\chi^2 - M_{Z'}^2)^2} \sqrt{1 - \frac{m_f^2}{m_\chi^2}} \, (2m_\chi^2 + m_f^2). \tag{B.153}$$

B.4.5 Spin-Independent Diffusion, Elastic Scattering

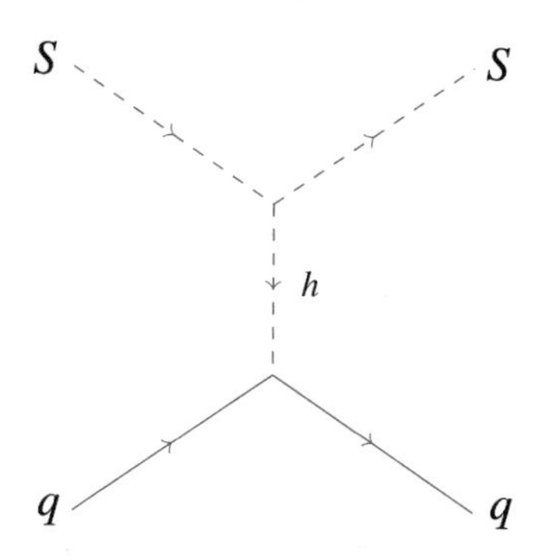

If one wants to compute the diffusion of a scalar particle of mass m_S and momentum p_S on a proton of momentum p_p and with momentum p'_p after the hit, following Eq. (B.110) (with $m_3 \to m_1$ and $m_4 \to m_2$) and noticing $\sqrt{(P_1.P_2)^2 - m_1^2 m_2^2} = \frac{1}{2}\sqrt{s^2 - 2m_1^2 s - 2m_2^2 s + (m_1^2 - m_2^2)^2}$, we can express

$$\frac{d\sigma^{SI}_{S-p}}{d\Omega} = \frac{|\mathcal{M}|^2}{64\pi^2 s} \Rightarrow \sigma^{SI}_{S-p} = \frac{|\mathcal{M}|^2}{16\pi s} \simeq \frac{|\mathcal{M}|^2}{16\pi (m_p + m_S)^2} \tag{B.154}$$

before averaging on the initial spins and before including the possible symmetry factors. Using Eq. (B.145), one can write

$$\mathcal{M} = -\frac{i}{2(P_H^2 - M_H^2)} \left[\sum_q \lambda_{HS} m_q \langle p | \bar{q} q | p \rangle \bar{u}(p'_p) u(p_p) \right]. \tag{B.155}$$

If one defines[10] $f_q = \frac{m_q}{m_p}\langle p|\bar{q}q|p\rangle$, $f = \sum_q f_q$, and noticing that $P_H^2 - M_H^2 \simeq -M_H^2$, Eq. (B.155) becomes

$$|\mathcal{M}|^2 = \frac{1}{2}\frac{m_p^2\lambda_{HS}^2 f^2}{M_H^4}(P'_p.P_p + m_p^2) \simeq \frac{m_p^4\lambda_{HS}^2 f^2}{M_H^4}$$

$$\Rightarrow \sigma_{S-p}^{SI} = \frac{m_p^4\lambda_{HS}^2 f^2}{16\pi(m_p + m_S)^2 M_H^4} \tag{B.156}$$

because the proton is mainly at rest[11] $P'_p.P_p + m_p^2 \simeq 2m_p^2$.

B.4.6 Decaying Particles

We have seen in Eq. (B.104) and more generally in Sect. B.4.3 how to compute the phase space for a n-body final state. Concentrating on decay processes, there exists some "tricks" to facilitate such computations that we will develop in this chapter.

B.4.6.1 2-Body Decay

The expression we obtained combining Eqs. (B.101) and (B.104) gives, for a particle ϕ of mass M_ϕ decaying into 2 quantum states $P_1 = (E_1, p_1)$ and $P_2 = (E_2, p_2)$:

$$d\Gamma_\phi = \frac{(2\pi)^4}{2M_\phi}\frac{d^3p_1}{(2\pi)^3 2E_1}\frac{d^3p_2}{(2\pi)^3 2E_2}\delta^{(4)}(P_1 + P_2 - P_\phi)|\mathcal{M}_{\phi\to 1,2}|^2 \tag{B.157}$$

with $P_\phi = (M_\phi, 0, 0, 0)$. The main difficulty is to express the phase space as function of the variables that can appear in $|\mathcal{M}_{\phi\to 1,2}|$, E_1, E_2 and $\cos\theta_{12}$. In Sect. B.4.3 we took the absolute momentum $|p_1|$ as the main variable, but we could also have taken E_1. Indeed, defining

$$d\Phi_2 = \frac{d^3p_1}{(2\pi)^3 2E_1}\frac{d^3p_2}{(2\pi)^3 2E_2}\delta^{(4)}(P_1 + P_2 - P_\phi) \tag{B.158}$$

and noticing that $E_1^2 = p_1^2 + m_1^2 \Rightarrow E_1 dE_1 = p_1 dp_1$, we can write within a solid angle $d\Omega_{12} = d\cos\theta_{12} d\phi_{12}$

$$d\Phi_2 = \frac{d^3p_1}{(2\pi)^6 2E_1 2E_2}\delta(E_1 + E_2 - M_\phi) = \frac{p_1 dE_1 d\Omega_{12}}{4(2\pi)^6 E_2}\delta(E_1 + E_2 - M_\phi),$$

[10]See Sect. 4.9.2 for the details to transform a scattering at the quark level to a scattering at a nucleon level.

[11]The first $\frac{1}{2}$ is coming from the mean on the spin of one fermion, the initial proton.

where E_2 is *not* a variable anymore as it is fixed by the condition $\delta^{(3)}(p_1 + p_2 = 0)$ which can be translated by

$$E_2 = \sqrt{E_1^2 - m_1^2 + m_2^2}$$

$$\Rightarrow d\Phi_2 = \frac{\sqrt{E_1^2 - m_1^2}\, dE_1\, d\Omega_{12}}{4(2\pi)^6\sqrt{E_1^2 - m_1^2 + m_2^2}}\delta(E_1 + \sqrt{E_1^2 - m_1^2 + m_2^2} - M_\phi),$$

where all the phase space is expressed uniquely as function[12] of E_1. Making use of the identity $\delta[f(x)] = \frac{1}{|f'(x)|}\delta(x)$ one can write

$$d\Phi_2 = \frac{\sqrt{E_1^2 - m_1^2}}{4(2\pi)^6 M_\phi}d\Omega_{12} \tag{B.159}$$

which is equivalent to Eq. (B.98) where we took p_1 as the fundamental parameter.

There exists a third way to compute the phase space, using a simple trick that will be useful once one needs to compute decays into 3-body, 4-body, or even more. The idea is to transform the last integration on the momentum space d^3p_2 into an integration on the 4-momentum space d^4P_2. Indeed, noticing that

$$\int \frac{d^3p_2}{2E_2} \quad \text{can be written} \quad \int d^3p_2 dE_2 \delta(E_2^2 - p_2^2 - m_2^2)\theta(E_2), \tag{B.160}$$

where we used the $\delta[f(x)]$ formula and $\theta(x)$ is the heaviside function. Eq. (B.158) then becomes

$$d\Phi_2 = \frac{d^3p_1}{(2\pi)^6 2E_1}d^4P_2\delta^{(4)}(P_1 + P_2 - M_\phi)\theta(E_2)\delta(E_2^2 - p_2^2 - m_2^2) \tag{B.161}$$

$$= \frac{\sqrt{E_1^2 - m_1^2}dE_1 d\Omega_{12}}{2(2\pi)^6}\delta((M_\phi - E_1)^2 - (E_1^2 - m_1^2) - m_2^2) = \frac{\sqrt{E_1^2 - m_1^2}d\Omega_{12}}{4(2\pi)^6 M_\phi}$$

with

$$E_1 = \frac{M_\phi^2 + m_1^2 - m_2^2}{2M_\phi}. \tag{B.162}$$

[12] It is common to see in the literature expressions as function of $x_i = \frac{E_i}{M_\phi}$ for convenience.

Combining Eq. (B.157) with Eq. (B.161), the 2-body decay can then be written as

$$d\Gamma_\phi = \frac{\sqrt{E_1^2 - m_1^2}}{32\pi^2 M_\phi^2} d\Omega_{12} |\mathcal{M}_{\phi\to 1,2}|^2 \quad \text{with } E_1 = \frac{M_\phi^2 + m_1^2 - m_2^2}{2M_\phi}. \tag{B.163}$$

We can also write

$$\sqrt{E_1^2 - m_1^2} = \frac{[(M_\phi^2 - (m_1+m_2)^2)(M_\phi^2 - (m_1 - m_2)^2)]^{1/2}}{2M_\phi}. \tag{B.164}$$

In the case where $m_1 = m_2$, the formula is greatly simplified, after integration on the phase space to

$$\Gamma = \left(\frac{1}{2}\right) \frac{|\mathcal{M}_{\phi\to 1,2}|^2}{16\pi M_\phi^2} \sqrt{M_\phi^2 - 4m_1^2} \tag{B.165}$$

the factor $\left(\frac{1}{2}\right)$ coming from the (anti)symmetrization of identical (fermions) bosons in the exchange $1 \leftrightarrow 2$. Notice also that $|\mathcal{M}|$ should be understood as a mean on the polarization state, which means care should be taken, notably dividing by the factor $\frac{1}{2S+1}$, S being the spin of the decaying particle.

B.4.6.2 3-Body Decay

One does not gain a lot using the trick explained in the previous section if one has to deal with 2-body decays only. This idea is more efficient concerning the 3-body decay. Indeed, rewriting the 3-body phase space and using (B.160) on p_3 :

$$\begin{aligned}
d\Phi_3 &= \frac{d^3p_1}{(2\pi)^3 2E_1}\frac{d^3p_2}{(2\pi)^3 2E_2}\frac{d^3p_3}{(2\pi)^3 2E_3}\delta^{(4)}(P - P_1 - P_2 - P_3) \\
&= \frac{d^3p_1 d^3p_2}{(2\pi)^9 2E_1 2E_2}\delta(E_3^2 - p_3^2 - m_3^2) = \frac{d^3p_1 d^3p_2}{(2\pi)^9 2E_1 2E_2}\delta(P_3^2 - m_3^2) \\
&= \frac{d^3p_1 d^3p_2}{(2\pi)^9 2E_1 2E_2}\delta[(P - P_1 - P_2)^2 - m_3^2] \\
&= \frac{E_1 dE_1 d\Omega_{12} E_2 dE_2 d\Omega_{23}}{4(2\pi)^9}\delta[M_\phi^2 - 2M_\phi(E_1 + E_2) + 2E_1E_2(1 - \cos\theta_{12})],
\end{aligned}$$

where, in the last expression, we put $m_1 = m_2 = m_3 = 0$ for simplicity. Having no particular axis of references, the integration over $d\Omega_{23}$ gives 4π, whereas $d\Omega_{12} = d\phi_{12} d\cos\theta_{12} = 2\pi d\cos\theta_{12}$. After integrating the δ function over $\cos\theta_{12}$, one obtains

$$d\Phi_3 = \frac{dE_1 dE_2}{4(2\pi)^7} \tag{B.166}$$

with

$$\cos\theta_{12} = \frac{M_\phi^2 - 2M_\phi(E_1 + E_2) + 2E_1E_2}{2E_1E_2}. \tag{B.167}$$

Whereas the limit for E_1 is $0 < E_1 < \frac{M_\phi}{2}$ the integration limits for E_2 can be obtained with the two extreme cases : 2 going in opposite direction of 1 and 3 (and then taking the maximum energy) corresponding to $\cos\theta_{12} = -1$ (which gives from Eq. (B.167) $E_2 = \frac{M_\phi}{2}$) and 2 going in the same direction of 1, whereas 3 takes all the energy ($\cos\theta = 1$, corresponding to $E_2 = \frac{M_\phi}{2} - E_1$);
we then can write

$$\Gamma_\phi = \frac{1}{64\pi^3 M_\phi} \int_0^{\frac{M_\phi}{2}} dE_1 \int_{\frac{M_\phi}{2}-E_1}^{\frac{M_\phi}{2}} dE_2 |\mathcal{M}_{\phi\to 1,2,3}|^2$$
$$\text{with} \quad \cos\theta_{12} = \frac{M_\phi^2 - 2M_\phi(E_1 + E_2) + 2E_1E_2}{2E_1E_2}. \tag{B.168}$$

B.4.6.3 Application to Muon Decay

We can directly apply our procedure above to compute the muon lifetime in the Fermi approximation of contact interactions. The Fermi Lagrangian can be written as

$$\mathcal{L} = \frac{4G_F}{\sqrt{2}} (\overline{\nu}_\mu \gamma^\mu P_L \mu)(\overline{e}\gamma_\mu P_L \nu_e) \tag{B.169}$$

which gives for the amplitude

$$|\mathcal{M}|^2 = 32G_F^2 (P.P_1)(P_2.P_3) = 32G_F^2 M_\mu^2 E_1(M_\mu - 2E_1), \tag{B.170}$$

where particle "1" is ν_μ, "2" ν_e and "3" the electron. The integration gives

$$\Gamma_\mu = \frac{G_F^2 M_\mu}{2\pi^3} \int_0^{\frac{M_\mu}{2}} E_1^2 (M_\mu - 2E_1) dE_1 = \frac{G_F^2 M_\mu^5}{192\pi^3}. \tag{B.171}$$

B.4.7 Higgs Lifetime

We can apply our generic discussions above to compute the Higgs branching ratios.

B.4.7.1 Higgs Lifetime from $H \to f(\mathbf{p_1})\bar{f}(\mathbf{p_2})$

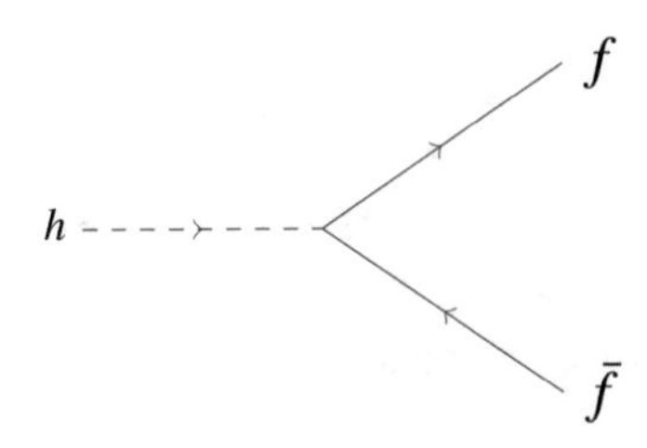

Combining Eq. (B.165) and the SM Lagrangian (Sect. B.9) one obtains

$$C_{Hf\bar{f}} = -i\frac{gm_f}{2M_W} \quad \Rightarrow \quad \mathcal{M} = \bar{u}(p_1)\frac{gm_f}{2M_W}u(p_2) \quad \Rightarrow \quad |\mathcal{M}|^2 = \frac{g^2 m_f^2}{4M_W^2}\mathrm{Tr}(\not{p}_1 + m_f)(\not{p}_2 - m_f). \tag{B.172}$$

Using Eq. (B.65) one can write

$$\Gamma_{H\to f\bar{f}} = \frac{g^2 m_f^2}{32\pi M_W^2 M_H^2}(M_H^2 - 4m_f^2)^{3/2}. \tag{B.173}$$

B.4.7.2 Higgs Lifetime from $H \to Z(p_1)Z(p_2)$

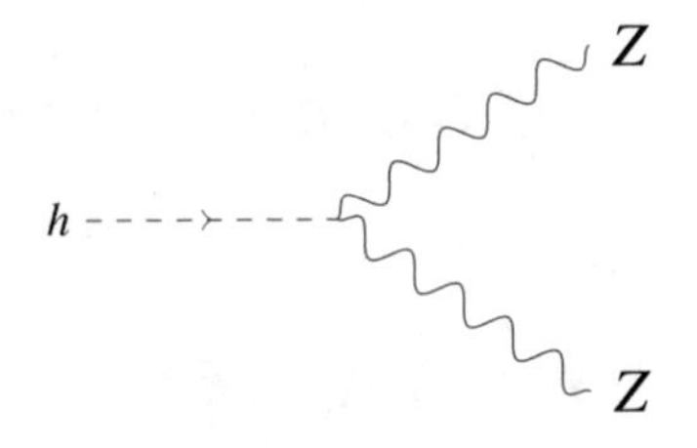

In this case, one uses the HZZ coupling from the Lagrangian given in Eq. (B.234), we obtain

$$\mathcal{M}_{\mu\nu} = i\frac{g}{\cos\theta_W} M_Z \eta_{\mu\nu} \epsilon^{\mu}_{1\lambda} \epsilon^{\nu}_{2\rho} \quad \mathcal{M}^{*}_{\mu\nu} = -i\frac{g}{\cos\theta_W} M_Z \eta_{\alpha\beta} \epsilon^{\alpha *}_{1\lambda} \epsilon^{\beta *}_{2\rho} \tag{B.174}$$

λ and ρ being the polarization of the $Z(p_1)$ and $Z(p_2)$, respectively. After the summation on the polarization, $\sum_\lambda \epsilon^{\mu}_{1\lambda} \epsilon^{\alpha *}_{1\lambda} = -\eta^{\mu\alpha} + p^{\mu}_1 p^{\alpha}_1 / M_Z^2$ one obtains

$$\begin{aligned} |\mathcal{M}|^2 &= \frac{g^2}{\cos^2\theta_W} M_Z^2 \eta_{\mu\nu} \left(-\eta^{\mu\alpha} + \frac{p_1^{\mu} p_1^{\alpha}}{M_Z^2} \right) \eta_{\alpha\beta} \left(-\eta^{\nu\beta} + \frac{p_1^{\nu} p_1^{\beta}}{M_Z^2} \right) \\ &= \frac{g^2 M_Z^2 M_H^4}{4 M_Z^4 \cos^2\theta_W} \left(1 - 4\frac{M_Z^2}{M_H^2} + 12\frac{M_Z^4}{M_H^4} \right) \end{aligned} \tag{B.175}$$

which gives

$$\Gamma_{H\to ZZ} = \frac{g^2 M_H^3}{128\pi M_W^2} \sqrt{1 - \frac{4M_Z^2}{M_H^2}} \left(1 - 4\frac{M_Z^2}{M_H^2} + 12\frac{M_Z^4}{M_H^4} \right). \tag{B.176}$$

B.4.7.3 Higgs Lifetime from $H \to W^+W^-$

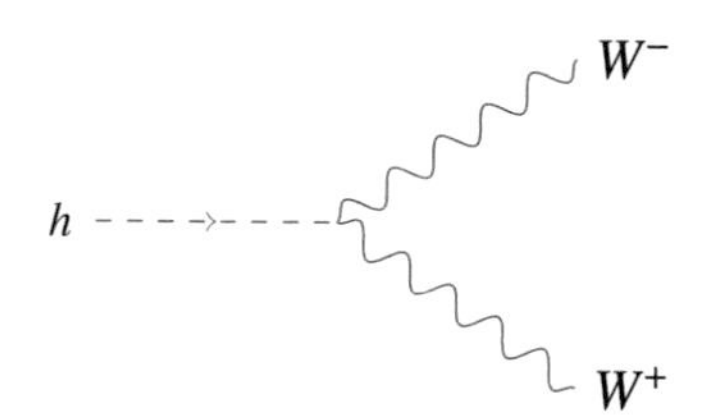

Similarly, on can compute

$$\Gamma_{H\to WW} = \frac{g^2 M_H^3}{64\pi M_W^2} \sqrt{1 - \frac{4M_W^2}{M_H^2}} \left(1 - 4\frac{M_W^2}{M_H^2} + 12\frac{M_W^4}{M_H^4} \right). \tag{B.177}$$

The difference between the WW and ZZ prefactor comes from the symmetry ($Z = Z^* \Rightarrow 1/2$ of WW final state.)

B.4.7.4 Higgs Lifetime from $H \to SS$ with Singlet Scalar Dark Matter

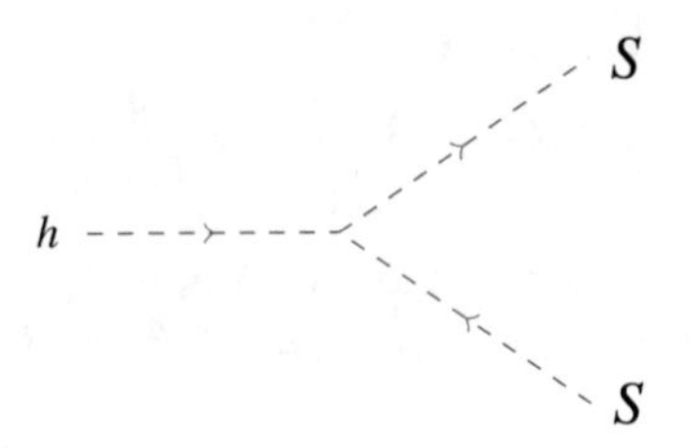

From the simplest singlet extension of the Standard Model, one can extract (see appendix) the HSS interaction term after the $SU(2) \times U(1)_Y$ breaking,[13]

$$\mathcal{L}_{HSS} = -\lambda_{HS}\frac{M_W}{2g}HSS \quad \to C_{HSS} = -i\frac{\lambda_{HS}M_W}{g} \quad \Rightarrow \quad |\mathcal{M}|^2 = \lambda_{HS}^2\frac{M_W^2}{g^2}. \tag{B.178}$$

From Eq. (B.165) one can compute

$$\Gamma_{H\to SS} = \frac{\lambda_{HS}^2 M_W^2}{32\pi g^2 M_H^2}\sqrt{M_H^2 - 4M_S^2}. \tag{B.179}$$

B.4.7.5 Higgs Lifetime from $H \to X_\mu X_\mu$

There are different ways of computing this process. The easiest way would be to see the vector as a 3-degree of freedom scalar. In this case, the sum on the 3 degrees of freedom convoluated with the spin averages gives

$$\Gamma_{X\to X_\mu X_\mu} = 3 \times \Gamma_{H\to SS}. \tag{B.180}$$

B.4.7.6 General Scalar Width: $T \to \chi\,\bar{\chi}$

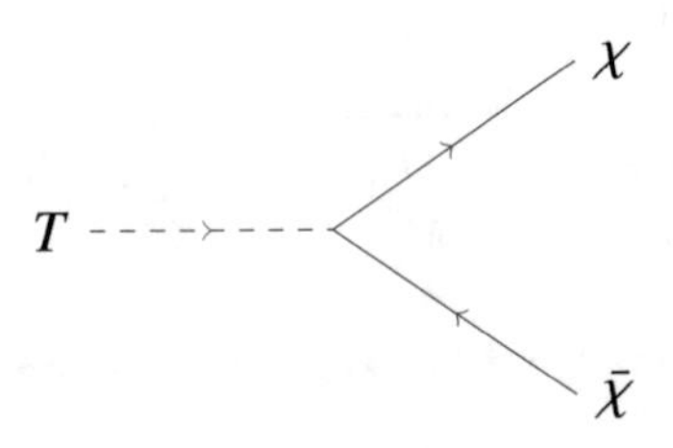

Generically, a scalar coupling to a fermion can be written as $y\,T\bar{\chi}\chi$. T can be a Higgs field or a moduli field decaying into 2 dark matter candidates for instance. In

[13]Notice the factor 2 between $\mathcal{L}_{HSS}$ and C_{HSS} coming from the fact that there are 2 identical particles in final states (it corresponds to the 2 possible contractions), see Eq. (B.239) for details.

this case,

$$\mathcal{M} = -iy\,\bar{u}(p_\chi)v(p_{\bar{\chi}}), \quad \Rightarrow |\mathcal{M}|^2 = 4y^2(p_\chi . p_{\bar{\chi}} - m_\chi^2)$$
$$\Rightarrow \Gamma_T = \frac{|\mathcal{M}|^2}{16\pi M_T^2}\sqrt{M_T^2 - 4m_\chi^2} = y^2\frac{(M_T^2 - 4m_\chi^2)^{3/2}}{8\pi M_T^2}. \quad \text{(B.181)}$$

T can also be the inflation field to compute the reheating temperature.

B.4.8 Majorana Case

To compute the decay of the scalar into a pair of Majorana particle, we let the reader have a look at Sects. B.2.2 and B.3.8 to understand the origin of the factors $\frac{1}{2}$ appearing in the Lagrangian and 2 in the Feynman rules, canceling each other at the end. Writing the Lagrangian $\mathcal{L} = y_\chi T\bar{\chi}\chi + \frac{1}{2}y_\lambda T\bar{\lambda}\lambda$, one obtains

$$\mathcal{M}_\lambda = -iy_\lambda\bar{u}(p_\lambda)v(p_\lambda) \quad \Rightarrow |\mathcal{M}_\lambda|^2 = 4y_\lambda^2(p_\lambda . p_{\bar{\lambda}} - m_\lambda^2)$$
$$\Rightarrow \Gamma_T^\lambda = \left(\frac{1}{2}\right)\frac{|\mathcal{M}_\lambda|^2}{16\pi M_T^2}\sqrt{M_T^2 - 4m_\lambda^2} = y_\lambda^2\frac{(M_T^2 - 4m_\lambda^2)^{3/2}}{16\pi M_T^2}. \quad \text{(B.182)}$$

The last factor $\left(\frac{1}{2}\right)$ is the symmetry factor (antisymmetrization because of identical final state). We can also apply it to the decay of the scalar into a spin 3/2 Majorana (like the gravitino for instance[14]).

B.4.9 Vector Lifetime

B.4.9.1 Generalities

We can write a general vectorial Lagrangian of a vector V_μ of mass M, momentum P, and polarization λ coupling to fermions ψ_i^1 and ψ_i^2 of masses m_1, m_2 and momentum p_1 and p_2,

$$\mathcal{L}_V = i\bar{\psi}_i^2\gamma^\mu(\partial_\mu - ig_V V_\mu)(\frac{V_i - A_i\gamma^5}{2})\psi_i^1 \quad \text{(B.183)}$$

[14]Be aware that the Lagrangian of supergravity exhibit more complex structures that the simplest one we propose to study here.

which gives the amplitude

$$\mathcal{M} = \frac{i}{2} g_V \bar{u}(p_1)\gamma^\mu (V_i - A_i\gamma^5) v(p_2)\epsilon^\lambda_\mu, \quad \mathcal{M}^* = -\frac{i}{2} g_V \bar{v}(p_2)\gamma^\nu (V_i - A_i\gamma^5) u(p_2)\epsilon^{\lambda *}_\nu. \tag{B.184}$$

After the summation on the polarization, $\sum_\lambda \epsilon^\mu_{1\lambda}\epsilon^{\nu *}_{1\lambda} = -\eta^{\mu\nu} + P^\mu P^\nu / M_V^2$ one obtains

$$|\mathcal{M}|^2 = C_i \frac{g_V^2}{4} Tr\Big[(\not{p}_1 + m_1)\gamma^\mu (V_i - A_i\gamma^5)(\not{p}_2 - m_2)\gamma^\nu (V_i - A_i\gamma^5)\Big]\Big[-\eta_{\mu\nu} + \frac{P_\mu P_\nu}{M^2}\Big]$$
$$= C_i \frac{g_V^2}{2}\Big[(|V_i|^2 + |A_i|^2)\left(2M^2 - m_1^2 - m_2^2\right) + 6(|V_i|^2 - |A_i|^2) m_1 m_2\Big]$$

C_i being internal degrees of freedom like the color charge of the fermions for instance. After a mean on the spin, $|\bar{\mathcal{M}}|^2 = \frac{1}{2J+1}|\mathcal{M}|^2 = \frac{1}{3}|\mathcal{M}|^2$ and using Eq. (B.163) we can write

$$\Gamma_V = \frac{g_V^2 M}{96\pi}\Big[(|V_i|^2 + |A_i|^2)\left(2 - x_1^2 - x_2^2\right) + 6(|V_i|^2 - |A_i|^2) x_1 x_2\Big]$$
$$\times \sqrt{[1 - (x_1 + x_2)^2][1 - (x_2 - x_1)^2]}, \quad \text{with } x_i = \frac{m_i}{M}. \tag{B.185}$$

B.4.9.2 $W^\pm$ and Z Widths

In the Standard Model, the Lagrangian involving $W^\pm$ is written following Eq. (B.183) with $g_V = g/\sqrt{2}$, $V_i = A_i = 1$ and neglecting the masses of the final states ($x_1 = x_2 = 0$) we obtain

$$\Gamma(W^+ \to e^+ \nu_e) = \frac{g^2 M_W}{48\pi}. \tag{B.186}$$

In the case of the Z boson $g_V = g/\cos\theta_W$the couplings A_i and V_i depend on the fermion ψ_i; we have

$$\Gamma(Z \to \psi_i \bar{\psi}_i) = C_i \frac{g^2 M_Z}{48\pi \cos^2\theta_W}\left(|V_i^Z|^2 + |A_i^Z|^2\right) \tag{B.187}$$

with $A_i^Z = T_i^3$ and $V_i^Z = -2q_i \sin^2\theta_W + T_i^3$, q_i being the electric charge of the particle i and $T_i^3 = \pm 1/2$ its isospin.

B.5 s-Wave, p-Wave, Helicity Suppression and All That

B.5.1 Velocity Suppression

We used frequently in the analysis of the annihilation cross section the development

$$\sigma v = a + bv^2. \tag{B.188}$$

This comes from the fact that, on a general ground, a cross section can be decomposed on the basis of the orbital momentum l : $\sigma v = \sum_l (\sigma v)_l \propto |\sum_l (2l + 1) a_l P_l(\cos\theta)|^2$, P_l being the Legendre polynomial function. The lth partial wave contribution to the annihilation rate is suppressed as v^{2l}, where v is the relative velocity between the annihilating particles. Indeed, the lth partial wave includes a factor $P_l(\cos\theta)$. Since the amplitude is an analytic function of the Lorentz invariants $s, t(\cos\theta)$ and $u(\cos\theta)$, the true argument of P_l is proportional to $\bar{p}_i \bar{p}_f \cos\theta$ for an initial particle of impulsion p_i and final particle of impulsion p_f, the latter scaling as $\bar{v}_i \bar{v}_f$. Integration over singularities picks out the highest power of P_l, and we get an amplitude scaling as $(\bar{v}_i \bar{v}_f)^l$. If annihilating particles have a non-zero spin, l is replaced by j and $P_l(\cos\theta)$ by the Wigner function $d^j_{\mu\lambda(\theta)}$. In the expression $a + bv^2$, a is then called s-wave ($l = 0$) term and bv^2 p-wave ($l = 1$) term.

The virial velocity in our galactic halo is about $\sim 300\text{km s}^{-1} \sim 10^{-3}c$, so even for $l = 1$ the suppression is considerable. Thus only the $l = 0$ partial wave gives an unsuppressed annihilation rate in today's Universe. The $l > 1$ states are too suppressed to give any observable rate. When computing the relic abundance the velocity is less suppressed. Indeed, the velocity of a dark matter particle χ of mass m_χ can be expressed as function of the temperature of the Plasma from Eq. (3.162), $\langle v^2 \rangle = 6\frac{T}{m}$. As we computed in Eq. (3.170), at the freeze out time T_{FO}, $\frac{m}{T_{FO}} \approx 20$ implying

$$\langle v^2 \rangle_{FO} = 6\frac{T_{FO}}{m_\chi} \approx 0.3. \tag{B.189}$$

B.5.2 Spin Selection

For any kind of process occurring at low velocity (like annihilation of dark matter in the halo for instance), one can have indications concerning the form of the annihilation cross section from selection rules. Indeed, fermion bilinears of momentum L have parity $\mathcal{P} = (-1)^{L+1}$. The $(-1)^L$ factor coming from the parity of the Legendre polynomials in the wave function decomposition (the polynomials are odd in $\cos\theta$ for odd values of L and even for even values of L, parity transforming $\cos\theta$ into $-\cos\theta$. The extra -1 factor exists for Majorana as for Dirac fermions (but not for scalar particles of course) but for different reasons. In the Dirac case, the u and v spinors (equivalently the positive and negative energy states) are independent and have opposite parity corresponding to the ± 1 eigenvalues of the parity operator γ^0 (see Sect. B.3.6). Reinterpreting the two spinor types, or positive and negative energy states, as particle and antiparticle, then leads automatically to opposite intrinsic parity for the particle–antiparticle pair. In the Majorana case, the fermion has intrinsic parity $\pm i$, and so the two-particle state has intrinsic parity $(\pm i)^2 = -1$.

As we discussed in the previous section, the Lth partial wave contribution to the annihilation rate is suppressed as v^{2L}. A Majorana pair is even under charge-conjugation (particle–antiparticle exchange), and so from the relation $C = (-1)^{L+S} = +1$ one infers that L and S must be either both even, or both odd for the pair. The origin of the $C = (-1)^{L+S}$ rule is as follows: under particle–antiparticle exchange, the spatial wave function contributes $(-1)^L$, and the spin wave function contributes (+1) if in symmetric triplet $S = 1$ state, and (-1) if in the antisymmetric $S = 0$ singlet state, i.e. $(-1)^{S+1}$. In addition, there is an overall (-1) for anticommutation of the two particle-creation operators $b^\dagger d^\dagger$ for the Dirac case, and $b^\dagger b^\dagger$ for the Majorana case. In the case of a bosonic initial state, the formula is the same $C = (-1)^{L+S}$ because the singlet/triplet spin cancels the anti-commutation relation.

B.5.3 Application to Specific Models

We can apply such reasoning to specific models to understand the behavior of the cross section we obtained. For instance, if a scalar dark matter as well as vectorial dark matter had no velocity dependence whereas fermionic dark matter is excluded as the annihilation is dominated by the p-wave and so is excluded by the direct detection experiments due to the large value of the dark matter coupling to the Higgs to still realize relic abundance constraint. Indeed, the scalar bilinear of the form $\bar{\Psi}\Psi$ is even under $\mathcal{P}$ (see Eq. B.43) which means for a fermionic dark matter, from $\mathcal{P} = (-1)^{L+1}$, that $L = 1$, and the wave function is velocity suppressed. On the contrary, terms of the form $\bar{\Psi}\gamma^5\Psi$ of $\bar{\Psi}\gamma^\mu\Psi$ are $\mathcal{P}$-odd and thus $L = 0$ is authorized. This is valid for Majorana as for Dirac cases. If the dark matter is a boson, the parity is $\mathcal{P} = (-1)^L = +1$; the S^*S or V^*V are thus not velocity suppressed contrarily to the fermion bilinear $\bar{\Psi}\Psi$. We then conclude that scalar operators for bosonic dark matter and pseudo-scalar ones for fermionic dark mater are the only ones not velocity suppressed.

At the more microscopic level, an amplitude from a Lagrangian containing a $\bar{\Psi}\Psi$ gives an amplitude proportional to $\mathcal{M} \propto \bar{v}u$, which when applied to the Eq. (B.19) we obtain $\mathcal{M} \propto |p|/m_\Psi$ and is thus velocity suppressed. On the other hand, since γ^5 is antidiagonal, annihilation through $\bar{\Psi}\gamma^5\Psi$ is not velocity suppressed. Concerning the direct detection, however, the difference is that we have now an initial state particle and final state particle, the amplitude is proportional to $\mathcal{M} \propto \bar{u}u$ which is unsuppressed. On the other hand, the $\bar{\Psi}\gamma^5\Psi$ is velocity suppressed due to the antidiagonal γ^5. So the operator $\bar{\Psi}\gamma^5\Psi$ is very efficient: it induces

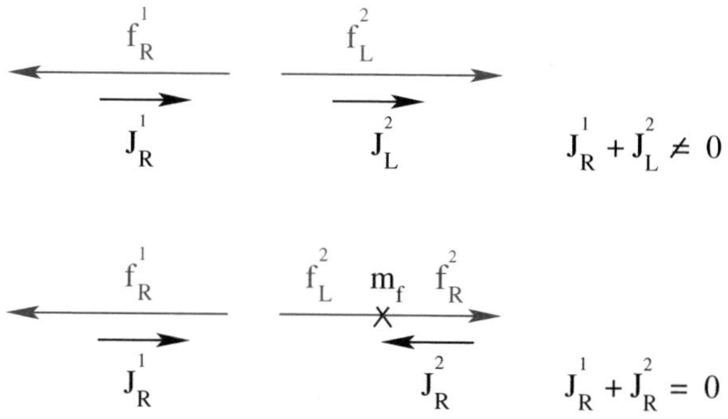

Fig. B.3 Illustration of dark matter annihilating into a pair of Standard Model Dirac fermions

f^1_R f^2_L

J^1_R J^2_L $J^1_R + J^2_L \neq 0$

f^1_R f^2_L m_f f^2_R

J^1_R J^2_R $J^1_R + J^2_R = 0$

unsuppressed annihilation and suppressed direct detection.[15] We observe this effect in supersymmetry where the Majorana neutralino annihilates mainly through the exchange of the pseudo-scalar A as the charge invariant term is $\bar{\chi}^0_1 \gamma^5 \chi^0_1 A$ (A being a pseudo-scalar, $A^* = -A$, ensuring that the Lagrangian is neutral with respect to electromagnetic charge). Terms of the type $\bar{\chi}^0_1 \chi^0_1 H/h$ exist but are velocity suppressed as we explained above.

B.5.4 Helicity Suppression

We just saw that only some configurations of the dark matter two-particle wave function survive from the velocity suppression effect. However, for some of them, there is another source of suppression called helicity suppression which acts on the s-wave function. Indeed, in the Majorana case, the operator $\bar{\Psi}\gamma^5\Psi$ is even under charge conjugate ($\Psi \rightarrow \Psi$ under C for a Majorana). As we discussed in the previous section, $C = (-1)^{L+S}$ which means that both L and S should be even/odd. In the s-wave part of the function, $L = 0 \Rightarrow S = 0 \Rightarrow J = L + S = 0$ (another way to see it is to observe that at a null velocity, for a spinless particle, no direction in space is preferred so $J = 0$). The helicity of the initial state is thus null. If the final state is made of Standard Model fermions, the bilinear operator should be of the form $\bar{f}_R f_L$, particles and antiparticles with opposite helicity, which imply $J_{\text{final}} \neq 0$, see Fig. B.3. The only way to conserve the total momentum J in the annihilation is to flip by mass insertion a flipping operator transforming the left-handed fermion f_L into a right-handed fermion f_R. The amplitude has thus to be proportional to m_f/m_Ψ. This is what we call the helicity suppression mechanism.
It is important to notice that this argument is valid for a Majorana fermion (argument of $J = 0$) but is also valid for a spin 0 particle. However, we cannot apply the conclusion to a Dirac fermion or fermions with vector-like coupling of the style

[15] We have to be careful as the operator $\bar{\Psi}\gamma^5\Psi$ is CP-odd and contribute to several processes, limiting the value of its coupling.

$\bar{\Psi}\gamma^{\mu}\Psi$, because the vectorial exchanged particle will discuss only to the left-handed fermions of the Standard Model.

B.5.5 Summary

We saw that for the annihilation process, it is the C and P quantum numbers that determine the nature of a final state, given an initial state. If one considers that the initial state and final state consist of a pair particle/antiparticle, the operators are given (for fermions) as

$$C = (-1)^{L+S} \quad P = (-1)^{L+1} \tag{B.190}$$

whereas for a bosonic pair

$$C = (-1)^{L+S} \quad P = (-1)^{L}. \tag{B.191}$$

The only allowed s-wave states are then $L = 0$, $S = 0$ ($J = 0$); $L = 0$, $S = 1$ ($J = 1$), and $L = 0$, $S = 2$ ($J = 2$). We list in the Table B.1 below the value of J, C and P for sermonic and bosonic final state. For any bosonic or sermonic bilinear, the transformation of the bilinear under rotations determines the total angular momentum of the state that this bilinear either creates or annihilates. This information along with the C and P quantum numbers of the bilinear are thus sufficient to determine the spin and orbital angular momentum of the initial and final state. The S and L quantum numbers of the states created (annihilated) by every lowest-dimension bilinear are listed in Table B.2.

One can recover all the results presented in Table B.3 noticing that for the annihilation, $|\bar{v}(p_1)u(p_2)|^2 \propto s - 4m_\chi^2 \propto v^2$ and for the direct detection, $|\bar{u}(p_1)u(p_2)|^2 \propto s + 4m_\chi^2$; whereas the roles are opposite in the case of pseudo-scalar couplings $|\bar{v}(p_1)\gamma^5 u(p_2)|^2$ as the introduction of a γ^5 matrix inverses the impulsion $p_2 \to -p_2$. Similar arguments are valid for $\bar{u}(p_1)\gamma^{\mu}u(p_2) = \frac{p_2^{\mu}}{m}\bar{u}(p_1)u(p_2)$ after applying the Dirac equation.

Table B.1 The C and P transformation properties of a fermion/anti-fermion (left) or boson/anti-boson (right) state for a given quantum number

S	L	J	C	P	S	L	J	C	P
0	0	0	+	−	0	0	0	+	+
0	1	1	−	+	0	1	1	−	−
1	0	1	−	−	1	0	1	−	+
1	1	0,1,2	+	+	1	1	0,1,2	+	−
1	2	1,2,3	−	−	1	2	1,2,3	−	+
1	3	2,3,4	+	+	2	0	2	+	+
					2	1	1,2,3	−	−
					2	2	0,1,2,3,4	+	+
					2	3	1,2,3,4,5	−	−
					2	4	2,3,4,5,6	+	+

Table B.2 The C and P and J quantum numbers of any state that can be either created or annihilated by the bilinear. For each possible state, the S and L quantum numbers are also given

Bilinear	C	P	J	State
$\bar{\psi}\psi$	+	+	0	$S=1, L=1$
$i\bar{\psi}\gamma^5\psi$	+	−	0	$S=0, L=0$
$\bar{\psi}\gamma^0\psi$	−	+	0	None
$\bar{\psi}\gamma^i\psi$	−	−	1	$S=1, L=0,2$
$\bar{\psi}\gamma^0\gamma^5\psi$	+	−	0	$S=0, L=0$
$\bar{\psi}\gamma^i\gamma^5\psi$	+	+	1	$S=1, L=1$
$\bar{\psi}\sigma^{0i}\psi$	−	−	1	$S=1, L=0,2$
$\bar{\psi}\sigma^{ij}\psi$	−	+	1	$S=0, L=1$
$\phi^\dagger\phi$	+	+	0	$S=0, L=0$
$iIm(\phi^\dagger\partial^0\phi)$	−	+	0	None
$iIm(\phi^\dagger\partial^i\phi)$	−	−	1	$S=0, L=1$
$B^\dagger_\mu B^\mu$	+	+	0	$S=0, L=0; S=2, L=2$
$iIm(B^\dagger_\nu\partial^0 B^\nu)$	−	+	0	None
$iIm(B^\dagger_\nu\partial^i B^\nu)$	−	−	1	$S=0, L=1; S=2, L=1,3$
$i(B^\dagger_i B_j - B^\dagger_j B_i)$	−	+	1	$S=1, L=0,2$
$i(B^\dagger_i B_0 - B^\dagger_0 B_i)$	−	−	1	$S=0, L=1; S=2, L=1,3$
$\epsilon^{0ijk}B_i\partial_j B_k$	+	−	0	$S=1, L=1$
$-\epsilon^{0ijk}B_0\partial_j B_k$	+	+	1	$S=2, L=2$
$B^\nu\partial_\nu B_0$	+	+	0	$S=0, L=0; S=2, L=2$
$B^\nu\partial_\nu B_i$	+	−	1	$S=1, L=1$

B.6 Schrodinger Equation

B.6.1 Generalities

The Schrodinger equation for a particle of mass M with an impulsion k described by a wave function ψ in a potential V is usually written as

Table B.3 Dependance of the annihilation cross section and the scattering cross section for different type of couplings

Operator	Annihilation $\langle\sigma v\rangle$	Direct detection [A,v_0]
$g\,\bar{\chi}\chi\bar{q}q$	$\propto g^2v^2$ [$\bar{\chi}\chi$: S=1, L=1]	SI $\propto g^2A^2\times f^N(q)$
$g\,\bar{\chi}\gamma^5\chi\bar{q}q$	$\propto g^2m_q^2$ [$\bar{\chi}\gamma^5\chi$: S=0, L=0 $\Rightarrow J=0$:h.f.]	0 [$\propto(v_0^\chi)^2$]
$g\,\bar{\chi}\chi\bar{q}\gamma^5 q$	$\propto g^2v^2$ [$\bar{\chi}\chi$: S=1, L=1]	0 [$\propto(v_0^q)^2$]
$g\,\bar{\chi}\gamma^5\chi\bar{q}\gamma^5 q$	$\propto g^2m_q^2$ [$\bar{\chi}\gamma^5\chi$: S=0, L=0 $\Rightarrow J=0$:h.f.]	0 [$\propto(v_0^q)^2(v_0^\chi)^2$]
$g\,\bar{\chi}\gamma^\mu\chi\bar{q}\gamma_\mu q$	$\propto g^2$	SI $\propto g^2\times(2u+d)\times p_\chi\cdot p_q$
$g\,\bar{\chi}\gamma^\mu\chi\bar{q}\gamma_\mu\gamma^5 q$	$\propto g^2$	0 [$\propto(v_0^q)^2$]
$g\,\bar{\chi}\gamma^\mu\gamma^5\chi\bar{q}\gamma_\mu q$	$\propto m_q^2$ [L=0,S=0(C=1) $\Rightarrow$ J=0:h.f.]	0 [$\propto(v_0^\chi)^2$]
$g\,\bar{\chi}\gamma^\mu\gamma^5\chi\bar{q}\gamma_\mu\gamma^5 q$	$\propto m_q^2$ [L=0,S=0(C=1) $\Rightarrow$ J=0:h.f.]	SD $\propto g^2\times(A-2Z)^2$

$$\frac{1}{2M}\Delta\psi + V\,\psi = E\,\psi = \frac{k^2}{2M}\psi \tag{B.192}$$

with Δ the Laplacian which can be expressed in spherical coordinates (r, θ, ϕ)

$$\Delta\psi = \frac{1}{r^2}\frac{\partial}{\partial r}\left(r^2\frac{\partial\psi}{\partial r}\right) + \frac{1}{r^2}\left[\frac{1}{\sin\theta}\frac{\partial}{\partial\theta}\left(\sin\theta\frac{\partial\psi}{\partial\theta}\right) + \frac{1}{\sin^2\theta}\frac{\partial^2\psi}{\partial\phi^2}\right]. \tag{B.193}$$

For a radial potential, it is possible to decompose the wave function

$$\psi = \psi_{kl}^m(r) = R_{kl}(r)\ \Theta_{lm}(\theta)\ \Phi_m(\phi). \tag{B.194}$$

In the above equation, we have defined ψ_{kl}^m as the wave function eigenvector of the momentum k, kinetic moment l, and its projection on z-axis m. The operator $\hat{l}_z\psi = m\psi = -i\frac{\partial}{\partial\phi}$, the solution for $\Phi(\phi)$ can then be solved directly after normalization $\int_0^{2\pi}\Phi_m^*(\phi)\ \Phi_{m'}(\phi)d\phi = \delta_{mm'}$

$$-i\,R_{kl}(r)\ \Theta_{lm}\ \frac{\partial\Phi_m(\phi)}{\partial\phi} = m\ R_{kl}(r)\ \Theta_{lm}(\theta)\ \Phi_m(\phi) \Rightarrow \Phi_m(\phi) = \frac{1}{\sqrt{2\pi}}e^{im\phi} \tag{B.195}$$

whereas the l dependance is included in the Θ function from the equation $\hat{l}^2\psi = l(l+1)\psi$ with $\hat{l}^2 = -\left[\frac{1}{\sin\theta}\frac{\partial}{\partial\theta}\left(\sin\theta\frac{\partial}{\partial\theta}\right) + \frac{1}{\sin^2\theta}\frac{\partial^2}{\partial\phi^2}\right]$. Implementing Eq. (B.194) in the preceding equation, we obtain

$$\frac{1}{\sin\theta}\frac{d}{d\theta}\left(\sin\theta\frac{d\Theta_{lm}}{d\theta}\right) - \frac{m^2}{\sin^2\theta}\Theta_{lm} + l(l+1)\Theta_{lm} = 0. \tag{B.196}$$

This equation is well known in the theory of spherical functions. Corresponding solutions are called *associated Legendre polynomials* $P_l^m(\cos\theta)$. After normalization, the eigenfunction solutions of the Eq. (B.196) can be written for $m > 0$,

$$\Theta_{lm} = (-1)^m i^l\sqrt{\frac{(2l+1)(l-m)!}{2(l+m)!}}P_l^m(\cos\theta). \tag{B.197}$$

For any m, we combined $\Theta(\theta)$ and $\Phi(\phi)$ in $Y_{lm}(\theta, \phi)$

$$Y_{lm}(\theta,\phi) = (-1)^{\frac{m+|m|}{2}}i^l\left[\frac{(2l+1)}{4\pi}\frac{(l-|m|)!}{(l+|m|)!}\right]^{1/2}P_l^{|m|}(\cos\theta)\ e^{im\phi}. \tag{B.198}$$

The $m = 0$ solution is often useful for problems with a z-axis symmetry (independence on the angle ϕ)

$$Y_{l0} = i^l \sqrt{\frac{2l+1}{4\pi}} P_l(\cos\theta) \tag{B.199}$$

with

$$P_n(x) = \frac{1}{2^n n!} \frac{d^n}{dx^n} \left[(x^2 - 1)^n \right]. \tag{B.200}$$

We then can summarize that the general solution of Schrodinger equation can be written as

$$\psi_k = \sum_l A_l \; P_l(\cos\theta) \; R_{kl}(r), \tag{B.201}$$

where $R_{kl}(r)$ are the continuum radial functions associated with angular momentum l satisfying

$$-\frac{1}{r^2} \frac{d}{dr} \left(r^2 \frac{d}{dr} R_{kl} \right) + \left(\frac{l(l+1)}{r^2} + 2M \; V(r) \right) R_{kl} = k^2 \; R_{kl}. \tag{B.202}$$

B.6.2 Solutions

Let us concentrate first on the case of the radial solution $Rkl^0(r)$ for a free particle $(V(r) = 0)$ Defining $\chi_{kl}(r) = r R^0_{kl}(r)$, Eq. (B.202) yields

$$\frac{d^2 \chi_{kl}(r)}{dr^2} + \left(k^2 - \frac{l(l+1)}{r^2} \right) \chi_{kl}(r) = 0. \tag{B.203}$$

For the simplest case $l = 0$, the two solutions are $\chi_{k0}(r) = \sin(kr)/\cos(kr)$ giving for $R^0_{k0}(r)$ the zeroth-order *Bessel* and *Neumann* functions, respectively :

$$R^0_{k0}(r) = \frac{\sin(kr)}{kr} \;\; \text{[Bessel]} \;\; \text{or} \;\; R^0_{k0}(r) = -\frac{\cos(kr)}{kr} \;\; \text{[Neumann]}. \tag{B.204}$$

Note that the Bessel function is the one well-behaved at the origin. To obtain the general solution for any value of the momentum l the idea is to write $R^0_{kl}(r) = (kr)^l \chi_{kl}(r)$ and replacing in Eq. (B.202), we can show that the function $\chi = \frac{1}{r}\frac{d}{dr}\chi_{kl}$ follows the equation for the momentum $l+1$. We can then write a recursion formula for generating all the χ_{kl} from χ_{k0}:

$$\chi_{k(l+1)}(r) = \frac{1}{k^2 r} \frac{d\chi_{kl}(r)}{dr} \;\; \Rightarrow \;\; R^0_{kl}(r) = (-kr)^l \left(\frac{1}{k^2 r} \frac{d}{dr} \right)^l \left(\frac{\sin kr}{kr} \right). \tag{B.205}$$

Developing[16] $\frac{\sin kr}{kr} = \sum_0^\infty (-1)^n \frac{(kr)^{2n}}{(2n+1)!}$, we can show that

$$R_{kl}^0(r) \simeq \frac{(kr)^l}{(2l+1)!!} \quad \text{as } r \to 0. \tag{B.206}$$

This r^l behavior near the origin is the usual well-behaved solution to Schrodinger's equation in the region where the centrifugal term dominates. Now, for large values or r, it is obvious that the dominant term is generated by differentiating *only* the trigonometric function at each step which gives

$$R_{kl}^0(r) \simeq \frac{1}{kr} \sin\left(kr - l\frac{\pi}{2}\right) \quad \text{as } r \to \infty. \tag{B.207}$$

In fact, the real wave function should be a combination of the Bessel *and* Neumann wave function. One should thus apply the recursive procedure also to the function $\frac{\cos kr}{kr}$ which is equivalent to introducing an extra phase factor called the *phase shift*, giving at the end the general solution

$$R_{kl}^0(r) \simeq \frac{1}{kr} \sin\left(kr - l\frac{\pi}{2} + \delta_l\right) \quad \text{as } r \to \infty. \tag{B.208}$$

In the case of a generic potential, the Schrodinger's equation can then be written as

$$\frac{d^2\chi_{kl}(r)}{dr^2} + \left(k^2 - \frac{l(l+1)}{r^2} - 2MV(r)\right)\chi_{kl}(r) = 0. \tag{B.209}$$

B.7 The Strong-CP Problem

B.7.1 QCD Lagrangian

The strong-CP problem is one of the oldest issue of the minimal versions of the Standard Model. In the *QCD* Lagrangian, there is the possibility to add a gauge invariant term of the form

$$\mathcal{L}_{QCD} = -\frac{1}{4}G^a_{\mu\nu}G^{a\mu\nu} + i\bar{q}_L\gamma^\mu D_\mu q_R + i\bar{q}_R\gamma^\mu D_\mu q_L + \theta_0\frac{g_3^2}{32\pi^2}G^a_{\mu\nu}\tilde{G}^{a\mu\nu}, \tag{B.210}$$

[16] The minus sign appearing in $(-kr)^l$ in Eq. (B.205) comes from the standard normalization of the function R_{kl}^0.

with

$$G^a_{\mu\nu} = \partial_\mu G^a_\nu - \partial_\nu G^a_\nu - g_3 f^{abc} G^b_\mu G^c_\nu, \qquad [G^a_\mu, G^b_\mu] = i f^{abc} G^c_\mu, \tag{B.211}$$

$$\tilde{G}^a_{\mu\nu} = \frac{1}{2}\epsilon_{\mu\nu\rho\sigma} G^{a\rho\sigma}, \tag{B.212}$$

is its dual tensor, and

$$f^{123} = 1; \; f^{147} = -f^{156} = f^{246} = f^{257} = f^{345} = -f^{367} = \frac{1}{2}; \; f^{458} = f^{678} = \frac{\sqrt{3}}{2}.$$

Exercise Show that in a parity P operation (B.39), $G^a_\mu \to G^{a\mu}$. Deduce then that $G^a_{\mu\nu} G^{a\mu\nu}$ does not violate P.

The term proportional to θ_0 in Eq. (B.210) violates P, CP, and T and produces a large neutron dipole moment which is in contradiction with experimental results (see the box). It is a total derivative, and therefore does not affect the equation of motion. However, the non-Abelian structure of the strong interactions, combined to the divergence of $g_3(\mu)_{\mu\to 0}$ means that it cannot be reduced to a boundary term in the action ($S = \int d^4x \mathcal{L}$). The gauge fields in this case do not necessarily vanish at infinity.

Exercise Show that the θ_0 term of Eq. (B.210) can be written $\partial_\mu K^\mu$, with

$$K^\mu = \epsilon^{\mu\nu\rho\sigma} G^a_\nu \partial_\rho G^a_\sigma + \frac{1}{3} f^{abc} G^a_\nu G^b_\rho G^c_\sigma. \tag{B.213}$$

The value of the electric dipole moment of the neutron expected from the θ_0-term is

$$d_n^{th} = 2.4 \times 10^{-16}\, \theta_0 \text{ e.cm},$$

corresponding to an experimental upper bound

$$|\theta_0| < 10^{-10}.$$

How to justify such a tiny parameter? This is called the strong-CP problem.

Electric Dipole Moment of the Neutron
The electric dipole moment can be roughly estimated by the difference of charge multiplied by the distance which separates them. If one takes the

(continued)

molecule of water for instance, H_20, the oxygen atom is separated from the two hydrogen atoms by a distance of ~ 0.1 nm. The dipole moment can then be approximated by

$$d^{th}_{H_2O} \simeq 10^{-8} \text{ e} \times \text{cm}, \tag{B.214}$$

e being the electric charge. The measured value $d^{exp}_{H_2O} = 0.5 \times 10^{-8}$ e.cm is in pretty good accordance with the prediction.

In the case of the neutron, the distance between the two down-type quarks and the up-type quark can be estimated at $\simeq 10^{-15}$ m, corresponding to a dipole moment $d_n \simeq 10^{-13}$ e.cm. The exact calculation gives

$$d^{th}_n = 10^{-15} \text{ e.cm}. \tag{B.215}$$

However, no dipole moment has been observed, the measurement giving an upper bound $d^{exp}_n < 10^{-26}$ e.cm, more than 10 orders of magnitude below the prediction.

Notice that, if we neglect the quark mass, a chiral transformation $q_L \to e^{-i\alpha} q_L$, $q_R \to e^{i\alpha} q_R$ will generate in the Lagrangian a term of the form

$$\mathcal{L} \supset -2\alpha \frac{g_3^2}{32\pi^2} G^a_{\mu\nu} \tilde{G}^{a\mu\nu}. \tag{B.216}$$

The reason for this anomalous symmetry is that the measure is not invariant under this transformation. We clearly see that there exists a shift symmetry $\theta_0 \to \theta_0 - 2\alpha$. Then, θ_0 is not physical, and there is no strong CP problem. However, quarks have masses. The full QCD Lagrangian should be written as

$$\mathcal{L}_{QCD} = -\frac{1}{4} G^a_{\mu\nu} G^{a\mu\nu} + \theta_0 \frac{g_3^2}{32\pi^2} G^a_{\mu\nu} \tilde{G}^{a\mu\nu} \tag{B.217}$$

$$+ i\bar{q}_L \gamma^\mu D_\mu q_L + i\bar{q}_R \gamma^\mu D_\mu q_R - m_q \bar{q}_L q_R - m_q^* \bar{q}_R q_L. \tag{B.218}$$

To render the mass matrices m_q real, one needs to rotate the spinors as follows:

$$q_L \to e^{i\phi_L} q_L\,; \quad q_R \to e^{i\phi_R} q_R. \tag{B.219}$$

If $m = |m| e^{i\phi}$, any rotation respecting $\phi_L - \phi_R = \phi$ should transform the masses into real parameters. However, these transformations on the quarks generate at quantum level a term

$$\mathcal{L}'_{QCD} = (\phi_L - \phi_R) \frac{g_3^2}{32\pi^2} G^a_{\mu\nu} \tilde{G}^{a\mu\nu}. \tag{B.220}$$

We give some arguments in the box below to catch the main argument to understand the form of $\mathcal{L}'_{QCD}$. The absence of observation of neutron dipole moment imposes an incredibly fine-tuned cancellation between θ_0 and ϕ, two parameters with a completely different nature : whereas the θ_0 contribution appears at tree level, even without the presence of any quark masses, the ϕ-term arises through the Yukawa interactions in the Higgs sector.[17] We will denote $\theta = \theta_0 + (\phi_L - \phi_R)$ from now on.

B.7.2 The Axionic Peccei–Quinn Solution

The idea proposed by Helen Quinn and Roberto Peccei in [8] is to introduce an axion field a such as, below a breaking scale f_a, under a chiral $U(1)_{PQ}$ transformation

$$q \to e^{-i\alpha\gamma^5} q \,; \quad \frac{a}{f_a} \to \frac{a}{f_a} + 2\alpha, \tag{B.221}$$

the Lagrangian

$$\mathcal{L}^c_{PQ} = [m_q \bar{q}_R e^{-i\frac{a}{f_a}} q_L + h.c.] + \theta \frac{g_3^2}{32\pi^2} G^a_{\mu\nu} \tilde{G}^{a\mu\nu} + \frac{1}{2}\partial^\mu a \partial_\mu a \tag{B.222}$$

stays invariant at the classical level.

Exercise Check that the Lagrangian (B.222) is invariant under the chiral $U(1)_{PQ}$ transformation (B.221).

Chiral Anomalies

To understand why transformation (B.219) generates terms of the form (B.220), we need to look at the figure below. A transformation of the type $q \to e^{i\gamma^5 \frac{\theta}{2}} q$ will generate in the Lagrangian a term of the type $im\theta\bar{q}\gamma^5 q$. At quantum level, this correspond to a $\delta\mathcal{L}_{eff}$

$$\delta\mathcal{L}_{eff} = -i\frac{\theta}{2} g_3^2 m Tr[T^a T^b] \int \frac{d^4p}{(2\pi)^4} Tr\, \gamma^5 \frac{1}{\not{p} + \not{k}_1 - m} \not{\epsilon}_1 \frac{1}{\not{p} - m} \not{\epsilon}_2 \frac{1}{\not{p} - \not{k}_2 - m},$$

(continued)

[17] Only the *up* and *down* quarks are important here, the heavier quarks giving a much more reduced contribution. ϕ should be understood as Arg(Det m_q).

T^i being the SU(3) generators. After some little algebra, the heavy quarks running in the loop give

$$\delta\mathcal{L}_{eff} = g_3^2 m^2 \theta Tr[T^a T^b]\epsilon_{\mu\nu\rho\sigma} k_1^\mu k_2^\nu \epsilon_1^\rho \epsilon_2^\sigma \int \frac{d^4 p}{(2\pi)^4} \frac{1}{(p^2 - m^2)^3} = \frac{\theta g_3^2}{32\pi^2} G^a_{\mu\nu}\tilde{G}^{a\mu\nu}$$

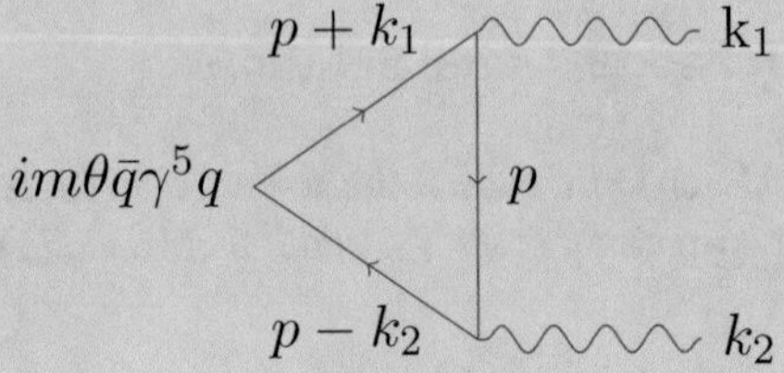

We saw that, at quantum level, redefining the quarks gives an equivalent Lagrangian

$$\mathcal{L}^Q_{PQ} = [m_q \bar{q}_R q_L + h.c.] + \left(\frac{a}{f_a} + \theta\right) \frac{g_3^2}{32\pi^2} G^a_{\mu\nu}\tilde{G}^{a\mu\nu} + \frac{1}{2}\partial^\mu a \partial_\mu a. \quad \text{(B.223)}$$

Clearly, the PQ symmetry is broken at the quantum level, and the axion is in fact a *pseudo*-Nambu–Goldstone boson. The miracle of this construction is that at a scale below $\Lambda_{QCD} \simeq 200$ MeV, the appearance of quark condensates under the form of mesons $\langle \bar{q} q \rangle$ breaks the chiral symmetry and generates an effective potential for the axion[18]

$$V(a) \sim \frac{1}{2}\left(\frac{a}{f_a} + \theta\right)^2 \frac{m_u m_d}{m_u + m_d} \langle \bar{q} q \rangle \sim \frac{1}{8}\left(\frac{a}{f_a} + \theta\right)^2 m_\pi^2 f_\pi^2, \quad \text{(B.224)}$$

with $m_\pi = 135$ MeV, $f_\pi = 93$ MeV are pion mass and pion decay constant, respectively. The potential admits a minimum at $\frac{a}{f_a} = -\theta$, eliminating the term responsible of the dipole moment in (B.223), solving in the most elegant way the strong-CP problem. In the meantime, once we develop a around its minimum $f_a\theta$,

[18]The potential should be periodic in $\frac{a}{f_a}$, the origin of the $\frac{1}{2}\left(\frac{a}{f_a} + \theta\right)^2$ term is in fact $1 - \cos(\frac{a}{f_a} + \theta)$.

$V(a)$ generates a mass term for the excitation of the axion field given by

$$m_a = \frac{m_\pi f_\pi}{2 f_a} \simeq 6.3\ \mu\text{eV} \left(\frac{10^{12}\ \text{GeV}}{f_a} \right). \tag{B.225}$$

B.8 Useful Spectrum

B.8.1 Gamma Spectrum

In order to determine these spectral functions, we generated 300,000 events of Standard Model particles decaying (directly or through secondary decays) into γ-rays using the PYTHIA [4] package, taking care in order to include all possible decay channels. Following the method of [5], we fitted the resulting spectra through functions of the form:

$$\frac{dN_\gamma^i}{dx} = \exp\left[F_i(\ln(x))\right], \tag{B.226}$$

where i represents the i-th WIMP annihilation channel, $i = WW, ZZ$, etc.; $x = E_\gamma/m_\chi$ with m_χ being the WIMP mass and F are seventh-order polynomial functions which were found to be the following:

$$\begin{aligned}
WW(x) &= -7.72088528 - 8.30185509\,x - 3.28835893\,x^2 - 1.12793422\,x^3 \\
&\quad - 0.266923457\,x^4 - 0.0393805951\,x^5 - 0.00324965152\,x^6 \\
&\quad - 0.000113626003\,x^7, \\
ZZ(x) &= -7.67132139 - 7.22257853\,x - 2.0053556\,x^2 - 0.446706623\,x^3 \\
&\quad - 0.0674006343\,x^4 - 0.00639245566\,x^5 - 0.000372241746\,x^6 \\
&\quad - 1.08050617 \cdot 10^{-5}\,x^7, \\
b\bar{b}(x) &= -11.4735403 - 17.4537277\,x - 11.5219269\,x^2 - 5.1085887\,x^3 \\
&\quad - 1.36697042\,x^4 - 0.211365134\,x^5 - 0.0174275134\,x^6 \\
&\quad - 0.000594830839\,x^7, \\
u\bar{u}(x) &= -4.56073856 - 8.13061428\,x - 4.98080492\,x^2 - 2.23044157\,x^3 \\
&\quad - 0.619205713\,x^4 - 0.100954451\,x^5 - 0.00879980996\,x^6 \\
&\quad - 0.00031573695\,x^7,
\end{aligned}$$

$$
\begin{aligned}
d\bar{d}(x) &= -4.77311611 - 10.6317139\,x - 8.33119583\,x^2 - 4.35085535\,x^3 \\
&\quad - 1.33376908\,x^4 - 0.232659817\,x^5 - 0.0213230457\,x^6 \\
&\quad - 0.000796017819\,x^7, \\
\tau^-\tau^+(x) &= -5.64725113 - 10.8949451\,x - 7.84473181\,x^2 - 3.50611639\,x^3 \\
&\quad - 0.942047119\,x^4 - 0.14691925\,x^5 - 0.0122521566\,x^6 \\
&\quad - 0.000422848301\,x^7.
\end{aligned}
$$

The case of WIMP annihilation into $\mu^+\mu^-$ pairs has a relatively small decay contribution, to the photon spectrum, coming from the $\mu \rightarrow e^-\bar{\nu}_e\nu_\mu\gamma$ channel, which has a small branching ratio. e^+e^- pair production contributes to the gamma-ray spectrum through different (not decay) processes, mainly Inverse Compton scattering and synchrotron radiation. These contributions depend crucially on the assumptions made concerning the intergalactic medium and will not be analyzed here. This means, practically, that the e^+e^- and $\mu^+\mu^-$ spectral functions are set equal to zero. A graphical representation of these functions can be seen in Fig. B.4. These functions can afterward be used in order to generate any gamma-ray flux according to Eq. (5.96). As we can see, all contributions are quite similar, apart from the $\tau^-\tau^+$ channel which has a characteristic hard form. Nevertheless, at high energies, the form of all contributions becomes almost identical.

B.8.2 Positron Spectrum

For the positron spectrum at the source, the same procedure gives us

$$
\begin{aligned}
WW(x) &= -1.470768 - 1.865343\,x - 1.865343\,x^2 - 1.940513\,x^3 \\
&\quad - 0.7538001\,x^4 - 01.432966\,x^5 - 0.01350088\,x^6 - 0.0005048965\,x^7, \\
ZZ(x) &= -2.909439 - 1.091164\,x - 1.184799\,x^2 - 1.621001\,x^3 \\
&\quad - 0.6838962\,x^4 - 0.1359553\,x^5 - 0.01320953\,x^6 - 5.058684 \cdot 10^{-4}\,x^7, \\
b\bar{b}(x) &= -12.33180 - 27.10877\,x - 29.62288\,x^2 - 18.59724\,x^3 \\
&\quad - 6.705565\,x^4 - 1.436652\,x^5 - 0.1813250\,x^6 - 0.01246600\,x^7, \\
c\bar{c}(x) &= -12.07344 - 17.57427\,x - 11.85548\,x^2 - 5.235151\,x^3 \\
&\quad - 1.391299\,x^4 - 0.2141368\,x^5 - 0.01755747\,x^6 - 0.0005935255\,x^7, \\
t\bar{t}(x) &= -8.90339 - 21.5246\,x - 27.6548\,x^2 - 18.8567\,x^3 \\
&\quad - 6.80256\,x^4 - 1.32199\,x^5 - 0.129489\,x^6 - 0.00494683\,x^7,
\end{aligned}
$$

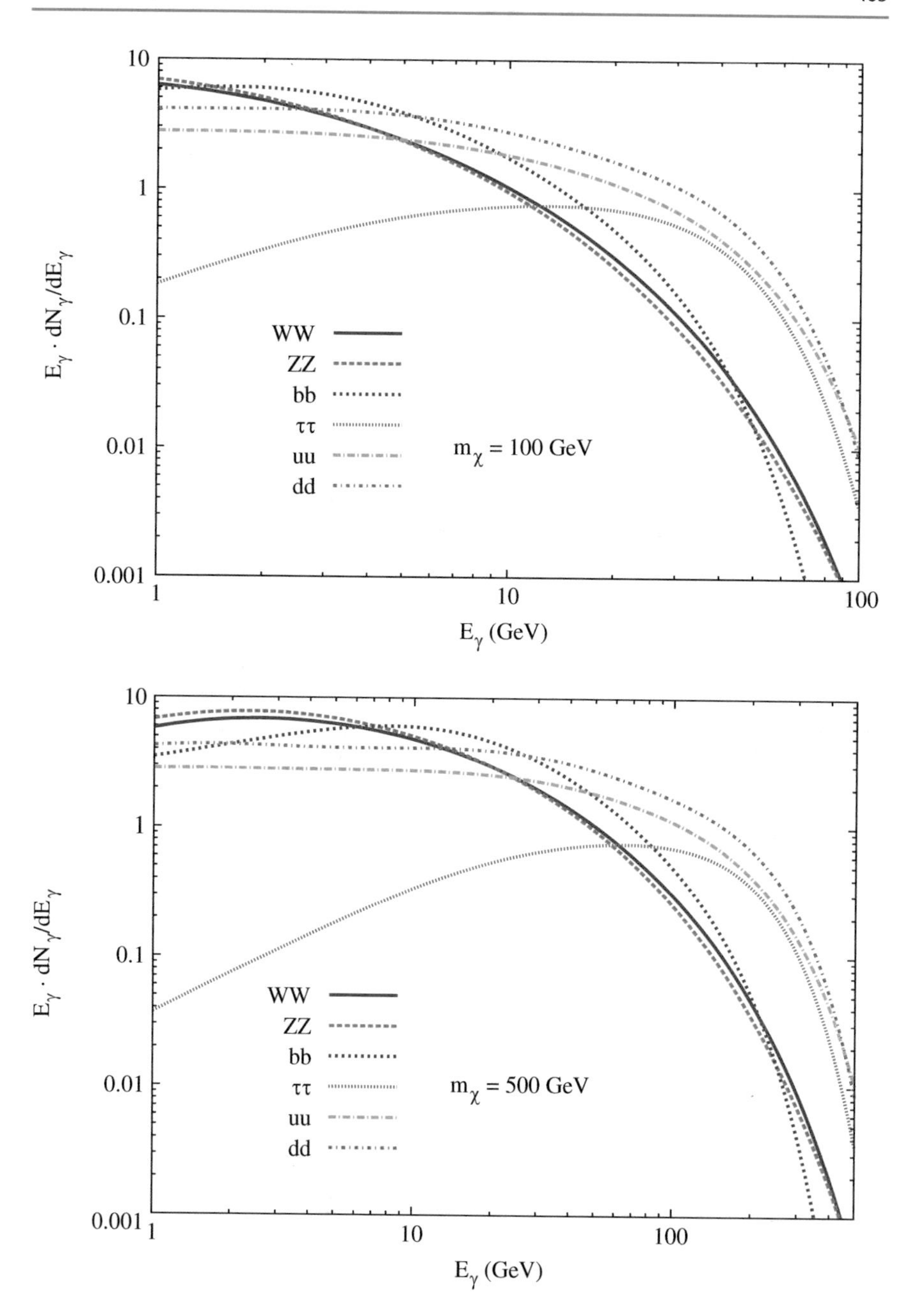

Fig. B.4 Separate contributions from Standard Model particles decaying into γ-rays for $m_\chi =$ 100 and 500 GeV. The PYTHIA result points have been suppressed for the sake of clarity

$$\tau^-\tau^+(x) = -4.004877 - 6.630655\,x - 5.779419\,x^2 - 3.504474\,x^3$$
$$- 1.200937\,x^4 - 0.2254182\,x^5 - 0.02172664\,x^6 - 0.0008412294\,x^7.$$
$$HH(x) = -8.731713 - 10.64431\,x - 6.924815\,x^2 - 3.131443\,x^3$$
$$- 0.7984188\,x^4 - 0.1135447\,x^5 - 0.0084599666\,x^6 - 0.00025748991\,x^7.$$

((for a 200 GeV Higgs)

B.8.3 Antiproton Spectrum

For the antiproton, we can parameterize the spectrum as follows :

$$\frac{dN^i_{\bar{p}}}{dE} = \frac{1}{m_\chi\left(p^i_1 x^{p_3} + p^i_2 |\log_{10}(x)|^{p^i_4}\right)}, \quad \text{(B.227)}$$

with

i	$1/p^i_1$	$1/p^i_2$	$1/p^i_3$	$1/p^i_4$
W	$306\, m_\chi^{0.28} + 7.4 \times 10^{-4} m_\chi^{2.25}$	$2.32\, m_\chi^{0.05}$	$-8.5\, m_\chi^{-0.31}$	$-0.39\, m_\chi^{-0.17} - 2 \times 10^{-2} m_\chi^{0.23}$
Z	$480\, m_\chi^{0.26} + 9.6 \times 10^{-4} m_\chi^{2.27}$	$2.17\, m_\chi^{0.05}$	$-8.5\, m_\chi^{-0.31}$	$-0.33\, m_\chi^{-0.075} - 1.5 \times 10^{-4} m_\chi^{0.71}$
t	$1.35\, m_\chi^{1.45}$	$1.18\, m_\chi^{0.15}$	-2.22	-0.21
b	$1.75\, m_\chi^{1.4}$	$1.54\, m_\chi^{0.11}$	-2.22	$-0.31\, m_\chi^{-0.052}$
c	$1.7\, m_\chi^{1.4}$	$3.12\, m_\chi^{0.04}$	-2.22	$-0.39\, m_\chi^{-0.076}$

B.9 Lagrangians

B.9.1 Standard Model

The SM Lagrangian can be written as

$$\mathcal{L}_{SM} = -\frac{1}{4}G^a_{\mu\nu}G^{\mu\nu a} - \frac{1}{4}F^a_{\mu\nu}F^{\mu\nu a} - \frac{1}{4}B_{\mu\nu}B^{\mu\nu} + |D_\mu H|^2 \quad \text{(B.228)}$$
$$+ i\bar{\Psi}\gamma^\mu D_\mu \Psi + \mu_H^2 |H|^2 - \lambda_H |H|^4 \quad \text{(B.229)}$$

with $T^a = \sigma^a/2$ (σ^a the Pauli matrices) and where D_μ represents the covariant derivative,

$$D_\mu = (\partial_\mu - i g_3 G^a T^a_3 - i g W^a_\mu T^a_2 - i g' \frac{Y}{2} B_\mu) \quad \text{(B.230)}$$

Ψ are fermionic multiplet $\begin{pmatrix} \nu_L \\ e_L \end{pmatrix} = \begin{pmatrix} \frac{1-\gamma^5}{2}\nu \\ \frac{1-\gamma^5}{2}e \end{pmatrix}$ in the case of the $SU(2)_L$ generators $W_\mu^a T^a$ for instance, and

$$F_{\mu\nu} = \partial_\mu B_\nu - \partial_\nu B_\mu \qquad W_{\mu\nu}^a = \partial_\mu W_\nu^a - \partial_\nu W_\mu^a + g f^{abc} W_\mu^b W_\nu^c, \tag{B.231}$$

where under the $SU(2) \times U(1)_Y$ of parameter α^a and β, W_μ, B_μ and H transforms as

$$\Psi_L/H \to e^{-i\alpha_a T^a - i\beta\frac{Y}{2}}\Psi_L/H \qquad \Psi_R \to e^{-i\beta\frac{Y}{2}}\Psi_R$$

$$B_\mu \to B_\mu - \frac{1}{g'}\partial_\mu\beta \qquad W_\mu^a \to W_\mu^a - \frac{1}{g}\partial_\mu\alpha^a - f^{abc}\alpha^b W_\mu^c.$$

B.9.1.1 Higgs Couplings

The minimum of the Higgs potential is obtained for

$$v = \mu_H/\sqrt{\lambda_H} \quad \text{and} \quad m_h^2 = 2\mu_H^2 \tag{B.232}$$

$$D_\mu H = \left(\partial_\mu - igW_\mu^a T^a - ig'\frac{Y_H}{2}B_\mu\right)H$$

$$= \begin{pmatrix} \partial_\mu - ig\frac{W_\mu^3}{2} - ig'\frac{Y_H}{2}B_\mu & -i\frac{g}{2}(W_\mu^1 - iW_\mu^2) \\ -i\frac{g}{2}(W_\mu^1 + iW_\mu^2) & \partial_\mu + ig\frac{W_\mu^3}{2} - ig'\frac{Y_H}{2}B_\mu \end{pmatrix} H.$$

If one develops around $H = \begin{pmatrix} 0 \\ (v+h)/\sqrt{2} \end{pmatrix}$, $|D_\mu H|^2 = \begin{pmatrix} -i\frac{g}{2\sqrt{2}}(W_\mu^1 - iW_\mu^2)(v+h) \\ \frac{\partial_\mu h}{\sqrt{2}} + i(\frac{gW_\mu^3}{2\sqrt{2}} - g'\frac{Y_H B_\mu}{2\sqrt{2}})(v+h) \end{pmatrix}^2$ if one defines $W_\mu^\pm = (W_\mu^1 \mp iW_\mu^2)/\sqrt{2}$, one can identify $M_W^2|W^\pm|^2 + \frac{M_Z^2}{2}|Z_\mu|^2$ which gives

$$M_W = \frac{gv}{2} \qquad M_Z = \frac{\sqrt{g^2 + g'^2}}{2}v = \frac{g}{2\cos\theta_W}v. \tag{B.233}$$

For the Higgs coupling, the development of $|D_\mu H|^2$ gives $\mathcal{L} \supset \frac{g^2 v}{2}HW_\mu^+ W_-^\mu + \frac{g^2 v}{4\cos^2\theta_W}HZ_\mu Z^\mu = gM_W HW_\mu^+ W_-^\mu + \frac{gM_Z}{2\cos\theta_W}HZ_\mu Z^\mu$. The couplings are then

$$C_{HWW} = igM_W, \quad C_{HZZ} = \frac{igM_Z^2}{2M_W} \times 2 \text{ (symmetry factor, Z} = \text{Z}^*) = ig\frac{M_Z^2}{M_W}. \tag{B.234}$$

B.9.1.2 Vectorial Couplings

Developing the Lagrangian (B.229) using (B.230) one can write

$$g\bar{\Psi}_L\gamma^\mu W^a_\mu T^a \Psi_L = \frac{g}{2\sqrt{2}}\bar{e}\gamma^\mu W^-_\mu(1-\gamma^5)\nu + \frac{g}{2\sqrt{2}}\bar{\nu}\gamma^\mu W^-_\mu(1-\gamma^5)e \quad \text{(B.235)}$$

B.9.2 Singlet Scalar

The simplest extension of the SM is the addition of a real singlet scalar field. Although it is possible to generalize to scenarios with more than one singlet, the simplest case of a single additional singlet scalar provides a useful framework to analyze the generic implications of an augmented scalar sector to the SM. The most general renormalizable potential involving the SM Higgs doublet H and the singlet S is

$$\begin{aligned}\mathcal{L} = \mathcal{L}_{SM} &+ (D_\mu H)^\dagger(D^\mu H) + \frac{1}{2}\mu_H^2 H^\dagger H - \frac{1}{4}\lambda_H|H|^4 \\ &+ \frac{1}{2}\partial_\mu S\partial^\mu S - \frac{\lambda_S}{4}S^4 - \frac{\mu_S^2}{2}S^2 - \frac{\lambda_{HS}}{4}S^2H^\dagger H \\ &- \frac{\kappa_1}{2}H^\dagger HS - \frac{\kappa_3}{3}S^3 - V_0,\end{aligned} \quad \text{(B.236)}$$

where D_μ represents the covariant derivative.

$$\begin{aligned}D_\mu H &= (\partial_\mu - igW^a_\mu T^a - ig'\frac{Y_H}{2}B_\mu)H \\ &= \begin{pmatrix} \partial_\mu - ig\frac{W^3_\mu}{2} - ig'\frac{Y_H}{2}B_\mu & -i\frac{g}{2}(W^1_\mu - iW^2_\mu) \\ -i\frac{g}{2}(W^1_\mu + iW^2_\mu) & \partial_\mu + ig\frac{W^3_\mu}{2} - ig'\frac{Y_H}{2}B_\mu \end{pmatrix} H\end{aligned}$$

and we have eliminated a possible linear term in S by a constant shift, absorbing the resulting S-independent term in the vacuum energy V_0. We require that the minimum of the potential occur at $v = 246$ GeV . Fluctuations around this vacuum expectation value are the SM Higgs boson. For the case of interest here for which S is stable and may be a dark matter candidate, we impose a Z_2 symmetry on the model, thereby eliminating the κ_1 and κ_3 terms. We also require that the true vacuum of the theory satisfies $\langle S\rangle = 0$, thereby precluding mixing of S and the SM Higgs boson and the existence of cosmologically problematic domain walls. After the electroweak breaking, the scalar Lagrangian can be written

$$\mathcal{L}_S = -\frac{\mu_S^2}{2}S^2 - \frac{1}{8}\lambda_{HS}S^2h^2 - \lambda_{HS}\frac{M_W}{2g}hS^2. \quad \text{(B.237)}$$

In this case, the masses of the scalars are

$$v = \frac{\sqrt{2}\mu_H}{\sqrt{\lambda_H}}, \qquad M_H = \mu_H, \qquad m_S = \sqrt{\mu_S^2 + \frac{\lambda_{HS}}{\lambda_H}\frac{\mu_H^2}{2}} \tag{B.238}$$

and the HSS coupling generated is

$$C_{HSS}: \quad -i\frac{\lambda_{HS} M_W}{2g} \times 2 \text{ (symmetry factor, S = S}^*) = -i\frac{\lambda_{HS} M_W}{g}. \tag{B.239}$$

B.9.3 Extra U(1) and Kinetic Mixing

The matter content of any *dark* $U(1)_D$ extension of the SM can be decomposed into three families of particles:

- The *Visible sector* is made of particles which are charged under the SM gauge group $SU(3) \times SU(2) \times U_Y(1)$ but not charged under $U_D(1)$ (hence the *dark* denomination for this gauge group).
- the *Dark sector* is composed of the particles charged under $U_D(1)$ but neutral with respect of the SM gauge symmetries. The dark matter (ψ_0) candidate is the lightest particle of the *dark sector*.
- The *Hybrid sector* contains states with SM *and* $U_D(1)$ quantum numbers. These states are fundamental because they act as a portal between the two previous sector through the kinetic mixing they induce at loop order.

From these considerations, it is easy to build the effective Lagrangian generated at one loop :

$$\begin{aligned} \mathcal{L} = \mathcal{L}_{\text{SM}} &- \frac{1}{4}\tilde{B}_{\mu\nu}\tilde{B}^{\mu\nu} - \frac{1}{4}\tilde{X}_{\mu\nu}\tilde{X}^{\mu\nu} - \frac{\delta}{2}\tilde{B}_{\mu\nu}\tilde{X}^{\mu\nu} \\ &+ i\sum_i \bar{\psi}_i\gamma^\mu D_\mu\psi_i + i\sum_j \bar{\Psi}_j\gamma^\mu D_\mu\Psi_j, \end{aligned} \tag{B.240}$$

$\tilde{B}_\mu$ being the gauge field for the hypercharge, $\tilde{X}_\mu$ the gauge field of $U_D(1)$ and ψ_i the particles from the hidden sector, Ψ_j the particles from the hybrid sector, $D_\mu = \partial_\mu - i(q_Y\tilde{g}_Y\tilde{B}_\mu + q_D\tilde{g}_D\tilde{X}_\mu + gT^aW_\mu^a)$, T^a being the $SU(2)$ generators, and

$$\delta = \frac{\tilde{g}_Y\tilde{g}_D}{16\pi^2}\sum_j q_Y^j q_D^j \log\left(\frac{m_j^2}{M_j^2}\right) \tag{B.241}$$

with m_j and M_j being hybrid mass states.

Notice that the sum is on all the hybrid states, as they are the only ones which can contribute to the $\tilde{B}_\mu \tilde{X}_\mu$ propagator. After diagonalization of the current eigenstates that makes the gauge kinetic terms of Eq. (B.240) diagonal and canonical, we can write after the $SU(2)_L \times U(1)_Y$ breaking:[19]

$$\begin{aligned} A_\mu &= \sin\theta_W W^3_\mu + \cos\theta_W B_\mu \\ Z_\mu &= \cos\phi(\cos\theta_W W^3_\mu - \sin\theta_W B_\mu) - \sin\phi X_\mu \\ (Z_D)_\mu &= \sin\phi(\cos\theta_W W^3_\mu - \sin\theta_W B_\mu) + \cos\phi X_\mu \end{aligned} \tag{B.242}$$

with, at the first order in δ:

$$\begin{aligned} \cos\phi &= \frac{\alpha}{\sqrt{\alpha^2 + 4\delta^2 \sin^2\theta_W}} \quad \sin\phi = \frac{2\delta \sin\theta_W}{\sqrt{\alpha^2 + 4\delta^2 \sin^2\theta_W}} \\ \alpha &= 1 - M^2_{Z_D}/M^2_Z - \delta^2 \sin^2\theta_W \\ &\pm \sqrt{(1 - M^2_{Z_D}/M^2_Z - \delta^2 \sin^2\theta_W)^2 + 4\delta^2 \sin^2\theta_W} \end{aligned} \tag{B.243}$$

and + (−) sign if $M_{Z_D} < (>) M_Z$. The kinetic mixing parameter δ generates an effective coupling of SM states ψ_{SM} to Z_D, and a coupling of ψ_0 to the SM Z boson which induces an interaction on nucleons. Developing the covariant derivative on SM and ψ_0 fermions state, we compute the effective $\psi_{SM}\psi_{SM}Z_D$ and $\psi_0\psi_0 Z$ couplings at first order in δ and obtain

$$\mathcal{L} = q_D \tilde{g}_D (\cos\phi\ Z'_\mu \bar{\psi}_0 \gamma^\mu \psi_0 + \sin\phi\ Z_\mu \bar{\psi}_0 \gamma^\mu \psi_0). \tag{B.244}$$

[19] Our notations for the gauge fields are $(\tilde{B}^\mu, \tilde{X}^\mu)$ before the diagonalization, (B^μ, X^μ) after diagonalization, and (Z^μ, Z^μ_D) after the electroweak breaking.

Neutrino Physics

C

C.1 Astrophysical and Cosmological Sources of Neutrino

The neutrino arriving on Earth has different origins. Some are produced in the Sun (*solar neutrino*) whereas others are produced via the interaction of cosmic rays, mainly composed of ultra-energetic proton, on the atmosphere (*atmospheric neutrinos*). Others can be directly generated via ultra-high energy (UHE) sources like blazars or Active Galactic Nuclei (AGN). These UHE neutrinos propagate and diffuse in the Universe, pinpointing their source clearly because, contrarily to the cosmic rays produced by the same objects, neutrinos are not affected by magnetic fields or the interstellar medium (ISM). Another production mechanism of neutrino is the interaction of the cosmic rays on the photons from the CMB *cosmogenic neutrino* : $p + \gamma_{\text{CMB}} \to \Delta \to n + \pi^+$, and then $\pi^+ \to \mu^+ \nu_\mu$ and $\mu^+ \to e^+ \bar{\nu}_\mu \nu_e$. This source is one of the dominant one and produces 2 μ-types neutrinos for 1 e-type one. The flavor ratio at the production is then (1:2:0), which will be an important point to distinguish them on Earth. Finally, the last source of neutrino, not yet detected, is obviously the relic ones from the decoupling of the primordial plasma. They have a temperature of 1.95 K and a density $n_\nu = 109$ cm^{-3}, see Sect. C.3, compared to the 2.5 K and $n_\gamma = 394$ cm^{-3} of the photons from the CMB (see Eq. 3.42). We will look into the details of each production and detection mechanism in this section.

C.1.1 Solar Neutrinos

Electron neutrinos are produced in the Sun as a product of nuclear fusion. By far the largest fraction of neutrinos passing through the Earth are Solar neutrinos. The main contribution comes from the proton–proton reaction. The reaction is

$$p + p \to d + e^+ + \nu_e.$$

Y. Mambrini, *Particles in the Dark Universe*,
https://doi.org/10.1007/978-3-030-78139-2

From this reaction, 86% of all solar neutrinos are produced. The deuterium will then fuse with another proton to create a ^{3}He atom and a gamma ray. This reaction can be seen as

$$d + p \rightarrow {}^3He + \gamma. \quad \text{(C.1)}$$

The isotope ^{4}He is then produced using two ^{3}He from the previous reaction

$${}^3He + {}^3He \rightarrow {}^4He + 2p. \quad \text{(C.2)}$$

With both helium-3 and helium-4 in the system now, beryllium can be fused by the reaction of one of each helium atom as seen in the reaction:

$${}^3He + {}^4He \rightarrow {}^7Be + \gamma. \quad \text{(C.3)}$$

Since there are four protons and only three neutrons, the beryllium can go down two different paths from here. The beryllium could capture an electron and produce a lithium-7 atom and an electron neutrino. It can also capture a proton due to the abundance in a star. This will create boron-8. Both reactions are as seen below respectively:

$${}^7Be + e^- \rightarrow {}^7Li + \nu_e. \quad \text{(C.4)}$$

This reaction produces 14% of the solar neutrinos. The lithium-7 will combine with a proton to produce 2 atoms of helium-4. On the other hand, the beryllium will capture a proton to form the boron-8 as follows:

$${}^7Be + p \rightarrow {}^8B + \gamma. \quad \text{(C.5)}$$

Then, the Boron-8 will beta decay into beryllium-8 due to the extra proton which can be seen below:

$${}^8B \rightarrow {}^8Be + e^+ + \nu_e. \quad \text{(C.6)}$$

The reaction produces about 0.02% of the solar neutrinos, but these few solar neutrinos have the larger energies. The highest flux of solar neutrinos comes directly from the proton–proton interaction, and has a low energy, up to 400 keV. From the Earth, the amount of neutrino flux at Earth is enormous (around 7×10^{10} particles cm^{-2}s^{-1}). Moreover, since neutrinos are insignificantly absorbed by the mass of the Earth, the surface area on the side of the Earth opposite the Sun receives about the same number of neutrinos as the side facing the Sun.

One should notice that the first solar neutrino generated by proton–proton fusion (constituting 99.77% of the solar neutrinos) was observed in August 2014 by the Borexino experiment [7]. The Borexino experiment was a joint collaboration of several European (Italy, Germany, France, Poland), USA, and Russian institutions, installed in the Gran Sasso Laboratory in central Italy. The reaction was harder to observe than the one generated by the Boron because of the very low energy of the escaping neutrino (around the keV compared to $\simeq$ 10 MeV for the ^{8}B ones). The fusion reactions occur deep within the Sun, and which then pass right through the Sun, taking just over eight minutes to reach the Earth. On the other hand, all solar energy measurements before were based on light from the Sun's photosphere (the familiar Sunlight which lights up our skies and warms the Earth). But the energy carried by this Sunlight was produced in solar fusion reactions about 100,000 years ago.[1] Comparison between the Borexino measurement and those of the Sun's radiant energy reveals that the solar power has not changed in quite a long time. The collaboration reported spectral observations of pp neutrinos, demonstrating that about 99% of the power of the Sun (3.84×10^{33} ergs per second) is generated by the proton–proton fusion process.

C.1.2 Atmospheric Neutrinos

Atmospheric neutrinos result from the interaction of cosmic rays with atomic nuclei in the Earth's atmosphere, creating showers of particles, many of which are unstable and produce neutrinos when they decay. A collaboration of particle physicists from Tata Institute of Fundamental Research (India), Osaka City University (Japan), and Durham University (UK) recorded the first cosmic ray neutrino interaction in an underground laboratory in Kolar Gold Fields in India in 1965.

C.2 Ultra-High Energetic Neutrinos

In the galaxy, or outer space, there are different sources of ultra-High Energy neutrinos ($>$ 1000 TeV = 1 PeV). It ranges from cosmic rays interaction with the Inter Stellar Medium (ISM) to direct production from Super Nova remnants. In any cases, these Ultra-High energy neutrino interacts with the CMB or the Earth atmosphere, producing specific phenomena, like a GZK-cut or the Glashow resonance. We will develop these two effects in this section and show how one can observe them experimentally.

C.3 Neutrino Mass

C.3.1 Dirac Mass

To give mass to neutrino, we face the same "issue" that we have concerning the mass term of "up" type quark. Indeed, the Lagrangian $\mathcal{L}_{down} = -\lambda_d \bar{L} \Phi d_R + h.c.$, where

[1] The average time for energy to percolate from the central regions of the Sun and reach its surface.

L represents the left doublet $(u, d)_L^T$, Φ the Higgs field, and d_R the right-handed part of the spinor $(d_L, d_R)^T$ [Weyl convention see Sect. B.3.1]. However, the Higgs field develops a *vev* v, $\Phi = (0, (v + \phi)\sqrt{2})^T$ which gives mass to down type particles like the bottom quarks, strange ones or the electrons or taus, but not to the top quarks or neutrino that are of the "up" type $SU(2)_L$. The idea is to introduce the charge conjugate of the Higgs field Φ, $\tilde{\Phi}^c = i\sigma_2\Phi^*$, σ_i being the Pauli matrices (see Sect. B.3.7). With this new field, one easily sees that the coupling $-\lambda_u \bar{L}\tilde{\Phi}^c u_R + h.c.$ generates a mass term for the up-type spinors once the Higgs develops a *vev*. The expression of the neutrino mass is then

$$\mathcal{L}_\nu^{Dirac} = -m_D \bar{\nu}\nu = -m_D(\bar{\nu}_L \nu_R + \bar{\nu}_R \nu_L), \quad \text{with } \nu = \begin{pmatrix} \nu_L \\ \nu_R \end{pmatrix} \tag{C.7}$$

and $m_D = \lambda_\nu v\sqrt{2}$. We know today that neutrinos (at least some of the neutrinos) have masses ($\lesssim 0.1$ eV). We can just introduce these very small masses via the Dirac mechanism, but of course this leaves completely unexplained the smallness of such masses with respect to other fermion masses. Alternatively, one could introduce the so-called Majorana mass term as we will see below.

C.3.2 Majorana Mass

C.3.2.1 Without Right-Handed Neutrino

It is also possible to generate mass to the left-handed neutrino through a Majorana mass term, without needing to ever introduce the right-handed degrees of freedom. Let us first recall that a fermion field ψ possess a charge conjugate counterpart $\psi^c = C\bar{\psi}^T$ where the role of particles and antiparticles are exchanged.[2] A Majorana fermion is one where $\psi^c = \psi$, i.e. a Majorana fermion is its own antiparticle. Let us assume that the left-handed neutrinos are particles of this kind. This is, in principle possible as they are not charged.[3] Thus let us define

$$\nu_L^c = C\bar{\nu}_L^T. \tag{C.8}$$

We cannot impose $\nu_L = \nu_L^c$ for the simple reason that ν_L^c is actually a right-handed field, but we can assume that ν_L and ν_L^c actually describe the left-handed and right-handed degrees of freedom of one fermion that is identical to its antifermion,

[2] See Sect. B.3.7 for more details.

[3] It is important to underline here that this Majorana mass term can only exist for neutrino as it is the only neutral fermion in the Standard Model, which is invariant under the electromagnetic $U(1)$ transformation. Indeed, a term of the form $\frac{1}{2}m_e^M(\bar{e}e^c + \bar{e}^c e)$ clearly violates the electric charge conservation.

respectively. We can then construct the Lorentz invariant mass term

$$-\frac{1}{2}m_M^L\bar{\nu}_L\nu_L^c + h.c. = -\frac{1}{2}m_M^L\bar{\nu}\nu \;\text{ with }\; \nu = \begin{pmatrix}\nu_L\\ \nu_L^c\end{pmatrix} = \nu^c. \tag{C.9}$$

This is the so-called Majorana mass term. It is not possible to make this gauge invariant with a dimension 4 operator. Indeed, this term changes weak hypercharge by 2 units—not possible with the standard Higgs interaction, requiring the Higgs field to be extended to include an extra triplet with weak hypercharge for instance—whereas for right chirality neutrinos as we will see, no Higgs extensions are necessary. For both left and right chirality cases, Majorana terms violate lepton number, but possibly at a level beyond the current sensitivity of experiments to detect such violations.

It is also interesting to notice that this minimal scheme to give mass to neutrino does not advocate right-handed states: is it indeed possible to give mass to neutrino (to the price of non-conserving lepton number and a Majorana nature for the neutrino). This is *the minimal scheme* which was the point of view adopted by Gribov and Pontecorvo [6].

Note that Majorana mass terms have nothing to do with the Majorana representation in spinor space. One can use any representation for the fields of which Majorana and Dirac mass terms are composed. Neither do Majorana mass terms imply the associated particles/fields are Majorana fermions, of which you may have heard. Majorana fermions are their own antiparticles. More on this in Sect. B.3.7. For now, we will assume that both Dirac and Majorana mass terms contain only Dirac type particles.

C.3.2.2 With Right-Handed Neutrino

We just saw that we can write a Majorana mass term for the left-handed neutrino, at the price to work in dimension 6 operators. However, if one introduces the right-handed neutrino from the Eq. (C.7), we clearly understand that nothing forbids us to write gauge invariant dimension 4 mass operators with the right-handed neutrino of the form:

$$-\frac{1}{2}m_M^R\left(\bar{\nu}_R\nu_R^c + \bar{\nu}_R^c\nu_R\right). \tag{C.10}$$

Note that in the expression (C.10), the second term destroys a right-handed particle and create a right-handed antiparticle (equivalent to a left-handed particle) as in the case of Eq. (C.7). The lepton number is however not conserved because two neutrino are created from the vacuum whereas their Dirac coupling attribute to them a leptonic number. Adding the three contributions (C.7, C.9, C.10) just discussed above, one obtains

$$\mathcal{L}_\nu = -m_D(\bar{\nu}_L\nu_R + \bar{\nu}_R\nu_L) - \frac{1}{2}m_M^L\left(\bar{\nu}_L\nu_L^c + \bar{\nu}_L^c\nu_L\right) - \frac{1}{2}m_M^R\left(\bar{\nu}_R\nu_R^c + \bar{\nu}_R^c\nu_R\right)$$

$$= -\frac{1}{2}\left(\bar{\nu}_L^c\bar{\nu}_R\right)\mathcal{M}\begin{pmatrix}\nu_L\\ \nu_R^c\end{pmatrix} + h.c. \text{ with } \mathcal{M} = \begin{pmatrix}m_M^L & m_D\\ m_D & m_M^R\end{pmatrix}, \tag{C.11}$$

where we used $\bar{\nu}_L\nu_R = \bar{\nu}_R^c\nu_L^c$ (Eq. B.52).

C.4 The See-Saw Mechanism

From the equation (C.11) one can easily diagonalize the mass matrix in the (ν_L^c, ν_R) space and then study the different possibilities or hierarchy between m_D, m_M^L, and m_M^R. However, to really understand the subtleties of the mechanism one should first study simple cases.

C.4.1 A Simple Example

Let us suppose that the Higgs does not couple to the neutrino; in this case, the Dirac mass is absent of the potential and one can write $m_D = 0$ in (C.11). One then needs to express the mass eigenstates (ν, N) as function of ν_L, ν_L^c, ν_R, ν_R^c from

$$-\frac{1}{2}(\bar{\nu}_L^c, \bar{\nu}_R)\begin{pmatrix}m_M^L & 0\\ 0 & m_M^R\end{pmatrix}\begin{pmatrix}\nu_L\\ \nu_R^c\end{pmatrix} + h.c. = -\frac{1}{2}(\bar{\nu}, \bar{N})\begin{pmatrix}m_\nu & 0\\ 0 & m_N\end{pmatrix}\begin{pmatrix}\nu\\ N\end{pmatrix} \tag{C.12}$$

which gives (noticing that $\bar{\nu}_R\nu_R = 0$)

$$\nu = \left(\nu_L + \nu_L^c\right) = \begin{pmatrix}\nu_L\\ \nu_L^c\end{pmatrix}; \quad N = \left(\nu_R + \nu_R^c\right) = \begin{pmatrix}\nu_R^c\\ \nu_R\end{pmatrix}, \tag{C.13}$$

where we have used the Dirac representation of the spinors ν and N which are 4-dimensional objects. As we can notice, in this case both ν and N are Majorana particles ($\nu^c = \nu$ and $N^c = N$), which can justify the term "Majorana masses" for m_M^L and m_M^R. Another important point concerns the notation, which is sometimes quite difficult to follow in the literature. The notation $\begin{pmatrix}\nu_L\\ \nu_L^c\end{pmatrix}$ should be read $\nu_L\begin{pmatrix}1\\ 0\end{pmatrix} + \nu_L^c\begin{pmatrix}0\\ 1\end{pmatrix}$, ν_L and ν_L^c being 4 dimensional spinors, even if as a left-handed spinor (ν_L) and right-handed spinor (ν_L^c), each of them possesses effectively only 2 degrees of freedom. $\begin{pmatrix}1\\ 0\end{pmatrix}$ and $\begin{pmatrix}0\\ 1\end{pmatrix}$ are the base vectors in the plane of the P_L and

P_R projectors. So, in a sense, it is formally correct to write $\nu_L = \begin{pmatrix} \nu_L \\ 0 \end{pmatrix}$, but on the left side, one should understand ν_L as a two–dimensional Weyl spinor.

C.4.2 Generalization

Now, if we assume that in the mass term enter the left–handed field ν_L, the right–handed sterile neutrino ν_R, and allowing for a violation of the lepton number L, from (C.11) we can write to the most general Dirac and Majorana mass term

$$\mathcal{L}^{\mathrm{D+M}} = -\frac{1}{2}\bar{n}_L\, \mathcal{M}\, n_L^c + h.c. \quad \text{with} \quad n_L = \begin{pmatrix} \nu_L \\ \nu_R^c \end{pmatrix} \quad \text{and} \quad \mathcal{M} = \begin{pmatrix} m_M^L & m_D \\ m_D & m_M^R \end{pmatrix}.$$

We denoted n_L as it is a 2 dimensional vector (in the mass basis) only formed by left-handed spinors. Noticing that a left-chiral spinor cannot mix with another left-chiral spinor ($\nu_L \nu_L = 0$), we can write

$$\mathcal{L}^{\mathrm{D+M}} = -\frac{1}{2}\bar{n}\, \mathcal{M}\, n \tag{C.14}$$

with $n = n_L + n_L^c$ which is a Majorana vector in the mass basis, which components are the sum of left and right chirality spinors. So we have reduced the problem of finding eigenstates to the one of diagonalizing M into the basis $n = \begin{pmatrix} \nu_L + \nu_L^c \\ \nu_R + \nu_R^c \end{pmatrix} = \begin{pmatrix} n_1 \\ n_2 \end{pmatrix}$. $\mathcal{M}$ being a complex symmetrical matrix, it can be diagonalized with the help of *one unitary matrix U*. We have $\mathcal{M} = UmU^T$ with

$$m = \begin{pmatrix} m_1 & 0 \\ 0 & m_2 \end{pmatrix}. \tag{C.15}$$

We skip the computation of the eigenvalues which are obtained solving the characteristic equation for the matrix $\mathcal{M}$: $|\mathcal{M} - \lambda Id| = 0$, given 2 values for the second order equation[4] for λ, m_1, and m_2:

$$\begin{aligned} m_1 &= \frac{1}{2}\left[(m_M^L + m_L^R) + \sqrt{(m_M^L + m_M^R)^2 + 4(m_D^2 - m_M^L m_M^R)}\right] \\ m_2 &= \frac{1}{2}\left[(m_M^L + m_L^R) - \sqrt{(m_M^L + m_M^R)^2 + 4(m_D^2 - m_M^L m_M^R)}\right]. \end{aligned} \tag{C.16}$$

[4]Notice that one eigenvalue is negative, which is not an issue as the field can be redefined by a global phase transformation.

C.4.3 The Specific Case $m_M^L = 0$

C.4.3.1 The Eigenvalues

We have seen in the preceding section that the Majorana mass term for the left-handed neutrino should appear only at two loops order, or generated by dimension 6 operator, and so should naturally be considered as much smaller than the right-handed Majorana term m_M^R which can be naturally present in the Lagrangian by gauge invariance. The mass values m_1 and m_1 then becomes

$$m_1 = \frac{1}{2}\left[m_L^R + \sqrt{(m_M^R)^2 + 4m_D^2}\right] \simeq m_M^R$$
$$m_2 = \frac{1}{2}\left[m_L^R - \sqrt{(m_M^R)^2 + 4m_D^2}\right] \simeq \frac{m_D^2}{m_M^R}, \qquad \text{(C.17)}$$

where we have developed at the first order in m_D/m_M^R, supposing $m_D \ll m_M^R$. This approximation is justified in a lot of extensions of the Standard Model (but not all) and can be understood as the fact that the Majorana mass term m_M^R being a free mass term, its natural scale should be the GUT or PLANCK scale (or at least an intermediate scale) whereas $m_D = y_{LR} v_H/\sqrt{2}$ ($v_H = 246$ GeV being the Higgs field vev and y_{LR} the Yukawa coupling between ν_L and ν_R, see below) is boded by the electroweak scale.

We then have a clear view on the famous see-saw mechanism (and on its name). Indeed, a quick glance at Eq. (C.17) shows us that a hierarchy between m_D and m_M^R will *naturally* drive one eigenvalue toward zero, keeping the second one relatively high. In other words, after diagonalization, the left-handed part of the multiplet (ν_L) will rotate and acquire a small admixture of ν_R which will generate its mass term, proportional to the admixture.

C.4.3.2 The Eigenvectors

If finding the mass terms m_1 and m_2 is an easy task, one should be careful when treating the compositions on the new states N_1 and N_2, mass eigenstates of the Lagrangian and thus, the *physical* states, the only one having really a meaning in any observable computation. To find them, one has to write down the equations

$$(\bar{N}_1\ \bar{N}_2)\begin{pmatrix} m_1 & 0 \\ 0 & m_2 \end{pmatrix}\begin{pmatrix} N_1 \\ N_2 \end{pmatrix} = (\bar{n}_1\ \bar{n}_2)\begin{pmatrix} 0 & m_D \\ m_D & m_M^R \end{pmatrix}\begin{pmatrix} n_1 \\ n_2 \end{pmatrix} \qquad \text{(C.18)}$$

with

$$\begin{pmatrix} N_1 \\ N_2 \end{pmatrix} = \mathcal{R}_\theta \begin{pmatrix} n_1 \\ n_2 \end{pmatrix} = \begin{pmatrix} \cos\theta & -\sin\theta \\ \sin\theta & \cos\theta \end{pmatrix}\begin{pmatrix} n_1 \\ n_2 \end{pmatrix}. \qquad \text{(C.19)}$$

Solving the Eq. (C.18) with the definition (C.19) we obtain

$$\tan 2\theta = \frac{2m_D}{m_M^R} \Rightarrow \theta \simeq \sin\theta \simeq \frac{m_D}{m_M^R} \tag{C.20}$$

which gives

$$\begin{pmatrix} N_1 \\ N_2 \end{pmatrix} \simeq \begin{pmatrix} n_1 - \theta\, n_2 \\ n_2 + \theta\, n_1 \end{pmatrix} = \begin{pmatrix} \nu_L + \nu_L^c - \theta\,(\nu_R + \nu_R^c) \\ \nu_R + \nu_R^c + \theta\,(\nu_L + \nu_L^c) \end{pmatrix}. \tag{C.21}$$

As we can notice, N_1 and N_2 are Majorana particles ($N_i = N_i^c$). N_1 represents the *physical* neutrino of the Standard Model, whose mass is constraint to be below 1 eV, whereas N_2 would be the (non-observed) heavy states responsible for the see-saw mechanism.

C.4.4 An Application: Coupling to a Scalar Field (Majoron)

It is always possible to work in a dynamical framework where the masses are given by the breaking of a symmetry "a la Higgs," with a new heavy scalar field (majoron field S) whose *vev* generates mass to the right-handed neutrino. The Lagrangian can be written, after the breaking of symmetry (assuming typical Higgs potential in the heavy scalar sector)

$$\begin{aligned}\mathcal{L} &= -\frac{h_1}{\sqrt{2}}\bar{\nu}_R^c(v_S + S)\nu_R - \frac{y_{LR}}{\sqrt{2}}\bar{\nu}_L(v_H + h)\nu_R \;+\; h.c.\\ &= -\frac{h_1}{\sqrt{2}}(\bar{\nu}_R^c + \bar{\nu}_R)(v_S + S)(\nu_R + \nu_R^c) - \frac{y_{LR}}{2\sqrt{2}}(\bar{\nu}_L + \bar{\nu}_L^c)(v_H + h)(\nu_R + \nu_R^c)\\ &\quad -\frac{y_{LR}}{2\sqrt{2}}(\bar{\nu}_R + \bar{\nu}_R^c)(v_H + h)(\nu_L + \nu_L^c),\end{aligned} \tag{C.22}$$

that we can express as function of the *physical* states N_1 and N_2

$$\begin{aligned}\mathcal{L} = &-\frac{h_1}{\sqrt{2}}S\bar{N}_2N_2 - \frac{h_1\theta^2}{\sqrt{2}}S\bar{N}_1N_1 + \frac{h_1\theta}{\sqrt{2}}S(\bar{N}_1N_2 + \bar{N}_2N_1)\\ &-\frac{y_{LR}}{2\sqrt{2}}h(\bar{N}_1N_2 + \bar{N}_2N_1) + \frac{y_{LR}\theta}{2\sqrt{2}}(\bar{N}_2N_2 + \bar{N}_1N_1) - m_1\bar{N}_1N_1 - m_2\bar{N}_2N_2\end{aligned} \tag{C.23}$$

with $m_1 \simeq \frac{y_{LR}^2 v_H^2}{2\sqrt{2}h_1 v_S}$ and $m_2 \simeq \sqrt{2}h_1 v_S$.
Such a dynamical model has a lot of interesting characteristics, like for instance the decay of the heavy scalar S into two (very energetic) light states N_1 which can be measured by the Icecube telescope for instance.

Useful Statistics

D

D.1 5σ and p-Value

D.1.1 5σ

An important example of constructing a confidence interval is when a single variable x follows a Gaussian distribution. This happens frequently when one needs to work with a great number ($\gtrsim 100$) of events. We then define the probability that a variable appear within an interval $\pm\delta$

$$1 - \alpha = \frac{1}{\sqrt{2\pi}\sigma} \int_{\mu-\delta}^{\mu+\delta} e^{-(x-\mu)^2/2\sigma^2} dx = \text{erf}\left(\frac{\delta}{\sqrt{2}\sigma}\right), \tag{D.1}$$

where μ is the mean of the variable x, and σ its variance. If one has to deal with a few number of events (dozens), it is more precise to use the Poisson distribution, $f(n; \nu) = \frac{\nu^n e^{-\nu}}{n!}$ with $n = 1, 2, 3..$ and $\nu > 0$. We illustrate the distribution behavior and its $\sigma limits$ in the Fig. D.1 and Table D.1. As a remark, the air below the Gaussian at 5σ correspond to a "null hypothesis" (see definition of the p-value below) of $\sim 6 \times 10^{-7}$ (see table) so a probability $\pm 3 \times 10^{-7}$. In other words, an event observed at 5 σ above a predicted background would have one chance on 3 millions to be part of the background (and thus 99.99997% of chance to be a "something else" beyond the Standard Model). We will discuss this case more in detail when examining the look-elsewhere effect.

Indeed, as we will see later, a 5σ confidence level—expressed in terms of Higgs search for instance—discovery corresponds to the condition that the probability of finding an excess rate of the measured size that is caused by Standard Model particles without Higgs particle is lower than 3×10^{-7}. Setting the limit for calling the collected data a discovery is, of course, a matter of convention. It is based on a trade-off between the advantage of calling a viable scientific claim empirically well-established and the potential damage of endorsing a false scientific

Y. Mambrini, *Particles in the Dark Universe*,
https://doi.org/10.1007/978-3-030-78139-2

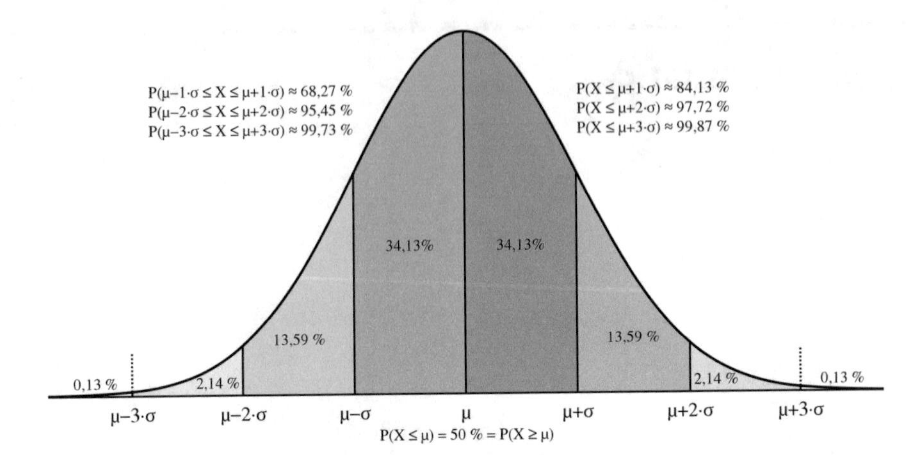

Fig. D.1 Gaussian distribution as function of standard σ deviations (Wolfgang Kowarschick, under the Creative Commons Attribution-Share Alike 4.0 International license)

Table D.1 Area of the tails α outside $\pm\delta$ from the mean of a Gaussian distribution

Confidence level δ	α	Probability
1σ	0.3173	68.27%
2σ	4.55×10^{-2}	95.45%
3σ	2.7×10^{-3}	99.73%
4σ	6.3×10^{-5}	99.9937%
5σ	5.7×10^{-7}	99.99994%
6σ	2.0×10^{-9}	99.9999998%

claim. In many scientific fields, a 3σ effect, that is a probability of less than 0.15% that the null hypothesis holds, is considered sufficient for establishing a phenomenon. In particle physics, the discovery of a new particle is taken to be of very high importance and is used in the analysis of the background in all future high energy scattering experiments. Therefore, the risk of erroneously acknowledging a discovery of a particle should be kept particularly low and a stronger criterion seems advisable. One should also take into account the look-elsewhere effect that can reduce considerably the 5σ effect (see below). Indeed, historically the 5σ limit was established based on largely pragmatic considerations. While statistical fluctuations at a 4σ level did and do occur from time to time in high energy physics experiments, no 5σ signal in a particle experiment has up to this point ever turned out to be a fluctuation. A 5σ limit therefore seemed plausible simply based on historical record. The fact that a 4σ fluctuations do occur can be statistically explained based on the number of experiments that are carried out in conjunction with the size of the look-elsewhere effects which usually apply in those contexts.

D.1.2 p-Value

The p-value is defined as the probability of obtaining a statistic at least as extreme as the one actually observed assuming that the null hypothesis is true. The smaller is the p-value, the less probable the "null hypothesis" is and so the results are "probably" a signal above a background. In other words, if the p-value computed by statistical tools (given by experimentalists) is below 0.05 or 0.01 (depending on the significance one is asking for), the researcher will reject the "null hypothesis": the observed result would be highly unlikely under the null hypothesis. The p-value can be seen as the probability that "something" observed happens assuming random coincidence.

D.2 Systematics vs Statistics

Any experimental analysis has to deal with two kind of uncertainties: the statistical uncertainties and systematics ones. In many cases, the systematics uncertainties are comparable to the statistical ones. However, consistent definition and practice are elusive which leads to confusion and in some cases incorrect interference. We will try to give some practical definitions in this section to help understand the importance of each of them.

Examples of statistical uncertainties include the finite resolution of an instrument, the Poisson fluctuations associated with measurements involving finite sample sizes, and random variations in the system one is examining. Systematic uncertainties, on the other hand, arise from uncertainties associated with the nature of the measurement apparatus, assumptions made by the experimenter, or the model used to make inferences based on observed data. Common examples of systematic uncertainty include uncertainties that arise from the calibration of the measurement device, the probability of detection of a given type of interaction (often called the "acceptance" of the detector), and parameters of the model used to make inferences that themselves are not precisely known.

Let us illustrate this correlation in a typical example, the measurement of the W boson cross section in hadronic colliders. The production rate of the W boson σ_W in $p\bar{p}$ collisions is theoretically known in the Standard Model. This cross section is also measured by a number of observed counts N_c (given some topologies and specific cuts depending on the channels analyzed by the experimenters), estimating the number of "background" events in this sample from other process N_b, estimating the acceptance of the apparatus including all selection requirements used to define the sample of events ϵ, and counting the number of $p\bar{p}$ annihilation, L. The cross section for W boson production is then

$$\sigma_W = \frac{N_c - N_b}{\epsilon L}. \tag{D.2}$$

For instance, a measurement performed by CDF collaboration where the transverse mass of a sample of candidate $W \to \epsilon \nu_e$ decays would be quoted as

$$\sigma_W = 2.64 \pm 0.01(\text{stat}) \pm 0.18(\text{syst})\ \text{nb}, \tag{D.3}$$

where the first uncertainty reflects the statistical uncertainty arising from the size of the candidate sample (approximately 38,000 candidates) and the second uncertainty arises from the background subtraction in Eq. (D.2). We can estimate these uncertainties as

$$\begin{aligned} \sigma_{stat} &= \sigma_0 \sqrt{\frac{1}{N_c}} \\ \sigma_{syst} &= \sigma_0 \sqrt{\left(\frac{\delta N_b}{N_b}\right)^2 + \left(\frac{\delta \epsilon}{\epsilon}\right)^2 + \left(\frac{\delta L}{L}\right)^2}. \end{aligned} \tag{D.4}$$

where the three terms in σ_{syst} are the uncertainties arising from the background estimate δN_b, the acceptance $\delta\epsilon$, and the integrated luminosity δL. The parameter σ_0 is the measured value.

The "paradox" that makes the definition of statistics and systematics so incertain is that in the *same* sample of events, experimenters also measured the acceptance ϵ through the Z boson observation from its decay into electron. The level of accuracy on ϵ becomes also a stochastic variable depending on the size of the sample. One can then also consider $\delta\epsilon$ as a statistical uncertainty and redefined

$$\begin{aligned} \sigma_{stat} &= \sigma_0 \sqrt{\frac{1}{N_c} + \left(\frac{\delta \epsilon}{\epsilon}\right)^2} \\ \sigma_{syst} &= \sigma_0 \sqrt{\left(\frac{\delta N_b}{N_b}\right)^2 + \left(\frac{\delta L}{L}\right)^2} \end{aligned} \tag{D.5}$$

resulting in a different assignment of statistical and systematics uncertainties. This matter in a sense that as a definition we took for "systematic" uncertainty, it should *not* depend on the size of the sample used to make the observation. It is supposed to come from an external measurement like a calibration for instance, and does not scale with the sample size. In the specific case of the measurement of the W cross section, ϵ does not fill this requirement. We can include in this family of systematics any degree of freedom which depends stochastically on an external measurement (which means any observable which is dominated by statistics uncertainties). If a threshold or an energy calibration (systematic) is dominated by a measurement from another experiment dominated by statistics, these uncertainties could be considered in fact as statistical uncertainties.

The background systematics however has a different nature. Indeed, the backgrounds events are gluon or quarks, produced in the collision but which fake an electron or a muon. A reliable estimate of this background is difficult to make from first principle, as the rate of such QCD events is many orders of magnitude larger than the W boson cross section. The way to reject background event is to discriminate between events having high missing transverse energy (taken by the neutrino in the W decay) and what experimenters called the "isolation" value: events from QCD background will have more particles produced at proximity of the electron or muon candidate. We remember of course also the "faster than light" neutrino results of OPERA that was claimed in 2011. This "faster-than-light" neutrino experiment suffered from a systematic error that affected all the data; faulty cables consistently gave the researchers bad readings. No matter how many times physicists repeated the experiments, they would get the same yet inaccurate results.

D.3 Look-Elsewhere Effect (LEE)

D.3.1 Generality

The look-elsewhere effect is common in scientific analysis of experiments studying models with a lot of degrees of freedom. Indeed, the more degrees of freedom a theory has, the more likely it is to make a statistically significant observation. This was very useful for the hunting of the Higgs boson in 2011.
What is the look-elsewhere effect in short? It is the main reason why the strong 5σ approach is applied in particle physics. Indeed, "5σ" corresponds to 3 events in 10 millions. But having such a signal which can be seen for instance as a bump at a given energy E in a diet spectrum corresponding to and invariant mass $M = E$. Naively speaking, suppose that such a bump is such high above the background that one has 1 chance on ten thousands for it to be a statistical fluctuation: it seems natural to claim for a potential discovery, corresponding to the production of a new particle of mass M. Now, suppose the study has been made in an energy range between 1 to 100 GeV, and the bump has a width of 2 GeV (see Fig. D.2). To know what is the real chance to observe such a bump at an energy E, say 60 GeV, one has to convolute the probability of observing such event in the 60 GeV bin to the general probability of having such a bump in another bin. In other words, if we do to "look-elsewhere" the probability of such event to be the background is only 1/10000 (and so we can consider it as an "event"); however, if one considers that this can statistically happen in any bin, we have 50 chances (100 GeV/2 GeV) that this event can occur in another bin which mean a total probability of $50 \times 1/10{,}000 = 1/200$ which cannot be considered anymore as a "discovery" signal. Put in another way, the probability of finding a signal *exactly* at 60 ± 2 GeV is one chance on 10,000, whereas the probability of finding a signal "somewhere between 1 and 100 GeV" is one chance on 200. In general, a good rule of thumb is the following: if the signal has a width W, and if one examines a spectrum spanning a mass range from M_1 to M_2, then the "boost factor" due to the LEE is $(M_2 - M_1)/W$. This may easily reach a factor 10 or 100 depending on the detail of the searches. Of course, this is a simplified version of the LEE effect because I have supposed that all 50 possibilities

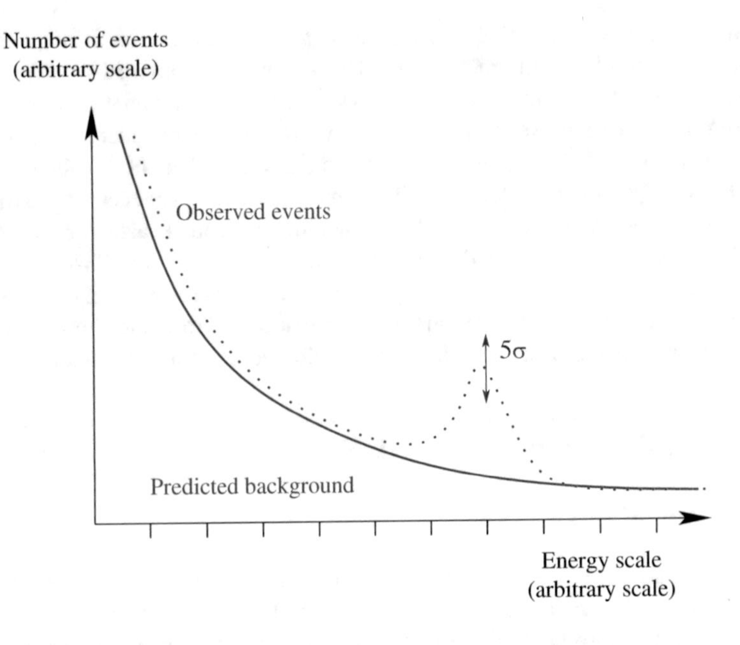

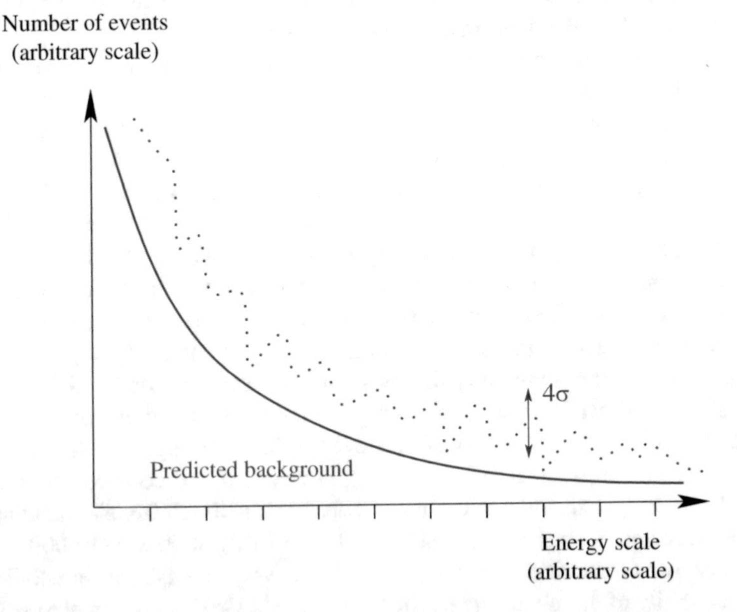

Fig. D.2 Illustration of the look-elsewhere effect: a 5σ signal observed above the background in a specific energy range (left figure) can be reduced to a 4σ signal (or even lower) if one take into consideration that, looking elsewhere, this signal has also a probability of appearing somewhere else (right figure). If one knows exactly the energy bin where the signal should appear (from other considerations or other experiments), this signal is effectively at 5σ of confidence level have the same chance to happen (a flat distribution), which is never the case in any complex studies. That explains the concrete difficulties to take into account this effect in modern analysis (especially at the LHC).

One can also find effects of such LEE effect in the common life: a Swedish study in 1992 tried to determine whether or not power lines caused some kind of poor health effects. The researchers surveyed everyone living within 300 meters of high-voltage power lines over a 25-year period and looked for statistically significant increases in rates of over 800 ailments. The study found that the incidence of childhood leukemia was four times higher among those that lived closest to the power lines, and it spurred calls to action by the Swedish government. The problem with the conclusion, however, was that they failed to compensate for the look-elsewhere effect; in any collection of 800 random samples, it is likely that at least one will be 3 standard deviations above the expected value, by chance alone. Subsequent studies failed to show any links between power lines and childhood leukemia, neither in causation nor even in correlation.

D.3.2 Applying the LEE Effect to the Higgs Discovery

Let us now look specifically at the Higgs detection at the LHC. The two main channels analyzed by the ATLAS and CMS collaborations are the $H \to \gamma\gamma$ and $H \to 4$ leptons final states, mainly because of the low background level of these two final states. In December 2011, the significance of the entire excess rate of both event types was close to 4σ. This corresponds to a probability close to 3×10^{-5} that the observed "signal" would arise as a fluctuation of Standard Model physics without a Higgs particle. This new boson was looked over a range of about 80 energy bins at the LHC. This corresponded to a probability of about 2.5×10^{-3} that an oscillation of the size of the December data could occur at the LHC experiment. This probability does not even amount a 3σ effect and therefore was clearly insufficient for establishing the existence of a new particle. The data of July 2012 then had a significance above the 5σ level, which corresponded to a probability of the null hypothesis of less than 3×10^{-5} after taking into account the LEE effect. This was sufficient for declaring the data a discovery.

What happened in December 2011, however, is that a lot of theoreticians (the author of the present book included) considered that a 4σ limit can be considered seriously as the look-elsewhere effect can be reduced by "theoretical" or "philosophical" argument. Indeed, there was strong evidence that (1) the Higgs field should exist and (2) that this mass range due to perturbative effects should lie below 1 TeV. Indeed, in this sense one can see the limit of the look-elsewhere effect, and its part of arbitrariness. Up to which scale one should take the possibilities of statistical fluctuations? Which kind of data from other experiments, or theoretical arguments should we take into account in the analysis? Or should we play the role of completely blind physicists, unaware of other experiments or theoretical constructions and history? For instance, before the Higgs signal of December 2011, there was other measurements from LEP or Tevatron, limiting the Higgs mass in the window 115 to 141 GeV, reducing thus the analysis to 26 bins, increasing at the same time the significance by a factor 3, the probability of a statistical fluctuation being then below 10^{-3}. But of course, that one should be taken into account if one believes that the Higgs exists.

D.3.3 Applying the LEE Effect to the Dark Matter Searches

In April 2012 was reported "A tentative gamma-ray line from Dark Matter annihilation at the Fermi Large Area Telescope" [9] with a statistical significance of 4.6σ at 130 GeV. When taking into account the look elsewhere effect, the signal dropped down to 3.2σ i.e. taking into account the fact that one should search for a line in multiple regions of the sky and at multiple energies. In May 2013 the Fermi LAT collaboration has reported their "Search for Gamma-ray Spectral Lines with the Fermi Large Area Telescope and Dark Matter Implications", computing the statistical significance of this line feature at 130 GeV at a lower statistical significance of 3.3σ, and only 1.6σ after taking the look-elsewhere effect into account. As we can see, the LEE effect was important in this case, as it seemed finally that such signal was not due to dark matter annihilation.

D.4 Bayesian vs Frequentist Approach

Since several decades, the fight exists between people sustaining the frequentist or Bayesian approach. In the supersymmetric communities several authors used to claim in every conferences that the Bayesian approach was *the one* to use in any supersymmetric study. We will remind you in this section the fundamentals of Bayesian approach.

To understand the Bayesian approach, we will take a very concrete example. Suppose a woman believes she may be pregnant after a single sexual encounter, but she is unsure. So she takes a pregnancy test that is known to be 90% accurate (meaning it gives positive result to positive cases 90% of the time) and the test produces a positive result. Ultimately, she would like to know the probability she is pregnant, given a positive test (p(preg|test +)); however, what she knows is the probability of obtaining a positive test result if she is pregnant (p(test + | preg)), and she knows the result of the test. Bayes' theorem offers a way to reverse conditional probabilities, and, hence, provides a way to answer this question.

Bayes' original theorem applied to point probabilities. The basic theorem states simply:

$$p(B|A) = \frac{p(A|B)p(B)}{p(A)}, \tag{D.6}$$

which can be translated by the (conditional) probability of having B given A is equal to the (conditional) probability of having A given B, multiplied by the (marginal) probability of having B divided by the (marginal) probability of having A.[1]
Proof: we understand easily that the probability of having A and B is $p(A, B) = p(A|B)p(B)$. With the same reasoning, we also can write that $p(B, A) = p(B|A)p(A)$. $p(A, B) = p(B, A)$ one then can write

$$p(B, A) = p(A, B) \;\Rightarrow\; p(B|A)p(A) = p(A|B)p(B) \tag{D.7}$$

which gives the Bayes' result.
Now, back to our example, suppose that, in addition to the 90% accuracy rate, we also know that the test gives false positive results 50% of the time. In other words, in cases in which a woman is *not* pregnant, she will test positive 50% of the time. Thus $p(test+|not\ preg) = 0.5$. With this information, combined with some "prior" information concerning the probability of becoming pregnant from a single sexual encounter, Bayes' theorem provides a prescription for determining the probability of interest.

The prior information we need, $p(B) = p(preg)$, is the marginal probability of being pregnant, not knowing anything beyond the fact that the woman has had a single sexual encounter. This information is considered prior information, because it is relevant information that exists prior to the test. We may know from previous research that, without any additional information (e.g. concerning the date of last menstrual cycle), the probability of conception for any single sexual encounter is approximately 15%. With this information, we can determine $p(B|A) \equiv p(preg|test+)$ as

$$p(preg|test+) = \frac{p(test+|preg)p(preg)}{p(test+|preg)p(preg) + p(test+|not\ preg)p(not\ preg)} \tag{D.8}$$

which gives

$$p(preg|test+) = \frac{0.90 \times 0.15}{0.90 \times 0.15 + 0.50 \times 0.85} = 0.241. \tag{D.9}$$

Thus, the probability that the woman is pregnant, given the positive test is only 0.241. Using Bayesian terminology, this probability is called "posterior probability" because it is the estimated probability of being pregnant, obtained *after* observing the data (the positive test). We can be surprised of such a low value given the "90%" accuracy of the test. This comes mainly from the 50% of probability (high) to obtain a positive test when the person is *not* pregnant.

[1] If anyone still believes that $p(A|B) = p(B|A)$ [probability of A given B = probability of B given A], remind them that the probability of being pregnant, given that the person is female is $\sim 3\%$, while the probability of being female, given that they are pregnant, is considerably larger.

If the woman is aware of these limitations, she can redo the test. But now, she can use the "updated" probability of being pregnant ($p = 0.241$) as the new $p(B)$; that is the prior probability for being pregnant has now been updated to reflect the results of the first test. If she repeats the test again and again observes a positive result, her new "posterior probability" of being pregnant is

$$p(preg|test+) = \frac{0.90 \times 0.241}{0.90 \times 0.241 + 0.50 \times 0.759} = 0.364. \tag{D.10}$$

This result is still not very convincing evidence that she is pregnant, but if she repeats the test again and finds positive result, her probability increases to 0.507 up to 0.984 at the tenth positive attempt. From a Bayesian perspective, we begin with some prior probability for some event, and we update this prior probability with new information to obtain a posterior probability. The posterior probability can then be used as a prior probability in a subsequent analysis. From a Bayesian point of view, this is an appropriate strategy for conducting scientific research.

We can then understand how one should apply such statistical analysis in supersymmetric models. Indeed, before that the LHC gave any limits, the aim was to know how a point in supersymmetric parameter space was "probable" or not. Usually, frequentists (the author of the present book included) make a scan on the par mater space, and compute how far, or near (from a frequentist approach, i.e. 2 or 3σ), this point is from the data measured. On the other hand, the bayesian researchers would take the complementary approach: *given* experimental data, what is the probability that a point in the parameter space is around the data.

Numbers

E

I have included in this appendix the main tables, conversion factors, and formulae I concretely regularly needed in my cosmological/astrophysical/particle computation. I hope you will enjoy using them as much as we enjoyed (sic!) writing them.

E.1 Useful Formulae

E.1.1 Cosmology

$$T_0 = 2.349 \times 10^{-4}\ \text{eV} = 2.725\ ^0\text{K}; \quad G = 1/M_{Pl}^2\,; \quad M_{Pl} = 1.22 \times 10^{19}\ \text{GeV}$$

$$H(T) = \sqrt{\frac{8\pi G}{3}\rho(T)} = \sqrt{g_\rho}\sqrt{\frac{4\pi^3}{45}}\frac{T^2}{M_{Pl}} \simeq 1.66 g_\rho^{1/2}\frac{T^2}{M_{Pl}} = 0.32 g_\rho^{1/2}\frac{T^2}{M_P}$$

$$= \left(\frac{g_\rho}{90}\right)^{1/2}\pi\frac{T^2}{M_P}\ \left[M_P = \frac{1}{\sqrt{8\pi}}M_{Pl}\right]$$

$$\rho_c^0 = \frac{3H_0^2}{8\pi G} = 10^{-5}\ h^2\ \text{GeV cm}^{-3} = 2 \times 10^{-29} h^2\ \ \text{g/cm}^3; \quad \Omega_A = \frac{\rho_A}{\rho_c^0}$$

$$n_\gamma^0 = 411\ \text{cm}^{-3};\ \rho_\gamma^0 = 2.62 \times 10^{-10}\text{GeV/cm}^3;\ \langle E_\gamma^0\rangle = \rho_\gamma^0/n_\gamma^0 = 6 \times 10^{-4}\text{eV};$$

$$T_\nu^0 = 1.95\ ^0\text{K} = 1.68 \times 10^{-4}\ \text{eV};\ n_\nu(T_\nu^0) = 112\ \text{cm}^{-3};\ g_\rho^0 = 3.36$$

$$s(T) = \frac{2\pi^2}{45}g_s(T)T^3; \quad g_s^0 = 3.91\,; \quad s(T_0) = 2.2 \times 10^{-38}\ \text{GeV}^3 = 2909\ \text{cm}^{-3}$$

$$Y(T_0) = \frac{\Omega\rho_c^0}{m s_0} = \left(\frac{\Omega}{0.1}\right)\left(\frac{100\ \text{GeV}}{m}\right)\frac{135 \times 10^{-3} H_0^2 M_P^2}{16\pi^3 g_s^0 T_0^3}$$

Y. Mambrini, *Particles in the Dark Universe*,
https://doi.org/10.1007/978-3-030-78139-2

$$Y(T_0) \simeq 5 \times 10^{-10} \left(\frac{1\ \text{GeV}}{m} \right)$$

$$\frac{\Omega_{dm} h^2}{0.1} = 5.9 \times 10^6 \left(\frac{n_{dm}(T)}{T^3} \right) \left(\frac{M_{dm}}{1\ \text{GeV}} \right) \quad [\text{for T} \gtrsim \text{EWSB}]$$

$$f_A = \frac{g_A}{e^{(E-\mu_A)/kT} \pm 1}; \qquad n_A(T) = \frac{g_A}{(2\pi)^3} \int f_A(\mathbf{p}) d^3 p$$

$$n_{EQ}^A(T \gg m) \left(\frac{3}{4} \right) \frac{\zeta(3)}{\pi^2} g_A T^3 \quad \text{(fermion) boson}$$

$$\rho_{EQ}^A(T \gg m) \left(\frac{7}{8} \right) \frac{\pi^2}{30} g_A T^4 \quad \text{(fermion) boson}$$

$$n_{EQ}^A(T \ll m) = g_A \left(\frac{mT}{2\pi} \right)^{3/2} e^{-m/T}; \quad \rho_{EQ}^A(T \ll m) = m \times n_{EQ}^A(T \ll m)$$

$$\Omega_\chi = \frac{n_\chi(T_0) m_\chi}{\rho_c} = \frac{n_\chi(T_0) m_\chi}{1.05 \times 10^{-5} h^2\ \text{GeV.cm}^{-3}} = \frac{1.6 \times 10^8}{h^2} \left(\frac{n_\chi(T_0)}{T_0^3} \right) \left(\frac{m_\chi}{1\ \text{GeV}} \right),$$

where we used $1\ \text{cm}^3 = \frac{1700}{T_0^3}$ (see table below). We then obtain

$$\Omega_\chi h^2 = 1.6 \times 10^8 \left(\frac{g_s^0}{g_s^{RH}} \right) \left(\frac{n_\chi(T_{RH})}{T_{RH}^3} \right) \left(\frac{m_\chi}{1\ \text{GeV}} \right) \tag{E.1}$$

where we used

$$n_\chi(T_0) = n_\chi(T_{RH}) \times \frac{a_{RH}^3}{a_0^3} \quad \text{and} \quad g_s^0 a_0^3 T_0^3 = g_s^{RH} a_{RH}^3 T_{RH}^3.$$

with $g_s^0 = 3.91$ and $g_s^{RH} = 106.75$ for $T_{RH} > m_t$.

E.1.2 Particle Physics

$$\langle H \rangle = v/\sqrt{2} \Rightarrow v = 246\ \text{GeV} \quad (174\ \text{GeV if } \langle H \rangle = v)$$

$$M_W = \frac{gv}{2}; \quad M_Z = \frac{\sqrt{g^2 + g'^2}}{2} v = \frac{g}{2\cos\theta_W} v; \quad M_Z = \frac{M_W}{\cos\theta_W}$$

$$\sigma_{EW} \simeq |M|^2 / 64\pi^2 s \simeq 10^{-9}\ \text{GeV}^{-2}$$

$$\cos\theta_W = \frac{g}{\sqrt{g^2+g'^2}}\,; \quad \sin\theta_W = \frac{g'}{\sqrt{g^2+g'^2}}\,; \quad e = g\sin\theta_W$$

$$A_\mu = \cos\theta_W\, B_\mu + \sin\theta_W\, W^3_\mu\,; \quad Z_\mu = -\sin\theta_W\, B_\mu + \cos\theta_W\, W^3_\mu$$

$$G_F = \frac{\sqrt{2}}{8}\frac{g^2}{M_W^2} \simeq 1.166 \times 10^{-5}\ \mathrm{GeV}^{-2}$$

$$W^+_\mu = \frac{1}{\sqrt{2}}(W^1_\mu - i W^2_\mu); \quad W^-_\mu = \frac{1}{\sqrt{2}}(W^1_\mu + i W^2_\mu);$$

$$\sigma_T = 6.65 \times 10^{-29}\ \mathrm{m}^2$$

Perturbativity: $\frac{\alpha}{4\pi} = \frac{g^2}{(4\pi)^2} < 1$ giving $g < 4\pi$

E.2 Tables

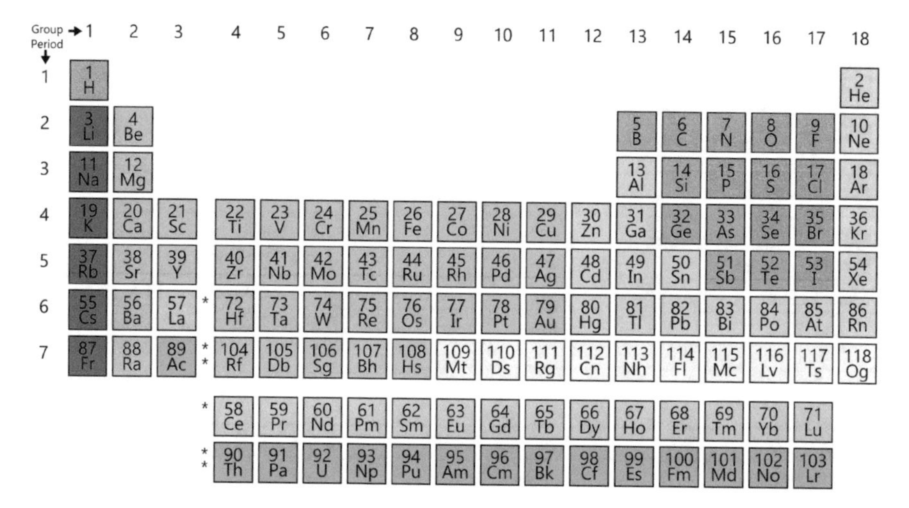

Fig. E.1 Periodic Table of the Elements (Double sharp, licensed under the Creative Commons Attribution-Share Alike 4.0 International license)

Quantity	Symbol	Value
Speed of light in vacuum	c	299 792 458 m/s
Planck constant	h	6.626×10^{-34} J s = 4.14×10^{-24} GeV
Planck constant reduced	$\hbar \equiv h/2\pi$	1.055×10^{-34} J s $= 6.58 \times 10^{-22}$MeV s
Planck Mass	M_{Pl}	1.2211×10^{19} GeV = 2.1768×10^{-5} g
Reduced Planck Mass	$M_P = M_{Pl}/\sqrt{8\pi}$	2.43×10^{18} GeV
Planck time	t_{Pl}	5.4×10^{-44} s
Planck size	l_{Pl}	1.6×10^{-35} m
Conversion factor	$\hbar c$	197.3 MeV fm
		1 GeV = 1.6×10^{-10} J
Boltzmann constant	k_B	1.38×10^{-23}J K^{-1} = 8.617×10^{-5} eV K^{-1}
Weinberg angle	$\sin\theta_W$ / $\sin^2\theta_W$	0.481/0.23122
Hypercharge coupling	g'	0.356
Weak coupling	g	0.65
Strong coupling	g_3	1.126
Electromagnetic coupling (M_Z)	$e = g(M_Z)\sin\theta_W$	0.313
Electromagnetic coupling (m_e)	$e = \sqrt{\frac{4\pi}{137}}$	0.303
$\alpha_1^{-1}(M_Z)$	$(g_1^2/4\pi)^{-1}$	58.98 ± 0.04
$\alpha_2^{-1}(M_Z)$	$(g_2^2/4\pi)^{-1}$	29.57 ± 0.03
$\alpha_3^{-1}(M_Z)$	$(g_3^2/4\pi)^{-1}$	8.40 ± 0.14
Fermi constant	$G_F = \frac{\sqrt{2}}{8}\frac{g^2}{M_W^2}$	1.166×10^{-5} GeV^{-2}
Typical EW cross section	σ_{EW}	10^{-9} GeV^{-2}
Electron charge	e	1.6×10^{-19} C
Electron mass	m_e	0.511 MeV/c^2 = 9.11×10^{-31} kg
Proton mass	m_p	938.27 MeV/c^2 = 1.672×10^{-27} kg
Neutron mass	m_n	939.57 MeV/c^2 = 1.675×10^{-27} kg
Electron charge	e	1.6×10^{-19} C
Electron radius	$r_0 = \frac{1}{4\pi\epsilon_0}\frac{e^2}{m_e c^2}$	2.8×10^{-15} m
Vacuum permittivity	ϵ_0	8.854×10^{-12} F m^{-1} [A^{2}s^{4}kg^{-1}m^{-3}]
Vacuum permeability	$\mu_0 = 1/\epsilon_0 c^2$	$4\pi \times 10^{-7} = 12.566 \times 10^{-7}$ N A^{-2}
Gauss	G	1 Gauss = 10^{-4} Tesla
Muon mass	m_μ	105.7 MeV/c^2
Pion mass	m_{π^0}	135 MeV/c^2
Tau mass	m_τ	1777 MeV/c^2
Deuteron mass	m_d	1876 MeV/c^2

Quantity	Symbol	Value
Higgs vev	$\langle H\rangle = \frac{v}{\sqrt{2}}$	$\langle H\rangle = 174$ GeV; $v = 246$ GeV
Higgs mass/width	m_h / Γ_h	125 GeV/c^2 / 4.07 MeV/c^2
W mass/width	$M_W = \frac{g}{2}v$ / Γ_W	80.385 ± 0.015 GeV/c^2 / 2.085 ± 0.042 GeV/c^2
Z mass/width	$M_Z = \frac{M_W}{\cos\theta_W}$ / Γ_Z	91.1876 ± 0.0021 GeV/c^2 / 2.4952 ± 0.0023 GeV/c^2
Up quark mass	m_u	1.5–4.5 MeV/c^2
Down quark mass	m_d	5–8.5 MeV/c^2
Strange quark mass	m_s	80–155 MeV/c^2
Charm quark mass	m_c	1–1.4 GeV/c^2
Bottom quark mass	m_b	4–4.5 GeV/c^2
Top quark mass	m_t	175 GeV/c^2
Conversion K/GeV		1K $=0.862 \times 10^{-4}$ eV
Conversion GeV^{-2}/cm^{3}s^{-1}	σv	1 GeV^{-2} $=1.2 \times 10^{-17}$cm^{3}s^{-1}
Conversion GeV^{-1}/L		1 GeV^{-1}$= 1.9733 \times 10^{-14}$ cm
Conversion GeV^{-1}/t		1 GeV^{-1}$= 6.67 \times 10^{-25}$ s
Conversion GeV^{-2}/barn		1 GeV^{-2}$= 4 \times 10^{-28}$ cm^2 = 4×10^{-4} barn
Conversion cm^2/barn		10^{-24}cm^2 = 1 barn
Conversion GHz/GeV		1GHz $=4 \times 10^{-6}$ eV
Conversion GeV/kg		1GeV $=1.78 \times 10^{-27}$ kg
Conversion barn/cm^2		1 barn = 10^{-24} cm^2 / 1 pbarn = 10^{-36} cm^2
Conversion cm^2 g^{-1}/GeV^{-3}		1 cm^2 g^{-1} $= 4.4 \times 10^3$ GeV^{-3}
Gravitational constant	G	6.7×10^{-11}m^3 kg^{-1} s^{-2} $6.7 \times 10^{-39}\hbar c(\text{GeV}/c^2)^{-2}$
Present Hubble expansion rate	H_0	100 h km/s/Mpc $=2.13 \times 10^{-42}$ h GeV
Hubble time	H_0^{-1}	$3.08 \times 10^{17} h^{-1}$ sec = 9.77×10^9 h^{-1} yrs
Hubble distance	cH_0^{-1}	2998 h^{-1} Mpc = $9.25 \times 10^{27} h^{-1}$ cm
Reduced Hubble expansion rate	h	H_0 /(100 km/s/Mpc)
Normalized Hubble expansion rate	h	0.677 ± 0.002
Horizon size	L_H	$\simeq$ 14 Gpc
Critical density	ρ_c	$1.05 \times 10^{-5} h^2$ GeV cm^{-3} $= 1.8 \times 10^{-32} h^2$ kg cm^{-3}
		$8 \times 10^{-47} h^2$ GeV^4 $\simeq 4 \times 10^{-47}$GeV^4
Age of the Universe	t_0	13.8 Gyr $= 4.3 \times 10^{17}$ s. $= 6.6 \times 10^{41}$ GeV
Mean free path of photon	λ_0	10^{26} m
CMB temperature	T_0	2.725 $\pm$0.001 K $= 2.349 \times 10^{-4}$ eV
(CMB temperature)^3	T_0^3	1700 cm^{-3}

Quantity	Symbol	Value
Zeta Riemann	$\zeta(3)$	1.20206
Parsec	pc	1 pc = 3.086×10^{16} m = 3.262 light-years
Solar velocity with respect to CMB		371 km/s
Jansky	Jy	10^{-26} W m^{-2} Hz^{-1}
Watt	W	J.s^{-1}=kg m^2 s^{-3}
Solar mass	$M_\odot$	1.98×10^{30} kg = 1.1×10^{57} GeV
Solar radius	$R_\odot$	696 340 km
Mass-to-light ratio	$M_\odot/L_\odot$	5133 kg/W
1 year		3.15×10^7 s
Light-years	lyrs	1 lyr = 9.46×10^{15} m
Binding energy of the Hydrogen	B_H	13.6 eV
Binding energy of the deuterium	B_D	2.2 MeV
Deuterium mass	m_D	1877.62 MeV

Quantity	z	Energy	Temperature	time
		0.862×10^{-4} eV	1 K	10^{17} s
Nucleosynthesis		0.1 MeV	10^9 K	100 s
Matter domination	3400	0.81 eV	9000 K	100,000 years
Decoupling (atoms formation)	1100	0.3 eV	3000K	380,000 years
Dark energy domination	0.66	3.9×10^{-4} eV	4.565 K	10^{10} years
Actual	0	2.35×10^{-4} eV	2.725K	13.8×10^9 years

Cross section	Value (in pb)
$\sigma_{e^+e^-\to\nu\nu\gamma}$	2.7×10^{-3}

Field	$q = T^3 + Y/2$	T^3	Y
u_L	$+2/3$	$+1/2$	$+1/3$
d_L	$-1/3$	$-1/2$	$+1/3$
u_R	$+2/3$	0	$+4/3$
d_R	$-1/3$	0	$-2/3$
ν_{eL}	0	$+1/2$	-1
e_L	-1	$-1/2$	-1
e_R	-1	0	-2
N_R	0	0	0
H	0	$-1/2$	$+1$

Quantity	Symbol	Value
Hubble Constant	$H_0; h$	67.77 ± 4 km/s/Mpc; 0.677 ± 0.002
Total density	Ω_t	1.02 ± 0.02
Critical density	ρ_c^0	$1.05 \times 10^{-5} h^2$ GeV cm^{-3} ; $8 \times 10^{-47} h^2$GeV^4
Dark matter energy density	ρ_{cdm}^0	1.28×10^{-6} GeV cm^{-3}
Dark energy density	ρ_Λ^0	3.32×10^{-6} GeV cm^{-3}
Dark Energy density	$\Omega_\Lambda; \Omega_\Lambda h^2$	0.689 ± 0.006; 0.316 ± 0.003
Total matter density	$\Omega_m; \Omega_m h^2$	0.311 ± 0.006; 0.142 ± 0.0009
CDM density	$\Omega_{cdm}; \Omega_{cdm} h^2$	0.265 ± 0.002; 0.119 ± 0.001
Baryonic density	$\Omega_b; \Omega_b h^2$	0.050 ± 0.0002; 0.0224 ± 0.0001
Photon density	$\Omega_\gamma; \Omega_\gamma h^2$	5.5×10^{-5}; 2.47×10^{-5}
Neutrino density	$\Omega_\nu; \Omega_\nu h^2$	$\lesssim 1.4 \times 10^{-3}$; $\frac{\sum m_\nu}{93.14} \lesssim \frac{0.06}{93.14} = 6.44 \times 10^{-4}$
Radiation density	Ω_R	$\Omega_R = \Omega_\gamma + \Omega_\nu = 1.68\ \Omega_\gamma$
Curvature density	Ω_K	$\Omega_K = 0.0007 \pm 0.0019$
Baryons to photons ratio	η_b	6.12×10^{-10}
Age of the Universe	t_0	13.787 ± 0.020 Gyr
Size of the Universe (radius)	R_0	~ 46 light years (14 Gpc)

References

1. A. Denner, H. Eck, O. Hahn, J. Kublbeck, Nucl. Phys. B **387**, 467–481 (1992). https://doi.org/10.1016/0550-3213(92)90169-C
2. R. d'Inverno, *Introducing Einstein's Relativity* (Clarendon, Oxford, 1992), 383 p.
3. C.S. Wu, E. Ambler, R.W. Hayward, D.D. Hoppes, R.P. Hudson, Phys. Rev. **105**, 1413–1414 (1957). https://doi.org/10.1103/PhysRev.105.1413
4. T. Sjostrand, S. Mrenna, P.Z. Skands, JHEP **05**, 026 (2006). https://doi.org/10.1088/1126-6708/2006/05/026. [arXiv:hep-ph/0603175 [hep-ph]]
5. J. Hisano, S. Matsumoto, O. Saito, M. Senami, Phys. Rev. D **73**, 055004 (2006). https://doi.org/10.1103/PhysRevD.73.055004. [arXiv:hep-ph/0511118 [hep-ph]]
6. V.N. Gribov, B. Pontecorvo, Phys. Lett. B **28**, 493 (1969). https://doi.org/10.1016/0370-2693(69)90525-5
7. G. Bellini et al. [BOREXINO], Nature **512**(7515), 383–386 (2014). https://doi.org/10.1038/nature13702
8. R.D. Peccei, H.R. Quinn, Phys. Rev. Lett. **38**, 1440–1443 (1977). https://doi.org/10.1103/PhysRevLett.38.1440
9. C. Weniger, JCAP **08**, 007 (2012). https://doi.org/10.1088/1475-7516/2012/08/007. [arXiv:1204.2797 [hep-ph]]

Index

Y. Mambrini, *Particles in the Dark Universe*,
https://doi.org/10.1007/978-3-030-78139-2

I

J

K

L

M

N

P

Printed in the United States
by Baker & Taylor Publisher Services